BIOLOGY OF BENTHIC ORGANISMS

BIOLOGY OF BENTHIC ORGANISMS

11th European Symposium on Marine Biology
Galway, October 1976

Edited by

B. F. KEEGAN and P. O. CEIDIGH

School of Marine Sciences, University College, Galway, Ireland

and

P. J. S. BOADEN

Marine Biology Station, Portaferry, Co Down, Northern Ireland

PERGAMON PRESS

OXFORD · NEW YORK · TORONTO · SYDNEY · PARIS · FRANKFURT

U.K.	Pergamon Press Ltd., Headington Hill Hall, Oxford OX3 0BW, England
U.S.A.	Pergamon Press Inc., Maxwell House, Fairview Park, Elmsford, New York 10523, U.S.A.
CANADA	Pergamon of Canada Ltd., 75 The East Mall, Toronto, Ontario, Canada
AUSTRALIA	Pergamon Press (Aust.) Pty. Ltd., 19a Boundary Street, Rushcutters Bay, N.S.W. 2011, Australia
FRANCE	Pergamon Press SARL, 24 rue des Ecoles, 75240 Paris, Cedex 05, France
WEST GERMANY	Pergamon Press GmbH, 6242 Kronberg-Taunus, Pferdstrasse 1, West Germany

Copyright © 1977 Pergamon Press Ltd.

All Rights Reserved. No part of this publication may be reproduced, stored in a retrieval system or transmitted in any form or by any means: electronic, electrostatic, magnetic tape, mechanical, photocopying, recording or otherwise, without permission in writing from the publishers

First edition 1977

Reprinted (with corrections) 1978

Library of Congress Cataloging in Publication Data

European Marine Biology Symposium, 11th, Galway,
Ire., 1976.
Biology of benthic organisms.

English or French.
1. Marine fauna--Congresses. 2. Benthos--
Congresses. I. Keegan, Brendon F. II. Ceidigh, P. O.
III. Boaden, P. J. S. IV. Title.
QL121.E94 1976   574.92   77-3482
ISBN 0-08-021378-2

In order to make this volume available as economically and rapidly as possible the authors' typescripts have been reproduced in their original form. This method unfortunately has its typographical limitations but it is hoped that they in no way distract the reader.

*Printed in Great Britain by William Clowes & Sons, Limited
London, Beccles and Colchester*

CONTENTS

FOREWORD — xi

OPENING ADDRESS — xiii

LIST OF PARTICIPANTS — xvii

PREVIOUS SYMPOSIA — xxxi

Daily Variation in Dichelopandalus bonnieri (Caullery) as a Component of the Epibenthos — 1
  A.H.Y. Al-Adhub and E. Naylor

Community Structures of Soft-bottom Macrofauna in Different Parts of the Baltic — 7
  A.B. Andersin, J. Lassig and H. Sandler

A Drop-trap Investigation of the Abundance of Fish in very shallow water in the Askö Area, Northern Baltic Proper — 21
  G. Aneer and S. Nellbring

Results and Problems of an "Unsuccessful" Benthos Cage Predation Experiment (Western Baltic) — 31
  W. E. Arntz

The Importance of being a 'Littoral' Nauplius — 45
  H. Barnes and M. Barnes

Physiological Measurements on Estaurine Bivalve Molluscs in the Field — 57
  B.L. Bayne, J. Widdows and R.I.E. Newell

Variability of Growth Rate of *Macoma balthica* (1.) in the Wadden Sea in relation to Availability of Food — 69
  J.J. Beukema, G.C. Cadee and J.J.M. Jansen

Studies on Anaerobic Nitrogen Fixation in the Sediments of Two Scottish Sea Lochs — 79
  D. Blake and J. Leftley

Production Primaire Saisonniere du Microphytobenthos des Sables Envases en Baie de Concarneau — 85
  D. Boucher

Modiolus modiolus (L.) - An Autecological Study — 93
  R.A. Brown and R. Seed

Distribution and Maintenance of a *Lanice conchilega* Association in the Weser Estuary (FRG), with special reference to the Suspension-Feeding Behaviour of *Lanice conchilega* — 101
  K.J. Buhr and J.E. Winter

Le Macrobenthos des Fonds Meubles de la Manche: Distribution
Generale et Ecologie
   L. Cabioch, F. Gentil, R. Glaçon and C. Retiere
   115

Colonisation et Distribution Spatiale des Copepodes dans des
Lagunes Semi-artificielles
   J. Castel and P. Lasserre
   129

Recrutement et Succession du Benthos Rocheux Sublittoral
   A. Castric
   147

Annual Macrofauna Production of a Soft-bottom in the Northern
Baltic Proper
   H. Cederwall
   155

Evolution dans les Temps des Peuplements des Sables Envases en
Baie de Concarneau (Bretagne)
   P. Chardy and M. Glémarec
   165

Distribution of Benthic Phyto- and Zoocoenoses along a Light
Gradient in a Superficial Marine Cave
   F. Cinelli, E. Fresi, L. Mazzella, M. Pansini, R. Pronzato and
A. Svoboda
   173

Bionomie Benthique du Plateau Continental des Iles Kerguelen.
8. Variations Spatiales et Temporelles dans le Peuplement des
Vases a Spicules.
   D. Desbruyères and A. Guille
   185

Some Observations on the Relative Abundance of Species in a
Benthic Community
   R.A. Eagle and P.A. Hardiman
   197

The Polychaete *Eulalia viridis* (O.F. Muller) as an Element in
the Energy Dynamics of Intertidal Mussel Clumps
   R.H. Emson
   209

A Diving Survey of Strangford Lough: the Benthic Communities
and their Relation to Substrate - a Preliminary Account
   D.G. Erwin
   215

Reproductive Strategies of the Winkle *Littorina rudis* in
Relation to Population Dynamics and Size Structure
   R.J. Faller-Fritsch
   225

Structural Features of a North Adriatic Benthic Community
   K. Fedra
   233

Structure of the Abyssal Macrobenthic Community in the Rockall
Trough
   J.D. Gage
   247

Ecology of the Pogonophore *Siboglinum fiordicum* Webb in a
Shallow-water Fjord Community
   J.D. George
   261

Contents

The Re-establishment of an *Amphiura filiformis* Population in the Inner Part of the German Bight ... 277
  D. Gerdes

An Ecophysiological Approach to the Microdistribution of Meiobenthic Oligochaeta, I. *Phallodrilus monospermathecus* (Tubificidae) from a Subtropical Beach at Bermuda ... 285
  O. Giere

Phosphoglucoisomerase Allele Frequency Data in *Mytilus edulis* from Irish Coastal Sites: its Ecological Significance ... 297
  E. Gosling and N.P. Wilkins

An "in situ" Study of Primary Production and the Metabolism of a Baltic *Fucus vesiculosus*-Community ... 311
  B. Guterstam

Reproductive Strategy in Two British Species of *Alcyonium* ... 321
  R.G. Hartnoll

Observations on the Behaviour and Distribution of *Virgularia mirabilis* (Coelenterata: Pennatulacea) in Holyhead Harbour, Anglesey ... 329
  B. Hoare

Meiobenthic Subcommunity Structure: Spatial vs. Temporal Variability ... 339
  W.D. Hummon and M.R. Hummon

Habitat Area and Development of Marine Epibenthic Communities ... 349
  J.B.C. Jackson

Quantitative Survey of Hard-bottom Communities in a Baltic Archipelago ... 359
  A.M. Jansson and N. Kautsky

Aspects of the Ecology of *Sargassum muticum* (Yendo) Fensholt, in the Solent Region of the British Isles. I. The Growth Cycle and Epiphytes ... 367
  N.A. Jephson and P.W.G. Gray

The Effect of Depth on Populations of *Laminaria hyperborea* ... 377
  J.M. Kain

Sublittoral Transects in the Menai Straits and Milford Haven ... 379
  E.W. Knight-Jones and A. Nelson-Smith

Epibenthic Assemblages as Indicators of Environmental Conditions ... 391
  G. Könnecker

Recherches sur le Régime Alimentaire et le Comportement Prédateur des Décapodes Benthiques de la Pente Continentale de l'Atlantique Nordoriental (Golfe de Gascogne et Maroc) ... 397
  J.P. Lagardère

Feedback and Structure in Deposit-feeding Marine Benthic
Communities 409
    J.S. Levinton, G.R. Lopez, H.H. Lassen and U. Rakn

The Role of Physical and Biological Factors in the Distribution
and Stability of Rocky Shore Communities 417
    J.R. Lewis

Meiofaunal Community Structure and Vertical Distribution: a
Comparison of Some Co. Down Beaches 425
    C. Maguire

Etude Comparative de L'Efficacite de Deux Bernes et d'une
Suceuse en Fonction de la Nature du Fond 433
    H. Masse, R. Plante and J.P. Rays

Organization in Simple Communities: Observations on the
Natural History of *Hyale nilssoni* (Amphipoda) in High Littoral
Seaweeds 443
    P.G. Moore

Dynamics and Production of *Pectinaria koreni* in Kiel Bay, West
Germany 453
    F.H. Nichols

The effects of Storms on the Dynamics of Shallow Water Benthic
Associations 465
    E.I.S. Rees, A. Nicolaidou and P. Laskaridou

Pachycerianthus multiplicatus - Biotope or Biocoenosis 475
    B. O'Conner, G. Konnecker, D. McGrath and B.F. Keegan

On the Ecology of a Suspension Feeding Benthic Community:
Filter Efficiency and Behaviour 483
    E.M. Ölscher and K. Fedra

Strategies of Energy Transfer from Marine Macrophytes to
Consumer Levels: The *Posidonia oceanica* Example 493
    J.A. Ott and L. Maurer

The Benthic Ecology of Some Shetland Voes 503
    T.H. Pearson and S.O. Stanley

Predation Pressure and Community Structure of an Intertidal
Soft-bottom Fauna 513
    K. Reise

Zonation in Deep-sea Gastropods: The Importance of Biological
Interactions to Rates of Zonation 521
    M.A. Rex

Molluscan Colonization of Different Sediments on Submerged
Platforms in the Western Baltic Sea 531
    W. Richter and M. Sarnthein

Contents

Epifaunal Ecology of Intertidal Algae 541
R. Seed and P.J.S. Boaden

The Hermit Crab Microbiocoenosis - The Role of Mobile Secondary Hard-bottom Elements in a North Adriatic Benthic Community 549
M. Stachowitsch

Sub-littoral Community Structure of Oxwich Bay, South Wales in Relation to Sedimentological, Physical Oceanographic and Biological Parameters 559
P. Tyler

On the Shapes of Passive Suspension Feeders 567
G.F. Warner

The Structure and Seasonal Fluctuations of Phytal Marine Nematode Associations on the Isles of Scilly 577
R.M. Warwick

Dissolved Organics in the Nutrition of Benthic Invertebrates 587
B. West, M. de Burgh and F. Jeal

Studies on the Shallow, Sublittoral Epibenthos of Langstone Harbour, Hampshire, Using Settlement Panels 595
R.G. Withers and C.H. Thorp

Modification of Association and Swarming in North Adriatic *Mysidacea* in Relation to Habitat and Interacting Species 605
K.J. Wittmann

Predator-prey Interactions Between the Crab *Pilumnus hirtellus* (leach) and the Brittle Star *Ophiothrix quinquemaculata* on a Mutual Sponge Substrate 613
R.S. Wurzian

Analysis of Ecological Equivalents among Littoral Fish 621
C.D. Zander and A. Heymer

FOREWORD

The 11th European Symposium on Marine Biology was held on campus at University College, Galway (Ireland) from October 5 through October 11, 1976 and was hosted by the College's School of Marine Sciences.

Some three hundred participants were welcomed by Colm Ó hEocha, President of University College, Galway and Chairman of the Irish National Science Council, and by Bruno Battaglia, President of the "International Committee for the European Symposia on Marine Biology". The opening address was given by Pádraig Ó Céidigh on behalf of the Organising Committee.

Sixty-three papers were presented on the topic: "Aspects of the Biology of Benthic Organisms".

The "camera-ready copy" system was employed in publishing the proceedings and most of the contributions remain, substantially, as they reached us. We thank the authors for their assistance in this attempt to reduce costs and to increase the speed of publication.

We extend our gratitude to all who contributed to the success of the symposium, particularly the staff and students of the University College Galway Department of Zoology, and gratefully acknowledge the financial support of our sponsors: Allied Irish Banks Ltd., Bank of Ireland Ltd., Corning Teo., Electricity Supply Board, Gaeltarra Eireann, Guinness Group Sales (Ireland) Ltd., Ireland West Tourism, Irish Cement and Texaco Ireland Ltd.

B.F. KEEGAN
P. Ó CÉIDIGH
P.J.S. BOADEN
Editors.

OPENING ADDRESS

A Cháirde, are mo shon féin agus ar son mo chompháid i gColaiste Ollscoile na Gaillimhe, i dteannta leis an Doctúir Boaden, stiúrtheoir an Stasún Muir-Bitheoluíochta, i dtuisceart na hEireann, ba mhaith lion céad míle fáilte do chur roibh go léir. Dear friends, as I have just said in Irish, on my own behalf and on behalf of my colleagues in University College, Galway, together with Dr. Boaden, director of the northern Ireland Marine Laboratory, I welcome you to Ireland and I hope that your stay will be profitable and enjoyable. I would like to emphasise that this symposium is an all-Ireland hosted symposium, organised here at Galway by Dr. Brendan Keegan in full collaboration with our northern colleague Dr. Pat Boaden.

Our Symposium President, Professor Battaglia, has requested me to outline briefly the development of marine studies in Ireland. Here of course, as in most other countries, the beginning of marine biology is a story of the last century and of the turn of the present one. Perhaps we could say it commenced with the work of the naturalist John Vaughan Thompson about whom Joe Hedgepeth says in his Treatise on Marine Ecology and Palaeoecology that "although Thompson published very little, his work was of the highest quality". Vaughan Thompson first collected and studied marine plankton in Cork Harbour in the eighteen twenties, long before the word plankton was invented, and demonstrated, in 1828, that zoeas were in fact crab larvae. I am glad to say that work on decapod larvae, and decapods in general, is still a feature of Irish marine biology. During the mid-eighteen hundreds many famous Irish and foreign (mainly English) naturalists made collections, lists and taxonomic studies of marine organisms from all around the Irish coast. To mention but a few of the many, names spring to mind such as the Corkman George Allman, appointed Professor of Botany in Trinity College Dublin in 1884, who wrote monographs on hydroids and polyzoa and contributed to the series of Challeger Reports; W.H. Harvey, the algologist from Limerick and also Professor at Trinity College Dublin; G.C. Hyndmann, the Belfast auctioneer and specialist in marine mollusca;

the Rev. Thomas Hincks, son of a Corkman and educated in Belfast, whose
treatises on hydroids and polyzoa are still standard reference works; and,
of course, William Thompson author of the "Natural History of Ireland"
which contains most of the early records of Irish marine animals.
Naturalists of this period who worked in Galway and in the western marine
area generally, included my predecessor Professor A.G. Melville, first
Professor of Natural History in this College, whose specimens of bivalves
and decapods, dredged from the Connemara bays, are still in the College
collection; J.R. Kinahan, specialist in crustacean distribution and taxonomy,
and the Roundstone school teacher, William Mac Calla, who pioneered the
knowledge of the occurrence and distribution of marine organisms in the
Connemara area in the eighteen forties, (W.H. Harvey named the seaweed
<u>Cladophora macollona</u> after him).

In the latter half of the last century teams of marine biologists
organised dredging expeditions and investigations of the seas around Ireland.
In 1869 the English Navy put the survey vessel <u>Porcupine</u> at the disposal of
the Royal Society for deep-sea research off the west coast and as far north
as Rockall. In the eighteen eighties and later, the Royal Dublin Society
and the Royal Irish Academy sponsored surveys of fishing grounds and deep-sea
dredging expeditions off west, south west and north west coasts of Ireland.
The published reports of these surveys are familiar to students of the
deeper water and coastal benthos of the eastern N. Atlantic to this day.

During the first two decades of the present century much important
fundamental marine biological research was carried out by the Irish Fisheries
Department. Such familiar names as Holt, Tattersall, Farran, Selbie, Massey,
Calman and Southern appear as contributors to the <u>Fisheries Ireland Scientific
Investigations</u> of this time. The Royal Academy sponsored Clare Island
survey, which included many works on marine ecology and taxonomy, took place
during this period and incidentally, on the historical scene, the English
Navy used the fisheries research vessel "Helga" to shell the Headquarters
of the Irish army during the Easter Rising.

After our war of Independence, and with the setting up of the new state
in 1922, finance was scarce and our small resources were ploughed into the
key development areas of agriculture and industry. This stood to us during
the emergency period 1939-45 but, unfortunately, it resulted in an almost

complete abandonment of government involvement in marine research for a
period of forty years.   During these lean years practically all marine
biological work was carried out by the Universities.   Professor Louis Renouf,
of University College Cork, recognised Lough Ine as a unique situation for
marine studies and set up a small field station there in the nineteen thirties.
This later attracted the attention of Professor Kitching of Bristol who also
established a small field laboratory on the Lough.   In the nineteen forties
Dr. Mairin de Valera commenced work at University College Galway on the
marine algae of the west coast and founded a school of marine botany here
that continues to flourish.   About the same time in the north, Dr. Williams,
lecturer in Zoology at Belfast, started a marine laboratory at Portaferry
of which Dr. Boaden is the present director.

In 1956 I joined the staff of the University College Galway department of
Natural History and shortly after coming here realised the great potential the
west coast has for marine biology.   I started field courses in marine zoology
in 1958 and persuaded the then professor of Natural History, Professor
T.J. Dinan, to buy and equip a fisherman's shed at Carna as a small field
laboratory.   We also purchased a small motor-launch for inshore dredging
and plankton work.   Carna, with its wide variety of littoral and sub-littoral
habitats, has a very rich and interesting fauna and the little laboratory
sited at the quay proved a very suitable centre for investigating the whole
area.   In 1962, I was appointed Professor of Zoology at Galway and adopted
a policy of specialisation in marine zoology for the new department.   A
resident biologist was appointed in the Carna laboratory and collection of
data on a regular basis was initiated.   In conjunction with Professor
de Valera and U.C.G. President O hEocha (then Professor of Biochemistry),
I persuaded the College authorities to institute a department of Oceanography
to which Professor McKay Bary was appointed in 1970.   Throughout the sixties
the numbers of postgraduate students reading Zoology and specialising in
marine studies continued to increase.   As the research school grew we
decided that in the national interest about a third of the effort of the
department should be directed towards assisting the fishing industry.   With
this in mind, Dr. Keegan and myself approached the government with a proposal
for setting up a special Shellfish Research Laboratory at Carna.   Due
mainly to the interest of Mr. Charles Haughey, then Minister for Finance,
the Shellfish Research Laboratory at Carna was built and equipped by a direct

grant of monies from the Department of Finance. Today, Dr. John Mercer as deputy director heads a team working there on several research programmes, including cultivation of commercially important shellfish.

Two years ago the Governing Body, recognising the marine interests of many departments in the College, officially established a School of Marine Sciences. One of the first activities of this school was a synecological survey of Killary Harbour, jointly carried out by University College Galway staff and scientists from the Station Biologique de Roscoff, France under the combined auspices of our National Science Council and the French C.N.R.S. Present marine biological programmes include microbiology (Dr. Patching), marine plankton (Dr. Fives, O'Céidigh and Mc Kay Bary), population genetics (Dr. Wilkins) and marine benthos (Dr. Keegan). The theme of the present symposium was chosen because it reflected the interest in benthic research here and research on meiobenthos being carried out by Dr. Boaden at Portaferry in the north. We regard your presence here as a recognition of our efforts and an encouragement and stimulus to future development of marine science in our country.

                                              Padraig Ó Céidigh,
                                              Professor of Zoology,
                                              University College,
                                              Galway,
                                              Ireland.

LIST OF PARTICIPANTS AND THEIR ADDRESSES

ALDRICH, J.C., Department of Zoology, Trinity College, Dublin 2, Ireland.

ALDRICH, (Mrs.), Same address.

ANDERSIN, A.B., Institute of Marine Research, Box 136, SF-00121 Helsinki/Helsingfors 12, Finland.

ANEER, G., Askö Laboratory, Box 58, S-150 13 Trosa, Sweden.

ANKAR, S., Askö Laboratory, Bos 58, S-150 13 Trosa, Sweden.

ARNTZ, W., Institut für Meereskunde an der Universität Kiel, Wischhofstr. 1, 23 Kiel 14, Fed. Rep. of Germany.

AUSTIN, J., School of Biological and Earth Sciences, the Northern Ireland Polytechnic, Jordanstown, Co. Antrim, Northern Ireland.

BAKKE, T., Biological Station, Espegrend, N-5065 Blomsterdalen, Norway.

BARNES, H., The Dunstaffnage Marine Research Laboratory, P.O. Box No. 3, Oban, Argyll PA34-4AD, Scotland.

BARNES, (Mrs.) Same address.

BARRY, M. Shellfish Research Laboratory, Carna, Co. Galway, Ireland.

BATTAGLIA, B., Instituto di biologia Animale dell'Universitá, 10 Via Loredan, Padova, Italy.

BATTAGLIA, (Mrs.) Same address.

BELSCHER, T., Station Biologique, 29211 Roscoff, France.

BERGERARD, J., Station Biologique, 29211 Roscoff, France.

BERGERARD, P., Station Biologique, 29211 Roscoff, France.

BERGMANS, M., Laboratorium voor Ecologie en Systematiek, Vrije Universiteit Brussel, Pleinlaan 2, 1050 Brussel, Belgium.

BEUKEMA, J.J., Netherlands Institute for Sea Research, P.O. Box 59, Texel, The Netherlands.

BILIO, M., Centro Ricerche Ittiologiche Di Comacchio, Fish Culture Research Institute, S.I.VAL.CO, Via Mazzini 200, 44022 Comacchio (Ferrara), Italy.

deBLOK, J.W., Nederlands Institut voor Onderzoek der Zee, Postbus 58, Texel, The Netherlands.

deBLOK, (Mrs.) Same address.

BOADEN, P.J., Marine Biology Station, Portaferry, Co. Down, Northern Ireland.

List of Participants and their Addresses

BOADEN, (Mrs.) Same address.

BODIN, P., Universite de Bretagne Occidentale, Faculté des Sciences, Laboratoire d'Oceanographie Biologique, Ave. Le Gorgeu, 29283, Brest Cédex, France.

BOOKOUT, C.G., Duke University Marine Laboratory, Beaufort, N.C. 28516, U.S.A.

BORGHOUTS, C.H., Delta Institute for Hydrobiological Research, Vierstraat 28, Yerseke, The Netherlands.

BOUCHER, D., Université de Bretagne Occidentale, Laboratoire d'Océanographie Biologique, Faculté des Sciences, Ave. Le Gorgeu, 29283 Brest, Cédex, France.

BOUCHER, M.J. Same address.

deBOVEE, F., Laboratoire Arago, 66650 Banyuls/mer, France.

BRIGGS, R.P., Fisheries Research Laboratory, 38 Casteror Road, Coleraine, Co. L'Derry, Northern Ireland.

BROWN, D., Shellfish Research Laboratory, Carna, Co. Galway, Ireland.

BROWN, R.A., Department of Zoology, Queens University Belfast, Belfast BT7 1NN, Northern Ireland.

BROWN, M.F., Same address.

deBROYER, C., Institut Royal des Sciences Naturelles de Belgique, Section d'Oceanographie, rue Vautier, 31, B-1040 Bruxelles, Belgium.

BRUCHMANN, Mrs. W., ll. Zoologisches Institut, Berliner Str. 28, D-34 Göttingen, Fed. Rep. of Germany.

BRUNEL, P., Départment des sciences biologiques, Université de Montréal, P.O. Box 6128, Station 'A', Montréal, Quebec, Canada H3C 3J7.

BUHR, K.J., Institut für Meeresforschung, Am Handelshafen 12, 285 Bremerhaven, Fed. Rep. of Germany.

CABIOCH, L., Station Biologique, 29211 Roscoff, France.

CALCOEN, J., Laboratorium voor Vergelykende Anatomie, Vrij Universiteit Brussel, Pleinlaan 2, Brussel, Belgium.

CASTEL, J., Institut de Biologie Marine, Université de Bordeaux, 2 rue du Professeur Jolyet, 33120 Arcachon, France.

CULLINANE, J., Department of Botany, University College, Cork, Ireland.

CASTRIC, A., Laboratoire de Biologie Marine, Concarneau 29110, France.

CEDERWALL, H., Askö Laboratory, Box 58, S-150 13 Trosa, Sweden.

CHASSE, P., Université de Bretagne Occidentale, Laboratoire d'Oceanographie Biologique, Faculté des Sciences, Ave. Le Gorgeu, 29283 Brest, Cédex, France.

List of Participants and their Addresses

CHRISTIANSEN, M.E., Zoological Museum, University of Oslo Sarsgt. 1, Oslo 5, Norway.

CINELLI, F., Reparto Ecologia Marina Stazione Zoologica di Napoli, Punta S. Pietro, 80070 Ischia Porto (Naples), Italy.

CINELLI, (Mrs.) Same address.

CONNEELY, M., Zoology Department, University College, Galway, Ireland.

COOKE, P., Botany Department, University College, Galway, Ireland.

CORRAL, J., Instituto Espanol de Oceanografia, Alcala 27, $4^{\underline{o}}$, Madrid, Spain.

COSTLOW, J.D., Duke University Marine Laboratory, Beaufort N.C. 28516, U.S.A.

COSTLOW, (Mrs.) Same address.

CRISP, D.J., N.E.R.C. Unit, Marine Science Laboratories, Menai Bridge, Gwynedd, North Wales.

CREUTZBERG, F., Netherlands Institute of Sea Research, P.O. Box 59, Texel, The Netherlands.

DAVIS, W.P., Bears Bluff Field Station, P.O. Box 368, Johns Island, S.C. 29455, U.S.A.

DELEPINE, R., Biologie Vegetale Marine, 7 Quai Saint-Bernard, 75230 Paris, Cedex 05, France.

DELEPINE, (Mrs.) Same address.

DINNEEN, P., Zoology Department, University College, Galway, Ireland.

DITIMER, J.D., Institut für Meeresforschung, 285 Bremerhaven-G, Am Handelshafen 12, Fed. Rep. of Germany.

DRINKWAARD, A.C., Netherlands Institute for Fishery Investigations, Mariculture Experimental Station, Polder't Horntje - H 47, Isle of Texel, The Netherlands.

DRINKWAARD, (Mrs.) Same address.

DUCHENE, J.C., Laboratoire Arago, 66650 Banyuls/mer, France.

DUNICAN, K. Microbiology Department, University College, Galway, Ireland.

DUNNE, J., Marine Zoological Station, Carna, Co. Galway, Ireland.

EAGLE, R.A., Fisheries Laboratory, (MAFF), Burnham-on-Crouch, Essex, England.

EDWARDS, J.C., Duke University Marine Laboratory, Beaufort, N.C. 28516, U.S.A.

ELLIOTT, A.J., Division of Biology, Department of Anatomy, Royal College of Surgeons in Ireland, St. Stephens Green, Dublin 2, Ireland.

List of Participants and their Addresses

ELMGREN, R., Askö Laboratory, Box 58, S-150 13 Trosa, Sweden.

EMSON, R.H., Zoology Department, Kings College, Strand, London WC2 R2LS, England.

ERSEUS, C., Department of Zoology, University of Gothenburg, Fack, S-400 33 Göteborg, Sweden.

ERWIN, D.G., Botany and Zoology Department, Ulster Museum, Botanic Gardens, Belfast BT9 5AB, Northern Ireland.

ESSINK, K., Rijksinstituut voor Zuivering van Afvalwater, Rijksweg 133, Sappemeer, The Netherlands.

EVANS, S., Institute of Zoology, Box 561, 751 22 Uppsala, Sweden.

EVANS, E.C., Naval Undersea Center/Hawaii Laboratory, P.O. Box 997, Kailua, Hawaii, 96734.

FALLER-FRITSCH, R., Department of Biological Sciences, University of Exeter, Hatherly Laboratories, Prince of Wales Road, Exeter EX4 4PS, England.

FEDRA, K., Zoologisches Institut d. Universität Wien, Wahringerstr. 17/VI, A 1090 Wien, Austria.

FIVES, J., Zoology Department, University College, Galway, Ireland.

FLÜGEL, H.J., Institut für meereskunde, Dusternbrooker Weg 20, 23 Kiel, Fed. Rep. of Germany.

FOREMAN, R.E., Department of Botany, University of British Columbia, Vancouver, B.C., Canada V6T 1W5.

FOTTRELL, P.F., Biochemistry Department, University College, Galway, Ireland.

FRESI, E., Villa Acquario, Ischia (Napoli), Italy.

FUREY, T., Oceanography Department, University College, Galway, Ireland.

GAGE, J.D., Scottish Marine Biological Assoc., Dunstaffnage Marine Research Lab., P.O. Box 3, Oban Argyll, PA34 4AD, Scotland.

GALVIN, J., Microbiology Department, University College, Galway, Ireland.

GARWOOD, P.R., 2 Oaklands, Ponteland, Northumberland, England.

GELDIAY, R., General Zoology Department, Institute of Hydrobiology, Fen Faculty, Ege University, Bornova-Izmir, Turkey.

GENTIL, F., Station Biologique de Roscoff, 29211 Roscoff, France.

GEORGE, C.L., Institute of Marine Environmental Research, 67 Citadel Road, Plymouth, Devon, England.

GEORGE, J.D., British Museum (Natural History), Cromwell Road, London SW7 5BD, England.

GEORGE, J.J., Same address.

GERDES, D., Institut für Meeresforschung, 285 Bremerhaven-G, Am Handelshafen 12, Fed. Rep. of Germany.

GHANNUDI, S.A., Zoology Department, Faculty of Science P.O. Box 656, University of Fateh, Tripoli, Libya.

GIERE, O., Universität Hamburg, Zoologisches Institut, und Zoologisches Museum, 2000 Hamburg 13, Martin-Luther-King-Platz 3, Fed. Rep. of Germany.

GLACON, R., Institut de Biologie Maritime, 28 Avenue Foch, 62930 Wimereux, France.

GLÉMAREC, M., Universite de Bretagne Occidentale, Laboratoire d'Oceanographie Biologique, Faculté des Sciences, Avenue Le Gorgeu, 29283 Brest, Cedex, France.

GLEMAREC, (Mrs.) Same address.

GOODBODY, I., Zoology Department, University of West Indies, P.O. Box 12, Kingston 7, Jamaica.

GOSLING, E., Zoology Department, University College, Galway, Ireland.

GOVAERE, J., Lab. v. Morfologie and Systematiek der Dieren, Ledeganckstraat 35, B-9000 Ghent, Belgium.

GRAINGER, J.N.R., Zoology Department, Trinity College, Dublin 2, Ireland.

GRASSLE, J., Marine Biological Laboratory, Woods Hole Mass. 02543, U.S.A.

GRASSLE, J.F., Marine Biological Laboratory, Woods Hole, Mass. 02543, U.S.A.

GRAY, P., Portsmouth Polytechnic, Marine Laboratory, Ferry Road, Hayling Island, Hants PO11 ODG, England.

GREEN, J.M., Department of Biology and M.S.R.L., Memorial University, St. John's, Newfoundland, Canada.

deGROOT, S.J., Netherlands Inst. Fish. Invest., P.O. Box 68, Ymuiden, The Netherlands.

GUDMUNDSSON, H., Department of Zoology, The University of Newcastle-upon-Tyne, NE1 7RU, England.

GUILLE, A., Laboratoire de Biologie des Invertebrés Marins, Museum National d'Histoire Naturelle, 55 Rue Buffon, 75005 Paris, France.

GULLIKSEN, B., University of Tromsö, Institute of Biology and Geology, Pb. 790, N-9001 Tromsö, Norway.

List of Participants and their Addresses

GUTERSTAM, B., Askö Laboratory, Box 58, S-150 13 Trosa, Sweden.

HAGERMAN, L., Marine Biological Laboratory, Strandpromenaden, Dk-3000 Helsingor, Denmark.

HARTNOLL, R.G., Department of Marine Biology, University of Liverpool, Port Erin, Isle of Man.

HARTNOLL, (Mrs.) Same address.

HARTWIG, E., Zoological Institute, University of Hamburg, 2 Hamburg 13, Martin-Luther-King-Platz 3, Fed. Rep. of Germany.

HEALY, B., Zoology Department, University College, Dublin, Ireland.

HEDGPETH, J.W., Marine Science Center, Marine Science Drive, Newport, Oregon 97365, U.S.A.

HEIP, C., Laboratorium Morfologie Systematiek, Ledeganckstraat 35, B-9000 Belgium.

HESTHAGEN, I.H., University of Oslo, Avd. marin zoologi og marin kjemi, P.O. Box 1064, Blindern, Oslo 3, Norway.

HISCOCK, K., Field Studies Council, Oil Pollution Research Unit, Orielton Field Centre, Pembroke, Dyfed, Wales.

HOARE, B., Department of Marine Biology, Marine Science Laboratories, Menai Bridge, Anglesey LL59 5EH, Wales.

HOLMES, J.M.C., Natural History Division, National Museum, Dublin 2, Ireland.

HOLMES, (Mrs.) Same address.

HOVGAARD, P., Biological Station, Espegrend, 5065 Blomsterdalen, Norway.

HUMMON, W.D., Department of Zoology and Microbiology, Ohio University, Athens, Ohio 45701, U.S.A.

HYNES, M., Chemistry Department, University College, Galway, Ireland.

IGLESIAS, J., Instituto Espanol de Oceanografia, Alcala 27, 40 Madrid, Spain.

JACKSON, J., The Johns Hopkins University, Baltimore, Maryland 21218, Department of Earth and Planetary Sciences, Baltimore, U.S.A.

JENSEN, P., Marine Biological Laboratorium, University of Copenhagen, Strandpromenaden DK-3000 Helsingør, Denmark.

JEPHSON, N.A., Portsmouth Polytechnic Marine Laboratory, Ferry Road, Hayling Island, Hants, PO11 0DG, England.

JONES, J.M., Department of Marine Biology, University of Liverpool, Port Erin, Isle of Man.

List of Participants and their Addresses

JONES, N.S., Department of Marine Biology, University of Liverpool, Port Erin, Isle of Man.

KANNEWORFF, E., Risby, 3100 Helsingør, Denmark.

KAUTSKY, N., Askö Laboratory, Box 58, S-150 13 Trosa, Sweden.

KEEGAN, B.F., Zoology Department, University College, Galway, Ireland.

KINNE, O., Biologische Anstalt Helgoland, Palmaille 9, D-2000 Hamburg 50, Fed. Rep. of Germany.

KLEPAL, W., 1. Zoologisches Institut, Lahrkanzel für Meeresbiologie, A-1090 Wien, Wahringer Strasse 17/6, Austria.

KNIGHT-JONES, E.W., Department of Zoology, University of Swansea, Singleton Park, Swansea SA2 8PP, Wales.

KNIGHT-JONES, P., Same address.

deKOCK, W.C., Central Laboratory TNO, P.O. Box 57, Den Helder, The Netherlands.

KØIE, M., Marine Biological Laboratory, DK-3000 Helsingør, Denmark.

KÖNNECKER, G., Zoology Department, University College, Galway, Ireland.

KUHL, H., 2190 Cuxhaven, Siedelhof 16, Fed. Rep. of Germany.

KUHL, (Mrs.) Same address.

LAGARDERE, J.P., Antenne de la Station Marine d'Endoume - C.R.E.O., Allee des Tamaris, 17 La Rochelle, France.

LAGARDERE, (Mrs.) Same address.

LAMBE, E., Botany Department, University College, Galway, Ireland.

LAMBECK, R.H.D., Delta Institute for Hydrobiological Research, Vierstraat 28, Yerseke, the Netherlands.

LAPPAILEN, A., Askö Laboratory, Box 58, S-150 13 Trosa, Sweden.

LASSEN, H., Institute of Ecology and Genetics, University of Aarhus, 8000 Aarhus C, Denmark.

LASSERRE, P., Institut de Biologie Marine, Université de Bordeaux, 2 rue du Professeur Jolyet, 33120 Arcachon, France.

LASSIG, J., Institute of Marine Research, Box 136, SF-00121 Helsinki/ Helsingfors 12, Finland.

LA TOUCHE, R.W., Zoology Department, Trinity College, Dublin 2, Ireland.

LAWLESS, A., Zoology Department, University College, Galway, Ireland.

List of Participants and their Addresses

LEE, J., The City College, Convent Avenue, 138 Street, New York, 10031, U.S.A.

LEFTLEY, J.W., Dunstaffnage Marine Research Laboratory, P.O. Box 3, Oban, Argyll PA34 4AD, Scotland.

LEVINTON, J., Department of Ecology and Evolution, Station University of New York, Stony Brook, N.Y. 11794, U.S.A.

LEWIS, J.R., Wellcome Marine Laboratory, Robin Hood's Bay, N. Yorkshire, England.

LINDAHL, K., Kristineberg Marine Biol. Station, S-450 34 Fiskebäckskil, Sweden.

LOPEZ, G.R., Instit. of Ecology and Genetics, University of Aarhus, DK-800 Aarhus C., Denmark.

LUNDALV, T., Kristineberg Marine Biol. Station, S-450 34 Fiskebäckskil, Sweden.

LUNDALV, (Mrs.) Same address.

MAGUIRE, C., Marine Biology Station, Portaferry, Co. Down, Northern Ireland.

MANGOLD, K., Laboratoire Arago, 66650 Banyuls/mer France.

MANGOLD, (Mrs.) Same address.

MATHERS, N.F., Zoology Department, University College, Galway, Ireland.

MAURER, L., 1. Zoologisches Institut der Universität Wien, Wahringerstrasse 17/6, A-1090 Vienna, Austria.

MAZZELIA, L., Reparto Ecologia Marina Stazione Zoologica, Napoli, Punta S. Pietro, 80070 Ischia Porto, (Naples), Italy.

McCARTHY, K., Zoology Department, University College, Galway, Ireland.

McCARTHY, M.F., Mathematical Physics Department, University College, Galway, Ireland.

McDONALD, M., Botany Department, University College, Galway, Ireland.

McGRATH, D., Zoology Department, University College, Galway, Ireland.

McK BARY, B., Oceanography Department, University College, Galway, Ireland.

McLUSKY, D.S., Department of Biology, The University, Stirling, Scotland.

MENA, O., St. George's Hall, Elmhurst Road, Reading RG1 5H2, Berkshire England.

MERCER, J.P., Shellfish Research Laboratory, Carna, Co. Galway, Ireland.

List of Participants and their Addresses   xxv

MERGNER, H., Ruhr-Universität, Lehrstuhl für Spezielle Zoologie, 463 Bochum, Universitätsstrasse ND F05, Fed. Rep. of Germany.

MICHAELIS, H., Forschungsstelle für Insel-und Küstenschutz, An der Mühle 5, 2982 Norderney, Fed. Rep. of Germany.

MONAGHAN, E.C., Oceanography Department, University College, Galway, Ireland.

MOORE, P.G., University Marine Biological Station, Millport, Isle of Cumbrae, KA28 OEG, Scotland.

MORAN, F., Institute for Industrial Research and Standards, Ballymun Road, Dublin 9, Ireland.

MOYNIHAN, E., Shellfish Research Laboratory, Carna, Co. Galway, Ireland.

MOYSE, J., Zoology Department, University of Swansea, Singleton Park, Swansea SA2 8PP, Wales.

MULDER, M., Netherlands Institute for Sea Research, Bova, P.O. Box 59, Texel, The Netherlands.

MULLIGAN, N., Zoology Department, University College, Galway, Ireland.

MURKEN, J., Institut für Meeresforschung, Am Handelshafen 12, 285 Bremerhaven, Fed. Rep. of Germany.

MWAISEJE, B., N.E.R.C. Unit, Marine Science Laboratories, Menai Bridge, Gwynedd, Wales.

MYERS, A.A., Department of Zoology, University College, Cork, Ireland.

NAYLOR, E., Department of Marine Biology, University of Liverpool, Port Erin, Isle of Man.

NICHOLS, F.H., Office of Marine Geology, U.S. Geological Survey, 345 Middlefield Road, Menlo Park, California 94025, U.S.A.

NICOLAIDOU, A., Department of Marine Biology, Marine Sc. Labs., Menai Bridge, Anglesey, LL59 5EH, Wales.

NICOLAISEN, W., Marine Biological Laboratory, Strandpromenaden, 3000 Helsingør, Denmark.

NIELSEN, C., Marine Biological Laboratory, DK-3000 Helsingør, Denmark.

NOODT, W., Zoologisches Institut der Universität, D23 Kiel, Hegewischstrasse 3, Fed. Rep. of Germany.

NOONAN, E., Shellfish Research Laboratory, Carna, Co. Galway, Ireland.

O'BRIEN, F.I., Kinsale Oysters Ltd., Castlepark Village, Kinsale, Co. Cork, Ireland.

O'CEIDIGH, P., Zoology Department, University College, Galway, Ireland.

O'CINNEIDE, S., Chemistry Department, University College, Galway, Ireland.

O'CONNOR, B., Zoology Department, University College, Galway, Ireland.

O'HALLORAN, G., Shellfish Research Laboratory, Carna, Co. Galway, Ireland.

O hEOCHA, C., University College, Galway, Ireland.

CLIFF, W.D., National Institute for Water Research, Council for Scientific and Industrial Research, P.O. Box 17001, Congella, 4013 Natal, South Africa.

OLIVER, G., Natural History Department, Royal Scottish Museum, Chambers Street, Edinburgh Eh1 1JF, Scotland.

OLSCHER, E., 1. Zoologisches Institut der Universität Wien, Währingerstr. 17/6, A-1090 Wien, Austria.

O'SULLIVAN, B., Shellfish Research Laboratory, Carna, Co. Galway, Ireland.

OTT, J., 1. Zoologisches Institut der Universität Wien, Wahringerstr. 17/6, A-1090 Vienna, Austria.

OTTWAY, B., Botany Department, University College, Galway, Ireland.

OUG, E., Biological Station, Espegrend, N-5065 Blomsterdalen, Norway.

PABST, B., Zoologisches Museum der Universität Zurich, Künstlerg 16, CH-8006 Zürich, Switzerland.

PANSINI, M., c/o Instituto di Zoologia dell'Università, Via Balbi 5, 16126 Genova, Italy.

PARKER, J.G., Department of Agriculture, Fisheries Research Laboratory, The Cutts, Coleraine, Co. Londonderry, Northern Ireland.

PARKER, M., Department of Agriculture and Fisheries, Fisheries Division, 3 Cathal Brugha Street, Dublin 1, Ireland.

PARMENTIER, J.H., Centraal Laboratorium TNO, P.O. Box 217, Delft, The Netherlands.

PARTRIDGE, K., Shellfish Research Laboratory, Carna, Co. Galway, Ireland.

PATCHING, J., Microbiology Department, University College, Galway, Ireland.

PAVICIC, J., Center for Marine Research, Institut "R. Boskovic", Rovinj, Yugoslavia.

PEARSON, T.H., Dunstaffnage Marine Research Laboratory, P.O. Box No. 3, Oban, Argyll, PA34 4AD, Scotland.

PEER, D.L., Marine Ecology Laboratory, Bedford Institute of Oceanography, Dartmouth, Nova Scotia, Canada.

List of Participants and their Addresses

PEER, (Mrs.) Same address.

PERSOONE, G., State University of Ghent, Lab. for Biological Research in Environmental Pollution, Plateaustraat 22, B 9000 Ghent, Belgium.

PFANNKUCHE, O., Zoologisches Inst. u. Mus., Martin-Luther-King-Platz 3, D-2000 Hamburg 13, Fed. Rep. of Germany.

PICKTON, B., Ulster Museum, Botanic Gardens, Belfast BT9 5AB, Northern Ireland.

PLANTE, R., Station Marine d'Endoume, Rue de la Batterie des Lions, 13007 Marseille, France.

PLATT, H.M., British Antarctic Survey, Madingley Road, Cambridge CB3 OET, England.

PRONZATO, R., c/o Instituto di Zoologia dell'Università, Via Balbi 5, 16126 Genova, Italy.

PYBUS, M., Oceanography Department, University College, Galway, Ireland.

PYBUS, C., Regional Technical College, Galway, Ireland.

RACHOR, E., Institut für Meeresforschung, Am Handelshafen 12, D-2850 Bremerhaven, Fed. Rep. of Germany.

RAINE, R., Microbiology Department, University College, Galway, Ireland.

REES, E.I.S., Marine Science Laboratories, Menai Bridge, Gwynedd, Wales.

REGILIO-TRONCONE, M., Stazione Zoologiga, Acquario, Villa Communale, 80121 Napoli, Italy.

REISE, K., II. Zoologisches Institut, Berliner Str. 28, D-34 Göttingen, Fed. Rep. of Germany.

RETIERE, C., Laboratoire Maritime, Museum d'Histoire Naturelle, 35800 Dinard, France.

REX, M.A., Department of Biology, University of Massachusetts/Boston, Boston, Massachusetts 02125, U.S.A.

REX, (Mrs.) Same address.

RIBI, G., 8006 Zürich, Nordstrasse 7, Switzerland.

RICHARDSON, M.G., Bannamin, Bridge End, Burra Isle, Shetland.

RICHTER, G., Senckenberg-Institut, 6000 Frankfurt, Senckenberganlage 25, Fed. Rep. of Germany.

RICHTER, (Mrs.) Same address.

RICHTER, W., Geologisch-Palaontologisches Institut, Olshausenstr. 40-60, D-2300 Kiel 1, Fed. Rep. of Germany.

ROBERTS, C., Ulster Museum, Botanic Gardens, Belfast BT9 5AB, Northern Ireland.

ROSENBERG, R., Swedish Water and Air Poll. Res. Lab., P.O. Box 5207, S-40224 Gothenburg, Sweden.

ROSENBERG, (Mrs.) Same address.

RUMOHR, H., Institut F. Meereskunde, D-23 Kiel, Düsternbrooker Weg. 20, Fed. Rep. of Germany.

RYLAND, J.S., Department of Zoology, University College of Swansea, Swansea SA2 8PP, Wales.

SALLING, P., A/S Københavns Pektinfabrik, DK-4623 Lille Skensved, Denmark.

SALZWEDEL, H., Institut für Meeresforschüng, Am Handelshafen 12, D-2850 Bremerhaven, Fed. Rep. of Germany.

SARA, M., Instituto di Zoologia dell'Universitá, Via Balbi 5, 16126 Genova, Italy.

SCHELTEMA, R.S., Woods Hole Oceanographic Institution, Woods Hole, Massachusetts, U.S.A. 02543.

SCHELTEMA, E.H., Same address.

SCHULTE, E.H., Cnen-Euratom, Lab. Cont. Mare, 1-19030 Tellara (SP) Via Fiascherino, Italy.

SCHULTE, (Mrs.) Same address.

SISULA, H., Zoological Station, SF-10850 Tvärminne, Finland.

SKULT, P., Zoological Station, 10850 Tvärminne, Finland.

SYLVAND, B., Laboratoire de Zoologie, Service du Pr. P. Lubet, Université de Caen, 14000, France.

SYLVAND, (Mrs.) Same address.

SMITH, S.M., Natural History Department, Royal Scottish Museum, Chambers Street, Edinburgh EH1 1JF, Scotland.

SMOL-KOCHANEK, N., State University of Ghent, Laboratorium voor Morfologie en Systematiek der Dieren, Ledeganckstraat, 35, 9000 Ghent, Belgium.

STACHOWITSCH, M., Lehrkanzel für Meeresbiologie, Wäringerstrasse 17/6, Vienna, 1090, Austria.

STEWART, M., Department of Biology, The Open University, Milton Keynes MK7 6AA, England.

STROMBERG, J.O., Kristineberg Marine Biological Laboratory, 450 34 Fiskebäckskil, Sweden.

List of Participants and their Addresses

SVOBODA, A., Ruhr-Universität Bochum, Lehrstuhl für Spezielle Zoologie, Postfech 2148, D-463 Bochum-Querenburg, Fed. Rep. of Germany.

TALLMARK, B., Institute of Zoology, Uppsala University, Box 256, S-75122, Sweden.

TESTER, P.A., School of Oceanography, Oregon State University, Corvallis, Oregon 97331, U.S.A.

THEEDE, H., Institut für Meereskunde an der Universität Kiel, 23 Kiel, Düsternbrooker Weg 20, Fed. Rep. of Germany.

THIRIOT-QUIÉVREUX, C., Station Zoologique, 06230 Villefranche sur Mer, France.

THOMAS, J.P., Sandy Hook Lab., Highlands, New Jersey 07732, U.S.A.

THORP, C.H., Portsmouth Polytechnic, The Marine Lab., Ferry Road, Hayling Island, PO11 ODG, England.

TINNBERG, L., Stockholms Universitet, Botaniska Institutionen, Lilla Frescati 104 05, Stockholm, Sweden.

TWIDE, P., A/S Københavns Pektinfabrik, DK-4623 Lille Skensved, Denmark.

TYLER, P.A., Department of Oceanography, University College of Swansea, Swansea, South Wales.

TYNDALL, P., Zoology Department, University College, Galway, Ireland.

URSIN, E., Danish Institute for Fishery and Marine Research, Charlottenlund Slot, DK-2920 Charlottenlund, Denmark.

URSIN, (Mrs.) Same address.

VAHL, O., Institute of Biology and Geology, University of Tromsø, P.O. Box 790, N-9001 Tromsø, Norway.

deVALERA, M., Botany Department, University College, Galway, Ireland.

VanARKEL, M.A., Netherlands Institute for Sea Research, Bova, P.O. Box 59, Texel, The Netherlands.

VELIMIROV, B., Department of Zoology, Marine Biology Unit, University of Capetown, Rondebosch 7700, Cape Town, South Africa.

WALKER, A.J.M., Marine Science Laboratories, Menai Bridge, Anglesey, Wales.

WARNER, G.F., Zoology Department, Reading University, Whitknights, Reading, Berks, RG6 2AJ, England.

WARWICK, R.M., I.M.E.R., 67 Citadel Road, Plymouth, PL1 3DH, England.

WEBB, J.E., Department of Zoology, Westfield College, Hampstead, London NW3 7ST, England.

List of Participants and thier Addresses

WEST, B., Department of Zoology, Trinity College, Dublin 2, Ireland.

WHITE, M.G., British Antarctic Survey, Madingley Road, Cambridge, England.

WHITTMANN, K., Lehrkanzel für Meeresbiologie, Währingerstrasse 17/6, A-1090 Wien, Austria.

WIDDOWS, J., I.M.E.R., 67 Citadel Road, Plymouth, PL1 3DH, England.

WILKINS, N.P. Zoology Department, University College, Galway, Ireland.

WILSON, J., Shellfish Research Laboratory, Carna, Co. Galway, Ireland.

WINTER, J., Institut für Meeresforschung, Am Handelshafen 12, 285 Bremerhaven, Fed. Rep. of Germany.

WITHERS, R.G., The Marine Laboratory, Ferry Road, Hayling Island, Hants, PO11 0DG, England.

deWOLF, P., Netherlands Inst. for Sea Research, P.O. Box 59, Texel, The Netherlands.

WOLFF, W., Netherlands Institute for Sea Research, P.O. Box 59, Texel, The Netherlands.

WURZIAN, R.S., Lehrkanzel, für Meeresbiologie, Währingerstr. 17/6, 1090 Wien, Austria.

ZANDER, C.D., Universität Hamburg, Zoologisches Institut und Zoologisches Museum, 200 Hamburg 13, Martin-Luther-King-Platz 3, Fed. Rep. of Germany.

ZANDER, R., Same address.

ZIJLSTRA, J., Netherland Inst. for Sea Research, Post Box 59, Texel, The Netherlands.

ZOUTENDYK, P., Oceanography Department, University of Cape Town, 7700 Rondebosch, South Africa.

ADDITIONAL PARTICIPANTS

BRACHI, R., 2 Coastguard, Kimmeridge. Wareham, Dorset, BHZO 5PF, England.

ERGEN, Z., Zoology Department, University College of Swansea, Swansea SA2 8PP, England.

GUELORGET, O., Universite des Sciences et Techniques du Languedoc, Laboratoire d'Hydrobiologie Marine, Place E. Bataillon, 34060 MONTPELLIER Cedex, France.

MICHEL, P., (address as in Geulorget above)

PREVIOUS EUROPEAN SYMPOSIA ON MARINE BIOLOGY

1st EUROPEAN SYMPOSIUM ON MARINE BIOLOGY

    Helgoland Fed. Rep. Germany, Sept. 26-Oct. 1, 1966.
    Biologische Anstait Helgoland.

    TOPICS - *Experimental ecology, its significance as a marine biological tool.*

           - *Subtidal ecology particularly as studied by diving techniques.*

           - *The food web in the sea.*

    Kinne, O., and H. Aurich (Editors).
    Helgoländer wissenschaftliche meeresuntersuchungen 15 (1967). 721 p.

2nd EUROPEAN SYMPOSIUM ON MARINE BIOLOGY

    Bergen, Norway, Aug. 24-28, 1967.
    Biological Station, Espegrend.

    TOPICS - *The importance of water movements for biology and distribution of marine organisms.*

    Brattström, H. (Editor).
    Sarsia, 34 (1968). 398 p.

3eme SYMPOSIUM EUROPEEN DE BIOLOGIE MARINE

    Arcachon, France, 2-7 Sept., 1968.
    Station Biologique d'Arcachon.

    THEMES - *Biologie des sediments meubles (Vol. 1).*

           - *Biologie des eaux a salinite variable (Vol. 11).*

    Soyer, J. (Editeur).
    Vie et Milieu (1971), Suppl. 22:   1-464.
    Vie et Milieu (1971), Suppl. 22:   464-857.

4th EUROPEAN MARINE BIOLOGY SYMPOSIUM

    Bangor, North Wales, Great Britain, Sept. 14-20, 1969.
    Marine Science Laboratories of the University College
    of North Wales, Menai Bridge.

    TOPICS - *Larval biology.*

            Light in the marine environment.

    Crisp, D.J. (Editor).
    Cambridge University Press (1971). 599 p.

5th EUROPEAN MARINE BIOLOGY SYMPOSIUM

Venice, Italy, Oct. 5-11, 1970.
Institute of Marine Biology, Venice.

TOPICS   - *Evolutionary aspects of marine biology.*

          - *Factors affecting biological equilibria in the Adriatic brackish water lagoons.*

Battaglia, B. (Editor).
Piccin Editore, Padova (1972). 348 p.

6th EUROPEAN SYMPOSIUM ON MARINE BIOLOGY

Rovinj, Yugoslavia, Sept. 27-Oct. 2, 1971.
Marine Biological Station (Center for Marine Research of the Rudjer Boskovic Institute), Rovinj.

TOPICS   - *Productivity in coastal areas of the sea.*

          - *Dynamics in benthic communities.*

Zavodnik, D. (Editor).
Thalassia Jugoslavica 7(1) (1971). 445 p.

7th EUROPEAN SYMPOSIUM ON MARINE BIOLOGY

Texel, The Netherlands, Sept. 11-16, 1972.
The Netherlands Institute for Sea Research, Texel.

TOPICS   - *Mechanisms of migration in the marine environment.*

          - *Respiratory gases and the marine organism.*

de Blok, J.W. (Editor).
Netherlands Journal of sea Research 7 (1973). 505 p.

8th EUROPEAN SYMPOSIUM ON MARINE BIOLOGY

Sorrento, Italy, Oct. 1-7, 1973.
Zoological Station of Naples.

TOPIC   - *Reproduction and sexuality in the marine environment.*

Bonaduce, G., and G.C. Carrada (Editors).
Pubblicazioni della Stazione Zoologica di Napoli 39, Suppl. 1(1975). 727 p.

9th EUROPEAN MARINE BIOLOGY SYMPOSIUM

    Oban, Scotland, Great Britain, Oct. 2-8, 1974.
    The Dunstaffnage Marine Research Laboratory, Oban.

    TOPIC   - *The biochemistry, physiology, and behaviour of marine organisms in relation to their ecology.*

    Barnes, H. (Editor).
    Aberdeen University Press (1975). 760 p.

10th EUROPEAN SYMPOSIUM ON MARINE BIOLOGY

    Ostend, Belgium, Sept. 17-23, 1975.
    Institute for Marine Research, Bredene, Belgium.

    TOPICS - *Research in mariculture at laboratory - and pilot scale (Vol. 1).*

             - *Population dynamics of marine organisms in relation with nutrient cycling in shallow waters (Vol. 11).*

    Personne, G. and E. Jaspers (Editors).
    Universa Press, Wettern, Belgium (1976), 620 p. (Vol. 1),
    710 p. (Vol. 2).

DAILY VARIATION IN *DICHELOPANDALUS BONNIERI* (CAULLERY) AS A COMPONENT OF THE EPIBENTHOS

A. H. Y. Al-Adhub* and E. Naylor

Department of Marine Biology, University of Liverpool, Port Erin, Isle of Man

ABSTRACT

Beam trawl catches of *Dichelopandalus bonnieri* in 75-80m depth off the S.W. Isle of Man show greatest numbers in the afternoon, sunset and sunrise. Evidence suggests that the shrimps swim up off the bottom at night and during the morning, possibly as a mechanism of dispersal and avoidance of bottom-feeding fish. Feeding appears to take place on the bottom, *D. bonnieri* probably competing for food with, and sometimes being eaten by *Nephrops norvegicus* which emerges from its burrows to feed at approximately the same time of the day.

INTRODUCTION

Though there is an extensive literature on rhythms and population dynamics of decapod crustaceans (see reviews by Allen 1966, 1972) it was not appreciated until fairly recently that diel variations in catches of epibenthic decapods in commercial trawls are in some cases attributable to rhythmic patterns of behaviour. For example it is now known that there are variations in trawl catches of *Nephrops norvegicus* (L.) which are determined by a daily rhythm of emergence from burrows (Chapman and Rice 1971; Naylor and Atkinson 1976). Similar diel changes occur in trawl catches of various caridean decapods such as *Pandalus borealis* (Kroyer) and *Pandalus jordani* Rathbun which, in contrast to the burrowing habit of *Nephrops*, show daily patterns of vertical migration off the bottom (Barr 1970; Pearcy 1970, 1972). Associated with populations of *Nephrops* in the W. Irish Sea another pandalid *Dichelopandalus bonnieri* (Caullery) also shows diel variations in trawls which, on preliminary evidence from another locality, Mason and Howard (1969) have suggested might be related to vertical migration. Present work was initiated to quantify the diel variations in trawl catches of *Dichelopandalus* and to attempt to assess whether these were attributable to vertical migration as a basis for further behavioural study.

METHODS

Samples were taken 15-20 km west of the Isle of Man in a depth of 75-80m at approximately 3 month intervals over a period of 15 months. On each occasion a 2.5m beam trawl was hauled for 15 min at 2hr intervals over a 24hr period, using mesh sizes of 20 and 9mm for the bag and cod-end respectively. In all samples the shrimps were sexed and, in samples taken in September, January, March and June, the presence or absence of food in the gut was also recorded. On two occasions in winter and summer, when it became apparent that the shrimps

*Present address: Department of Biology, College of Science, University of Basrah, Iraq.

probably made vertical excursions off the bottom, an attempt was made to sample throughout the water column. To do this a 1m plankton net with the can replaced by a 1mm brass mesh and with a 25kg weight attached was lowered to about 10m above the bottom and towed obliquely at slow speed.

RESULTS

Trawl to trawl variance was high but there was some patterning in the 24hr samples which showed no apparent sequence of change over the 15 month sampling period. Fig. 1. therefore illustrates average catches over a representative 24hr period based on all samples, standardized in 2hr periods around the times

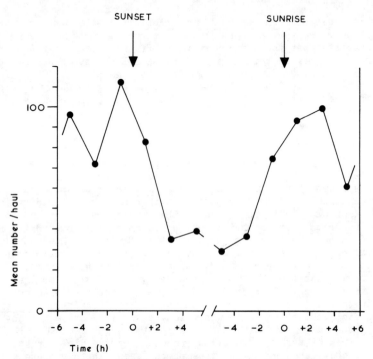

Fig. 1. Catches of <u>Dichelopandalus bonnieri</u> taken in standard 15 min hauls throughout a 24hr period at various times of year. Catches are averaged in 2hr periods centred on the times of sunset and sunrise.

of sunset and sunrise. The most often repeated high catches tended to occur during a period of 6-8hr up to and including sunset, and again around sunrise, with the most repeated occurrence of low numbers occurring at night. Numbers in standard 15 min hauls by day ($\bar{x} = 92$) were significantly greater than at night ($\bar{x} = 53$) ($0.5 > p > 0.1$, 57 d.f.).

In view of the repeated fall in numbers caught at night it was necessary to consider whether the shrimps avoided capture by burrowing or by swimming above the bottom. During the course of several days of observations of <u>Dichelopandalus</u> in laboratory tanks provided with habitat mud, no animals were observed

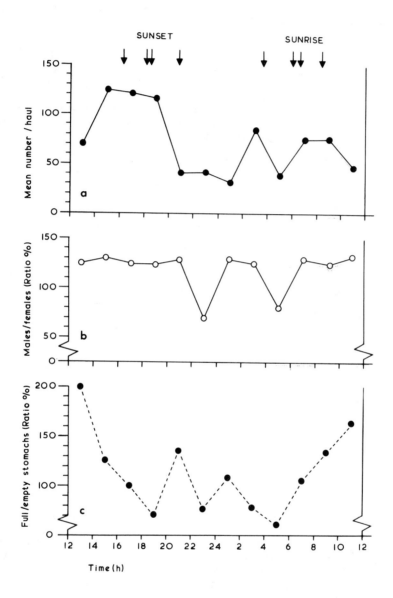

Fig. 2. Analysis of catches of Dichelopandalus bonnieri taken in standard 15 min hauls throughout periods of 24hr during September, January, March and July : (a) mean number per haul (b) sex ratio (c) feeding condition (Arrows indicate times of sunset and sunrise).

to burrow but they were observed to swim up from the bottom for prolonged
periods particularly at night. Swimming above the bottom is also supported by
two records at sea when single specimens were taken in oblique hauls by plankton net at night but not by day.

Fig. 2. illustrates the average catch in standard 15 min hauls throughout a
standard 24hr period based on the four sets of samples in which stomach contents
were also examined. In this case the samples are not standardized around the
times of sunset and sunrise, the timings of which are indicated by arrows. In
addition, since it is known that the male/female sex ratio remains fairly
constant throughout the year at this depth of sampling (Al-Adhub and Bowers
In prep.), the possibility of diel changes in sex ratio was investigated. In
Fig. 2. it is noteworthy that low numbers of shrimp in trawls at night were
accompanied by falls in the male/female ratio(Fig. 2b.) in which at other times
and, in the samples overall, males outnumbered females. The total number of
animals caught at night was low but despite this the results suggest that, if
the fall in numbers was related to upward swimming, then females tended to be
less involved in vertical migration and spent more of their time on the bottom.

Fig. 2c. illustrates the average feeding condition throughout the day from
which it is clear that at times when shrimp numbers increased in the trawls
(Fig. 2a.) then the proportion of those with full stomachs was reduced, particularly from 1400-2000hr and before sunrise. If, as seems likely, Dichelopandalus migrated upwards off the bottom around midnight and before midday it is
evident that when they returned to the bottom they had empty stomachs and therefore did not appear to have fed when in the water column. Stomach contents
were to a great extent unidentifiable in present samples but the presence of
polychaete setae and the shells of ostracods and bivalves further supports the
suggestion of bottom feeding.

DISCUSSION

It seems reasonable to conclude that the present beam trawl catches are of
Dichelopandalus walking on the substratum, as has been clearly shown for
Pandalus jordani by Pearcy (1972). Periodical reductions in numbers in the
trawls are best explained by the hypothesis of swimming upwards off the bottom.
This substantiates the provisional suggestion by Mason and Howard (1969) who
reported greater catches of D. bonnieri by bottom trawling in bright sunshine
than in dull weather and that large numbers of this species were often taken
in midwater trawls as a by-catch when fishing for clupeoids at night in Scottish
waters.

Similar diel vertical movement up from the bottom has been suggested for Pandalus borealis by various workers (see Hjort and Rudd, 1938; Horstedt and Smidt
1956) and substantiated for that species by Barr (1970) and for Pandalus jordani
by Pearcy (1970, 1972). However, whereas both these species are considered to
migrate upwards into the water column to feed on planktonic animals, present
work suggests that Dichelopandalus bonnieri feeds predominantly on bottom-
living organisms. Moreover, this is confirmed by Lagardère (1973) who showed
that Dichelopandalus bonnieri in the Bay of Biscay fed predominantly on the
bottom, eating dead fish in winter and benthic invertebrates in spring, though
bathypelagic crustaceans were taken in summer. Indeed, since in the present study
it appears that the most pronounced upward swimming occurs after sunset when
truly planktonic species also migrate towards the surface, it seems unlikely
that as a general occurrence Dichelopandalus moves up from the bottom to feed
on plankton organisms. In that event it seems most likely that movement off

the bottom at night permits some degree of dispersion and avoidance of bottom feeding fish, which are additional advantages of vertical migration suggested for Pandalus jordani by Pearcy (1970). The latter interpretation would be consistent with preliminary evidence available from the present sampling area which suggests that cod (Gadus morhua L.) shoal above the bottom by day and disperse downwards to feed on the bottom at night as has been reported elsewhere (see Woodhead 1966).

Another factor to be considered is that peak catches of D. bonnieri occur at the same times of day as the burrowing prawn Nephrops norvegicus in the same locality (Naylor and Atkinson 1976), which is recorded as feeding on Dichelopandalus as one of several food organisms (Farmer 1974). In other localities it is known that adult natantian decapods such as Dichelopandalus form only a small proportion of the diet of Nephrops (Thomas and Davidson 1962) but where they are abundant the descent of vertically migrating pandalids to feed on the bottom could at times play a significant role in timing the emergence activity of Nephrops (see Naylor and Atkinson 1976).

As yet there have been no experimental studies of the behaviour of Dichelopandalus such as have been carried out on Nephrops (Chapman, Johnstone and Rice, 1975; Atkinson and Naylor, 1977; Hammond and Naylor 1977). However further work on the light responses and rhythmic behaviour of Dichelopandalus bonnieri would clearly be worthwhile. As Barr (1970) points out, an understanding of such behaviour is of importance to fishery studies of deep water shrimp populations where sampling is by bottom trawling and the catches are subjected to daily variations in abundance and size composition.

REFERENCES

Al-Adhub, A.H.Y. and Bowers, A.B. Growth and breeding of Dichelopandalus bonnieri in Isle of Man waters. (In prep.).

Allen, J.A. 1966. The rhythms and population dynamics of Decapod Crustacea. Oceanogr. Mar. Biol. Ann. Rev. 4, 247-265.

Allen, J.A. 1972. Recent studies on the rhythms of post-larval Decapod Crustacea. Ibid, 10, 415-436.

Atkinson, R.J.A. and Naylor, E. Endogenous activity rhythm and rhythmicity of catches of Nephrops norvegicus (L.). Exp. mar. biol. Ecol. (In press).

Chapman, C.J. and Rice, A.L. 1971. Some direct observations on the ecology and behaviour of the Norway lobster Nephrops norvegicus. Mar. Biol. 10, 321-329.

Chapman, C.J., Johnstone, A.D.F. and Rice, A.L. 1975. The behaviour and ecology of the Norway lobster, Nephrops norvegicus (L.). Proc. Eur. mar. Biol. Symp. 9, 59-74.

Farmer, A.S.D. 1974. Field assessments of diurnal activity in Irish Sea populations of the Norway lobster, Nephrops norvegicus (L.) (Decapoda: Nephropidae). Estuar. Coast. Mar. Sci. 2, 37-47.

Hammond, R.D. and Naylor, E. Effects of sunset and sunrise on locomotor activity rhythms in the Norway lobster, Neprhops norvegicus. Mar. Biol.

(In press).

Hjort, J. and Rudd, J. 1938. Deep-sea prawn fisheries and their problems. Hvalrad. Skr., 17, 144pp.

Horsted, S.A. and Smidt, E. 1956. The deep sea prawn (Pandalus borealis Kr.) in Greenland waters. Meddr. Danm. Fisk.-og Havunders., Kbh., (N.S.) 1 (11), 118pp.

Lagardère, J.P. 1973. Données sur la biologie et sur l'alimentation de Dichelopandalus bonnieri (Crustacé-Natantia) dans le Golfe de Gascogne. Tethys, 5 (1), 155-166.

Mason, J. and Howard, F.G. 1969). Notes on the distribution and biology of Pandalus bonnieri Caullery off the west of Scotland. I.C.E.S., C.M. 1969. Shellfish Comm. No. K. 34. 3pp (Mimeo).

Naylor, E. and Atkinson, R.J.A. 1976. Rhythmic behaviour of Nephrops and some other marine curstaceans. Perspect. Exp. Biol. 1, 135-143, Pergamon Press, Oxford and New York.

Pearcy, W.G. 1970. Vertical migration of the ocean shrimp, Pandalus jordani : a feeding and dispersal mechanism. Calif. Fish & Game 56, 125-129.

Pearcy, W.G. 1972. Distribution and diel changes in the behaviour of pink shrimp, Pandalus jordani off Oregon. Proc. Natl. Shellfish Assoc. 62, 15-20.

Thomas, H.J. and Davidson, C. 1962. The food of the Norway lobster. Mar. Res. 3, 1-15. H.M.S.O. Edinburgh.

Woodhead, P.M.J. 1966. The behaviour of fish in relation to light in the sea. Oceanogr. Mar. Biol. Ann. Rev. 4, 337-403.

COMMUNITY STRUCTURES OF SOFT-BOTTOM MACROFAUNA IN DIFFERENT PARTS OF THE BALTIC*)

Ann-Britt Andersin, Julius Lassig and Henrik Sandler

Institute of Marine Research, P.O. Box 136, SF-00121, Helsinki/Helsingfors 12, Finland

ABSTRACT

The bottom fauna communities in the open Baltic are mainly influenced by the decrease in salinity from about 20‰ (above the bottom) in the south to about 4‰ at the heads of the northern gulfs, and by the poor oxygen conditions below the permanent halocline.

In 1961 the Institute of Marine Research in Helsinki started a long-term study on the occurrence of benthic macrofauna in the seas around Finland. Some years later the study was widened to comprise the whole of the Baltic Sea inside the transect Trelleborg-Arkona. The study is unique in so far as it is the first attempt to cover the whole Baltic with the same methods and gear.

In the present paper some of the results of this investigation, chiefly data on abundance, biomass and species composition, have been used to describe various types of benthic communities in the Baltic Sea in summer 1967.

In the deepest part of the southern Baltic, the former mollusc-dominated community has been replaced by a polychaete community. In the Central Basin and the Gulf of Finland, an area totally devoid of macrofauna has developed since the early part of this century. Comparison of the results from summer 1967 with data from the beginning of this century does not reveal any drastic changes in the Gulf of Bothnia.

---

*) The treatment of the material for this study was supported by the National Swedish Environmental Protection Board.

INTRODUCTION

The Baltic Sea is a unique brackish-water area, with a low but fairly stable salinity, which decreases from about 20‰ (above the bottom) in the south to about 4‰ at the heads of the northern gulfs. Its hydrographical pecularities are mainly caused by the narrow and shallow connection with the North Sea, the strong inflow of fresh water from rivers, and the climate, which affects many of the processes operating in it. The hydrography of the deeper parts of the Baltic basins is strongly dependent on the amount and density of the saline water flowing over the sills which separate the Baltic from the ocean and the Baltic basins from each other. In the Bornholm Basin, the Central Basin and the Gulf of Finland, the bottom layer is separated from the surface layer by a permanent halocline. During periods of stagnation this separation gives rise to an oxygen deficit and periodically to complete oxygen depletion and formation of hydrogen sulphide. The result is a total disappearance of macrofauna in the deepest parts of these basins. (For additional information on the hydrography of the Baltic, cf. e.g. (Segerstråle 1957),(Fonselius 1962, 1969, 1971),(Magaard and Rheinheimer 1974).

In (1961) the Institute of Marine Research in Helsinki started a long-term study on the occurrence of benthic macrofauna in the seas around Finland. Some years later the study was widened to comprise the whole Baltic inside the transect Trelleborg-Arkona (TA). This study is unique in so far as it is the first attempt to cover the whole Baltic with the same method and gear. The advantage of this is obvious, since an intercalibration between a Finnish and a Swedish laboratory in (1974) showed that their results were not fully comparable, although both laboratories followed the same recommendations (Dybern et al.1976) regarding methodology.

The present paper is based mainly on material from summer 1967, which has been used to exemplify the community structure at the present time in different parts of the Baltic.

MATERIAL AND METHODS

Since the main aim of the study was to follow the situation in the deeper layers of the Baltic basins, the stations have primarily been chosen in deep localities, and only a few represent shallower areas, mainly at the margin of unfavourable regions, or in transitional areas between the basins (Fig. 1).

In 1967 the stations in the Gulf of Bothnia were visited in June, those in the Gulf of Finland in June and July, and the stations in the Baltic proper in July. Although all sampling was carried out within about two months, the great climatic differences between different parts of the Baltic affect the seasonal course of the abundance and biomass values, and thus somewhat restrict the possibility of comparing the values from different regions.

The sampling method corresponds closely to the recommendations given by the Baltic Marine Biologists (BMB) (Dybern et al. 1976). The samples were taken with a 0.1 m^2 van Veen grab, sieved through a 1-mm sieve, handpicked and stored in 4% buffered formalin. In most cases, three samples were taken at each station.

Species composition, abundance and biomass were determined from the samples. The animals were identified to species level when possible, but the oligochaetes and the nemertineans, except for *Prostoma obscurum* (Schultze), were treated as groups. The biomass was determined as formalin wet weight after at least one month's preservation, in order to obtain constant values. The species diversity was calculated according to Shannon's formula (Pielou 1966):

$$H' = -\sum_{i=1}^{S} p_i \ln p_i,$$

where S is the number of species and p_i is the proportion of individuals in the ith species (i = 1,2,3...S). This index was calculated for stations where more than 50 individuals were sampled.

RESULTS

Total abundance (Fig. 2)

In the Arkona Basin fairly high values, mostly around 2000 ind./m^2, were found. In the Bornholm Basin, at depths below 70m, the values were very low. However, none of the stations in this area were found to be totally devoid of macrofauna in 1967. In the Slupsk Furrow, the oxygen conditions are favourable, which is mirrored in the high density values recorded there. On the long transect between Oeland and the Gulf of Gdansk, the values did not generally exceed 1000 ind./m^2. In the Gdansk Deep none of the stations were devoid of benthic macrofauna.

However, at depths below 60m the greater part of the Central Basin, including the Gulf of Finland, was characterized by very low density values, or, over vast areas, by a total absence of macrofauna. In shallower parts of the Central Basin and Gulf of Finland, the density varied with the topographic, hydrographic and sediment conditions.

The highest density values in the entire Baltic occurred in the Bothnian Sea. In 1967 values exceeding 3000 ind./m^2 were not unusual, while the maximum values recorded were c. 5000 ind./m^2. The situation in the northernmost part of the Baltic, the Bothnian Bay, differs remarkably from that of the Bothnian Sea.

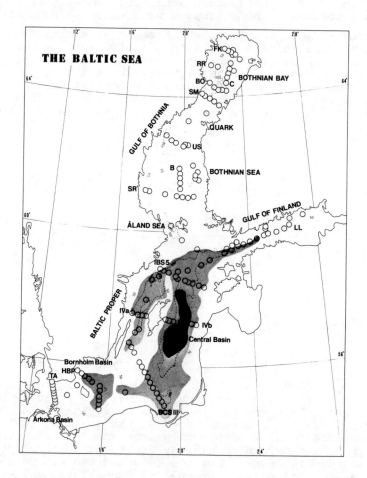

Fig. 1. The main basins in the Baltic, with the stations visited in 1967 inserted. The light shaded shows the areas in which oxygen values below 2 ml/l were recorded in the bottom water in 1967. The dark shading shows the area where hydrogen sulphide occurred. (data from Havsfiskelaboratoriet, Lysekil, 1967a, 1967b, 1968a, 1968b, and Inst. Mar. Res., Helsinki, unpublished).

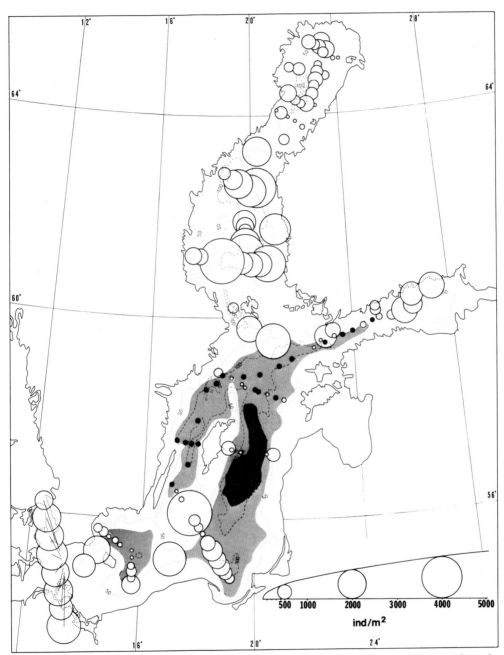

Fig. 2. Density of macrofauna in the Baltic Sea in June and July 1967. Filled circles indicate stations devoid of macrofauna. Shaded areas as in Fig. 1.

Here the density values were much lower, typically ranging from 200 to 500 ind./m^2.

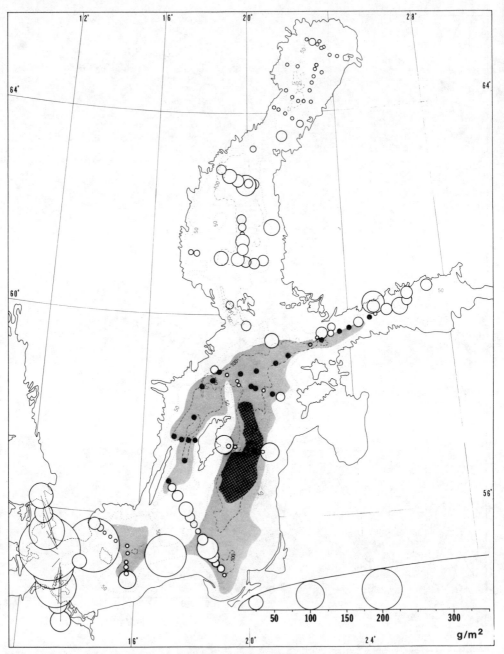

Fig. 3. Biomass (wet weight) of macrofauna in the Baltic Sea in June and July 1967. Shaded areas as in Fig. 1. Filled circles as in Fig. 2.

Total Biomass (Fig. 3)

The biomasses were higher in the Arkona Basin than in any other part of the Baltic; at several stations the values exceeded 200g/m^2. The biomass values recorded below 70m in the Bornholm Basin were all under 2.5g/m^2, at some stations even below 1 g/m^2. In the Central Basin the values varied widely (cf. total abundance). As expected, they showed a tendency to decrease towards the north. In the areas with critical oxygen conditions the values were mostly very low.

The difference between the Bothnian Sea and the Bothnian Bay is even more striking as regards the biomass than in the case of abundance. In the Bothnian Sea the values varied between 10 and 25 g/m^2 at the majority of the stations; in the Bothnian Bay, almost all the biomass values were under 2g/m^2, and the majority even under 1g/m^2. This difference is partly due to the higher abundance values in the Bothnian Sea of the numerically dominant *Pontoporeia affinis* Lindstroem (Crustacea), and by the higher frequency of the large isopod *Mesidotea entomon* (Linné) (Crustacea), which forms 30-80% of the total biomass in that region. Another contributory factor is the clear difference in the weight of *P. affinis*. The average weight of *Pontoporeia* individuals in the Bothnian Sea was 4.5mg., whereas in the Bothnian Bay it was only 2.0mg.

This difference is due partly to differences in size distribution and partly to differences in the weights of animals of the same size groups (length). The results of further studies on seasonal fluctuations in the population of *P. affinis* in the Gulf of Bothnia from spring to autumn 1974 showed that the majority of the individuals in the oldest year class measured 7mm in the Bothnian Sea, while the corresponding value for the Bothnian Bay was only 5mm. It could also be shown that animals of the same size group were clearly heavier in the Bothnian Sea than in the Bothnian Bay. It has not yet been determined whether this reduction in size and weight should be attributed to salinity, temperature, nutrient conditions or to the combined effect of these factors.

Faunal Composition

The distinct difference in the number of species between the Arkona Basin and the rest of the Baltic Sea is well illustrated by the material of 1967. The total number of taxa recorded was 43, and of these 41 occurred in the Arkona Basin, and 20 were found there alone. From the Bornholm Basin and the southern part of the Central Basin 20 taxa were recorded, from the northern half of the Central Basin and the Gulf of Finland 14 taxa, and from the Gulf of Bothnia only 8 taxa.

The species diversity calculated for the soft-bottom macrofauna in 1967 is shown in Fig. 4. The Arkona Basin and the transition areas between the Arkona Basin and the Bornholm Basin, and between the Bornholm Basin and the Central Basin show the highest diversity, while the diversity is strikingly low in the Gulf of Bothnia, where the fauna consists mainly of one single very

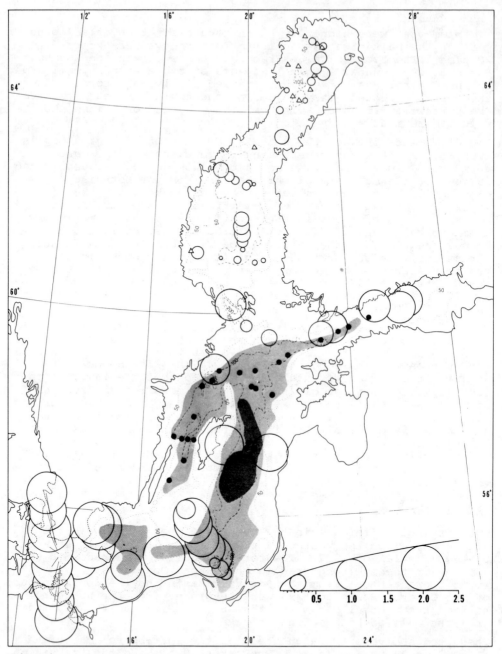

Fig. 4. Species diversity of macrofauna in the Baltic Sea in June and July 1967. Triangles indicate stations where only one species was recorded. Shaded areas as in Fig. 1. Filled circles as in Fig. 2.

numerous species, P. affinis.

As regards the numbers of individuals the dominance relations of the macrofauna change towards the north in the Baltic, dominance being concentrated in a decreasing number of species (Fig. 5). In the Arkona Basin the three most dominant species accounted for about 70% of the total number of individuals; in the Gulf of Finland the corresponding value was about 97% and in the Gulf of Bothnia more than 99%.

In the southern part of the Baltic, on bottoms with good oxygen conditions, no particular taxonomic group was dominant in respect of the density values. At most stations 60-80% of the total number of individuals belonged to various combinations of the following species: *Scoloplos armiger* (Mueller) (Polychaeta), *Diastylis rathkei* (Kroeyer) and *Pontoporeia gemorata* Kroeyer (Crustacea), *Astarte borealis* (Chemnitz) and *Macoma baltica* (Linné) (Lamellibranchiata) (Table 1). The most frequent species in this part of the Baltic were *Halicryptus spinulosis* v. Siebold (Priapulida), *Harmothoe sarsi* (Kinberg) (Polychaeta), *Diastylis rathkei*, *Pontoporeia femorata* and *Macoma baltica*, which occurred at all or nearly all stations.

As regards biomass, the molluscs dominated clearly in regions with good oxygen conditions in the southern part of the Baltic, constituting 70-90% of the total biomass at most stations (Table 1). This overwhelming dominance was partly due to the inc-lusion of the shell weights in the biomass values.

In the northern half of the Central Basin and in the Gulf of Finland, bottoms with good oxygen conditions typically showed a very pronounced dominance of crustaceans as regards both abundance and biomass. Almost 90% of the fauna on soft bottoms consisted of *Pontoporeia affinis*. The most important accompanying species were *Mesidotea entomon*, *Pontoporeia femorata* and *Macoma baltica*.

In the Gulf of Bothnia, where the oxygen conditions are always good, the dominance of crustaceans was still more distinct. The community was almost entirely made up of one single species, P. affinis. In two exceptional cases M. baltica was dominant in respect of both abundance and biomass. M. entomon, which occurs frequently, especially in the Bothnian Sea, had very small numbers of individuals, but generally made a major contribution to the biomass. (cf.p. 5).

In the Baltic proper and in the Gulf of Finland, in areas where the oxygen conditions are unfavourable, but still allow a macrofauna community to develop, the polychaetes clearly dominate. In the Bornholm Basin and in the southern half of the Central Basin, the main species were *Harmothoe sarsi*, *Scoloplos armiger* and *Capitella capitata* (Fabricius)(Table 1). In the northern part of the Central Basin and in the Gulf of Finland, the only polychaete occurring in such areas was *Harmothoe sarsi*, which mostly formed the entire macrofauna community in this region. However, at some of the critical stations in the Gulf of Finland the macrofauna community consisted mainly of *Macoma baltica*.

Table 1. Contributions to total abundance and total biomass of the three most dominant species in the southern part of the Baltic Sea in June and July 1967.

Station	Species	Percentage distribution Abundance	Biomass
Arkona Basin			
TA 1	Pontoporeia affinis/Diastylis rathkei/Macoma baltica Macoma baltica/Mya sp/Mytilus edulis	40/64/77	78/88/94
TA 2	Pontoporeia femorata/Scoloplos armiger/Diastylis rathkei Macoma baltica/Scoloplos armiger/Pontoporeia femorata	47/81/84	68/83/89
TA 3	Scoloplos armiger/Pontoporeia femorata/Macoma baltica Macoma baltica/Halicryptus spinulosus/Scoloplos armiger	28/50/65	86/92/95
TA 4	Astarte borealis/Aricidea suecica/Scoloplos armiger Astarte borealis/Buccinum undatum/Dendrodoa grossularia	31/44/56	86/91/94
TA 5	Pontoporeia femorata/Diastylis rathkei/Scoloplos armiger Cyprina islandica/Macoma baltica/Nephtys ciliata	31/58/60	57/89/92
TA 6	Diastylis rathkei/Pontoporeia femorata/Scoloplos armiger Cyprina islandica/Macoma baltica/Scoloplos armiger	28/54/74	84/94/96
TA 7	Scoloplos armiger/Capitella capitata/Macoma baltica Macoma baltica/Nephtys ciliata/Scoloplos armiger	50/74/85	72/91/97
TA 8	Scoloplos armiger/Pontoporeia femorata/Macoma baltica Macoma baltica/Scoloplos armiger/Mya arenaria	33/58/67	51/70/80
W 19a	Aricidea suecica/Scoloplos armiger/Terebellides stroemi Nephtys ciliata/Nemertini sp/Macoma baltica	48/69/83	24/38/52
SB 2	Astarte borealis/Dulichia monacantha/Terebellides stroemi Astarte borealis/A. elliptica/Scoloplos armiger	19/34/48	81/87/87
Bornholm Basin			
HBP 217	Clitellio arenarius/Pontoporeia femorata/Diastylis rathkei Macoma baltica/Mytilus edulis/Clitellio arenarius	27/49/69	95/98/99
HBP 216	Pontoporeia femorata/Scoloplos armiger/Terebellides stroemi Astarte elliptica/Macoma baltica/Scoloplos armiger	20/36/52	51/68/80
HBP 215	Harmothoe sarsi/Scoloplos armiger/Diastylis rathkei Harmothoe sarsi/Scoloplos armiger/Pontoporeia femorata	54/74/87	63/81/90
HBP 213	Harmothoe sarsi/Scoloplos armiger/Capitella capitata Harmothoe sarsi/Scoloplos armiger/Capitella capitata	37/68/100	89/100/100
HBP 212	Harmothoe sarsi/Scoloplos armiger Harmothoe sarsi/Scoloplos armiger	94/100	99/100
HBP 133	Harmothoe sarsi Harmothoe sarsi	100	100
HBP 134	Harmothoe sarsi Harmothoe sarsi	100	100
HBP 115	Capitella capitata/Harmothoe sarsi/Scoloplos armiger Harmothoe sarsi/Capitella capitata/Scoloplos armiger	41/74/100	69/96/100
HBP 135	Capitella capitata/Scoloplos armiger/Harmothoe sarsi Harmothoe sarsi/Capitella capitata/Scoloplos armiger	60/72/99	50/93/100
HBP 136	Scoloplos armiger/Harmothoe sarsi/Halicryptus spinulosus Scoloplos armiger/Harmothoe sarsi/Halicryptus spinulosus	76/96/98	46/78/99
HBP 137	Aricidea suecica/Scoloplos armiger/Diastylis rathkei Astarte borealis/Macoma baltica/Scoloplos armiger	38/68/80	81/88/93
Slupsk Furrow			
DBS 2	Pholoe minuta/Astarte borealis/Pontoporeia femorata Astarte borealis/A. elliptica/Diastylis rathkei	32/48/61	64/92/94
Gdansk Deep (transect BCS-III)			
12	Pontoporeia femorata/Scoloplos armiger/Diastylis rathkei Macoma baltica/Scoloplos armiger/Halicryptus spinulosus	45/65/75	32/51/69
13	Scoloplos armiger/Harmothoe sarsi/Diastylis rathkei Scoloplos armiger/Harmothoe sarsi/Diastylis rathkei	96/99/100	75/99/100
14	Scoloplos armiger/Harmothoe sarsi/Pontoporeia femorata Scoloplos armiger/Harmothoe sarsi/Pontoporeia femorata	93/99/100	63/97/100
15	Scoloplos armiger/Harmothoe sarsi/Aricidea suecica Scoloplos armiger/Harmothoe sarsi/Aricidea suecica	96/98/100	93/99/100

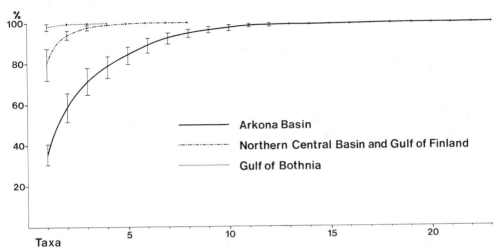

Fig. 5. Increase in percentage of total number of individuals with increasing number of taxa in different parts of the Baltic Sea in June and July 1967. Means and standard deviations (vertical lines) of all the stations sampled in each region.

DISCUSSION

It is difficult to compare recent results, especially quantitative ones, with older data. The sampling stations are mostly different, and the methods have changed. However, some general features of the situation at the present time can be compared with observations from the past.

In the deepest parts of the southern Baltic, the former mollusc-dominated community, (*Macoma calcarea* (Chemnitz), *M. baltica*, *Astarte borealis*), seen at the beginning of this century (Thulin 1922, Hessle 1923, Hagmeier 1930), and still present in the early 50's (Demel and Mulicki 1954, Mulicki 1957), has been replaced by a community dominated by polychaetes. This was noted during Leppaekoski's studies (1969, 1971, 1975), and was still the case in 1974, when we took our last samples in that area. Like Leppaekoski (1975), we are inclined to suppose that the sudden appearance of polychaetes in the Bornholm Basin indicates passive transport of juvenile specimens by currents, which enables these species to rapidly colonize the dead bottoms when the oxygen conditions have improved. However, Leppaekoski's (1975) description of the oceanization of this area may be slightly exaggerated, since at least one of the species character-ized by him as a recent immigrant in the Bornholm area, was recorded there already in the 20's (Thulin 1922), and may have been overlooked earlier, being a tiny species (*Pholoe minuta* Fabricius). The present polychaete-dominated community may rather be the first step in a succession which is wiped out by unfavourable oxygen conditions before it can develop further.

Thus the present community would primarily indicate unfavourable oxygen conditions rather than an oceanization caused by higher salinity.

In the Central Basin, comparison with the few older data available shows that the area devoid of macrofauna has grown since the early decades of this century, when macrofauna was found much deeper than in our days (cf. Hessle 1923, Sjoeblom 1955, Segerstråle 1962).

In the Gulf of Bothnia no drastic changes can be seen when the results from 1967 are compared with those of Hessle (1924), although he did not find such an overwhelming difference between the Bothnian Sea and the Bothnian Bay as is indicated by our values.

A comparison of the persistence ("stability") of the soft-bottom macrofauna in the deeper parts of the Baltic shows that the communities of the Gulf of Bothnia are the most stable, while those of the Bornholm and Gdansk Deeps vary strongly, mainly owing to the periodic oxygen deficit. Fluctuations attributable to variations in salinity are also much more marked in the Arkona and Bornholm Basins than in the northern region of the Baltic, but are less dramatic than those caused by oxygen deficits. The area with periodically unfavourable oxygen conditions, often referred to in discussions on the health of the Baltic, was at most about 100 000 km^2 in 1967, which is c. 25% of the total area of the Baltic Sea.

Almost from the beginning of research on Baltic biology, it has been known that, when the bottom type and depth are compared, variations in the community structure of benthic animals are primarily attributable to differences in salinity. The importance of the effect of deficiency of oxygen was also realized at an early date (e.g. Johansen 1918), and oxygen deficits are not themselves a new problem. Consequently it may seem unnecessary to carry out an investigation of the type presented here, since the main features of soft-bottom macrofauna communities were already known at the beginning of this century.

However, the salinity has risen since that time, increasing the stability of the halocline (Fonselius 1969), and restricting still further the water exchange between the more saline bottom layer and the overlying water masses. As this makes the Baltic deep basins more sensitive to all kinds of overfertilization, it is certainly worthwhile to trace the possible effect of such natural environmental changes on the benthic communities.

Studies of this type will therefore surely have value as a basis of reference in future analyses of man-made changes in the Baltic environment.

REFERENCES

Demel, K. and Mulicki, Z. 1954 Quantitative investigations on the biological bottom productivity of the South Baltic. Pr.morsk. Inst. ryb. Gdyni 7:75-126. (In Polish, summary in English).

Dybern, B.I. Ackefors, H. and Elmgren, R. (ed.) 1976 Recommendations on methods for marine biological studies in the Baltic Sea. The Baltic Marine Biologists, 1:3-98.

Fonselius, S.H. 1962. Hydrography of the Baltic deep basins. Rep. Fishery Bd. Swed. 13:5-41.

- 1969 Hydrography of the Baltic deep basins III. Rep. Fishery Bd. Swed. 23:5-97.

- 1971 Om Oestersjoens och speciellt Bottniska vikens hydrografi. Vatten 3:309-324.

Hagmeier, A. 1930 Die Bodenfauna der Ostee im April 1929 nebst einigen Vergleichen mit April 1925 und Juli 1926. Ber. dt. wiss. Kommn. Meeresforsch. V:78-95 (157-173).

Havsfiskelaboratoriet 1967a Hydrographical Data January-June 1967 R.V. Skagerack. Meddn Havsfiskelab. Lysekil 38.

- 1967b Hydrographical Data January-June 1967. R.V. Thetis. Meddn. Havsfiskelab. Lysekil 41.

- 1968a Hydrographical Data July-December 1967. R.V. Thetis. Meddn. Havsfiskelab. Lysekil 51.

- 1968b Hydrographical Data July-December 1967. R.V. Skagerack. Meddn. Havsfiskelab. Lysekil 52.

Hessle, Chr. 1923. Undersoekningar roerande bottnen och bottenfaunan i farvattnen vid Oeland och Gotland. Meddn K. Lantbr-Styr. 243 (2):143-156.

Johansen, A.C. 1918. Om hydrografiske faktorers inflydelse paa Molluskernes udbredelse i Østersøen. Forh. skand. Naturf. Møte 16:633-654. Kristiania.

Leppaekoski, E. 1969. Transitory return of the benthic fauna of the Bornholm Basin after extermination by oxygen insufficiency. Cah. Biol. mar. X(2): 163-172.

- 1971 Benthic recolonization of the Bornholm Basin (Southern Baltic) in 1969-1971. Thalassia jugosl. 7(1):171-179.

- 1975 Macrobenthic fauna as indicator of oceanization in the Southern Baltic. Merentutkimuslait. Julk./Havsforskningsinst. Skr. 239:280-288.

Magaard, L.and Rheinheimer, G. (ed.) 1974 Meereskunde der Ostsee. Heidelberg. 1-269.

Mulicki, Z. 1957 Ecology of the more important Baltic invertebrates. Pr. morsk. Inst. ryb. Gdyni 9:313-379.(In Polish, summary in English).

Pielou, E.C. 1966. The measurement of diversity in different types of biological collections. J. Theoret. Biol. 13:131-144.

Segerstråle, S.G. 1957 Baltic Sea. In: Treatise on Marine Ecology and Paleoecology. Mem. geol. Soc. Am. 67 (1):751-800.

- 1962 Investigations on Baltic populations of the bivalve Macoma baltica (L). Comment. biol.XXIV (7):1-26.

Sjoeblom, V. 1955 Bottom fauna. (In: Granqvist, G., The summer cruise with m/s Aranda in the northern Baltic 1954) Merentutkimuslait. Julk./Havsforskningsinst. Skr. 166:37-40.

Thulin, G. 1922. Bottenboniteringar i soedra Oesterjoen i samband med fisktrålningar. Sv. Hydrogr.biol.Komm.Skr. VII:1-9

A DROP-TRAP INVESTIGATION OF THE ABUNDANCE OF FISH IN VERY SHALLOW WATER IN THE ASKÖ AREA, NORTHERN BALTIC PROPER

Gunnar Aneer and Sture Nellbring

Department of Zoology and the Askö Laboratory, University of Stockholm, Box 6801, S-113 86 Stockholm, Sweden

ABSTRACT

An investigation of the abundance of fish in very shallow water in the northern Baltic proper was started in July 1975 and terminated in August 1976. Drop traps enclosing a bottom area of 1 m^2 and the water column above have been used down to a depth of 0.5 m. Out of 83 drops only six were completely without any trapped fish. 16 species were caught but only a few of these occurred frequently. On average, the total abundance was 15.7 fish $\cdot m^{-2}$ with a mean biomass of 4.3 $g \cdot m^{-2}$ (2.98 $\cdot m^{-2}$ if a school of rudd is excluded). The sand goby (*Pomatoschistus minutus* Pallas) dominated and was caught in 61 per cent of the drops. Its mean abundance was 4.9 fish $\cdot m^{-2}$ and the corresponding biomass 0.9 $g \cdot m^{-2}$. A school of rudd (*Scardinius erythrophthalmus* L.) caught in one drop affected the mean biomass value considerably and the mean for rudd was 39 percent of the total although this species was present in two drops only. Shannon-Weaver indices for abundance and biomass have been calculated and both averaged 0.47.

INTRODUCTION

In ecosystem work, information on abundance and amounts of living matter are most essential for the understanding of energy flow within the systems and for subsequent modelling of the systems as well as of their parts. For many organisms, as for example benthic macro-fauna, values of these parameters are fairly easy to obtain at least when compared to fish and especially marine fish.

The problem of how to obtain good estimates of true fish abundance has puzzled fishery scientists all over the world for many years, especially in the marine environment where the problems are even greater than in the limnic biotopes. Many different approaches have been tested but not many have been very successful. Trials with tags, poison, seines, hydro-acoustics, electrofishing and traps (Hellier 1959, Kahl 1963, Higer and Kolipinski 1967, Moseley and Copeland 1969, Wegener et al. 1973 and Kushlan 1974 to mention a few) among other things have been carried out but they all lack the universal fitness that would make them the ideal method for most environments. The present study has been an attempt to quantify the fish fauna in very shallow water with the aid of drop-traps. The investigation formed part of the project "Dynamics and energy flows through Baltic ecosystems" which is being carried out at the Askö Laboratory in the northern Baltic proper.

MATERIAL AND METHODS

A total of 83 drops with two drop-traps of slightly different construction (Fig. 1) were made in the vicinity of the Askö Laboratory (58°49'N, 17°39'E) in the northern Baltic Sea where

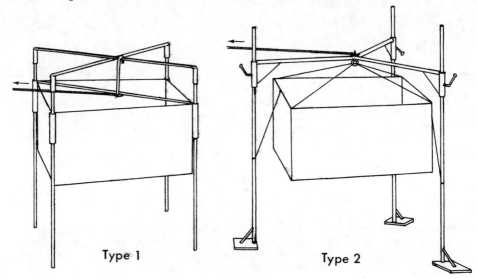

Fig. 1. The types of drop-traps used. Type 1, slide-fall model and type 2, modified free-fall model.

the salinity is around 7 °/oo. Sampling was carried out between July 1, 1975 and August 31, 1976. The construction of the type 1 trap is largely that described by Kahl (1963) but lacks the bottom frame and slides inside the supporting vertical poles. Type 2 is a modified free-fall version of type 1. The dropping frames in both types are 1x1x0.5 m and made of 1 mm sheet iron. The traps were used in three similar bays near to the laboratory in depths down to 0.5 m. They were used in day-time only, on most occasions after noon. A new place was selected for each drop and the traps were left at least over night before a new drop was performed in order to obtain as natural conditions as possible. The collection of trapped fish was carried out by means of fine-meshed dipnets. Several dips were carried out after the last specimen was caught to make sure that no fish still remained in the trap.

The bottoms consisted of clay covered with coarse and fine sand and small stones. The dominant macro-phytes were *Scirpus maritimus* L., *Sc. Tabernaemontani* C.C. Gmel, *Phragmites communis* Trin., *Fucus vesiculosus* L., *Chorda filum* L., *Potamogeton pectinatus* L., and low grown *Chara* spp. The bottom was normally sparsely covered

by vegetation but on some occasions dense covers of *Chara* spp. and filamentous algae were common in the whole area. Due to strong wave action, high water levels, ice-problems and other factors, sampling was limited to August, October and December 1975, i.e. only one sample per month. Ice-problems in January-March 1976 stopped all sampling.
A total of 1291 specimens of 14 identified species and 16 larvae of two unidentified species were caught. Lengths and wetweights for each specimen were measured on each occasion.
Numbers behind $\pm$ signs, after presented means in the text, are standard errors of the mean. The Shannon-Weaver diversity index (H') estimated by $H" = -\sum n_i/N \times \ln n_i/N$ for samples out of populations have been used for the diversity calculations.

RESULTS

Abundance

1307 specimens were caught during the investigation period. On the average this was 15.7 fish$\cdot m^{-2}$ (Table 1) but there was a strong seasonal variation ranging from a mean of 2 fish per m^2 in April 1976 to 29.6 in August 1976. The monthly averages were higher in summer and autumn than in spring (Fig. 2). Only six of the 83 drops were empty. The range in numbers per m^2 was 0-89.

TABLE 1 Total numbers and wetweights for all species as well as abundance and biomass per m^2 and their ranges.

Scientific name	ABUNDANCE				BIOMASS (g)			
	n	$\bar{n}\cdot m^{-2}$	st.err.	range	weight	$\bar{w}\cdot m^{-2}$	st.err.	range
Pomatoschistus minutus Pallas	406	4.89	0.77	0-38	76.61	0.92	0.16	0- 6.98
Gasterosteus aculeatus L.	371	4.47	1.47	0-84	40.47	0.48	0.11	0- 4.31
Pygosteus pungitius L.	158	1.90	0.57	0-39	33.39	0.40	0.17	0- 13.62
Gobius niger L.	139	1.67	0.52	0-26	9.11	0.10	0.04	0- 2.14
Phoxinus phoxinus L.	131	1.58	1.02	0-82	7.65	0.09	0.05	0- 3.29
Pleuronectes flesus L.	46	0.55	0.15	0- 9	24.88	0.29	0.08	0- 3.86
Scardinius erythrophthalmus L.	20	0.24	0.22	0-18	138.47	1.66	1.38	0-112.02
Unidentified Cyprinidae	11	0.13	0.08	0- 6	0.02	3×10^{-4}	2×10^{-4}	0- 0.01
Ammodytes tobianus L.	6	0.07	0.03	0- 2	3.97	0.05	0.03	0- 1.71
Unidentified sp.	5	0.06	0.06	0- 5	0.01	1×10^{-4}	1×10^{-4}	0- 0.01
Siphonostoma typhle L.	4	0.05	0.03	0- 2	0.46	0.01	0.01	0- 0.45
Zoarces viviparus L.	3	0.04	0.03	0- 2	4.32	0.05	0.04	0- 2.57
Cottus gobio L.	3	0.04	0.03	0- 2	0.57	0.01	0.01	0- 0.49
Sprattus sprattus L.	2	0.02	0.02	0- 2	2.00	0.02	0.02	0- 2.00
Perca fluviatilis L.	1	0.01	0.01	0- 1	12.39	0.14	0.15	0- 12.39
Rutilus rutilus L.	1	0.01	0.01	0- 1	0.06	7×10^{-4}	7×10^{-4}	0- 0.06
Total	1307	15.75	1.96	0-89	354.38	4.27	1.52	0-124.41

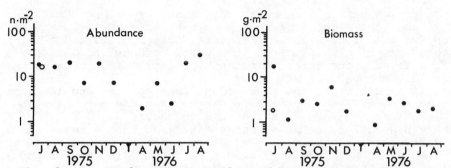

Fig. 2. Seasonal variation of monthly means for abundance and biomass, total material. O = The school of rudd excluded.

The most dominant species was the sand goby (*Pomatoschistus minutus* Pallas) of which a total of 406 specimens were caught which corresponds to 31 per cent of the material (Table 1). As shown in Fig. 3, the average number of sand gobies per month per m^2 was highest in late summer and autumn. This species was the only one present in all months and was caught in 61 per cent of the drops. Table 2 shows its percentages of the total numbers per month. Sand gobies were dominant in six of the eleven months.

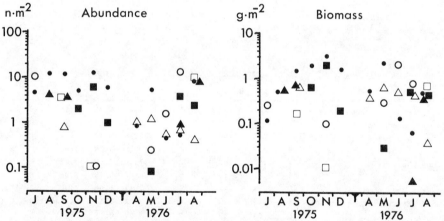

Fig. 3. Seasonal variation in abundance and biomass for more important species. ● = Sand goby, O = Three-spined stickleback, ■ = Nine-spined stickleback, □ = Black goby, △ = Flounder and ▲ = Minnow.

The second dominant species where total numbers are concerned was the three-spined stickleback (*Gasterosteus aculeatus* L.). 371 specimens were caught and thereby made up 28 per cent of the total number. This species was present in 38 per cent of the drops. Their occurrence was predominantly in the period May-July. Table 2 shows that they were dominant in July 1975, June and July 1976.

Nine-spined sticklebacks (*Pygosteus pungitius* L.), black gobies (*Gobius niger* L.) and minnows (*Phoxinus phoxinus* L.) all made up more than 10 per cent of the total catch (Table 1). None of the remaining species exceeded 3.5 per cent.

Nine-spined sticklebacks were present in 35 per cent of the drops. They were not dominant in any month (Table 2). Figure 3 shows their monthly mean abundance.

The two other species, black gobies and minnows, were present in 18 and 9 per cent of the drops respectively. Table 1 shows their mean abundances and Table 2 their percentages for the monthly total abundances. None of these species was dominant at any time. Figure 3 shows their mean abundances per m^2 per month.

TABLE 2 Percentage composition of monthly catches for more important species. Numbers in normal typing = Per cent of monthly total numbers. Numbers in italics = Per cent of monthly total biomass.

	1975						1976				
	J	A	S	O	N	D	A	M	J	J	A
Pomatischistus minutus	0.7 *25.6*	46.8 *75.0*	47.6 *57.5*	75.2 *71.4*	52.8 *65.1*	89.1 *85.7*	57.3 *41.7*	63.9 *74.1*	5.0 *17.4*	3.6 *2.6*	23.8 *27.9*
Gasterosteus aculeatus	1.5 *55.5*	-	-	-	1.6 *0.6*	-	-	8.7 *3.5*	76.7 *60.9*	44.6 *68.4*	-
Pygosteus pungitius	-	-	-	24.8 *28.6*	32.8 *30.3*	10.9 *14.3*	-	0.8 *1.2*	-	28.0 *18.8*	21.0 *8.1*
Gobius niger	-	-	5.2 *17.5*	-	0.2 *0.6*	-	-	-	-	-	34.3 *34.9*
Phoxinus phoxinus	-	53.2 *25.0*	22.9 *17.5*	-	-	-	-	-	-	0.3 *4.4*	17.2 *27.0*
Pleuronectes flesus	-	-	20.1 *3.8*	-	-	-	39.4 *50.0*	18.2 *16.5*	18.3 *21.7*	23.3 *3.4*	1.8 *1.4*
Scardinius erythrophthalmus	89.8 *12.2*	-	-	-	-	-	-	-	-	-	-
Others	8.0 *6.7*	-	4.1 *3.8*	-	12.5 *3.4*	-	3.4 *8.3*	8.4 *4.7*	-	0.2 *2.4*	2.0 *0.6*
Number of drops	9	1	4	1	9	1	6	12	9	19	12

Of the remaining species (Table 1) the flounder (*Pleuronectes flesus* L.) is worth mentioning as it occurred in 23 per cent of the drops. It was dominant with regard to number in April 1976 (Table 2) and present every month of sampling in 1976 but only in September 1975.

The rest of the species (Table 1) were less numerous and none was caught in more than five drops.

Biomass

The seasonal variation in mean biomass per m^2 is shown in Fig. 2 and percentage composition with regard to species in Table 2. The relatively high mean biomass in July 1975 (17.1 $g \cdot m^{-2}$) was the result of the capture of a school of rudd (*Scardinius erythrophthalmus* L.). This species, although present in only two drops,

was responsible for 39 per cent of the total biomass. Mean biomass values for each species as well as for the entire material are shown in Table 1.
A ranking of dominant species according to total biomass for each species results in a picture somewhat different from that based on abundance (Table 1). Rudd with its 39 per cent would be dominant. Sand gobies were second dominant but where monthly percental compositions of biomass are considered (Table 2) it is shown that sand gobies dominate from September through May. The variation in monthly mean biomass is shown in Fig. 3 where it is also seen that their highest biomass per m^2 and month was found in November. Three-spined sticklebacks made up 11 per cent of the total biomass. They were dominant only in June-July 1976 (Table 2).
The seasonal variations in biomass are shown in Fig. 3.
The minnows made up 53 per cent of the biomass in August 1975 but were on the whole of relative little importance as their total biomass was only 2 per cent of the total. Flounders were present throughout 1976 but contrary to abundance they were never dominant with regard to biomass.

Diversity

Shannon-Weaver diversity indices have been calculated for each sampling occasion for both numbers and biomasses. The means for each month are presented in Fig. 4. The diversity was highest in August and the same patterns are found for the biomass and abundance diversity.

The number of species caught was low. A total of 16 species was found but more than 6 species were never present in a single drop. The most frequent number of species per drop was 2 and on the average 2.1 ± 0.1 species were present per m^2.

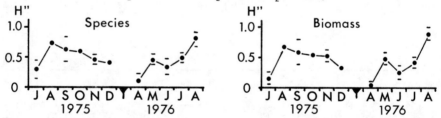

Fig. 4. Seasonal variation in deversity index (H") (Shannon-Weaver) for species and biomass. Horisontal bars = standard errors of means.

Abundance and biomass in relation to depth

Table 3 shows the result of number and biomass in relation to depth for 58 of the drops. The abundance increased with depth and almost doubled per dm whereas the biomass remained almost constant.

TABLE 3 Abundance and biomass in relation depth.

Depth range (cm)	Number of drops	Abundance		Biomass	
		$\bar{n} \cdot m^{-2}$	st.err.	$\bar{w} \cdot m^{-2}$	st.err.
20 - 29	11	7.27	2.42	2.18	0.56
30 - 39	30	15.40	2.93	2.66	0.63
40 - 50	17	25.29	6.39	2.39	0.39

Discussion

The presented method seems to offer reliable results on the type of soft shallow bottoms where it has been used. According to Kushlan (1974) the dipnet removal efficiency is approximately 99 per cent. A limited test performed with our traps gave a result of 97 per cent but it is obviously a matter of personal efficiency. At most the traps' working range may be extended to slightly greater depths. The problems of emptying the traps increase with increments in height and area as they offer trapped fish greater possibilities of keeping away from the dipnets (Kushlan 1974). Factors to be considered are noise, shading effects, alteration of the bottom structure as well as dipnetting efficiency. The type 1 (Fig. 1) trap produces noise and vibrations as it slides down the vertical poles. Even if greased the noise and vibrations may cause escape reactions in fish within the working area. To overcome this problem the type 2 trap was constructed. Noise and vibrations during the fall are practically nil but the stay-strings (Fig. 1) may transfer frame oscillations caused by wind-action to the rest of the construction and thus affect fish in the drop area. It is therefore to be somewhat modified for future sampling.
Shading effects of the frame may bias day-time sampling. The presented results might be somewhat too low as Higer and Kolipinski (1967) and Moseley and Copeland (1969) showed that night catches often resulted in a higher biomass. Kjelson and Johnson (1973), however, found no significant difference between night and day catches for some species.
If the traps are used extensively in a limited area, alterations of the bottom may cause the results to be severely biased. The dipnets destroy the area within the frame if all fish are to be caught but as the traps are easily portable it is easy to select a new undisturbed area for each drop.

A point of interest is that larger fish were missing in the catches but this is most probably an effect of the limited numbers of drops. As abundance is negatively correlated to size, theoretically, the number of drops needed to get a good picture of the abundance within a certain relatively large area is higher than the number of drops presented in this paper. The species low down in the food web, have a higher catch probability, being more numerous per unit area, compared to the catch probability for predatory fish. For example, pike (*Esox lucius* L.) are known to be present in the area but were never caught.

If the school of rudd is excluded the mean biomass would be only
2.9 ± 0.4 g·m^{-2}. The values in Table 3 are all of this magnitude
and if compared to available literature data the presented biomass
figures are rather low. Ackefors (1975) has theoretically calculated the biomass of fish in the Baltic Sea to 5-9 g·m^{-2}. The average for this study is low as the littoral zone is regarded as a
productive area where a great number of invertebrates and young
fish abound. When compared to values for demersal fish presented
by Oviatt and Nixon (1973) for marine areas our values still fall
in lower parts of the ranges presented (0.6-19.6 g·m^{-2}). Their
data were obtained from trawl catch data of demersal fish and this
kind of data often gives low numbers because of low trawling efficiencies. It is likely that the actual density might be somewhat higher. Hellier (1959) found a mean of 16.6 g·m^{-2}, Laguna
Madre, Texas, using a drop net. Hall and Woodwell (unpublished)
presented a value of 39.7 g·m^{-2} for a productive Long Island estuary and Moseley and Copeland (1969) on the average about 35 g·m^{-2}
for Guadalupe Bay, Texas. Nixon and Oviatt (1973) presented biomass values for some aquatic ecosystems and only a few were lower
than our result.

On the contrary, the mean abundance found in our study (15.7 fish
per m^2) was high compared to the values of Oviatt and Nixon (1973),
0.05 fish per m^2, and more of the same magnitude as those presented for an area in the Everglades by Higer and Kolipinski (1967),
5 and 7.6 · m^{-2} for day and night catches respectively. The investigated area is an important nursery area for juvenile fish and
therefore explains the high abundance values.

Wahlberg (1969) worked in some of the areas we used. He seined and
found both sand gobies and common gobies (*Pomatoschistus microps*
Krøyer). The latter was not found by us but according to him there
was a relationship in occurrence of roughly 10:1 in favour of the
sand gobies. The highest abundance for them found by him was 1.5
·m^{-2} in October which is low compared to our mean for the species
(Table 1) and very low compared to our maximum mean found in
November, 12.7·m^{-2}. The seasonal pattern in occurrence of sand
gobies showed by him was also found in our material.

Abundance for sand gobies in the Dutch Wadden Sea and the adjacent
North Sea coastal area has been examined by Fonds (1973) and a
maximum abundance of 0.4 gobies per m^2 for the whole area was
reported for October. It is a low value compared to our results
but it must be kept in mind that our results were from the littoral zone while his were mostly from deeper areas. He also found
that sand gobies were more abundant in shallow water in summer and
autumn which is in agreement with our findings. He showed that
most of their spawning took place between depths of 10-25 m and
that an offshore migration in April-May to suitable spawning areas
was probably the cause of the decrease in numbers in the shallow
areas. This seems in general to be valid for this area although
Wahlberg (1969) noticed spawning in shallow areas around the
laboratory.

The number of species found was low but one would expect more
species to be found in the area. A continuation of the investiga-

tion with more traps would probably result in more species being caught and a better picture formed of the seasonal abundance and biomass fluctuations as well as of total production.

Conclusions

The presented results have given a preliminary picture of the most abundant and important fish species on soft bottoms in very shallow water of the northern Baltic proper. The drop-trap method offers reliable results but more samples are needed before a clear picture and a good understanding can be obtained of the importance of fish in this part of the Baltic ecosystem.

Acknowledgements

Financial support from Carl Tryggers Stiftelse and from Längmanska Kulturfondens nämnd have enabled us to construct the traps and their support is greatfully acknowledged.

References

Ackefors, H. 1975. Production studies of zooplankton in relation to the total production of the Baltic proper. Medd. Havsfiske-lab., Lysekil, 181 22 pp (mimeo).

Fonds, M. 1973. Sand gobies in the Dutch Wadden Sea (*Pomatoschistus, Gobiidae, Pisces*). Neth. J. Sea Res. 6 (4) 417-478.

Hall, C.A.S. and Woodwell, G.M. Unpublished but submitted to J. Fish. Res. Bd. Can. Spatial and seasonal patterns of standing crop, productivity, and diversity of fishes in Flax pond, a Long Island estuary.

Hellier, T.R. 1959. The dropnet quadrat, a new population sampling device. Publ. Inst. Mar. Sci. Texas 6 165-168.

Higer, A.L. and Kolipinski, M.C. 1967. Pull-up trap: A quantitative device for sampling shallow-water animals. Ecology 48 (6) 1008-1009.

Kahl, M.P. Jr. 1963. Technique for sampling population density of small shallow-water fish. Limnol. Oceanogr. 8 302-304.

Kjelson, M.A. and Johnson, G.N. 1973. Description and evaluation of a portable drop-net for sampling nekton populations. Proc. 27th Ann. Conf. S.E. Assoc. Game and Fish Comm. 653-662.

Kushlan, J.A. 1974. Quantitative sampling of fish populations in shallow, fresh-water environments. Trans. Am. Fish. Soc. 103 (2) 348-352.

Moseley, F.N. and Copeland, B.J. 1969. A portable drop-net for representative sampling of nekton. Publ. Inst. Mar. Sci. Texas 14 37-45.

Nixon, S.W. and Oviatt, C.A. 1973. Ecology of a New England salt marsh. Ecol. Monogr. 43 (4) 463-498.

Oviatt, C.A. and Nixon, S.W. 1973. The demersal fish of Narragansett Bay: An analysis of community structure, distribution and abundance. Estuar. Coast. Mar. Sci 1 361-378.

Wahlberg, B. 1969. Smörbultar. Zoologisk Revy årg. 31 (1-2) 36-37 (in Swedish).

Wegener, W., Holcomb, D. and Williams, V. 1973. Sampling shallow water fish populations using the Wegener ring. Proc. 27th Ann. Conf. S.E. Assoc. Game and Fish Comm. 663-674.

RESULTS AND PROBLEMS OF AN "UNSUCCESSFUL" BENTHOS CAGE PREDATION EXPERIMENT (Western Baltic) +

Wolf E. Arntz

Institut für Meereskunde an der Universität Kiel, FRG

ABSTRACT

In the Protected Research Area of the Joint Research Programme "Interaction Sea - Sea Bottom" of Kiel University, at 20 m depth, a number of in situ macrobenthos experiments were carried out to follow up the establishment of a faunal community as well as specific community mechanisms like competition, displacement and predation. This paper deals with an experiment carried out in 9 benthos cages, each 1 m^2, between August 1972 and May 1974.

Originally the cages were intended to provide data on the amount of predation by demersal fish (i.e. to show the development of a community that was not preyed upon by fish larger than 10 cm). Towards the end of the experiment however the fauna was no richer inside than outside the cages. One of the main reasons was that a "secondary hard bottom" in a muddy sand area attracts large numbers of starfish (Asterias rubens). The starfish, when young, enter the cages and prey upon the fauna much more heavily than on a "normal", 20 m depth, bottom. The same applies to the shore-crab, Carcinus maenas, which for lack of shelter does not normally appear at this depth, and to gobies which were protected from large predators and thus appeared in unusually large numbers. Another reason was that drifting macroalgae caught on the cages and started decaying after a while, increasing O_2 depletion and resulting in adverse H_2S conditions.

Apart from a number of useful experiences which were employed in designing a new experiment, the "unsuccessful" experiment yielded some results on the impact of oxygen decrease (which is a usual feature in the investigation area during late summer) on different faunal groups and species. While the mollusc populations collapsed with the exception of a few old specimens, mainly Mya truncata and Astarte elliptica, the polychaetes reacted in a very distinctive way. Some species were killed off whereas a few others (especially Capitella capitata) developed very well. With crustaceans (mostly Diastylis rathkei), as with the young stages of molluscs and polychaetes, predation, inside the cages, must be considered as having influenced the process of recolonization. Contrary to the original idea, this would give the benthic species even less chance to recover than under natural conditions.

+ Publication No 106 of the Joint Research Programme "Interaction Sea - Sea Bottom" of Kiel University.

INTRODUCTION

Within the Joint Research Programme (SFB 95) "Interaction Sea - Sea Bottom" of Kiel Univeristy, one working group deals with the energy transfer between the higher members of the benthic food web in the western Baltic. One of the basic problems studied is the question of how much is produced by macrozoobenthos and to what extent this production is utilized by the different demersal fish populations. To this end, besides a number of field studies on macrobenthos biomass and production, demersal fish biology and the food of these fish (for review of literature see Arntz & Brunswig, 1976 and Arntz, in press), a number of field experiments have been started on submerged substrate platforms (Sarnthein & Richter, 1974; Richter & Sarnthein, 1977) and in containers designed to study productivity on the bottom (Brunswig et al., 1976). This paper deals with a benthos cage experiment planned to exclude predation by larger fish in the Abra alba community of Kiel Bay.

The idea of comparing areas sheltered by cages with the open environment is quite old. To my knowledge, Blegvad (1928) was the first to carry out an experiment of this type in very shallow water of the Limfjord. He reports a sixtyfold greater standing crop under the cages after one summer. While similar shallow water studies in the investigation area are in preparation, the decision to start at 20 m resulted from the fact that at this depth the bottom fauna is richest and the share of good fish food is highest over wide areas of Kiel Bay (Arntz & Brunswig, in press). Although the experiment turned out to be a near failure, it yielded data which are of some value in understanding the factors controlling benthic community structure.

EXPERIMENTAL DESIGN, MATERIAL AND METHODS

From August 1972 to May 1974, 9 stainless steel cages (without a bottom of 1 m^2 each (Fig. 1) were placed on an existing benthos association at 20 m depth, in the "Hausgarten" research area of SFB 95. The mesh size of the cages was 30 mm, designed to prevent fish larger than about 10 cm (mainly cod, Gadus morhua L., and different flatfish species) from entering the

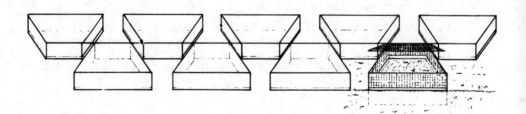

Fig. 1. The "benthos cages".

cages and large enough to avoid clogging by hydrozoans and small algae.

The investigation area lies on a slope which in a distance of 600 m falls from 13 m to 27 m, and where the sediments change from medium sand to H_2S mud (Wefer & WTGK Kiel, 1974). The conditions in this area, although considerably removed from the influx zone of the Great Belt providing most of Kiel Bay with more saline bottom water rich in oxygen, have been shown to be still fairly typical for the whole of the bay (Arntz et al., 1976; Arntz & Brunswig, in press). However, the deeper part of the protected research area is more often exposed to oxygen deficiency, and the well.colonized belt characterized by muddy sand - sandy mud sediments is narrower. At 20 m depth, the sediment is a muddy fine sand with median sediment values ranging between 0.4 and 0.08 mm, and colonized by a rich meiofauna with an annual biomass mean of 0.56 g m^{-2} dry weight (Scheibel, 1976).

At approximately two-monthly intervals, divers took nine 192 cm^2 samples from inside the cages (one from each cage; for sampling scheme see Fig. 2) and another nine samples from the unsheltered area between the cages. The sampler, a mechanical push box,

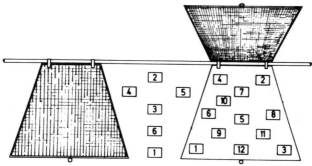

Fig. 2. Sampling sequence in- and outside the cages.

penerated about 10 cm into the sediment (cf. Brunswig et al., 1976), thus yielding quantitative results (Table 1) with the exception of large Mya truncata. There was no statistically significant difference between the inside and outside samples at the beginning of the experiment (for comparisons the Mann-Whitney U-Test, P = 0.05, is used (Elliott, 1971)).

Since the sampling area was insufficient for a comparison of the single boxes and because there is the probability that the divers mixed up the cage numbers, the data from the 9 cage samples are combined and compared with the data from the 9 outside samples. Between June and July 1973, the space between the cages was clogged by drifting algae; as these samples were no longer considered quantitative, subsequent samples were taken a few metres away. This incident has been marked on the figures with two lines (II). Additional sampling carried out in June 1973 (marked with a (·) for the new samples) indicates, however, higher densities within the new sampling scheme only for the molluscs.

TABLE 1a. Mean, standard deviation and variation coefficient of specimen numbers inside and outside the cages (October, 1972)

	Bivalves		Polychaetes		Crustaceans		Others		Total	
	inside	outside	inside	outside	inside	outside	inside	outside	inside	outside
Cage 1	14	19	63	29	4	27	6	0	87	75
2	19	24	37	33	12	45	0	1	69	103
3	24	14	68	30	5	5	4	0	101	49
4	17	19	21	19	12	2	1	0	51	40
5	39	12	44	17	5	0	5	2	93	41
6	12	17	38	15	8	15	0	2	58	49
7	14	7	21	25	6	8	3	0	44	40
8	31	4	21	17	9	15	5	4	66	40
9	23	14	15	40	21	28	4	2	63	84
$\bar{x}$	21.4	14.4	36.4	25.0	9.1	16.1	3.1	1.2	70.2	57.9
s	9.0	5.6	19.2	8.6	5.4	14.8	2.3	1.4	19.6	23.5

VAR. COEFF.

$V\% = \dfrac{s}{\bar{x}} \cdot 100$

	42.1	18.1	52.8	34.4	59.3	91.9	74.2	116.7	27.9	40.6

TABLE 1b. Mean, standard deviation and variation coefficient of biomass inside and outside the cages (October, 1972; molluscs excl. individuals 5 g).

	Bivalves		Polychaetes		Crustaceans		Others (excl. Asterias)		Total	
	inside	outside	inside	outside	inside	outside	inside	outside	inside	outside
Cage 1	4.14	6.46	0.99	0.35	0.04	0.39	0.10	0.00	5.27	7.20
2	7.79	3.53	0.63	0.29	0.16	0.63	0.00	0.03	8.58	4.48
3	1.14	5.12	0.30	0.48	0.07	0.06	0.05	0.00	1.56	5.66
4	3.85	9.71	0.87	1.64	0.16	0.04	0.02	0.00	4.90	11.39
5	7.46	18.56	1.39	0.31	0.06	0.00	0.05	0.03	8.96	18.90
6	2.56	4.25	2.20	0.48	0.09	0.21	0.00	0.05	4.85	4.99
7	4.99	0.45	0.74	0.34	0.07	0.10	0.02	0.00	5.82	0.89
8	12.40	2.63	0.66	0.31	0.12	0.19	0.13	0.15	13.31	3.28
9	11.55	1.96	0.40	1.76	0.20	0.32	0.24	0.03	12.39	4.07
$\bar{x}$	6.21	5.85	0.91	0.50	0.11	0.22	0.07	0.03	7.29	6.60
s	3.89	5.48	0.58	0.69	0.09	0.20	0.10	0.03	3.84	5.50

VAR. COEFF.

$V \% = \dfrac{s}{\bar{x}} \cdot 100$

| 62.6 | 93.7 | 63.7 | 138.0 | 81.8 | 90.9 | 142.9 | 100.0 | 52.7 | 83.3 |

All the samples were washed on an O.5 mm screen on board the research vessel and fixed individually in 4% formalin. The molluscs were preserved in 70% alcohol after sorting which was carried out within one week of sampling. For illustration, abundance of specimens was chosen rather than biomass since the density figures are not as much influenced by a few old specimens as are the weight data.

RESULTS AND DISCUSSION

A first glance at the data showing the variations in specimen number inside and outside the cages (Figs. 3-5) reveals that the protected fauna by no means exceeded the fauna outside as was expected, but that on the contrary the unprotected fauna became much more numerous at certain times during the experiment. Besides, it showed seasonal variations which are typical for the western Baltic (Arntz 1971a) with a strong decline of many species towards the winter and distinct recruitment in late spring, whereas the cage fauna never recovered from its original decline.

In the development of the different faunal groups during the experiment there is one basic pattern: From originally similar densities at the beginning (except for Diastylis at the first sampline date, where the inside numbers were twice the value outside) there is a general seasonal decline until March (Diastylis, Nephthys) or June 1973 (most other species). Although the inside numbers of molluscs and polychaetes seem to be a little lower from the start, they are significantly lower only from December, 1972 onwards. Then with spring spatfall, the outside values climb to distinct peaks between July and September, while under the cages there is practically no change. From September to October a sudden breakdown is seen, only in some cases (Nephthys, Pectinaria) extending over more than a month. There are only few exceptions to this general pattern: Scoloplos armiger shows a continuous rise inside the cages between the beginning of the experiment and its breakdown between July and September 1974: Capitella capitata reveals a strong population development restricted to the cages only after most of the other fauna died (Fig. 6), confirming once again its "opportunistic" properties. Large Ophiura albida were significantly more abundant in the cages until the general breakdown in October 1973.

There are two main reasons for unexpected development of the experiment: 1) disturbance by O_2 deficiency and development of H_2S, and 2) disturbances by the experimental design itself. Although it will become apparent that these influences cannot really be separated, they will be treated individually for reasons of greater clarity.

1) Influence of O_2 deficiency

It has been shown in a number of publications (Arntz et al., 1976; Brunswig et al., 1976; Arntz & Brunswig, in press; Nichols, this symposium) that the O_2 conditions are of outstanding importance in the deeper part of the investigation area. Von Bodungen's (1975) data (Fig. 6) show that oxygen declines - especially in warm summers - every year between the months of August and Novem-

Benthos Cage Predation Experiment 37

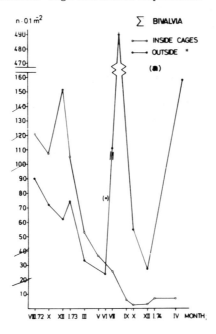

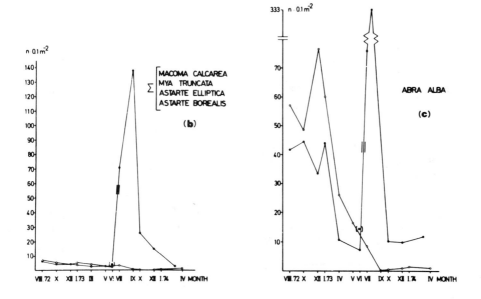

Fig. 3. Changes is abundance of total bivalves (a), several long-lived species (b) and Abra alba (c) in- and outside the cages.

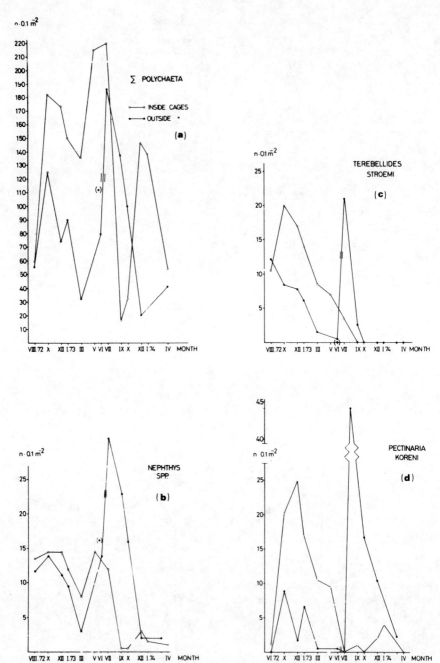

Fig. 4. Changes in abundance of total polychaetes (a), Nephthys spp. (b), Terebellides stroemi (c) and Pectinaria koreni (d) in- and outside the cages.

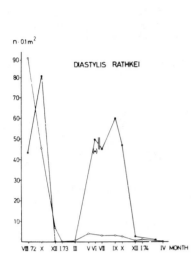

Fig. 5
Changes in abundance of *Diastylis rathkei* in- and outside the cages.

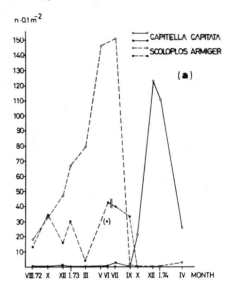

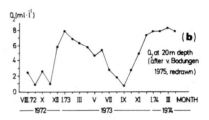

Fig. 6. Changes in abundance of *Capitella capitata* and *Scoloplos armiger* (a) and oxygen concentration at 20 m depth (b).

ber. The result is oxygen deficiency and development of H_2S not only in the sediment but also in the bottom water. In exceptionally warm summers with little wind movement, e.g. 1973, O_2 values of 1 ml l^{-1} (measured 0.5-1 m above the bottom) are reached, and the effect of the fauna can be deleterious even above 20 m. Figs 3 and 4 reveal that there was an almost complete breakdown of the fauna in the surrounding area in 1973. While areas below 22 m in southern Kiel Bay, where oxygen dificiency is a normal feature are colonized only by a few species tolerant to H_2S (e.g. by the polychaete *Capitella capitata* and the priapulid *Halicryptus spinulosus*) and - during periods with improved conditions - by species able to immigrate quickly (e.g. by the cumacean *Diastylis rathkei*, the mysid *Gastrosaccus spinifer* and the polychaete *Harmothoe sarsi*, only a few benthic organisms survive a breakdown such as that in 1973 at 20 m. They are mainly old bivalve specimens of <u>Astarte</u>, <u>Mya</u> and <u>Cyprina</u>, which are possibly able to break a thin H_2S bottom water layer by strong filtration (Schulz 1968). That the rest of the fauna is not extinguished from the area for sometime but shows considerable recruitment in the spring following a catastrophy is due to the fact that the bottom water and sediment normally recover during the winter, that spatfall is heaviest in spring, and that a large number of meroplanktonic larvae are carried in by the channel currents.

So why did the fauna only recover outside the cages? The reason has probably to be sought in the experimental design itself.

2)   Influence from the experimental design

Stainless steel cages on a level sea bottom alter the normal conditions to such an extent that this may have caused the failure of the predation experiment from the very beginning.   There are two main reasons which can be associated with a)   an increase of O_2 deficiency (and, perhaps, other abiotic parameters) brought about by the experimental design, and b) biotic influences caused by the existence of a secondary hard bottom in a soft bottom area.

a)   In spite of the rather large mesh size of 30 mm it cannot be completely excluded that the water and sediment conditions inside the cages were less natural than outside.   Hydrozoans were never observed to clog the meshes, algae did not grow on the cages, and occasional ascidians were of no importance; however, rather large amounts of drifting macroalgae collected between the cages and may have influenced the water exchange to some extent.   The quantity of these red and brown algae had never become apparent to divers because normally there is nothing at 20 m depth to stop them.   They were always removed, but some of them had already started to decay.   Unfortunately changes in O_2, light conditions and sediment chemistry were not recorded during the experiment, so there is no clear indication about the effect of these algae.

Another possibly negative effect may have come from the cages themselves.   Although made of "V4A" stainless steel, they became corroded and were more or less rotten towards the end of the experiment.   This applies especially to the lower part of the walls covered by sediment.   The only explanation so far at hand is that stainless steel is affected by H_2S.

b)   The secondary hard bottom on the sea floor attracted great numbers of predators and scavengers which normally do not or only rarely appear at this depth.

<u>Asterias rubens</u>:   The common starfish, which in the western Baltic is much more abundant in shallow water with a stony bottom, was found in all sizes under the cages, with an abundance of larger animals (6 cm diameter) of over 10 per square metre.   At the first two sampling dates divers removed the starfish but later they were left, as their removal apparently had no effect at all.

<u>Carcinus maenas</u>:   The shorecrab is normally never found at this depth because of lack of shelter from predation by large cod.

However, throughout the experiment, divers noted the presence of adult Carcinus in the cages, never exceeding 2 per cage and 10 for all cages. They were always removed.

Buccinum undatum: The whelk lives in the Abra alba community of Kiel Bay in low abundance. This scavenger, obviously living on remains left by predators, appeared in numbers of up to three per cage.

Harmothoe imbricata, Lepidonotus squamatus: These carnivorous polychaetes were found in rather large numbers (50 per cage) when the experiment was finished and the cages were lifted. In contrast to Harmothoe sarsi they do normally not appear because most of the time they live firmly attached to walls, stones etc.

Asterias and Carcinus must have immigrated into the cages as young animals or even as larvae, since as large specimens they were no longer able to leave. Apparently Carcinus could only survive because the cages sheltered them against predation. This is also true for gobies:

Pomatoschistus minutus, Chaparrudo flavescens   Gobies are a very important food for cod and dab, Limanda limanda (L.) in Kiel Bay (Arntz, 1971b, 1974), but only during the winter when they appear in large numbers in the deeper parts. In the cages, they were in fact most abundant during December and January, when the divers reported numbers of 20 per m^2. However, there were some gobies also during summer, amounting up to 5 per cage. The figures may have been higher since it has to be assumed that some of the fish hide when the divers open the cages.

A favourable effect of the cage shelter was only noticed during the first phase of the experiment, in the case of Ophiura albida large specimens of which could obviously not be taken by the small predators inhabiting the cages.

All these observations indicate that beside a possible increase of the unfavourable oxygen conditions during the experiment due to macroalgae clogging some of the cage walls, predation may have negatively influenced recruitment under the cages or hindered immigration of some more motile species which were found to be common in the open surroundings. The rise of Capitella capitata after the breakdown might be taken as an indication that oxygen deficiency was on the whole the more important factor. On the other hand the rise of this population did not last more than three months before it also became subject to predation (as the samples show there were no competitors in the cages until April 1974). It is most unsatisfying that the two main influencing factors cannot be separated.

In planning field experiments in the sea, even in a protected research area, certain disturbances can be forseen from the beginning. Among these are bad weather conditions preventing the divers working or resulting in insufficient samples, loss of buoys by ship propellers, gales or theft, and loss of other equipment due to anchoring ships, fishing gear etc. In fact in

May 1974 most of the cages were destroyed by an anchor of the research vessel, leading to somewhat premature end of the experiment.

Many of the problems shown in this paper can be excluded in future experiments: The vessel can be anchored at mooring buoys, drifting algae can be kept away by underwater fences, and even the mesh size could be reduced to some extent if the lids were more often changed by the divers. Water and sediment conditions under the cages would have to be measured in future, and the outside samples might be taken a few metres away from the cages from the very beginning. What is much worse is the enormous attraction an artifical hard bottom has for some benthic animals, and the shelter it provides for a number of small predators which may have the same (or greater) grazing effect as the few large fish design wanted to exclude.

It would not be a good solution simply to place the experimental cages into shallow water next time although there the problems do not seem to be as grave. Steele (1966) reports removal of sand by wave action from tanks exposed 2 m below the level of low water springs. Nichols (pers. comm.) and Virnstein (pers. comm.) were quite successful with shallow water experiments. Even with a smaller mesh size, there was no indication that recruitment was less good in their containers than outside, showing that as such our cages might have been satisfactory (excluding the stainless steel material which can easily be altered). However, our aim was to measure the grazing effect on species actually eaten by fish, under conditions where they really live. Although it seems sad I must conclude that this type of deeper water experiment is not likely to be a success in future, at least not on a Baltic soft bottom.

ACKNOWLEDGEMENT

I would like to thank my colleagues of the Scientific Diving Group of Kiel University for their help in taking the samples under conditions which were everything but favourable.

REFERENCES

Arntz, W.E. 1971a. Biomasse und Produktion des Makrobenthos in den tieferen Teilen der Kieler Bucht im Jahr 1968. Kieler Meeresforsch. 27, 36-72.

Arntz, W.E. 1971b. Die Nahrung der Kliesche (Limanda limanda (L.)) in der Kieler Bucht. Ber. dt. wiss. Kommn. Meeresforsch 22, 129-183.

Arntz, W.E. 1974. Die Nahrung juveniler Dorsche in der Kieler Bucht. Ber. dt. wiss. Kommn. Meeresforsch, 23, 97-120.

Arntz, W.E. (in press). The "upper part" of the benthic food web: The role of macrobenthos in the western Baltic. JOA Conference, Edinburgh (U.K.), Special Symp. C8, Sept. 13-23, 1976.

Arntz, W.E. and D. Brunswig 1976. Studies on structure and dynamics of macrobenthos in the western Baltic carried out by the Joint Research Programme "Interaction Sea - sea bottom" (SFB 95-Kiel). 10th European Symposium on Marine Biology, Ostend, Belgium.

Arntz, W.E. and D. Brunswig (in press). Zonation of macrobenthos in the Kiel Bay channel system and its implications for demersal fish. 4th Baltic Symp. on Marine Biology, Gdansk, Oct. 1975, Prace Morskiego Instytutu Rybackiego.

Arntz, W.E., D. Brunswig and M. Sarnthein 1976. Zonierung von Mollusken und Schill im Rinnensystem der Kieler Bucht. Senckenbergiana maritima 00, (in press).

Blegvad, H. 1928. Quantitative investigations of bottom invertebrates in the Limfjord 1910-1927 with special reference to the plaice food. Rep. Danish biol. Sta. 34, 33-52.

Bodungen, B. v. 1975. Der Jahresgang der Nährsalze und der Primärproduktion des Planktons in der Kieler Bucht unter Berücksichtigung der Hydrographie. Diss. Univ. Kiel, 116pp

Brunswig, D., W.E. Arntz and H. Rumohr 1976. A tentative field experiment on population dynamics of macrobenthos in the western Baltic. Kieler Meeresforsch. 00, (in press).

Elliott, J.M. 1971. Some methods for the statistical analysis of samples of benthic invertebrates. Freshwater Biol. Ass. Scientific Publ. No. 25, 148 pp.

Nichols, F.H. 1977. The dynamics and production of Cistera cylindraria (Pallas) = Pectinaria koreni (Malmgren) in Kiel Bay, West Germany. In Proceedings of the 11th European Symposium on Marine Biology ed. Keegan, B.F., P. O Ceidigh and P.J.S. Boaden.

Richter, W. and M. Sarnthein 1977. Benthic colonization of different sediments on submerged platforms in the western Baltic. In Proceedings of the 11th European Symposium on Marine Biology ed. Keegan, B.F., P. O Ceidigh and P.J.S. Boaden.

Sarthein, M. and W. Richter 1974. Submarine experiments on benthic colonization of sediments in the western Baltic Sea. I. Technical layout. Mar. Biol. 28, 159-164.

Scheibel, W. 1976. Quantitative Untersuchungen am Meiobenthos eines Profils unterschiedlicher Sedimente in der westlichen Ostsee. Helgoländer wiss. Meeresunters. 28, 31-42.

Schulz, S. 1968. Rückgang des Benthos in der Lübecker Bucht. Monatsber. Dt. Akad. Wissensch. Berlin 10, 748-754.

Steele, J.H. 1966. Experiments on O-group plaice in underwater tanks. <u>ICES C.M. 1966/C:</u> 9, 4 pp.

Wefer, G. and Wiss. Tauchgruppe Kiel 1974. Topographie und Sedimente im "Hausgarten" des Sonderforschungsbereichs 95 der Universität Kiel (Eckernförder Bucht, westliche Ostsee). <u>Meyniana</u> 26, 3-7.

THE IMPORTANCE OF BEING A 'LITTORAL' NAUPLIUS

H. Barnes and Margaret Barnes

The Dunstaffnage Marine Research Laboratory, Oban, Argyll, Scotland

ABSTRACT

The possible contribution of littoral organisms to the zooplankton of inshore waters is largely neglected - particularly in studies of productivity. The relation between the physical dimensions of an inshore system and the relative input from the bottom, the water mass, and the littoral will be considered - largely relative to some common cirripedes. An assessment of the importance of this contribution in terms of both seasonal and non-seasonal events will be considered. A few calculations will show the possible quantitative importance, in some situations at least, of littoral organisms in this respect, particularly when compared with published estimates of the amount of "permanent plankton". Apart from its general character, "littoral" plankton may be especially important at certain seasons and in particular areas not only in terms of its quantity but also as regards the "quality" of the nutriment it provides for other organisms; it is probably an integral part of some temporary ecosystems.

INTRODUCTION

There is general agreement that the title of a scientific paper should give some guide as to its contents; the somewhat facetious title of the present one demands, therefore, an excuse if not an explanation. The reason, as for many facetious statements, is simply to attract attention; but, since this in itself is - or should be - deplored by scientists, further comment seems in order. We believe that the contribution of littoral organisms to the ecosystems of inshore waters has been distinctly neglected and that this state of affairs should be rectified. Littoral ecologists are largely interested in what remains on the shore and planktologists in what stays in the plankton. The so-called meroplankton or 'non-permanent' part of the plankton whether originating from the littoral or the benthos tends to be neglected - except when it is being considered as some part of the life cycle of an organism. It may also be pointed out that while 'non-permanent' in terms of the biology of the individual species this category of plankton makes a continual 'permanent' contribution. In terms of the world ocean the area of the littoral and its contribution to the marine biomass is, of course, trivial in the extreme. Nevertheless, there is ample evidence, that in terms of unit area the littoral contribution is extremely high and some of this - particularly in the form of reproductive and excretory products - finds its way directly into inshore waters where it makes a substantial contribution to the over-all marine productivity.

We would like to illustrate this by reference to *Balanus balanoides*. Crisp (1976) has recently considered the importance of a pelagic stage and consequent dissemination in regard to variable strategies of reproduction.

Some morphometric calculations

For simplicity, we will consider first a closed circular basin - neglecting any discontinuity at the entrance - of radius r at the water surface at mean high tide, with a mean tidal range of h and with the shore sloping at an angle of θ. Calculations will be made for h = 4 m, r = 500 and 1000 m, and θ = 60°, 20° and 50°. The plane area of the littoral in terms of h and θ is:

$$A_L = \pi(2r - h \cot \theta) \sqrt{h^2 + 2r h \cot \theta - h^2 \cot^2 \theta}$$

The appropriate values are given in Table I.

Table 1

Littoral plane area, A_L ($m^2 \times 10^5$), for a circular basin of radius r (m) and slope of shore θ.

Radius of basin	Slope of shore		
	60°	20°	5°
500	1.51	3.24	6.26
1000	4.27	9.24	18.34

The surface area of the water at mean tide (A) is 7.85×10^5 and 31.42×10^5 for r = 500 and 1000, respectively.

The output of cirripede larvae

The mean population density of *Balanus balanoides* will be taken as $1/cm^2$. The size and age distribution of the population will vary with the locality and particularly with the exposure; barnacle cover is much reduced by algae in quiet-water situations. If 50% of the shore is covered by barnacles at this population density, this gives 5×10^3/animals m^2. In some situations this may well be an over-estimate but it must be remembered that on shores covered by boulders the true surface area available for settlement is far greater than the plane area; for example, if the boulders are treated as open-packed spheres the ratio of the true area to the plane area is 3:1 (the ratio would be greater for close packing). *Balanus balanoides* releases its stage I nauplii once each year in the spring, this release taking place - or the major part of it - over a very restricted period. A mean value of 2000 stage I nauplii/individual (equivalent to one with a body weight of ≈ 1.5 mg and a basal diameter of ≈ 8 mm) will be assumed (Barnes & Barnes, 1965, 1968). This gives an estimated output of 10^7 stage I larvae/m^2. The dry weight of a stage I nauplius is 1.0μg (Barnes, unpubl.). The littoral production of these nauplii is, therefore, 10 g dry wt/m^2. The total larval contribution (g dry wt) and per unit surface area of the 'basin', assuming the water is completely mixed, are given in Table II, for the same three values of θ and two values of r.

Table II

Balanus balanoides: total contribution (g dry wt $\times 10^5$) from littoral with variable angle of shore (θ) and contribution per unit surface area of water (g dry wt/m^2).

Radius of basin	Slope of shore					
	60°	20°	5°	60°	20°	5°
	Total			Areal (water surface)		
500	15.1	32.4	62.6	1.93	4.13	7.97
1000	42.7	92.7	184	1.36	2.95	5.84

The biochemical composition of a stage I nauplius may be taken as carbohydrate 2%, lipid 32%, and protein 62%; taking conventional values of 4.2 kcal/g for carbohydrate and protein and 9.5 kcal/g for lipid gives a value of 5.72 kcal/g dry wt stage I nauplii. In terms of kcal/m^2 the areal values in Table II gives those of Table III.

Because the tidal rise and fall remains constant, the smaller the basin the greater the contribution of larvae when expressed in terms of surface area. There are data - of very variable accuracy and kind - on the standing crop and rate of production of zooplankton from a wide variety of localities and in order to emphasize the present thesis it is not necessary to compare each estimate with the present 'model'; a comparison with some of the data given by Steele (1974) will suffice. Steele gives a value of 175 kcal/m^2/yr as the annual production of copepods - the main contributors to the planktonic

Table III

Balanus balanoides: areal (water surface) production in kcal/m^2.

Radius of basin	Slope of shore		
	60°	20°	5°
500	11.0	23.7	45.6
1000	7.8	36.9	33.4

biomass - of the North Sea. This is higher than the values for *Balanus* with r = 500 or 1000 even at a rocky shore slope of 5° (Table III), but when it is remembered that the *Balanus* input takes place over only a few weeks its impact on planktonic events at that time of the year must clearly be very considerable. It may also be noted that if it is assumed that the larvae are all present in the upper 3-5 m of the water column the values are of the same order as given by Wiborg (1976) for commercially fishable *Calanus finmarchicus*, namely 2-25 g wet wt/m^3 ($\simeq$ 0.5 - 6.3 g dry wt/m^3). We may also note that Rzepishevsky (1962) found *B. balanoides* larvae often form more than 99% of all the zooplankton in the Balne-Zelenetskaya inlet of the Eastern Murman sea; it is known that in this region they form the main food of herring larvae. He states that the larvae reach concentrations of 3 x 10^4/m^3, i.e., 0.15 g dry wt/m^2 assuming the larvae are present in the upper 5 metres. The larvae will enter the planktonic food web; some will pass directly to the carnivores; some will die and be added to the particulate and soluble fraction; some will reach the benthos. All larvae living by the time the cyprids settle will, unless returned to the shore, be added to the latter two fractions. Even at extremely high settlement densities, say, 100/cm^2 (10^6/m^2) only 10% of the input returns to the shore - and subsequently most of this is returned to the water as dead juveniles since only one larvae/adult needs ltimately to survive to maintain a steady-state shore population (such return may be indirect if account is taken of predation in the littoral. These values may be compared with those of Barnes (1971) on the annual production of *Elminius modestus* in the semi-enclosed Bassin d'Arcachon. The shores are sandy and do not carry a significant barnacle population but the latter settle on the numerous pignots which surround the famous oyster grounds; because of the regular structure of these, reasonable estimates of the area and number of animals can be made. The estimated egg production (which will be similar to nauplius production) was found to be 6.26 g dry wt/m^2 surface area/yr. It has also been shown by Barnes & Barnes (1968) that no matter how many broods are produced per year - many in a warm-water species, one in a boreo-arctic species such as *Balanus balanoides* - the amount/yr in cirripedes is relatively species-independent. The value of 6.26 g dry wt/m^2 is of the same order of magnitude as that calculated for *Balanus balanoides* (5.84 d dry wt) for the model with θ = 5° and r = 1000 (Table II); the estimated area of pignots and collectors was 7.3 x 10^5/m^2 which is equivalent to the model for r = 1000 and θ between 20° and 60°.

Clearly, by treating a bay with a narrow entrance as a closed system the importance of any littoral contribution to production is maximized and, indeed, for many natural systems grossly over-estimated. Yet situations which approximate to the one considered do exist. Lewis & Powell (1960) have described the littoral of Loch Sween, Scotland; the entrance to the sea is extremely narrow and because of the small exchange with open water a well-developed population of *Chthamalus stellatus* is present on the shores of the Loch even though this species is very infrequent on the adjacent open coast; *Balanus balanoides* is also present on the shores of the Loch and must make a distinct contribution to the zooplankton in the spring.

The possible contribution in two other cases which are the least favourable for larval retention may be considered. Off a straight open coast the simplest model is to consider all the larvae to be instantaneously released along an infinite distance with a uniform population of *B. balanoides* which will be assumed to be of the same density as above; it will be assumed that N larvae are immediately uniformly distributed through a depth H and then uniformly spread offshore by horizontal diffusion alone. Then the concentration C (per m³) is given by

$$\frac{\partial C}{\partial t} = \frac{\partial}{\partial x}(K_x \cdot \frac{\partial C}{\partial x})$$

where t is the time, x the distance offshore, and K and eddy diffusion constant (see Csanady, 1973). Assuming that K remains constant the solution is,

$$C = \frac{N}{H\sqrt{\pi K t}} \cdot e^{-\frac{x^2}{4Kt}}$$

K may be taken of the order of 10^3 to 10^4 cm²/sec. The value for the larval output per m² of the littoral will again be taken as $10^7/m^2$ (= 10.0 g dry wt/m²), the intertidal height as 4 m and θ = 20°. The same conversion of dry wt to kcal as that used previously will be assumed. The amounts present relative to the distance offshore are shown in Table IV.

Table IV

Balanus balanoides: amounts present in g dry wt/m² and kcal/m² at various distances offshore assuming larvae instantaneously released: x, distance (m) offshore; t, 28 h and 12 days; K, 10^3 cm²/sec; angle, θ, offshore, 20°; 10^7 larvae released/m² littoral plane area.

		x, distance offshore (m)				
		50	100	200	500	1000
g dry wt/m²	28 h	0.62	0.51	0.24	-	-
	12 days	-	0.21	-	0.11	0.02
kcal/m²	28 h	3.55	3.03	1.38	-	-
	12 days	-	1.17	-	0.65	0.11

Under the above assumptions the contribution of these 'littoral' larvae/m² soon becomes very much reduced compared with that found for the enclosed basin (Table III) - but nevertheless at moderate distances from the shore cannot be regarded as completely negligible. During such diffusion there will

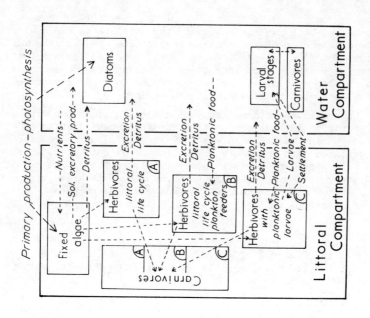

Fig. 2. Certain aspects of relation between littoral and water compartments: grossly simplified as regards many aspects in order to emphasize others.

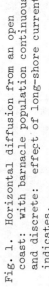

Fig. 1. Horizontal diffusion from an open coast: with barnacle population continuous and discrete: effect of long-shore current indicates.

be larval development and mortality.

Severe loss from the immediate environment of the shore may also take place in an estuary where complete mixing takes place on each tidal cycle. Assuming such complete mixing at each tidal cycle and no return of exchange water on successive tides then,

$$P_n = P_o (1-R)^n$$

(Ketchum, 1951) where P_o is the original concentration, P_n that after n tides, and R the exchange ratio. Bousfield (1955) has given some detailed data for the Miramichi Estuary (Canada) with particular reference to barnacle larvae. He estimated r to be 0.15 so that after 18 days < 1% of the original population should remain within the estuary. But these 'losses' were partially reduced by the interaction between other physical factors and larval behaviour; seaward drift and a landward counter-current together with differences in the vertical distribution of the several larval stages led to a retention of 10% rather than the calculated < 1% in this estuary.

The three examples given above only serve to illustrate the possible range of magnitudes involved. Coastlines have an almost infinite variety of topographies - bays with different kinds of entrances, estuaries, and so forth; shorelines vary in slope even over short distances, as do the population densities of the littoral animals; the effect of wind force and direction during the planktonic phase of 'littoral' nauplii is of extreme importance in their dispersal. Moreover, there is interaction between these variables, e.g., sheltered shores, where dispersal might be expected to be minimal, have low population densities of some littoral animals, particularly *B. balanoides*, as a result of the increased algal cover; and the situation may vary considerably from year to year. Barnes (unpubl.) has shown that the intensity of settlement of *Balanus balanoides* (which may be taken as inversely related to its dispersal) from year to year in lochs around the Clyde Sea Area is directly related to the orientation of the loch relative to the direction and speed of the wind.

Littoral and planktonic systems

Once shed into the water the nauplii become part of the planktonic system - both as predator and prey - as they develop through successive larval stages. Because of a lack of quantitative data estimates become largely a matter of speculation. The shore is easily accessible and the population density of the adults - at least in certain cases - may be assessed if the appropriate techniques are used; for example, on a rocky reef accurate estimates are possible; on a shore covered with variable sized boulders it is much more difficult. But it is not so as regards the plankton: extensive sampling programmes are essential and due respect must be paid to a wide variety of other species - both plant and animals. During their planktonic life the nauplii will consume diatoms and in view of their density in certain situations, as has already been demonstrated, must make a direct impact. In addition there will be a loss by mortality and predation. The organic content of the various stages is known and according to the assumptions made regarding losses it is possible to reach almost any desired conclusion as to the effect upon the planktonic system; speculation really ceases to be profitable. There are, nevertheless, some general and even specific aspects of the effect upon the water compartment which can be profitably considered.

In oceanic, and even far offshore waters, plankton ecology is largely that of forms which pass all stages of their life-history in the planktonic phase; there is some contribution from species in which the adults are free-living (e.g., fish) and the larval stages usually considered planktonic as well as some from the larvae of benthic organisms. Most of the components of this planktonic system are subject to the same physical and chemical variables and only actions and interactions within this total system have to be considered. The whole may be considered as a single compartment. In inshore waters, however, the input of planktonic larvae of littoral animals is not controlled by the same factors. The breeding adults are subject to quite different and often more fluctuating conditions; the standing stocks of the littoral populations vary according to quite different factors and the littoral can be considered as a compartment distinct from the planktonic one. Furthermore, primary production is quite independent of that in the water mass and in some cases provides the food of littoral herbivores; in other cases e.g., barnacles, the adult food is derived from the water mass. When the food of the littoral organisms with planktonic larvae is derived from littoral algae then their input of larvae is an 'absolute' gain i.e., no phytoplankton components are utilized by the adults in the production of the initial larval stages.

In warm waters the planktonic input from the littoral may vary little with season and may be considered as a steady transport from one compartment to another. In boreo-arctic waters where the larvae, of, say, *B. balanoides* are often released into the plankton over a very short period (Barnes, 1962) there is a transfer between littoral and free water compartments but it must take the form of a marked perturbation of the water compartment. In the evolution of the whole ecosystem the perturbation has, so to speak, 'been taken care of'. Thus if larvae were introduced in the absence of an adequate phytoplankton population the results would be disastrous for the survival of the 'littoral' larvae themselves. In the case of *B. balanoides* the larvae are retained in the mantle cavity of the adult until release is stimulated by the diatom bloom itself (Crisp, 1956; Barnes, 1957; Crisp & Spencer, 1958); in the case of *B. balanus,* a low littoral but largely sublittoral species, the eggs are fertilized at such a time that embryonic development is completed by the time of the diatom increase - with possibly a final triggering mechanism of the latter. The distinction between the two compartments is here very clear: the factors which determine the timing of the spring increase seem to be quite different from those which regulate cirripede reproductive activity. The degree of any perturbation such as just described will be dependent upon the input of larvae and this, in turn, will be determined by the population density of the adults which in its turn, will be determined by variations in the littoral environment. For example, a succession of warm years in southwest Britain increased the density of *Chthamalus stellatus,* a warm-temperate species, at the expense of *Balanus balanoides* - in places almost to the extinction of the latter (Southward & Crisp, 1954, 1956). Elsewhere in Britain the Australasian immigrant *Elminius modestus* has decimated the local *Balanus balanoides* population - particularly in estuarine situations (Crisp, 1958). In both cases the spring perturbation of *B. balanoides* must have been markedly reduced to by replaced by a more extended period of moderate larval input.

The perturbation is only of a short duration. The planktonic larvae have, in most cases, a very limited life-span before 'returning' to the shore as sedentary forms to grow into the adult. During their planktonic existence they must have an effect on the food web via competition and consumption. Since they pass quickly through a size-range of 0.3-1.1mm (total length) they could provide a spectrum of prey to a variety of carnivores. The variations in the timing of the input could markedly effect the success of any predators relying wholly or in part on their presence.

One particular aspect of the food web related to cirripede nauplii is the possible importance with respect to larval and young herring. Rosenthal & Hempel (1970) have found that herring larvae > 15 mm take largely copepod nauplii and those < 15 mm largely copepodid stages. Taking the size of *Calanus helgolandicus* as representative of these stages (from Marshall & Orr, 1955) the two groups correspond to lengths of 0.2-0.6 and 0.9 to 3.0 mm respectively; the size of *Balanus balanoides* larvae cover the range 0.3 to 1.1 mm and are, therefore, particularly suitable as regards their size for the smaller herring larvae. It is essential that predators – if they are to grow – should find food sufficiently rapidly to ensure that the energy spent in searching does not exceed that of their intake. The sudden large impact of cirripede larvae could, therefore, provide a favourable situation for carnivorous predators such as herring larvae. The values for $r = 1000$, $\theta = 60°$ in Table II when reconverted to no./unit volume are far greater (270/l) than those given by Rosenthal & Hempel as optimally required by 10-11 mm herring larvae i.e., mean 42 larvae/h (their maximal estimate). Even the value for 100 metres offshore on an open coast after 12 days (Table IV) gives 41 larvae/l and at 1000 metres $\simeq$ 4 larvae/l. These values will, of course, represent the original stage I nauplii and take no account of losses and changes to later stages during diffusion. There is, however, conflicting evidence as to the part played by cirripede nauplii in the nutrition of larval herring. In the White Sea the importance of the larval *B. balanoides* as the main food of larval herring is well known. In the coastal waters of the Gulf of Maine, Sherman & Honey (1971) found that autumn spawned herring reached a length of 30-40 mm by the spring and then cirripede nauplii formed a constant component of the diet. Cirripede nauplii were not present in the larval guts in autumn and winter – they were probably not present in any quantity in the plankton – but in the spring, when *B. balanoides* larvae would be abundant in these coastal waters, they become a constant and important component of the larval diet: the regular appearance of *B. balanoides* in the spring would seem to represent an important source of food for these larval herring. In the Clyde there is a well known spring spawning of herring on the Ballantrae Banks at the same time as the input of *B. balanoides*. Bainbridge & Forsyth (1971) examined the variability of the zooplankton on the Banks over three successive years; they found cirripede nauplii to form a considerable proportion of the zooplankton and, perhaps more important for its relation to an annual predator, more constant in its abundance. Since water movements would, during any dispersal, tend to keep the larval herring and cirripede nauplii input together a direct predator-prey relation could be important to larval success. In spite of the fact that Blaxter (1962, 1968) has found early stage larvae of *B. balanoides* to be extremely successful as food for larval herring of 10-20 mm length and that Sherman & Honey found a mean size of $\simeq$ 0.5 mm to be the size of prey (non-selective) taken by the Maine (10-20 mm) larval herring, Bainbridge & Forsyth could find no cirripede nauplii in the guts of larval herring. For the earlier stages of herring larvae Blaxter used stage I larvae hatched from littoral samples and the size would be 0.33 m; stage II is 0.53. stage III 0.63, stage IV 0.73, stage V 0.94, and stage VI 1.13 (total length in mm). Stage I rapidly passes to stage II and this would seem to be the most suitable as regards size for the 7-10 mm larvae investigated by Bainbridge & Forsyth (1971). Stage II nauplii are, however, delicate – perhaps more so than copepod nauplii – and may have been too completely digested to be recognized in gut contents. Rosenthal & Hempel

(1970) have shown that, at least under experimental conditions, larval herring become imprinted with a particular kind of food; imprinting to plankton other than cirripede nauplii may explain some of the above discrepancies.

The larvae as a source of 'essential' food

There is now much evidence to suggest that the 'quality' of the food required by a variety of marine invertebrate larvae is as important as the quantity and that the desideratum as regards quality may vary even within a species according to the larval stage (Sulkin, 1975, q.v., for further references). The ease of capture, and ability to be digested are, of course, essential and this probably involves both physical factors such as size, shape, nature of external covering as well as behavioural characteristics such as swimming speed and type of activity: that of cirripede nauplii is very different from that of copepods. The prey may be too large to be ingested whole or too tough to be broken down by some particular larval stage. There is evidence that a wide variety of the smaller holo - and meroplanktonic organisms can probably serve as prey and provide an adequate diet for early zoea stages, but that late zoea stages require a more specialized diet high in lipids — perhaps particularly in esterified sterols which cannot be synthesized by some larvae and which ultimately originate in the phytoplanktonic components (Whitney, 1969, 1970): Sulkin (loc. cit.) has shown that a diet rich in lipids is essential for the successful rearing of the late larvae of *Callinectes sapidus*. There is evidence, too, that diets low in lipids delay metamorphosis in some fish larva. Might not the cirripede larvae provide just such an adequate diet? They are certainly rich in lipids. Lawinski & Pautsch (1969) have successfully reared the larvae of *Rithropanopeus harrisii* on *Balanus improvisus* nauplii and Reed (1969) used *B. glandula* nauplii to rear the brachyuran *Cancer magister*; in both cases, under natural conditions, the cirripede nauplii concerned would be available to the crab larvae. In the waters of the Firth of Clyde, Marshall (1925) found zoea and cirripede nauplii commonly to occur together in plankton hauls.

We have largely illustrated the thesis that the larvae of littoral animals may be of considerable importance to events in the plankton by reference to barnacles; there are, however, many other littoral animals with planktonic larvae, and the situation will be similar - but no less complex. Some of the simpler relations are given in Fig. 1 which makes it evident that more data are needed to understand these complex relations some of which will be discussed elsewhere.

ACKNOWLEDGEMENT

We wish to thank Professor K.F. Bowden for valuable discussions regarding diffusion problems in relation to larval distribution.

REFERENCES

Bainbridge, V. & D.C.T. Forsyth, 1971. The feeding of herring larvae in the Clyde. *Rapp. P.-V. Reun. Cons. Int. Exp. Mer.* Vol. 160, pp.104-113.

Barnes, H., 1957. Processes of restoration and synchronization in marine ecology; the spring diatom increase and the 'spawning' of the common barnacles, *Balanus balanoides* (L.). *Annee Biol.*, T. 33, pp. 67-85.

Barnes, H., 1962. Note on variations in the release of nauplii of *Balanus balanoides* with special reference to the spring diatom outburst. *Crustaceana*, Vol., 4, pp. 118-122.

Barnes, H., 1971. Organic production by *Elminius modestus* in an enclosed basin. *J. exp. mar. Biol. Ecol.*, Vol. 6, pp. 79-82.

Barnes, H. & M. Barnes, 1965. Egg size, nauplius size and their variation with local, geographical and specific factors in some common cirripedes. *J. Anim. Ecol.*, Vol. 34, pp. 391-402.

Barnes, H. & M. Barnes, 1968. Egg numbers, metabolic efficiency of egg production and fecundity- local and regional variations in a number of common cirripedes. *J. exp. mar. Biol. Ecol.*, Vol. 2, pp. 135-153.

Blaxter, J.H.S., 1962. Herring rearing. IV. Rearing beyond the yolk-sac stage. *Mar. Res.*, No. 1, 18 pp.

Blaxter, J.H.S., 1968. Rearing herring larvae to metamorphosis and beyond. *J. mar. biol. Ass. U.K.*, Vol. 48, pp. 17-28.

Bousfield, E.L., 1955. Ecological control of the occurrence of barnacles in the Miramichi estuary. *Natl. Mus. Can.*, Bulletin, No. 137, 69 pp.

Crisp, D.J., 1956. A substance promoting hatching and liberation of yound in cirripedes. *Nature, Lond.*, Vol. 178, p.263 only.

Crisp, D.J. 1976. The role of pelagic larvae. In, *Perspectives in Experimental Biology*, Vol. 1, Zoology, Edited by P. Spencer Davies. Permanon Press, Oxford, pp. 145-155.

Crisp, D.J., 1958. The spread of *Elminius modestus* Darwin in north-west Europe. *J. mar. biol. Ass. U.K.*, Vol. 37, pp. 483-520.

Crisp, D.J. & C.P. Spencer, 1958. The control of the hatching process in barnacles. *Proc. R. Soc. B.*, Vol. 148, pp. 278-299.

Csanady, G.T., 1973. *Turbulent diffusion in the environment.* D. Reidel Pub.Co.

Ketchum, B.H., 1951. The exchange of fresh and salt waters in estuaries. *J. mar. Res.*, Vol. 10, pp. 18-38.

Lawinski, L. & F. Pautsch, 1969. A successful trial to rear larvae of the crab *Rhithropanopeus harrisii* (Gould) subsp. *tridentatus*. *Zoologica Poloniae*, Vol. 19, pp. 495-506.

Lewis, J.R. & H.T. Powell, 1960. Aspects of the intertidal ecology of rocky shores in Argyll, Scotland. II. The distribution of *Chthamalus stellatus* and *Balanus balanoides* in Kintyre. *Trans. Roy. Soc. Edinb.*, Vol. 64, pp. 75-100.

Marshall, S.M., 1925. A survey of Clyde plankton. *Proc. R. Soc. Edinb.*, Vol. 45, pp. 117-141.

Marshall, S.M. & A.P. Orr, 1955. *The biology of a marine copepod.* Oliver & Boyd, Edinburgh, 188 pp.

Reed, P.H., 1969. Culture methods and effects of temperature and salinity on survival and growth of the Dunganess crab (Cancer magister) in the laboratory. *J. Fish. Res. Bd. Can.,* Vol. 26, pp. 389-397.

Rosenthal, H. & G. Hempel, 1970. Experimental studies in feeding and food requirements of herring larvae (*Clupea harengus* L.). In, *Marine food chains* edited by J.H. Steele, Oliver & Boyd, Edinburgh, pp. 344-364.

Rsepishevsky, I.K., 1962. Conditions of mass liberation of the nauplii of the common barnacle, *Balanus balanoides* (L.), in the Eastern Murman. *Int. Revue. ges. Hydrobiol. Hydrogr.,* Bd 47, S. 471-479.

Sherman, K. & K.A. Honey, 1971. Seasonal variations in the food of larval herring in coastal waters of central Maine. *Rapp. P.-V. Reun. Cons. Int. Exp. Mer,* Vol. 160, pp. 121-124.

Southward, A.J. & D.J. Crisp, 1954. Recent changes in the distribution of the intertidal barnacles *Chthamalus stellatus* and *Balanus balanoides* in the British Isles. *J. Anim. Ecol.,* Vol. 23, pp. 163-177.

Southward, A.J. & D.J. Crisp, 1956. Fluctuations in distribution and abundance of intertidal barnacles. *J. mar. biol. Ass. U.K.,* Vol. 35, pp. 211-219.

Steele, J.H., 1974. *The structure of marine ecosystems.* Harvard University Press, Cambridge, Mass., 128 pp.

Sulkin, S.D., 1975. The significance of diet in the growth and development of the blue crab, *Callinectes sapidus* Rathbun, under laboratory conditions. *J. exp. mar. Biol. Ecol.,* Vol. 20, pp. 119-135.

Whitney, J.O., 1969. Absence of sterol synthesis in larvae of the mud crab *Rhithropanopeus harrisii* and of the spider crab *Libina emarginata. Mar. Biol.,* Vol. 3, pp. 134-135.

Whitney, J.O., 1970. Absence of sterol biosynthesis in the blue crab *Callinectes sapidus* Rathbun and in the barnacle, *Balanus nubilus* Darwin. *J. exp. mar. Biol. Ecol.,* Vol. 4, pp. 229-237.

Wiborg, K.F., 1976. Fishery and commercial exploitation of *Calanus finmarchicus* in Norway. *J. Cons. perm. int. Explor. Mer,* Vol. 36, pp. 251-258.

PHYSIOLOGICAL MEASUREMENTS ON ESTUARINE BIVALVE MOLLUSCS IN THE FIELD

B. L. Bayne, J. Widdows and R.I.E. Newell

Institute for Marine Environmental Research, 67 Citadel Road, Plymouth, Devon PL1 3DH, England

ABSTRACT

Techniques for measuring the rates of feeding and of oxygen consumption by bivalve molluscs have been tested and applied under various conditions in the field, including working from a research vessel and from a small mobile laboratory on the shore. The animals are held in flowing water pumped from their natural environment, and conditions of temperature, salinity and ration are kept at ambient values. Rates of feeding are determined either by particle counts (by Coulter Counter) on water samples taken before and after the water has passed over the animal, or by passing the water through a fluorimeter for the continuous estimation of living particulate plant material. Rates of oxygen consumption are determined by recording the decline in oxygen tension in respirometer vessels which are isolated for brief periods of time from the flowing seawater. These techniques are described in more detail, and their uses in the field illustrated with data on two aspects of the environmental physiology of bivalves. 1: Rates of oxygen consumption by the mussel, Mytilus edulis over two annual cycles, to illustrate seasonal changes in metabolic rate under ambient conditions. 2: Feeding rates by Mytilus and by the cockle, Cardium edule, at various concentrations of suspended particulate matter, including a discussion of particle-size selection.

INTRODUCTION

The extrapolation of the results of physiological measurements made in the laboratory for the interpretation of physiological response in natural populations is often fraught with difficulties, both conceptual and technical. We have recently approached this problem by adapting certain laboratory techniques for the measurement of the metabolic and feeding rates of bivalve molluscs to use on a small research vessel at sea, or from a mini-bus, converted to a mobile laboratory, on the shore. In this paper we describe these procedures, and illustrate their use in examining two aspects of the physiological ecology of bivalves viz. the annual cycle in their rates of oxygen consumption, and their capacity to select particles of certain size during feeding.

METHODS

An essential requirement for any physiological experiment with bivalves is that the individuals are kept undisturbed not only during the measurements, but for some time preceding them. This precludes the use of closed experimental chambers for metabolic and feeding rate measurements, and also renders inadvisable the use of any attachments, such as rubber sleeves, to the shell valves. In our experiments in the field, individuals are introduced into the

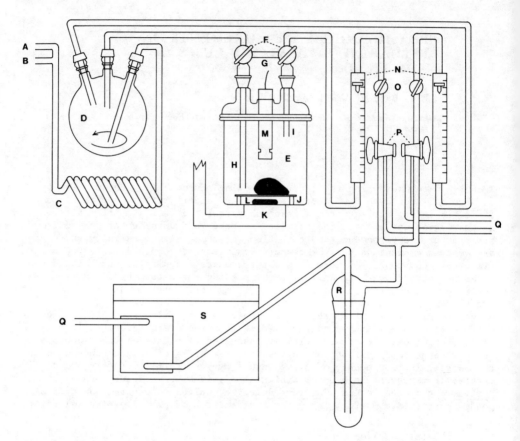

Fig. 1. Apparatus for the measurement of rates of oxygen consumption and feeding in bivalves. A, seawater inflow; B, algal cell inflow; C, temperature equilibration coil; D, mixing chamber; E, experimental chamber; F, two-way taps; G, by-pass; H, inflow; I, outflow; J, perforated base plate; K, magnetic stirrer; L, stirring bar; M, oxygen electrode; N, flow meter; O, flow regulating taps; P, two-way taps; Q, to waste; R, bubble trap; S, fluorimeter.

experimental chamber (Fig. 1) and left undisturbed for at least 45-60 min before physiological measurements are started. During the entire experimental period water is pumped from that directly over the mussel- or cockle-bed, and passed at a rate of 70 ml min^{-1} through the experimental apparatus. In this way all physiological determinations are made under conditions of ambient temperature, salinity and the quantity and quality of suspended particulate matter.

Experiments were carried out with Mytilus edulis and Cardium (Cerastoderma) edule from populations in the Tamar estuary at Plymouth. Rates of oxygen consumption and of feeding were measured in the apparatus shown in Fig. 1. Mytilus were introduced directly into the experimental chamber (volume 500 ml) and allowed to attach by byssus threads to a perforated glass plate over a stirring bar. Cardium were first allowed to burrow into sterilised sand in small plastic containers, and these containers placed in a larger experimental chamber (volume 850 ml).

The rate of oxygen consumption was measured by isolating the chamber from the flowing water (by turning the taps F) for periods of 30-60 min while monitoring the decline in oxygen tension in the chamber with a Radiometer oxygen electrode coupled to a chart recorder. Oxygen tension was not allowed to fall below 120 mm Hg. This is similar to the technique described by Bayne (1971). For the results recorded in this paper, the rates of oxygen consumption ($\dot{V}O_2$) of twelve mussels, ranging in size from 0.3 to 2.9 g dry tissue weight, were measured at approximately bi-monthly intervals over two years. On each occasion, temperature, salinity and the concentrations of suspended particulate material were also recorded.

Two techniques were used to determine rates of feeding (= filtration rate), the one using a Coulter Counter, the other using a Turner fluorimeter.

Filtration rate by particle counts. A period of at least 45-60 min was allowed for the animal to come to equilibrium with the conditions in the experimental chamber. Samples of the water flowing into and out of the chamber were then obtained. The Coulter Counter was used to determine the concentration in the sample of all suspended particles greater than 2 μm in diameter. The filtration rate (FR), defined as the volume of water cleared of suspended particulates >2 μm per unit time, was calculated as follows:

$$FR = F \cdot \left(\frac{C_1 - C_2}{C_1}\right) \qquad (1)$$

where FR is in litres per hour, F is the flow rate of water through the chamber (litres per hour), and C_1 and C_2 are concentrations of particles in the inflowing and outflowing water, respectively. Recently, Hildreth and Crisp (1976) have criticised this method of calculating filtration rate, arguing that the inflow concentration of particles does not represent the true particle concentration which actually impinges on the animal's ctenidia; they recommended that C_0 be substituted for C_1, where C_0 is the concentration of particles surrounding the animal within the experimental chamber. Hildreth and Crisp (1976) are entirely correct in their theoretical treatment, a point noted by us in all our earlier studies on filtration by Mytilus (see Bayne, Thompson and Widdows, 1976). However, the error that is introduced by using C_1 and not C_0 in the calculation of FR is reduced to an insignificant level in a chamber of the size and design illustrated in Fig. 1. The experimental chamber has a volume of 500 ml, compared with a tank of 2.5 litres used by Hildreth and Crisp (1976). In addition, the inflowing water is taken to the base of the flask, near to the animal's inhalent mantle edge; the outflow from the chamber is at the top. The water surrounding the animal is not stirred during the measurement of filtration. This arrangement of water flow and the use of relatively high flow rates (70 ml min^{-1}) ensures that the particle concentration at the inhalent edge (C_0) is within 3% of the inflow concentration (C_1) and that recirculation of the water by the animal is minimal.

When measuring the filtration rate of Cardium, the positioning of the animal's inhalent and exhalent siphons, and the presence of the plastic beaker, prevent a concentration gradient developing within the flask. Therefore the water was stirred, so that the concentration of particles at the outflow (C_2) was similar to the concentration around the cockle. The filtration rate is then calculated as:

$$FR = F \cdot \left(\frac{C_1 - C_2}{C_2}\right) \qquad (2)$$

Hildreth and Crisp (1976) recommend the use of this equation when C_0 is not determined directly.

Filtration rate by fluorimetry. Use of the Coulter Counter provides discrete measurements of filtration rate, which are adequate in many circumstances. A continuous record of filtration rate is often preferable, however, especially with the many bivalve species which ventilate the mantle cavity, and therefore feed, sporadically (Brand and Taylor, 1974), and which show marked short-term changes in feeding rate. We have therefore employed a constant flow fluorimeter (Turner model 111) to provide a continuous record of feeding activity (Fig. 1). Seawater, carrying algal cells, is flowed through the experimental chamber at 70 ml min^{-1}, into a bubble trap, and thence into the flow cuvette of the fluorimeter. By use of 2-way taps (P) it is also possible to by-pass the chamber and flow the "inflow" water directly through the fluorimeter. The chlorophyll in the algal cells is excited by light at 430 nm and the emitted light at 650 nm is detected and amplified (Lorenzen, 1966). The volume of water within the system is kept to a minimum to reduce the response time to change in concentration of algal cells.

The linearity of the response of the fluorimeter was confirmed initially over a wide range of algal cell concentrations, by calibrating the fluorimeter output against cell concentrations determined with a Coulter Counter. At the start of each experimental determination, the fluorimeter is adjusted so that the inflow concentration of cells results in a full-scale deflection of the output on the chart recorder. In the observations recorded here we also determined occasional particle counts by Coulter Counter to provide a check for the calculations of feeding rate taken from the fluorimeter readings.

Particle selection

In the determination of filtration rates, the Coulter Counter was used to measure total particle concentrations. We also used the Counter to count separately particles of different size in an attempt to determine whether Mytilus is capable of the differential selection of particles within a limited size range. Two series of experiments are recorded here, both of which were carried out from a research vessel using equipment described earlier. In the first experiment, eight mussels of similar size (6-7 cm shell length) were placed individually in flasks and two flasks were left without animals, to act as controls. Water was pumped through the flasks for two days during which time eight samples were taken of inflowing and outflowing water. These water samples were then immediately counted by a Coulter Counter (model ZB) fitted with a 140 µm orifice tube, providing discrete counts, converted to relative volume, for the following spherical particle diameters: 2.5, 3.1, 3.9, 5.0, 6.2, 7.9, 9.9, 12.5, 15.7, 19.8, 25.0 and 31.5 µm. This experiment, called the 'variance' trial for size-selection, was used to calculate the natural variance to be expected in estimates of an index of particle selection, D, as described later.

In a second series of experiments, Mytilus were sampled on 10 February, 1 April and 11 May, 1976, from the mouth of the Lynher river in the Tamar estuary. On each occasion six animals of similar size (6-7 cm) were placed individually in experimental flasks and left undisturbed for 1.5 hours in flowing water bearing a natural load of suspended particles. Inflow and outflow samples were then taken, and the various size fractions counted as before (the 'natural particulates' sample). Cells from a laboratory culture of the diatom Phaeodactylum tricornutum were then added to the natural particulates in the inflow water, by continuous pumping into the inflow water via the inlet B (Fig. 1). After

1.5 hours, inflow and outflow samples were taken and counted as before (the 'natural particulates plus Phaeodactylum' sample). In this way particle selection by Mytilus was determined at three different times of the year, a) under ambient conditions with the natural concentration of suspended matter in the water, and b) with a 'pulse' of one cell-type added to the natural particulates, providing both a higher concentration of particles and a domination of the complete size-spectrum of particles with a single particle type. These determinations are called the 'experimental' trial for size selection.

In order to determine whether differential size-selection had occurred, we have used the electivity index, D, of Jacobs (1974), which is a modification of the index originally described by Ivlev (1961). Jacobs defines

$$D = \frac{r-p}{r+p - 2rp} \tag{3}$$

where r is the fraction of a particular food type in the ingested ration, and p is the fraction in the environment (the available ration). From the Coulter Counter measurements of particle concentrations, we designate the inflow concentrations as p, and the inflow minus outflow concentrations as r. It is important to recognise that the electivity index D is not the same as the more commonly estimated "retention efficiency" (e.g. Vahl, 1972), since D is dependent upon the proportional concentration of particles of a particular size in the available and the ingested ration, whereas the retention efficiency is a function only of the numbers of those particles in the inhalent and exhalent water flows of the animal. D takes values from -1 to +1; negative values signify rejection of a particle type, positive values signify selection of a particle type, and zero indicates the absence of any active selection (Jacobs, 1974).

RESULTS

Seasonal changes in the rate of oxygen consumption

The relationship between the rate of oxygen consumption ($\dot{V}O_2$; ml O_2 hr^{-1} by Mytilus edulis and its dry tissue weight (W; grams) is decribed by the equation:

$$\dot{V}O_2 = aW^b \tag{4}$$

The best fit parameters (a = intercept; b = gradient) for each regression equation, calculated by least squares regression analysis, are listed in Table 1. An analysis of covariance for these data showed that values for a, but not for the gradient b, were significantly heterogeneous at $P = 0.01$. The data were therefore re-analysed with a common regression coefficient b = 0.445 and the intercept values recalculated. These recalculated values illustrate the seasonal changes in $\dot{V}O_2$ by an animal of 1 g dry tissue weight at ambient temperature, salinity and ration level (Fig. 2). $\dot{V}O_2$ was at a maximum in spring (April-May) and then declined to a minimum in late autumn and winter. The seasonal changes in water temperature are also shown in Fig. 2.

Measurements of filtration rate by fluorimetry

Traces representative of many obtained with the fluorimeter are presented in Fig. 3, together with occasional cell counts made with a Coulter Counter. Trace 1, recorded in the laboratory, shows the steady inflow reading (A) obtained with a monoculture of Phaeodactylum tricornutum. When the system was switched to monitor the outflow from a chamber containing a specimen of Cardium (B) the position of the recorder trace was altered in proportion to the number of cells filtered from suspension by the animal. At (C) the fluorimeter was

TABLE 1 Regression constants (a = intercept; b = slope) for equations describing the relationship between rate of oxygen consumption and dry tissue weight of Mytilus edulis during two annual cycles.

Regression Equation: $\dot{V}O_2 = aW^b$

Date	a = intercept	b = slope	Correlation coefficient	Recalculated intercept (b = 0.445)
5. 5.73	0.570	0.51	0.62	0.570
10. 7.73	0.558	0.36	0.67	0.574
4. 9.73	0.516	0.38	0.74	0.519
29.11.73	0.436	0.35	0.85	0.443
13. 2.74	0.300	0.72	0.84	0.296
1. 4.74	0.326	0.49	0.88	0.325
2. 5.74	0.694	0.44	0.66	0.696
13. 6.74	0.628	0.84	0.94	0.579
16. 7.74	0.423	0.28	0.37	0.445
13. 8.74	0.468	0.66	0.84	0.458
19. 9.74	0.345	0.50	0.84	0.348
14.11.74	0.252	0.81	0.91	0.265
12. 2.75	0.350	0.77	0.85	0.341
3. 3.75	0.546	0.59	0.71	0.541
2. 5.75	0.581	0.46	0.89	0.577

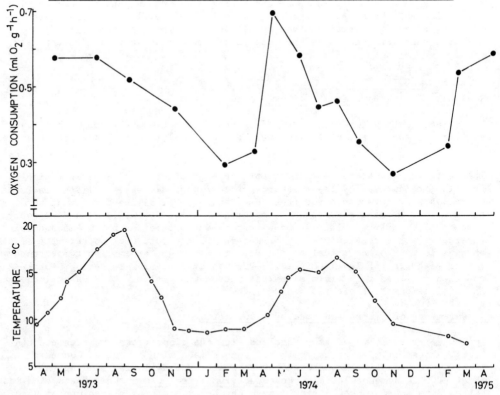

Fig. 2. The rate of oxygen consumption by Mytilus edulis over two years, and the seawater temperature.

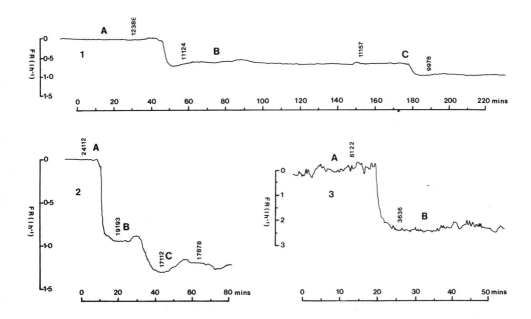

Fig. 3. Traces from a fluorimeter recording the feeding of Mytilus and Cardium together with occasional particle counts (numbers of cells per ml). See text for details.

switched to monitor the outflow from a different chamber, also containing one Cardium; this animal had a higher rate of filtration (1.1 $l.hr^{-1}$ as against 0.7 $l.hr^{-1}$) which shows as a further depression of the recorder trace.

Trace 2 was also obtained in the laboratory, with cells of Phaeodactylum, and illustrates the rapid response time of the system. Within 3 min of switching from inflow to outflow (A) the recorder trace had stabilised. In this experiment the chamber contained a specimen of Mytilus which increased its rate of filtration over a period of 10 min, from 0.95 $l.hr^{-1}$ (B) to 1.3 $l.hr^{-1}$ (C).

Trace 3 was recorded in the field from the research vessel, using natural seawater. This trace was not as stable as those obtained in the laboratory, probably due to a greater variation in the fluorescent material present, but the same response is apparent as the taps are turned from monitoring the inflow (A) to the outflow (B).

Particle selection by Mytilus

When Mytilus were left undisturbed for some hours in flowing seawater (the 'variance' trial for size-selection: Fig. 4a), there was evidence of rejection of particles < 4-5 μm in diameter (D<0) and either no selection (D=0) or slight positive selection (D>0) of particles greater than 6 μm. The calculation of D is based upon the proportions of numbers of particles of different sizes in the available ration and the ingested ration (see Methods). It follows that negative D values for small particles (resulting, most probably, from reduced retention efficiency) must be reflected in positive values for larger particles. The data in Fig. 4a suggest that this positive component of D is spread

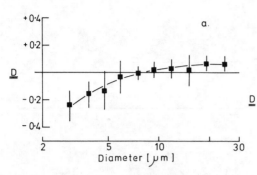
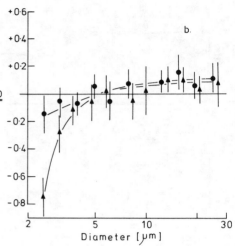
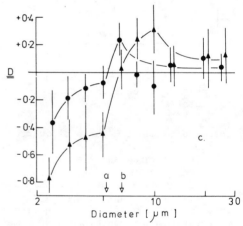

Fig. 4. The electivity index, D, related to particle diameter: a, the 'variance' trial for size selection; b, the 'experimental' trial, natural particulates; c, the 'experimental' trial, natural particulates plus Phaeodactylum. In b and c, ● refers to February and May, ▲ to April. Values are means ± SD. In c, the arrows refer to the modal size of Phaeodactylum offered in February and May (a) and in April (b).

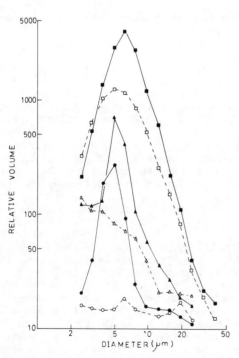

Fig. 5. Size frequency of particles in the 'experimental' trials for size selection: ---, natural particulates; ——, natural particulates plus Phaeodactylum. △, February; □, April; ○, May.

uniformly over all sizes >8-10 µm, when the mussels are feeding on natural suspended particles. This result was reproduced whenever we measured particle retention by <u>Mytilus</u> at ambient particle concentrations (Fig. 4b) although the magnitude of the negative component of D varied between different trials.

However, when cells of <u>Phaeodactylum</u> were added to the natural particles and then fed to the mussels (the 'experimental' trial for size selection) there was positive selection of the algal cells (D>0; Fig. 4c). The particle diameters at which these marked positive values of D were calculated were: February and May, 6.2 µm; April, 7.9 and 9.9 µm. In Fig. 5 are plotted typical size-frequency data, for both natural particles and natural particles with added <u>Phaeodactylum</u>, in February, April and May. The modal size of the <u>Phaeodactylum</u> altered from 5.3 µm in February to 6.4 µm in April and 5.0 µm in May. The increase in size of the <u>Phaeodactylum</u> used in April is consistent with an increase in size at which marked positive selection occurred. However, in all cases the main peak of positive D values occurred at particle diameters 1-2 µm greater than the modal size of the added <u>Phaeodactylum</u>. Comparing the data in Fig. 4b and 4c also suggests that, when <u>Phaeodactylum</u> was added to the natural particles, there was a depression of the D values for particles <5 µm, i.e. a more marked rejection of particles in this size range.

There was also a further seasonal difference in the data shown in Fig. 4b and 4c. In February and in May particles up to 4 um were rejected, but in April particles as large as 5-6 µm were rejected and D values for particles <5 µm diameter were much lower than on the other two occasions. In April there were very high concentrations of natural particulate matter in the water (Fig. 5), but this was largely inorganic in origin and occurred before the main phytoplankton bloom in May (own unpublished observations). Marked selection against small particles (resulting from low retention efficiency) may have been a response to these higher concentrations of inorganic particles.

DISCUSSION

Seasonal changes in rates of oxygen consumption

There is a seasonal cycle in the rate of oxygen consumption by <u>Mytilus edulis</u> when measured in the field at ambient conditions. This confirms the presence of seasonal patterns previously recorded in laboratory studies, where $\dot{V}O_2$ was measured either at a constant acclimation temperature (Kruger, 1960; Bayne, 1973) or at the seasonal ambient temperature (Widdows and Bayne, 1971; de Vooys, 1976). In all cases the seasonal cycle was not closely correlated with temperature. For example, the results of this study show that in October the rates of oxygen consumption were 50% lower than in May, although measured at the same temperature ($12^\circ C$). The seasonal metabolic cycle cannot be explained as a response to temperature increase in the spring or decrease in the autumn, since previous studies have shown that <u>Mytilus</u> can acclimate fully to temperatures between 5 and $20^\circ C$ and that the rate of acclimation to changes in temperature is usually more rapid than the rate of change in the mean seawater temperature (Widdows and Bayne, 1971; Widdows, 1976).

The seasonal changes in oxygen demand appear rather to be correlated with the cycles of gametogenesis and the storage and utilisation of body reserves (Bruce, 1926; Kruger, 1960, Bayne, 1973; Gabbott and Bayne, 1973). The storage cycle (Gabbott, 1975) and gametogenesis are, in turn, dependent at least in part upon temperature and the availability of food (Bayne, 1975; Seed, 1976). In a recent paper, de Vooys (1976) has questioned these conclusions. He suggested that high rates of oxygen uptake at low seasonal temperatures were an artefact, and that $\overline{VO_2}$ is largely temperature-dependent. In commenting upon results from a North Sea population during the winter (Widdows and Bayne, 1971), de Vooys (1976) failed to appreciate that the determinations were made at ambient seasonal temperatures and not, as he suggested, at temperatures above the ambient mean.

There is now ample evidence of a considerable variability in the seasonal metabolic cycle of Mytilus edulis (Bruce, 1926; Kruger, 1960; Widdows and Bayne, 1971; Bayne, Thompson and Widdows, 1976; de Vooys, 1976) which cannot be accounted for simply in terms of the different experimental techniques, geographical distribution or annual temperature cycles. In view of the known complex relationships between metabolic rate and a variety of endogenous and exogenous factors (reviewed by Bayne et al., 1976), this variability is hardly surprising. We wish to make the point here that, if viewed simply in terms of a response to temperature, resulting interpretations are misleading and oversimplified.

Feeding and particle selection

Reports in the literature suggest that the minimum size of particle which can be filtered by Mytilus with maximum efficiency is between 5-6 μm (Davids, 1964) and 2 μm diameter (Vahl, 1972). Jørgensen (1974) records that retention efficiency decreases gradually with diminishing particle size between 3 and <1 μm. Negative values for the selection coefficient D for particles between 2 and 5 μm are most readily explained in terms of low retention efficiency by the gill. A shift in the range of negative D values, such as we observed from February to April, may then be due to control by the animal of the porosity of the gill. Equally, both long-term and short-term changes in the magnitude of the negative D values at small particle sizes signify quantitative changes in porosity in response to changing populations of particles in the water. Minimum porosity of the gill would appear not to be a fixed condition, therefore, but subject to control by the animal. When the concentration of small particles of inorganic silt in the water column is high, such control might be adaptive.

When undisturbed, Mytilus rejects small particles, but retains larger particles with a high and a constant efficiency (Vahl, 1972; Jørgensen, 1975). Data recorded here for February, April and May 1976, and unpublished data for 1975, confirm that mussels normally graze the available suspended particles above a variable minimum size strictly in proportion to particle concentration. However, when Phaeodactylum was made to dominate the population of suspended particles, Mytilus was apparently able actively to select these cells in preference to other particles, as evidenced by a peak of positive D values in Fig. 4c. This might be due either to selection on the gill/labial palp, or to a more passive retention which is a function of the shape of the cell. Phaeodactylum may be either elongated or tri-radiate and may therefore not be retained by the bivalve gill in proportion to any single linear dimension. This possibility is masked in our present data by the electronic properties of the Coulter Counter, which measures particle sizes as the volume of spherical equivalents (Sheldon and Parsons, 1967). The simplest explanation for our finding a preferential selection for Phaeodactylum, bearing in mind our present understanding of the functioning of the bivalve gill (Jørgensen, 1975; Bayne et al., 1976), is that the shape of the diatom cell results in a higher retention efficiency than might be expected for a spherical particle of similar volume. Further experiments, with algal cells of different shapes and volumes, should clarify this finding. In considering particle retention by copepods, Boyd (1976) pointed out that the mechanical properties of a filter will naturally result in the modal size of particles filtered being larger than the mode of the particles available, so long as the most numerous particles are collected with a lower efficiency than are larger particles. Our data suggest that a similar phenomenon may occur in mussels.

ACKNOWLEDGEMENTS

This work forms part of the experimental ecology programme of the Institute for Marine Environmental Research, a component of the Natural Environment Research Council. It was commissioned in part by the Department of the Environment (Contract No. DGR 480/47).

REFERENCES

Bayne, B.L. 1971. Oxygen consumption by three species of lamellibranch mollusc in declining ambient oxygen tension. Comp. Biochem. Physiol. 40A, 955-970.

Bayne, B.L. 1973. Physiological changes in Mytilus edulis L. induced by temperature and nutritive stress. J. mar. biol. Ass. U.K., 53, 39-58.

Bayne, B.L. 1975. Reproduction in bivalva molluscs under environmental stress. In: Physiological ecology of estuarine organisms (ed. F.J. Vernberg) pp 259-277. University of South Carolina Press.

Bayne, B.L., Thompson, R.J. and Widdows, J. 1976. Physiology : 1. In: Marine mussels: their ecology and physiology (ed. B.L. Bayne) pp 121-206. Cambridge University Press.

Boyd, C.M. 1976. Selection of particle sizes by filter feeding copepods: A plea for reason. Limnol. and Oceanogr. 21, 175-179.

Brand, A.R. and Taylor, A.C., 1974. Pumping activity of Arctica islandica (L.) and some other common bivalves. Mar. Behav. Physiol. 3, 1-15.

Bruce, J.R. 1926. The respiratory exchange of the mussel (Mytilus edulis L.) Biochem J. 20, 829-846.

Davids, C. 1964. The influence of suspensions of micro-organisms of different concentrations on the pumping and retention of food by the mussel (Mytilus edulis L.) Neth. J. Sea Res. 2, 233-249.

Gabbott, P. 1975. Storage cycles in marine bivalve molluscs: a hypothesis concerning the relationship between glycogen metabolism and gametogenesis. In: Proceedings of the ninth European marine biology symposium (ed. H. Barnes)pp 191-211. Aberdeen University Press.

Gabbott, P.A. and Bayne, B.L. 1973. Biochemical effects of temperature and nutritive stress on Mytilus edulis L. J. mar. biol. Ass. U.K. 53, 269-286.

Hildreth, D.I. and Crisp, D.J. 1976. A corrected formula for calculation of filtration rate of bivalve molluscs in an experimental flowing system. J. mar. biol. Ass. U.K. 56, 111-120.

Ivlev, V.S. 1961. Experimental ecology of the feeding of fishes. New Haven. Yale University Press.

Jacobs, J. 1974. Quantitative measurement of food selection. Oecologia 14, 413-417.

Jørgensen, C.B. 1974. On gill function in the mussel Mytilus edulis L. Ophelia, 13, 187-232.

Jørgensen, C.B. 1975. Comparative physiology of suspension feeding. Ann. Rev. Physiol. 37, 57-79.

Kruger, F. 1960. Zür Frage der Grössenabhäugigkeit des Sauerstoffverbranchs von Mytilus edulis L. Helg. wissen Meeresunters 7, 125-198.

Lorenzen C.J. 1966. A method for the continuous measurement of "in vivo" chlorophyll concentration. Deep Sea Res. 13, 223-227.

Seed, R. 1976. Ecology. In: Marine mussels: their ecology and physiology (ed. B.L. Bayne) pp 13-66. Cambridge University Press.

Sheldor, R.W. and Parsons, T.R. 1967. A practical manual on the use of the Coulter Counter in marine research. Coulter Electronics Sales Co. Toronto, Canada.

Vahl, O. 1972. Efficiency of particle retention in Mytilus edulis L. of different sizes. Ophelia 12, 45-52.

Vooys, C.G.N. de, 1976. The influence of temperature and time of year on the oxygen uptake of the sea mussel Mytilus edulis. Mar. Biol. 36, 25-30.

Widdows, J. 1976. Physiological adaptations of Mytilus edulis to cyclic temperatures. J. Comp. Physiol. 105, 115-128.

Widdows, J. and Bayne B.L. 1971. Temperature acclimation of Mytilus edulis with reference to its energy budget. J. mar. biol. Ass. U.K. 51, 827-843.

VARIABILITY OF GROWTH RATE OF *Macoma balthica* (L.) IN THE WADDEN SEA IN RELATION TO AVAILABILITY OF FOOD

J. J. Beukema, G. C. Cadee and J. J. M. Jansen

Netherlands Institute for Sea Research, P.O. Box 59, Texel, the Netherlands

ABSTRACT

Annual length increments during the second growing season were studied in the deposit-feeding bivalva Macoma balthica (L.) over 8 years at 15 intertidal stations in the westernmost part of the Dutch Wadden Sea. In the same area distribution of primary production and functional chlorophyll-a content in the top layer of the sediment were estimated over 1 year and over 7 years at a nearby station.

Both the place-to-place and year-to-year variation in the growth rate of Macoma were significant. The observed variations in growth rates were related to measures of food supply and to the time available for feeding. Correlations suggest that shortage of food, at the lowest intertidal stations, and short periods of immersion, at the highest stations, both limit growth rate.

INTRODUCTION

Macoma balthica (L.), a deposit-feeding tellinid bivalve with a mainly boreal distribution, is one of the most abundant and common macrobenthic species of the tidal flats of the Dutch Wadden Sea (Beukema, 1976). Its occurrence in a wide range of environmental conditions offers an opportunity to carry out in situ studies on the variability of its growth rate in relation to conditions like the length of the twice daily immersion period and the food density.

Such conditions vary from year to year. Durations of immersion are affected by the direction of the prevailing wind. Production rates of microphytobenthos, constituting probably the main food of Macoma, also varies considerably from year to year (Cadee & Hegeman, 1974). Therefore, we estimated growth rates not only at a number of stations, but also over a number of years.

MATERIALS AND METHODS

Shells of Macoma were obtained from macrobenthos samples taken at each of 15 places scattered over Balgzand (1 to 15 in Fig. 1), a tidal flat area in the westernmost part of the Dutch Wadden Sea. See Beukema (1974) for further details on sampling. From samples taken during the late winters of 1969 up to and including 1976, we selected the shells of all living Macoma that had completed two growing seasons. These shells provided estimates of growth rates at 15 stations during the growing seasons of 8 foregoing years.

Measurements of length increments were preferred to estimates of weight change of soft parts, because a seasonal weight loss begins immediately at the end of the growing season (Beukema & De Bruin, in prep.). This makes it difficult to record maximum weight. Length increments, on the other hand, leave a permanent record on the shells which can be read at any time after the end of the growing season.

For each individual, two distances were measured (accuracy ± 0.1 mm) using sliding callipers, viz. the total length (L) and the length at the end of the first growing season (L_1). Length is defined as the longest distance between the anterior and posterior margin of the shell. The length at the end of the first growing season was always clearly marked by a narrow dark year-ring formed during late summer (see plate I in Lammens, 1967). Length increment during the second growing season is denoted by LIN, and equals L - L_1. All data on lengths are in mm.

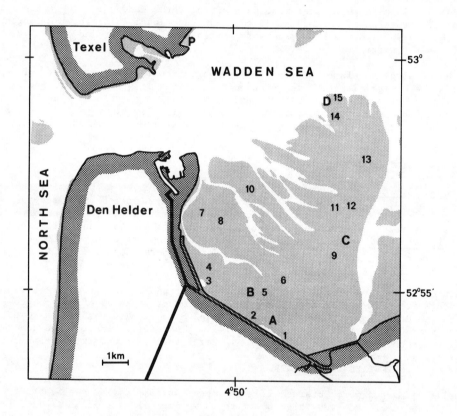

Fig. 1. Map of the westernmost part of the Wadden Sea, showing the location of the 15 sampling stations (1 to 15) for Macoma and the 5 stations (A, B, C, D, and P) for measurements of primary production of microphytobenthos and chlorophyll-a content of the sediment.

Primary production of the benthic microflora and functional/chlorophyll-a content of the top cm of the sediment were concurrently estimated, over the period 1968 - 1975, at one station in the western part of the Wadden Sea (P in Fig. 1). During 1974 such estimates were also made at 14 stations, along 4 transects, on Balgzand (A to D in Fig. 1). For primary production measurements, the ^{14}C-method, as described in Cadee & Hegeman (1974), was used with some minor alterations (Cadee & Hegeman, in prep.). For functional chlorophyll-a estimates, the method of Lorenzen (1967) was employed.

Wind records were obtained from synopses issued monthly by the Royal Dutch Meteorological Institute (KNMI). The levels of the intertidal sampling stations were obtained from unpublished sounding charts kindly provided by the Rijkswaterstaat, Studiedienst Hoorn.

RESULTS

Spatial variability

Environmental conditions. Balgzand is a 50 km^2 area of intertidal sand and mud flats and shallow channels (Fig. 1). The tidal range in this area is small: MLW is about 80 cm below and MHW about 50 cm above MTL. Except for a narrow strip along the dikes in the Southwest, the tidal flats are below MTL. The content of fine material (silt and clay) in the top layer of the sediment gradually decreases from about 40% in the South and Southwest part to less than 1% in the Northwest part of Balgzand (Ente, 1969). Thus with increasing distance from the Southwest coast, both level and silt content decrease. Daily time of immersion (Fig. 3b) and degree of exposure to waves and tidal streams increase with distance from the shore.

Functional chlorophyll-a content of the top layer of the sediment decreases with increasing distance from the Southwest coast (Fig. 3d). Primary production of the benthic microflora shows a similar decrease (Fig. 3c). Thus food supply is highest along the Southwest coast and decreases gradually with increasing distance from shore to reach values well below half of those found in the near-shore area.

Growth rates. From the 120 samples (15 stations, 8 years) we obtained 119 estimates of mean figures for L, L_1, and LIN. Macoma of the proper age were found on all but one occasion. Figure 2a shows the (7-or) 8-year averages for each of the 15 stations of both L_1 and LIN. The contour lines enclose points representing stations situated at about the same distance from the SW-coast, characterized by about the same environmental conditions like level in the intertidal zone, silt content in the sediment, and food density. Such stations are represented by nearby points in Fig. 2a, pointing to similar starting sizes (L_1) as well as growth rates (LIN) at similar environmental conditions.

Macoma from two stations within 1 km of the coast show a slow rate of growth (mean LIN only about 4½ mm), starting from a mean L_1 that is close to average (6 mm). Growth is most rapid (mean LIN about 6½ mm) at stations situated at distances between 2 and 4 km from the coast and is intermediate at stations in between (about 5½ mm at 1 to 2 km). At distances farther than 4 km from the coast, no further increase of growth rates are observed. Note, however, that the mean values of L_1 at these stations far off shore are all below the average of 6 mm.

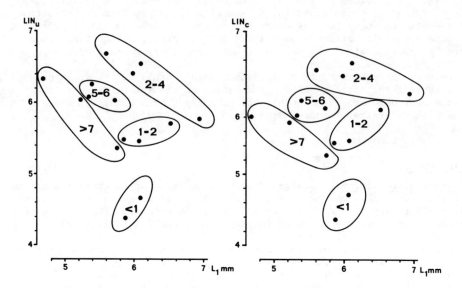

Fig. 2. Mean length increments during the second growing season (LIN in mm, vertical axis) at each of 15 stations plotted against mean lengths at the start of the growing season (L_1 in mm, horizontal axis). Means of 7 or 8 estimates for separate years (1968 to 1975, inclusive). Lines surround points for neighbouring stations. Distance from the Southwest coast for such groups indicated in km. a. uncorrected LIN, b. LIN corrected for influence of L_1 (see text).

Does the starting length L_1 affect the following increment LIN? Lammens (1967) described a "catching-up phenomenon" in <u>Macoma</u>, i.e. a faster growth in individuals starting at a smaller size. We observed essentially the same at all stations and in all years and could quantify it: on average LIN increased by 0.4 mm for each 1 mm L_1 decreased. Thus figures for LIN from samples with a L_1 figure differing from the average of 6 mm can be corrected using the following relationship:

$$LIN_c = LIN_u + 0.4(L_1 - 6),$$

wherein LIN_u is the observed uncorredted value of LIN, and LIN_c is the corrected one. LIN_c may be considered as the (mean) length increment during the second growing season of a standard animal with a shell length of exactly 6 mm at the start of that growing season. Such corrections are essential for a comparison of values of LIN between years or stations, as a Friedman two-way analysis of variance by ranks (Siegel, 1956) of the 119 mean values of L_1 showed that these means probably do not belong to a common population. Both

differences between stations (n = 8, k = 15, p< 0.01) and between years
(n = 15, k = 8, p<0.01) were signified.

Figure 2b shows the results of the correction procedure, plotted as in Fig. 2a.
A Friedman analysis of the 119 values of mean LIN showed highly significant
(p< 0.001) differences between the stations.   Thus at some stations growth
was relatively rapid (almost) all years, whereas, at some other stations it was
(almost) always relatively slow.   These high- and low-growth rate stations
are grouped as to distance from the coast.   More clearly than in Fig. 2a, the
corrected results included in Fig. 2b show that conditions for growth were
optimal in the central part of the area with a decline in both directions,
viz. to the coast and away from it.

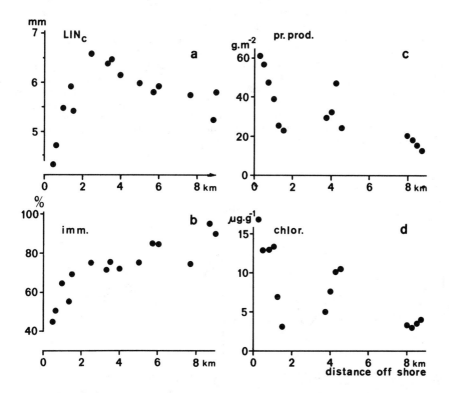

Fig. 3.   Measures of growth rate, time of immersion and food supply
at increasing distances from the Southwest coast (in km,
horizontal axis).   (A)   Growth rates as 8-year averages of
LIN_c,   (b)   Immersion time as percentages of total time,
(c)   Rate of primary production (in g C per m^2) of the
Microphytobenthos during the April-May-June period of 1974,
and (d)   Concentrations of functional chlorophyll-a in the
top-cm of the sediment during the same period (in μg
Chlorophyll-a per g of sediment).

Relationship with availability of food. Figure 3a shows the relationship of growth rate to distance from the coast more directly. Food density does not show a similar relationship with distance from the coast (Fig. 3c and 3d). Both the rate of primary production of the microphytobenthos and concentration of chlorophyll-a in the top centimeter of the sediment decrease continually with increasing distance from the shore. Sedimentation of allochthonous organic material, being another potential food for a deposit feeder like Macoma, will also be higher in near-shore areas (Cadee & Hegeman, in prep.). The distribution of food may thus account only for the decrease of growth rates beyond 3 km off shore. But in the near coastal zone the situation appears paradoxical at first sight: decreasing growth rates going with increasing food densities. By estimating amounts and freshness of food in the stomachs of Macoma collected at various stages of the tide, Hummel (pers. comm.) observed that feeding rates at high tide, far exceeded those at low tide. When immersed Macoma rapidly augment the amounts of chlorophyll-a in their stomachs. The longer the periods of immersion, the higher these amounts and the longer the periods with high contents of fresh algal material. During the the two daily periods when the tidal flats were drained, the amounts of chlorophyll-a in the stomachs gradually declined, pointing to retardation or even cessation of food intake during these low tide periods. Thus, long daily immersion times will favour food intake, whereas too short immersion times may be expected to limit food intake and consequently rate of growth. Immersion times increase with increasing distances from the SW-coast (Fig.3b). They are particularly short (about 50% of the total time) at the two near-shore stations, situated near MTL. At these two high-level stations, growth rates were slowest, suggesting short daily feeding periods. The duration of the feeding period rather than the (high) food density of (high) food production appears to limit rate of growth in the near-shore area.

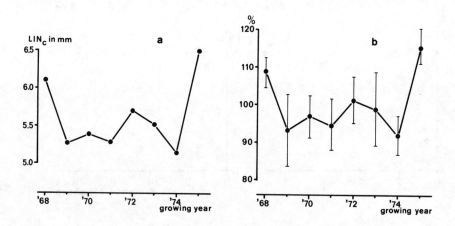

Fig. 4   Mean length increments during each of the growing seasons of 8 years.   (a) Means of 15 station-values of LIN_c in mm, (b) Means and 95%-confidence limits of 15 values of relative growth rates in percentages of 8-year station-averages.

Year-to-year-variability

<u>Growth rates</u>. Mean length increments were not equal for the 8 years of observation. At almost all stations, growth was more rapid during some years (1968, 1975) than others. A Friedman analysis of the 119 values of LIN_c showed highly significant ($p < 0.001$) differences between the years.

Figure 4a shows the mean values of LIN_c of the 15 stations for each of the 8 years. Figure 4b shows similar means of relative growth rates, calculated after conversion of the 119 absolute values to percentages of the 8-year averages at each station. This conversion was necessary to exclude from the confidence limits the significant place-to-place variability. The 95%-confidence limits of the relative growth rates show that the values for 1968 and 1975 clearly stand out above those for the other 6 years.

<u>Relationships with environmental factors</u>. When dealing with spatial variability, we suggested that growth rates might be controlled by different factors in the upper as compared to the lower parts of the intertidal zone. Food density as such might be limiting growth in the lower tidal range, whereas time available for feeding might be decisive in the upper range. Therefore, data on year-to-year variation obtained at the 5 higher stations will be dealt with separately from those of the other 10 stations that were immersed more than 70% of the time (Fig 3b). We predict that: 1. at the lower stations, the growth rates for the various years will be related positively to measures of food supply, and 2. growth rates at the higher stations will be related positively to time of immersion.

As a measure for the food supply in each of the years, estimates of primary production of microphytobenthos at the nearby station P (Fig. 1) are available for the 7 growing seasons of 1969 to 1975, inclusive. They varied from 24 to 69 g C per m^2 per 3 months. A ranking test for trends in several arranged series of observations described by De Jonge (1963: 333-337) with $k = 7$ (years) and $n = 10$ (stations) revealed a significantly ($p < 0.01$) positive correlation with the estimates of growth rate at the 10 lower stations. At the 5 higher stations, such correlation was weakly negative and non-significant (same test with $k = 7$ and $n = 5$, $p > 0.1$). Thus only in the low intertidal area, where food production might limit growth rate (as predicted from the positively correlated <u>spatial</u> variability of both, see Fig. 3), did high rates of food production in certain years go with high rates of growth in <u>Macoma</u> in those years.

Our second prediction concerns the time available for feeding, i.e. the duration of the immersion. Year-to-year variability will be caused especially by the prevailing directions of the wind: easterly winds lower the water levels in the Wadden Sea and thus shorten the immersion periods at intertidal stations, whereas westerly winds do the opposite. The proportions of easterly winds (expressed as a percentage of the total duration of easterly plus westerly winds) varied from 34 to 58% during the April-May-June periods of 1968 to 1975, inclusive. The same ranking test as used above with $k = 8$ (years) and $n = 5$ (stations) revealed a significantly ($p < 0.02$) negative correlation with growth rates of <u>Macoma</u> at the 5 higher stations. For the growth rates at the other 10 stations a similar correlation was found ($p < 0.01$). Thus, years with a high proportion of easterly winds during the growing season, causing lowered water levels and shortened immersion periods, are characterized by relatively slow growth in <u>Macoma</u> at all stations.

The second prediction is thus also confirmed, but should be extended to the stations low in the intertidal zone. This is not contradictory to the original second prediction. It will simply mean that the effect of a reduction of feeding time adds to that of shortage of food supply.

DISCUSSION

Some obvious risk is inherent in any correlation method when cause-effect relationships are looked for. In the present case, a causative dependence of growth rate on food supply and the time this supply is available would be a plausible one. More importantly, such dependence was suggested from the mere data on place-to-place variation. The two resulting predictions could be confirmed by additional independent data on year-to-year variation. Therefore, the observed correlations will point to causative relationships. The risk appears to be remote that the variability of some other (unmeasured and intercorrelated) factor could offer a better explanation for the two types of variability in the growth rate of Macoma.

Of course, an unknown number of other environmental factors will affect growth rate in Macoma. In laboratory conditions, temperature has been found to be such a factor (De Wilde, 1975). However, for the 8-year period of 1968 to 1975, the correlations between temperatures and growth rates of Macoma were only slight and statictically non-significant (De Jonge - test with $k = 8$ and n = either5, 10, or 15 for the groups of 5, 10 and all 15 stations, respectively). It should be added, however, that mean temperatures during the growing seasons of the particular years of this study happened to deviate little from long term averages. Data on Macoma growth rates gathered by Lammens (1967) during years with more extreme temperature conditions point to a negative influence of high temperature during the growing season.

Unexpectedly, in an apparently food-limited situation, highest growth rates in Macoma were observed at stations where also highest numerical densities of adult Macoma were found. Data on numbers per m^2 at the start of the growing season are available for the 7 years 1969 to 1975, inclusive. The 7-year averages vary from about 20 both at the stations closest to,as well as those farthest away from,the coast to 50 to 100 at the midway stations. The correlation of these density estimates with growth rates at the same stations (7-year means of LIN_c) is positive and significant (Spearman's $r = 0.56$, $n = 15$, $p < 0.05$). A similar correlation of growth rate with Macoma biomass (7-year means of ashfree dry weight of soft parts per m^2) at the start of the growing season is also significantly positive ($r = + 0.72$, $n = 15$, $p < 0.01$).

The central area, showing highest values for both growth and numerical density for Macoma, also shows highest estimates for the biomass and species density of total macrozoobenthos. Total biomass at the end of winter is about 30 $g.m^{-2}$ ashfree dry weight here, as compared to less than 10 $g.m^{-2}$ at the two types of extreme stations as to distance from the coast. Number of macrozoobenthic species per m^2 varies from about 14 in the central parts of Balgzand to about 10 at the edges. Apparently, the central area is an optimal one not only for Macoma. Possible competition for food in this area will not have been severe enough to reduce growth rates to the low levels observed in the other areas, where either lower food densities or shorter daily periods of food intake limit growth rates.

ACKNOWLEDGEMENTS

We are indebted to Mr. H. Hummel for kindly placing at our disposal results of his unpublished work on stomach contents in Macoma. We express our thanks to Mr. W. de Bruin and Mr. J. Hegeman for assistance in sampling.

REFERENCES.

Beukema, J.J. 1974. Seasonal changes in the biomass of the macro-benthos of a tidal flat area in the Dutch Wadden Sea. Neth. J. Sea Res. 8, 94-107.

Beukema, J.J. 1976. Biomass and species richness of the macro-benthic animals living on the tidal flats of the Dutch Wadden Sea. Neth. J. Sea Res. 10, 236-261.

Cadee, G.C. and J. Hegeman 1974. Primary production of the benthic microflora living on tidal flats in the Dutch Wadden Sea. Neth. J. Sea Res. 8, 260-291.

Ente, P.J. 1969. De bodemgesteldheid en de bodemgeschiktheid van het Balgzand en de Breehorn. Flevobericht (Rijksdienst Ijsselmeerpolders, Zwolle) 66, 1-29.

Gilbert, M.A. 1973. Growth rate, longevity and maximum size of Macoma balthica (L.). Biol. Bull. 145, 119-126.

Jonge, H. de 1963. Inleiding tot de medische statistiek, deel I. Verh. Inst. prev. Geneesk. 41, 1-421.

Lammens, J.J. 1967. Growth and reproduction of a tidal flat population of Macoma balthica (L.) Neth. J. Sea Res. 3, 315-382.

Lorenzen, C.J. 1967. Determination of chlorophyll and phaeopigments. Spectrophotometric equations. Limnol. Oceanogr. 12, 343-346.

Siegel, S. 1956. Nonparametric statistics for the behavioral sciences. McGraw-Hill, New York.

Wilde, P.A.W.J. de 1975. Influence of temperature on behaviour, energy metabolism, and growth of Macoma balthica (L.). Proc. 9th Europ. mar. biol. Symp. (H. Barnes, ed) Aberdeen Univ. Press, 239-256.

STUDIES ON ANAEROBIC NITROGEN FIXATION IN THE SEDIMENTS OF TWO SCOTTISH SEA-LOCHS

D. Blake* and J. W. Leftley**

*Department of Biological Sciences, The University, Dundee, DD1 4HN, Scotland. Present address: North West Water Authority, Manchester, England
**Dunstaffnage Marine Research Laboratory, P.O. Box 3, Oban, PA34 4AD, Argyll, Scotland

ABSTRACT

Anaerobic nitrogen fixation has been studied in sediments from two Scottish sea-lochs, Loch Eil and Loch Etive. The problem was approached in three ways: 1. Estimation of nitrogen fixation (acetylene reduction) rates. 2. A characterisation of the bacteria involved. 3. An assessment of some chemical and physical factors likely to influence nitrogen fixation.

Aerobic nitrogen fixation was not detected. Anaerobic nitrogen fixation rates in the range 0.06-4.10 n-moles ethylene produced. g dry weight^{-1} hr^{-1} were measured.

In these sea-lochs nitrogen fixation in the sediments appears to be due principally to sulphate-reducing bacteria of the genus _Desulfovibrio_, which are associated with sulphide deposits and negative Eh. _Clostridium_ spp. are also present in the sediments but do not appear to be active.

INTRODUCTION

The availability of nitrogen may be a factor limiting primary productivity in the sea and it is pertinent to study processes involved in the cycling of this element, particularly those likely to result in a net input of nitrogen to the system. Nitrogen fixation by micro-organisms is such a process: Nitrogen-fixing autotrophic cyanobacteria have been well studied in the laboratory and in the field (Stewart 1973) but less attention has been paid to the role of nitrogen-fixing heterotrophic bacteria in sediments, particularly in the marine environment. Several genera of N-fixing heterotrophic bacteria have been reported to be present in marine sediments: _Azotobacter_ (Pshenin 1963), _Clostridium_ (Pshenin 1963, Patriquin and Knowles 1972, Herbert 1975), _Desulfovibrio_ (Patriquin and Knowles 1972, Herbert 1975), _Enterobacter_ (Werner et al., 1974) and _Methanobacterium_ (Barker 1940).

With the exception of _Azotobacter_, the above organisms are facultative or obligate anaerobes. Consequently, where marine sediments are anaerobic, or contain anaerobic microenvironments (Patriquin and Knowles 1975) and receive an input of organic carbon, of natural or man-made origin, a potential for nitrogen fixation exists.

AREAS STUDIED

Loch Eil and Loch Etive are sea-lochs situated on the west coast of Scotland

(Fig. 1). L. Eil receives an input of organic carbon, in the form of cellulose fibres discharged from a nearby paper mill, at a rate of about 10 to 14 tonnes suspended solids/day (Pearson and Rosenberg 1976), as well as the natural input. L. Etive receives only natural runoff.

Sediments from both lochs are characterised by being rich in sulphide and rapidly become anaerobic below the surface as indicated by negative Eh values. The carbon content of these sediments is quite high (3.5 to 7%) and C:N ratios are between 10-15. All these characteristics are indicators of organically-rich reduced sediment (Pearson and Stanley 1976 - this symposium). Three stations were sampled in L. Eil: E24 (depth 30 m), E70 (54 m) and E2 (60 m) and two stations in L. Etive: E24 (23 m) and E6 (55 m).

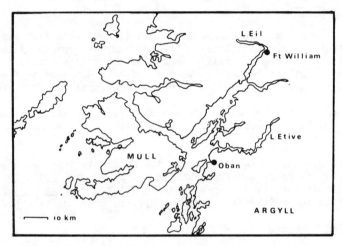

Fig. 1. Part of the Mainland of Scotland showing the location of the lochs studied.

MATERIALS AND METHODS

Undisturbed sediment samples were obtained using a Craib corer (Craib 1963) fitted with 245 mm x 56 mm o.d. acrylic plastic core tubes. The tubes were stoppered, placed in ice and returned to the laboratory for analysis. All manipulations were carried out in a glove-bag (Vickers Medical Ltd), filled with oxygen-free nitrogen which was further purified by passage over R3-11 catalyst (BASF-AG) at 160°C.

The overlying water was removed from the sediment and the first 5 cm of core extruded and well mixed.

Enumeration of Bacteria. Viable counts of Desulfovibrio, Clostridium and facultative anaerobes were determined using the most-probable-number (MPN) method (Alexander 1965) with the following modification: 10 ml samples of sediment were taken with a cut-down 20 ml syringe and serial dilutions made from these. MPN counts were calculated per unit volume and then converted to a dry-weight basis. Media used were essentially those used by Herbert (1975). Filter-sterilized sodium dithionite and resazurin were added to all media to a final concentration of 0.03% and 0.0001% respectively. A chelating agent, trisodium citrate, was added to the Desulfovibrio medium to a final concentration of 0.5%.

MPN counts were made in duplicate sets. One series from each set of tubes was supplemented with 0.4 M NaCl. The tubes were incubated at 15°C under nitrogen in anaerobic jars for up to 21 days. The tubes were then examined for growth and most-probable-number counts determined from statistical tables (Alexander 1965). Positive tubes were tested for acetylene reduction as described below.

Measurement of nitrogen fixation. This was measured indirectly by the acetylene-reduction method (Postgate 1972). Using a cut-off 5 ml disposable syringe, 2 ml (approx. 0.8 g dry-weight) portions of sediment were placed in the bottom of each of a series of universal bottles which were then closed with Suba Seal rubber stoppers. After injection of any amendments, the sediment was spread around the wall of each bottle by means of a vortex mixer, thus increasing the surface-area of the sediment. The bottles were then evacuated and back-filled three times with oxygen-free argon. The internal pressure of each bottle was equalised to atmospheric and then 3 ml of argon injected to produce a slight positive pressure (Postgate 1972). 1 ml of acetylene was injected into each bottle and a sample of gas (0.6 ml) withdrawn immediately into a plastic syringe for analysis, and then at various intervals thereafter. The bottles were incubated in the dark at 10°C.

The gas samples were analysed for ethylene by gas-chromatography (Pye-Unicam, Series 104). Separation of gases was carried out on a 960 mm long, 2 mm diameter column packed with 'Porapak T' maintained at 80°C, and then measured with a hydrogen flame ionization detector. Nitrogen was used as carrier gas; flow rate 40 ml min^{-1}. Ethylene peak areas were integrated and calibrated against known standards.

Ethylene production was not detected in samples previously held at 80°C for 20 minutes, or in samples which received no acetylene. The time-course of ethylene production was followed and the rate calculated from this (Fig. 2).

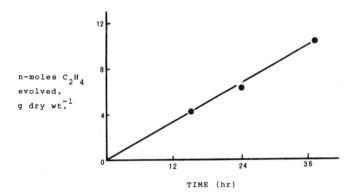

Fig. 2. Time-course of acetylene reduction by sediment from L. Eil (Station 70, July, 1976). Points are means of 3 replicates.

RESULTS

TABLES 1(a) and 1(b) Range of MPN counts in N-free medium and effect of sodium chloride. Except where indicated by an asterisk, all cultures gave positive results with the acetylene reduction test.

(a) L. Eil

Bacterial Group		Sample Site		
		E24	E70	E2
			(Count/g dry wt)	
Desufovibrio spp.	+ 0.4 M NaCl	$10^3 - 10^4$	$10^4 - 10^6$	$10^4 - 10^5$
	- NaCl	$10^2 - 10^3$	$10^3 - 10^4$	$10^3 - 10^4$
Clostridium spp.	+ 0.4 M NaCl	$10^1 - 10^2$*	$10^1 - 10^2$*	$10^1 - 10^2$*
	- NaCl	$10^2 - 10^3$	$10^2 - 10^3$	$10^2 - 10^3$
Facultative anaerobes	+ 0.4 M NaCl	0	0	0
	- NaCl	0	0	0

(b) L. Etive

Bacterial Group		Sample Site	
		Etive 6	Etive 24
		(Count/g dry wt)	
Desulfovibrio spp.	+ 0.4 M NaCl	$10^3 - 10^4$	$10^2 - 10^3$
	- NaCl	10^2	10^2
Clostridium spp.	+ 0.4 M NaCl	0	0
	-NaCl	$10^3 - 10^4$	10^3
Facultative anaerobes	+ 0.4 M NaCl	0	0
	- NaCl	0	0

TABLE 2 Maximum and minimum rates of acetylene reduction by sediments from the two lochs.

Sample Site	nmoles $C_2H_4 \cdot$ g dry wt$^{-1} \cdot$ hr^{-1}
Loch Eil	
E24	0.07-0.70
E70	0.07-4.10
E2	0.06-0.53
Loch Etive	
E6	0.08-0.28
E24	0.06-0.19

L. Eil was sampled monthly over a twelve month period, L. Etive over a four month period.

DISCUSSION

Only heterotrophic N-fixing bacteria belonging to the genera *Desulfovibrio* and *Clostridium* could be isolated in significant numbers from the sampling areas. In the presence of 0.4 M NaCl the clostridia showed little or no growth and did not reduce acetylene. In medium without salt these bacteria grew and fixed nitrogen. The MPN count of clostridia in sediment which was held at 80°C for 20 minutes and then inoculated into salt-free medium was similar to the count in unheated sediment, indicating that the clostridia are present as spores. It seems likely that the *Clostridium* spp. are of terrestrial origin and are present in sediments as dormant spores. Numerically dominant in all the sediments studied are sulphate-reducing bacteria belonging to the genus *Desulfovibrio*. These organisms will grow in nitrogen-free media in presence of 0.4 M NaCl and these isolates actively reduce acetylene.

Nitrogen fixation, as determined by acetylene reduction, has been consistently detected at all the sampling sites. These rates are low but comparable to those measured by other workers (Table 3).

There is no clear relationship between MPN counts of sulphate-reducers and acetylene reduction rates. A tenfold difference in numbers between samples does not necessarily mean a tenfold difference in the rate of acetylene-reduction. This may indicate that some factor, possibly carbon, is limiting. Sulphate-reducing bacteria can utilise only a limited number of carbon compounds, principally lactate which under natural conditions is produced by the fermentative metabolism of other micro-organisms: It is possible that our method of assaying acetylene reduction disrupts this syntrophic relationship, thus depriving the sulphate-reducers of a carbon supply.

TABLE 3   Reported rates of nitrogen fixation, as measured by acetylene reduction, in marine sediments.

Authors	Location	nmoles $C_2H_4 \cdot g^{-1} \cdot day^{-1}$
Patriquin & Knowles (1972)	Barbados	0.3 – 1.37
	New Brunswick	2.3 – 2.93
Maruyama et al (1974)	Tokyo Bay	1.45
Knowles (1975)	Beaufort Sea	0.1 – 1.0
Herbert (1975)	Tay Estuary	1.7 – 3.4
Marsho et al (1975)	Chesapeake Bay (Marsh)	4.8 – 300
This Study	L. Eil & L. Etive	1.3 – 98.4

Addition of lactate to sediment from L. Eil does, after a lag period of 24 – 30 hrs stimulate the rate of acetylene reduction (unpublished observations), though this effect is somewhat variable. Enhancement of acetylene reduction in marine sediments by various other carbon compounds has also been reported by other workers (Maruyama et al., 1974, Patriquin and Knowles 1975, Herbert 1975).

Finally, it can be said that much work needs to be done in order to obtain a better understanding of the interrelationship between N-fixation and carbon metabolism in sulphate-reducing bacteria, both in the laboratory and *in situ*, and also on the interrelationship between these bacteria and other micro organisms in sediments.

CONCLUSIONS

Anaerobic nitrogen fixation (acetylene reduction) can be detected in sediments from Loch Eil and Loch Etive. Desulfovibrio spp. appear to be the bacteria responsible for nitrogen fixation since they are the only group of heterotrophic nitrogen fixers so far isolated from these sediments which can function in a marine environment.

ACKNOWLEDGEMENTS

Some of the work described here was carried out as part of the Loch Eil project, a co-operative study involving personnel from the S.M.B.A., and the Universities of Dundee and Strathclyde. D.B. acknowledges the receipt of a studentship from the Natural Environment Research Council.

REFERENCES

Alexander, M. 1965. Most-probable-number method for microbial populations. In Methods of Soil Analysis, Part 2. Ed. by C.A. Black. American Society for Agronomy, Madison. 1467-1472.

Barker, H.A. 1940. Studies upon the methane formation. IV. The isolation and cultivation of Methanobacterium omelianskii. Antonie van Leeuwenhoek J. Microbiol. Serol. 6, 201-220.

Craib, J.S. 1965. A sampler for taking short undisturbed marine cores. J. Cons. perm. int. Explor. Mer. 30, 34-39.

Herbert, R.A. 1975. Heterotrophic nitrogen fixation in shallow estuarine sediments. J. exp. mar. Biol. Ecol. 18, 215-225.

Knowles, R. 1975. Nitrogen fixation in Arctic marine sediments. Beaufort Sea Technical Report No. 9. Dept. of the Environment, Canada. 44 pp.

Marsho, T.V., Burchard, R.P. and Fleming R. 1975. Nitrogen fixation in the Rhode River estuary of Chesapeake Bay. Can. J. Microbiol. 21, 1348-1356.

Maruyama, Y., Susuki, T. and Otobe, K. 1974. Nitrogen fixation in the marine environment: the effect of organic substrates on acetylene reduction. In Effect of the Ocean Environment on Microbial Activities. Ed. by R.R. Colwell and R.Y. Morita. University Park Press, Baltimore. 341-353.

Patriquin D.G. and Knowles, R. 1972. Nitrogen fixation in the rhizosphere of marine angiosperms. Mar. Biol. (Berlin), 16, 49-58.

Patriquin, D.G. and Knowles, R. 1975. Effects of oxygen, mannitol and ammonium concentrations on nitrogenase (C_2H_2) activity in a marine skeletal carbonate sand. Mar. Biol. (Berlin) 32, 49-62.

Pearson, T.H. and Rosenberg, R. 1976. A comparative study of the effects on the marine environment of wastes from cellulose industries in Scotland and Sweden. Ambio. 5, 77-79.

Postgate, J.R. 1972. The acetylene reduction test for nitrogen fixation. In Methods in Microbiology. Ed. by J.R. Norris and D.W. Ribbons. Vol. 6B. Academic Press, London. 343-356.

Pshenin, L.N. 1963. Distribution and ecology of Azotobacter in the Black Sea. In Symposium on Marine Microbiology. Ed. by C.H. Oppenheimer. Thomas, Illinois. 383-391.

Stewart, W.D.P. 1973. Nitrogen fixation by photosynthetic microorganisms. Ann. Rev. Microbiol. 27, 283-316.

Werner, D., Evans, H.J. and Seidler, R.J., 1974. Facultatively anaerobic nitrogen-fixing bacteria from the marine environment. Can. J. Microbiol. 20, 59-64.

PRODUCTION PRIMAIRE SAISONNIERE DU MICROPHYTOBENTHOS DES SABLES ENVASES EN BAIE DE CONCARNEAU

Denise Boucher

Université de Bretagne Occidentale - Laboratoire, d'Océanographie Biologique - 29283 BREST Cédex - FRANCE

ABSTRACT

The annual cycle of primary production and biomass has been studied in sublittoral muddy sands at 5, 10 and 15 metres. Biomass is measured by Chlorophyll a and Pheophytin a at the superficial layer of the sediment. Production is estimated by a ^{14}C method: *in situ* incubation and liquid scintillation counting. The seasonal cycle of biomass presents two maxima, the first in spring, the second in autumn. They are markedly apparent at the 10 and 15 metre stations. The annual variation ranges between 1 and 50 µg Chl.a/g of dry sediment. Sediment stability and internal sedimentary surface govern this biomass. The seasonal variations of production - in the range of 0.1 to 30 µg C/g of sediment / day - are connected with biomass variations, except for the 5 metre depth station where the biomass does not vary during the year. Production rate is connected with light, decreasing with depth: the maximum values of biomass and production were recorded at the 10 metre station with an annual average of 15 µg Chl.a/g and 5.5 µg C/g of sediment / day. Stability of the upper layer of the sediment and light are the main limiting factors for microphytobenthos production.

INTRODUCTION

Cette étude a pour but d'évaluer la biomasse et la production du microphytobenthos de l'étage infralittoral en fonction des facteurs climatiques, au cours d'un cycle annuel réalisé d'octobre 1971 à octobre 1972, au niveau de peuplements benthiques bien définis et étudiés par ailleurs. Afin d'éliminer les variations liées aux facteurs édaphiques, les trois stations choisies sont situées par 5, 10 et 15 m de fond (Fig. 1) sur des sables fins envasés, peuplés

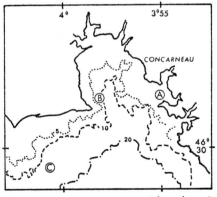

Fig. 1. Baie de Concarneau, localisation des stations

par la communauté à "*Amphiura*". Ce type sédimentaire est bien répandu en baie de Concarneau; les populations macrofauniques y présentent des biomasses élevées.

MATERIEL ET METHODES

Le prélèvement des échantillons de sédiment effectué aux trois stations selon trois jours consécutifs, est répété périodiquement une fois par mois pendant 12 mois. Il a lieu pour chaque station, entre 10 h et 11 h du matin; nous n'avons pas tenu compte des coefficients et des heures de marée. Les stations étant accessibles en scaphandre autonome, le sédiment est prélevé à l'aide d'un tube cylindrique de plexiglass de 20 cm de longueur et de 21 mm de diamètre intérieur. La zone de prélèvement correspond à un cercle d'1 m de rayon environ, choisie en tenant compte de la répartition de la macrofaune et des grosses particules. Douze carottes de sédiment sont prélevées. Les mesures de biomasse et de production sont appliquées au centimètre superficiel. La biomasse est évaluée par la teneur en chlorophylle a mesurée sur l'extrait acétonique d'un broyat de sédiment séché à 40°C d'après la méthode de Steele & Baird (1968) sur trois carottes par station. Toutefois, cette opération de séchage peut entraîner une perte de chlorophylle a et Plante-Cuny (1973) recommande l'extraction sur sédiment humide avec correction de la concentration finale d'acétone. Les densités optiques de l'extrait sont lues à 750, 665, 430 et 410 nm. La chlorophylle a "totale" est exprimée en µg/g de sédiment sec à partir de la formule: Chl.a T = $(10^3/84).(OD665/1).(v/p)$ où OD665: densité optique lue à 665 nm corrigée de la densité optique lue à 750 nm, v : volume d'extrait acétonique en ml, 84 : coefficient d'absorption de la chlorophylle a à 665 nm dans l'acétone à 90% ml/mg/cm (Talling & Driver, 1963), p: poids de sédiment sec en gramme, l : longueur de la cuve en cm. Les proportions de chlorophylle a fonctionnelle et de phéophytine sont obtenues par la méthode de Moss (1967) que nous avons choisie du fait des faibles valeurs de OD665 nm souvent mesurées.

La production est mesurée par la méthode du ^{14}C avec incubation *in situ*. Le sédiment provenant de 5 à 8 carottes, est réuni et mélangé. Un échantillon de 0.5 à 1.5 g est prélevé et placé dans un piluliier de verre de 96 ml, rempli d'eau de mer filtrée sur Whatman GF/C; 1 ml de Na_2 $^{14}CO_3$ d'activité 4 µCi est ajouté au dernier moment. Cinq incubateurs (2 blancs - 2 noirs - 1 formolé) ainsi préparés, sont placés, couvercle en bas, sur un portoir qui est descendu sur le fond. L'incubation dure une demi-journée. Le sédiment est alors recueilli par filtration sur Whatman GF/C, puis rincé. Après séchage à 40°C, il est broyé et la radioactivité fixée et mesurée sur des aliquotes de 100 mg en scintillation liquide (compteur Tricarb-Packard 314 ex). La composition de la solution scintillante utilisée (5 g PPO, 100 mg POPOP, 40 g de "Cab.O.Sil" pour 1 l de toluène) permet le maintien du sédiment en suspension. Le rendement est mesuré par la méthode du standard interne. Les résultats sont exprimés en µg de carbone assimilés par g de sédiment sec et par jour. Les mesures de l'énergie lumineuse arrivant sur le fond sont effectuées au moyen d'un photomètre à cellule au sélénium équipé de filtres neutres, la réponse est donnée en microampères. La mesure est effectuée au midi solaire, au moment de la mise à l'eau des incubateurs.

RESULTATS

Les variations des principaux facteurs écologiques sone résumés ici: Ces sédiments présentent une médiane granulométrique de la fraction sableuse respectivement égale à 100 - 86 et 100µ en A, B et C. Le taux de pélites (mesuré sur le centimètre superficiel), égal en moyenne à 37% en B et 29% en C varie entre 27 et 52% en B et entre 21 et 54% en C, les valeurs maximales étant atteintes en mars et avril, les minimales en hiver, seules des mesures ponctuelles ont été faites en A, on a mesuré 14.5% en mars, 10% en octobre. L'amplitude maximale des variations saisonnières de la température est de 10.5°C, elle est minimale en janvier (6°C), maximale en août aux stations A et B (16.5°C), en septembre (14°C) en C. La plus faible température mesurée sur le fond en C entre juin et septembre correspond à une hétérothermie. Les variations de l'éclairement mesuré sur le fond sont reproduites sur la Fig. 2. Il faut noter que les mesures ne sont pas effectuées le même jour aux mêmes stations. L'éclairement relatif mesuré sur le fond est égal en moyenne entre mai et octobre à 14.4% en A, 6.1% en B et 3.3% en C.

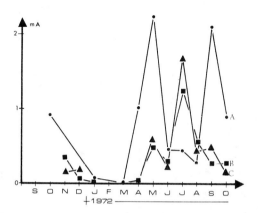

Fig. 2. Variations saisonnières de l'éclairement (mA) mesuré sur le fond, aux trois stations (● — ●)A, (■ — ■)B, (▲ — ▲)C.

Les variations saisonnières de la biomasse et de la production aux stations A, B et C sont représentées sur la Fig. 3.
C'est à la station A que l'on observe la plus faible amplitude de variation de la biomasse chlorophyllienne. A cette station, la teneur en chlorophylle a fonctionnelle est stable au cours de l'année, oscillant entre 0.7 et 2.5 µg/g de sédiment sec, c'est la phéophytine a qui est responsable des variations observées sur la chlorophylle a totale. Aux stations B et C, un cycle saisonnier apparaît présentant un minimum en mars, immédiatement suivi du maximum printanier en avril, un second pic de plus faible importance est observé en septembre. Les teneurs moyennes annuelles aux stations A, B et C sont respectivement égales à 1.3 - 5.8 et 5.4 µg/g de Chl.a fonctionnelle et à 4.6 - 9.3 et 6.7 µg/g de phéophytine.
Le rapport de la chlorophylle a fonctionnelle à la phéophytine ou rapport pigmentaire (Rp), est faible dans les sables fins envasés indiquant une présence importante de matériel détritique: ainsi les moyennes annuelles sont respectivement égales à 0.38 - 0.53 - 0.49 aux stations A, B et C alors que pour des sables fins propres de l'étage infralittoral méditerranéen Colocoloff M (1972) obtient des Rp compris entre 1.77 et 2.68. Les forts accroissements de ce rapport coincident aux "poussées de la biomasse, ce qui apparaît nettement au mois d'avril.

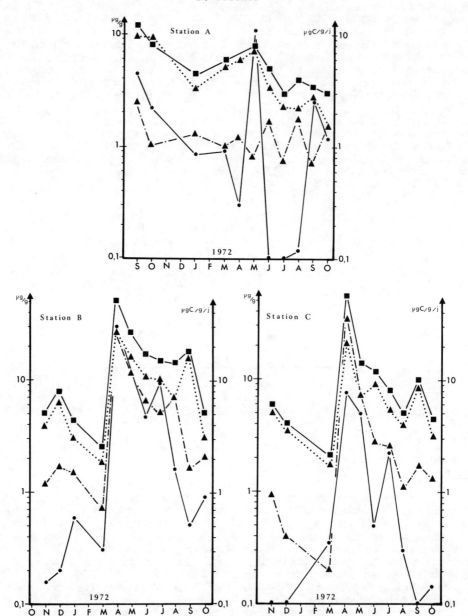

Fig. 3. Variations saisonnières de la production: (●─●), exprimée en µgC produit par g de sédiment sec et par jour (échelle de droite) et de la chlorophylle a totale: (■─■), de la chlorophylle a fonctionnelle: (▲─·─·▲) de la phéophytine a: (▲······▲) exprimée en µg/g de sédiment sec (échelle de gauche), pour chaque station (A, B et C). Coordonnées semi-logarithmiques.

L'indice de diversité pigmentaire OD430/OD665 est compris entre 2.6 et 6. Il diminue brutalement aux stations B et C entre mars et avril, où il atteint sa valeur minimale (2.9 et 2.6), et reste inférieur ou égal à 3 jusqu'au mois de juillet. Une décroissance de plus faible amplitude est observée entre août et septembre à ces deux stations. La brusque diminution de cet indice coincide avec une forte augmentation de la production, comme ce qui est connu pour le phytoplancton. Cet indice a été appliqué aux sédiments par Chassé (1972) et Plante-Cuny (1973), ce dernier auteur obtient une corrélation forte et négative avec la production ($r = -0.871$ $p \leq 0.001$).

Les cycles saisonniers de la production primaire aux stations B et C sont semblables. La production, faible en hiver, augmente brutalement entre mars et avril, où elle atteint son maximum respectivement égal à 32 et 7.7 µgC/g/j aux stations B et C. Elle decroit ensuite en présentant un second pic, moins élevé, en juillet. Le cycle à la station A présente des variations opposées: les deux périodes de forte production apparaissant en mai et septembre encadrent les valeurs minimales observées en été. La production moyenne hivernale est la plus élevée en A: 1 µgC/g/j pour 0.4 et 0.15 µgC/g/j en B et C. Les productions annuelles sont estimées à 0.7 - 1.8 et 0.5 mgC/g soit 6-15 et 4 gC/m^2 respectivement pour les stations A, B et C (valeurs étroitement liées à la méthodologie).

Le taux de production, ou rapport de la production à la biomasse exprimée ici en chlorophylle a fonctionnelle, est nul aux stations B et C au cours des mois d'hiver, tandis qu'il est relativement élevé an A au cours de cette période (0.65 µgC/ µgChl.a/j en janvier). Le taux maximal (13.2 µgC/ µgChl.a /j) est mesuré au mois de mai à la station A. Le taux de production montre également des variations saisonnières identiques à celles de la production à la station B. Les maxima (1.3 et 2 µgC/ µgChl.a/j) apparaissent respectivement en avril et juillet, mais alors que la production est maximale en avril coincidant avec le maximum de biomasse, le taux de production le plus élevé est mesuré en juillet, correspondant à l'éclairement maximum sur le fond. A la station C le taux de production est élevé en mars (1.7 µgC/ µgChl.a /j) et en mai (0.7 µgC/ µgChl.a/j) après une brutale diminution en juin, on note en deuxième pic en juillet (0.9 µgC/ µgChl.a/j) correspondant, comme en B, à un maximum d'éclairement. Les taux moyens de production calculés sur une année, respectivement égaux à 2.18 - 0.58 et 0.40 µgC/ µgChl.a/j soit 0.3 - 0.2 et 0.1 µgC/ µgChl.a "totale"/j, aux stations A, B et C, diminuent avec la profondeur d'immersion croissante.

DISCUSSION

Un cycle saisonnier de la biomasse chlorophyllienne, en relation avec les variations des facteurs climatiques est généralement observable en mode calme. Leach (1970) observe en été un enrichissement en chlorophylle a fonctionnelle des vases d'un estuaire. Colocoloff (1972) met en évidence dans des sables fins infralittoraux en Méditerranée deux maxima de la chlorophylle situés au printemps et en automne. Par contre, en milieu instable, Steele & Baird (1968) remarquent que les actions à court terme de l'hydrodynamisme en zone intertidale masquent ou détruisent une possibilité d'action à long terme des saisons sur la biomasse, alors que celle-ci est observée pour des sédiments plus abrités de l'étage infralittoral. De même Pamatmat (1968) qui n'observe pas de variations saisonnières de la biomasse chlorophyllienne des hauts niveaux en zone intertidale, note une augmentation significative de celle-ci à une station plus profonde, au mois de mai.

Outre ces variations saisonnières, la biomasse chlorophyllienne d'un sédiment est liée à sa granulométrie. Un rôle important est joué par la surface sédimentaire interne (Chassé, 1972) qui conditionne les possibilité de fixation des éléments de la microflore. Parmi les Diatomées d'un sable, les formes liées aux grains sont 10 à 100 fois plus nombreuses que les formes libres (Moss & Round, 1967, Colocoloff , 1971). La stabilité du sédiment est un autre facteur limitant de la biomasse microphytobenthique, le brassage du sédiment sous l'action des forces hydrodynamiques provoquant une érosion de la population par enfouissement ou mise en suspension.

La comparaison des biomasses hivernales (moyennes d'octobre à mars) respectivement égales à 1.2 - 1.5 et 0.7 µg/gChl.a en A, B et C concordent avec les surfaces sédimentaires internes calculées pour la fraction sableuse à ces trois stations (220 - 270 - 280 cm^2/g). A ces biomasses minimales correspond une stabilité minimale, les actions des forces hydrodynamiques étant très fortes pendant cette période aux trois stations. Le cycle saisonnier nettement marqué aux stations B et C, les plus profondes, reflète l'action conjuguée des facteurs climatiques et de l'accroissement de la stabilité apparaissant à ces stations dès le mois d'avril avec la diminution de la force des vents et de la fréquence des tempêtes. Le sédiment de la station A, peu profonde, reste soumis à l'action régulière des vagues et courants de marée. La faible stabilité du substrat serait le facteur limitant la biomasse estivale à cette station. Si on utilise la teneur en pélites de la couche superficielle comme critère de stabilité du sédiment, elle est la plus faible en A, par ailleurs les valeurs maximales obtenues en mars-avril aux stations B et C respectivement égales à 50.5% et 52%, diffèrent peu et coincident avec les valeurs maximales de la chlorophylle a "totale" (51 et 56 µg/g) qui sont également du même ordre de grandeur à ces deux stations.

Production et chlorophylle a fonctionnelle décrivent un cycle saisonnier similaire aux stations B et C. La corrélation calculée entre ces deux termes: ($r = 0.69$ $p \leq 0.1\%$) est meilleure que celle obtenue entre production et rapport pigmentaire ($r = 0.45$ $p \leq 2\%$) ou indice de diversité pigmentaire ($r = 0.48$ p 2%). En A, par contre, le cycle de production est lié à celui de l'énergie lumineuse disponible sur le fond ($r = 0.62$ $p \leq 5\%$). Entre les trois stations, une corrélation est obtenue entre le taux de production et l'énergie lumineuse mesurée sur le fond ($r = 0.48$ $p \leq 1\%$).

Une relation entre la production et l'irradiation mise en évidence par Pamatmat (1968), Leach (1970) n'est cependant pas toujours observée, l'action de la lumière pouvant soit être secondaire à celle de la température (Hargrave 1969), soit être marquée par les variations de la biomasse (Steele & Baird, 1968).

Dans notre étude, les incubations toutes réalisées *in situ* n'ont révélé aucune corrélation entre la production et la température, alors que l'action de ce facteur apparaît dans les expériences menées en lumière constante par Grøntved (1960), qui l'explique par l'accélération de la minéralisation sur le fond provoqué par le réchauffement.

Le rôle de la lumière comme facteur limitant la production explique les très faibles valeurs mesurées en A de juin à août, qui correspondent à une période de très forte turbidité des eaux à cette station. Les taux de production mesurés en mars (1.7 µgC/ µgChl.a/j) en C, en avril (1.3 µgC/ µgChl.a/j) en B ne sont pas reliés à de forts éclairements; il est reconnu toutefois que les Diatomées benthiques ont une forte efficacité photosynthétique à de très faibles intensités lumineuses (Pomeroy, 1959) et d'autre part que l'intensité lumineuse de saturation est liée à la température (Hunding, 1971). Cette adaptation est le fait soit d'un accroissement de la teneur en pigments soit d'une modification de la répartition de l'ensemble de ces pigments (Round, 1971). La concentration en sels nutritifs de l'eau intersticielle n'a pas été mesurée, mais le rôle de ce facteur est peut-être important au moment de la poussée printanière.

CONCLUSION

Le cycle du microphytobenthos est étudié sur une année en trois stations de la baie de Concarneau possédant chacune leurs caractères propres, l'action des facteurs écologiques comme la stabilité de la couche superficielle et la lumière étant mise en évidence. Un doute subsiste quant au facteur principal responsable du déclenchement de la poussée printanière.

REMERCIEMENTS

Ce travail a été réalisé avec l'aide financière et matériel du Centre National pour l'Exploitation des Océans - Contrats N° 71-363/72-527/73-755.

REFERENCES

Chassé, C. 1972. Economie sédimentaire et biologique (production des estrans meubles des côtes de Bretagne. Thèse Dr. Sc. Nat. Fac. Sci. Paris, 293p.

Colocoloff, C. 1971. Recherches sur la production primaire d'un fond sableux. 1. Ecologie quantitative et qualitative des Diatomées. Thèse spécialité Marseille.

Colocoloff, M. 1972. Recherches sur la production primaire d'un fond sableux. 2. Biomasse et production. Thèse spécialité Marseille.

Grøntved, J; 1960. On the productivity of microbenthos and phytoplankton in some Danish Fjords. Mddr. Danm. Havunders., (NY ser.), 3, 55-92.

Hargrave, B.T. 1969. Epibenthic algal production and community respiration in the sediments of Marion Lake. J. Fish. Res. Bd. Can., 26 (8), 2003 - 2026.

Hunding, C. 1971. Production of benthic microalgae in the littoral zone of a eutrophic lake. Oikos, 22 (3), 389 - 397.

Leach, J.H. 1970. Epibenthic algal production in an intertidal mud-flat. Limnol. Oceanogr., 15 (4), 514-521.

Moss, B., 1967a. Spectrophotometric method for the estimation of percentage degradation of chlorophylls to pheo-pigments in extracts of algae. Limnol. Oceanogr., 12 (2), 335-340.

Moss, B. and Round F.E., 1967. Observations on standing-crops of epipelic an epipsammic algal communities in shear water, Wilts. Br. phycol. Bull., 3 (2), 241-248.

Pamatmat, M.M., 1968. Ecology and metabolism of a benthic community on an intertidal sandflat. Int. Rev. Ges. Hydrobiol., 53, 211-298.

Plante-Cuny, M.R. 1973. Recherches sur la production primaire benthique en milieu marin tropical. I. Variations de la production primaire et des teneurs en pigments photosynthétiques sur quelques fonds sableux. Valeur des résultats obtenus par la méthode du ^{14}C. Cah. O.R.S.T.O.M., ser. Oceanogr., 11, 317-348.

Pomeroy, L.R. 1959. Algal productivity in salt marshes of Georgia. Limnol. Oceanogr., 4 (4), 386-397.

Round, F.E. 1971. Benthic marine diatoms. Oceanogr. Mar. Biol. Ann. Rev., 9, 83-139.

Steele, J.H. and Baird, I.E. 1968. Production ecology of a sandy beach Limnol. Oceanogr., 13 (1), 14-25.

Talling, J.F. and Driver, D., 1963. Some problems in the estimation of chlorophyll a in phytoplankton. In Doty, M. S. ed. Proc. conf. primary productivity measurement, marine and freshwater. Univ. Hawaii 1961. U.S. Atomic Energy Commission T.I.D. 7633, 142-146.

MODIOLUS MODIOLUS (L.) - AN AUTECOLOGICAL STUDY

R. A. Brown and R. Seed

Departments of Zoology, Queen's University, Belfast, Northern Ireland, and University College of North Wales, Bangor, Wales

ABSTRACT

This paper examines the ecology of the horse mussel Modiolus modiolus (L.) in Northern Ireland. Most of the work relates to the sublittoral beds in Strangford Lough though more limited data are also available for a much less extensive intertidal population in Belfast Lough. Reproductive cycles have been examined in some detail and estimates of protein, lipid, glycogen and calorific value have been monitored throughout one complete annual cycle. Growth has been estimated from ring analysis and from experimentally caged populations. Relative growth in the intertidal and subtidal populations has also been compared. Population structure and the major community associates of Modiolus are briefly considered.

INTRODUCTION

Modiolus (Fam. Mytilidae) which dates back to Silurian or Devonian times is exceedingly variable and probably includes several phyletic lines. The simple modioliform shell with subterminal umbones, lack of sculpturing and absence of hinge teeth is almost certainly convergent. The lineage from Modiolus to Mytilus shows a transition from an endobyssate (infaunal) to an epibyssate (epifaunal) habit. The functional and evolutionary significance of this change is discussed in detail by Stanley (1972).

Despite its widespread distribution and potentially large size Modiolus (= Volsella) modiolus has received far less attention than its widely exploited relative Mytilus edulis L. Perhaps a predominantly sublittoral distribution, generally slower growth and poorer recruitment, together with its apparently greater intolerance of inshore waters of reduced salinity are at least partly responsible.

The most comprehensive account of M. modiolus is that by Wiborg (1946) for mussels in Norwegian waters where this species has a certain limited value in the commercial fisheries. Apart from this work only scattered accounts are available concerning the ecology (eg. Rowell 1967; Roberts 1975; Seed & Brown 1975) and physiology (eg. Winter 1970; Coleman & Trueman 1971; Coleman 1973; Schlieper et al. 1958) of this species. More recently its larval development has also been investigated in some detail (Schweinitz & Lutz 1976).

Modiolus (= Arcuatula) demissus Dillwyn has likewise been rather poorly investigated even though this species is abundant along the Atlantic coast of America where it is important in the energy and biogeochemical cycles of salt marsh ecosystems (Kuenzler 1961a, b; Lent 1968, 1969). This paper briefly examines our existing and rather limited knowledge of the ecology of M. modiolus particularly in relation to our findings for populations of this species in Northern Ireland.

MATERIALS AND METHODS

The subtidal Modiolus population in Strangford Lough has been sampled monthly since December 1971, the less extensive low shore population at Bangor, Belfast Lough since May 1975. Samples were carefully sorted and the major community associates removed and identified. The length of each mussel was measured and the population size frequency distribution established. Reproductive tissue from the fleshy mesosoma (the gonad does not penetrate the mantle as in Mytilus) was removed, fixed in Bouin's fluid and stored in 70% alcohol. Sections cut at 10 - 15 μm were stained in haematoxylin. From such sections the reproductive condition of each animal and the gonad index of the whole population could be established. Details of the scheme used are given elsewhere (Seed & Brown 1975). Animals used for biochemical analyses and calorific estimations were homogenised and deep frozen until required. Total lipid was estimated after ether extraction whilst protein and glycogen were estimated colorimetrically. Absolute growth was established from analyses of annual rings and from measurements of marked animals of differing initial size retained in experimental cages. These were set out on the beds and the animals measured every six weeks. To check whether repeated disturbance had any marked effect on growth, one cage was left in situ for a period of seven months. Differential growth between selected size parameters was examined by testing each pair of variables y and x for their fit to the allometric equation $y = Ax^b$ where A and b are constants. These constants were estimated by least squares regressions. Comparisons of variability between the two populations and between different size groups were made by analysis of relative variation i.e. by comparing coefficients of variation - standard deviations expressed as percentages of the mean. Further details are given in our earlier paper (Brown, et al. 1976)

RESULTS

Distribution

From data kindly supplied to us by various Marine Laboratories and Universities the distribution of Modiolus in N.W. Europe has been established. It occurs in low intertidal pools and crevices to depths of over 100 m. on substrata ranging from rock through sand and shell gravel to mud. It is especially abundant in the Barent and White Seas, Norway, Iceland and the northern coasts of Britain but is apparently absent from much of the Baltic. Although less common in the southern North Sea and Channel it does extend south as far as the Bay of Biscay.

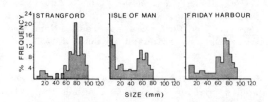

Fig. 1 Length frequency distributions of three Modiolus populations.

Population and community structure. Size frequency distributions of Strangford Modiolus are typically bimodal the major part of the population consisting of large individuals (> 50 mm) with smaller mussels forming a secondary more variable peak. This pattern seems to be characteristic of Modiolus populations elsewhere (Fig. 1). The species is extremely gregarious, clumps of mussels living partially buried in the sediment. Their shells provide a stable substratum for numerous epifaunal species including juvenile mussels, whilst the weft of byssus threads and large amount of mud and shell debris which accumulates on the bed, creates a suitable habitat for numerous infaunal organisms. This increase in species diversity produces a more complex food network which in turn results in enhanced stability of the whole community. To date 90 invertebrate taxa have been found associated with the Strangford population with most of the major groups well represented (Fig. 2). Ophiuroids and Chlamys spp. are especially abundant along with

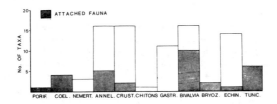

Fig. 2 Associated fauna of the Strangford L. Modiolus community.

numerous tunicates, hydroids and serpulids. Although not present in large numbers, predators such as Cancer and Asterias are unquestionably important functional members of this community. Pea crabs, which often inhabit the mantle cavities of bivalves are not uncommon. Mussel samples from Strangford (n = 1702) and Port Erin (n = 200) had low levels of infection

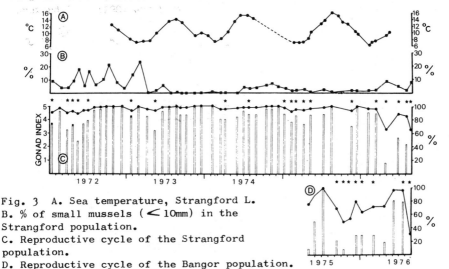

Fig. 3 A. Sea temperature, Strangford L.
B. % of small mussels (< 10mm) in the Strangford population.
C. Reproductive cycle of the Strangford population.
D. Reproductive cycle of the Bangor population.
% Ripe, open columns; % Spent, closed columns;
* over 20% spawning; Gonad index●

with Pinnotheres - 2.23% and 2.00% respectively whilst the level of infestation by Fabia in a Friday Harbor population (n = 126) was somewhat higher at 8.73%. The presence of these "parasites" may cause loss of condition though this effect is not especially pronounced.

Reproduction. The majority of Strangford mussels first became sexually mature when measuring between 30 - 40 mm in length. Only animals over this size were therefore used in our investigation of the reproductive cycle. Results (Fig. 3) reveal an apparent absence of any marked pattern. Gonads were ripe virtually throughout the period of investigation. Gonad indices fell slightly during February - March when temperatures were at their lowest but with the exception of 1976 this was not significant. A more marked cycle was evident in the intertidal population, development in late spring and summer being followed by a pronounced spawning. This commenced as temperatures fell in September and apparently continued until late March. Completely spent individuals were rarely encountered though reripening gonads with residual and developing gametes were not uncommon. The absence of any obvious cycle in the subtidal population is interpreted as evidence of a slow, continual release of gametes. This should result in a similarly protracted period of recruitment and the presence of a small though variable proportion of small mussels ($<$ 10 mm) throughout the year (Fig. 3) argues in favour of this interpretation. No small mussels were found in the intertidal population. Fig. 4 shows that tissue calorific values through 1975 generally varied between 4 - 5 K cals/g dry weight with little evidence of any seasonal trend. Similarly tissue weights of mussels of similar size changed little with the exception of a slight increase in June. This coincided with the local phytoplankton bloom. Lipid and protein levels again remained steady throughout the year and although glycogen did increase between May and July it never accounted for more than a small proportion of total tissue weight.

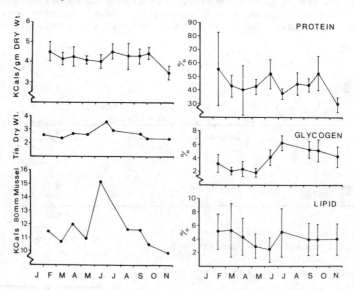

Fig. 4 Changes in calorific value and biochemical content (% dry wt.) in Strangford mussels during 1975.

Growth. Size frequency distributions of Strangford Modiolus changed little
throughout the year and could not therefore be used to monitor the progress
of individual size classes through the population. Ring analyses and caged
populations showed that growth in shell length was most rapid in small
mussels (Fig. 5). Little or no further increase occurred in the largest
animals. Estimates from caged populations however, suggested a rather slower
growth rate and this was more pronounced in older mussels. Such results
could possibly be explained in terms of the very localised nature of the beds
themselves. Whilst every effort was made to dredge from exactly the same
location minor shifts in position could produce mussels grown under quite
different conditions. Repeated disturbance and fouling by algae could
adversely influence growth in caged animals though no significant differences
were observed between those mussels sampled regularly and those left undisturbed for longer periods. Ring analyses suggest that subtidal mussels in
the 60 - 100 mm range could vary from 6 to over 20 years of age. Furthermore

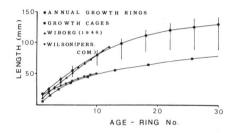

Fig. 5 Growth rates of Modiolus. ● and Strangford L.; ▲ Norway; ▼ N. Wales.

the mean value was distinctly skewed towards the maximum size in all age
classes possibly indicating that selection occurs in favour of faster growing
animals. Growth in intertidal cages was much slower ceasing completely
during winter when subtidal animals of comparable size were still growing
albeit rather slowly. In summer (June - August) growth of intertidal (0.50
± 0.52 mm, n = 14) and subtidal animals (0.52 ± 0.57 mm, n = 57) was similar.
Details of relative growth in the Strangford population are given elsewhere
(Brown, et al. 1976). In this account comparisons are made between the two
local populations. All the parameters measured in the subtidal population
showed a decline in variability with increasing size (Fig. 6). Intertidal
animals were more variable and the trend towards declining variability with
increased size was less evident. Regressions of tissue wet weight on shell
weight suggest that shell growth is slightly faster than tissue growth in
intertidal (b = 0.975) than in subtidal (b = 1.033) animals. Growth in shell
weight relative to length gave similar results (b = 2.938 and 2.901 for
intertidal and subtidal mussels respectively) suggesting that littoral
animals have somewhat heavier shells than their subtidal conspecifics of
similar size.

DISCUSSION

M. modiolus the largest British mussel can grow up to 15 - 20 cm in length.
Essentially a northern species it occurs in suitable habitats from North
Russia to the Bay of Biscay. On the Atlantic coast of America it is present
from Labrador to North Carolina and in the Pacific from the Bering Sea to
Japan and California. Tolerance of salinities down to only 25 - 30°/oo
(cf Mytilus edulis 5 - 6°/oo) probably explains its absence in the Baltic

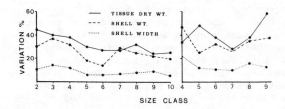

Fig. 6 Coefficients of variation in subtidal (left) and intertidal (right) Modiolus populations. Size class corresponds to 10 mm steps.

(Schlieper et al. 1961). Penetration of the intertidal is restricted by its physiological limitations and air gaping habit (Coleman & Trueman 1971; Coleman 1973). Air gaping, however, need not necessarily prevent successful intertidal colonisation as witnessed by M. demissus an extremely successful shore dweller. Controlled gaping and the more extreme infaunal nature of this species reduces water loss whilst its upper thermal tolerance ($37^{\circ}C$) is considerably higher than that of M. modiolus (Read & Cumming 1967). Modiolus can frequently dominate large areas of seabed and the communities which they support are amongst the most luxuriant in the cold temperate waters of the N.E. Atlantic. In this respect the Strangford mussel community is apparently no exception.

The absence of any obvious annual pattern in the breeding cycle of Strangford mussels is curious. The sheltered stable nature of the habitat together with ideal feeding conditions through much of the year could possibly be responsible since a more marked cycle is evident in the intertidal population where temperature, salinity and food supply fluctuate widely. Spawning, especially in the Bangor population seems to be most intense over winter when temperatures are low. This might be expected in these waters for a species which is principally northern in its distribution. Wiborg (1946) and Rowell (1967) suggest that an interval of several years may elapse between successive spawnings and there seems to be some evidence for this in the Strangford population where an apparently marked spawning occurred only once during the past five years. Wiborg (1946) also comments on the poor spatfalls in Modiolus compared with Mytilus whilst Kuenzler (1961a) explains the continued presence of small mussels in M. demissus populations in terms of variable growth rates since spawning was confined to late summer. The absence of any marked seasonal change in calorific value is consistent with previous findings for both subtidal (Ansell 1974) and intertidal (Dare & Edwards 1975) bivalves which do exhibit more seasonal reproductive patterns. Changes in food availability and breeding condition do not therefore seem to be paralleled to any great extent by changes in tissue calorific value though the actual volume of tissue may vary seasonally. Food scarcity during winter is often reflected in loss of tissue weight in bivalves though no such loss is apparent in Modiolus. The slight increase in weight during June could be attributed to the local phytoplankton bloom at that time. Gonad development generally results in increased weight in many bivalves but this does not seem to occur in Modiolus. Absence of seasonal variation in protein and lipid is again compatible with the relatively stable conditions experienced by the Strangford population though a seasonal trend in glycogen content is evident (see also Dare & Edwards). Depressed glycogen levels during winter reported for other bivalves may be due to utilisation when energy reserves are low. The stable conditons experienced by the Strangford population could, however, render such reserves unnecessary.

Growth in Modiolus is most rapid in the smaller size groups and declines
rapidly with increasing age until little or no further increase occurs in the
largest animals. The changeover from a period of fast growth to a period of
slower growth possibly coincides with the interval between the two modes in
the size frequency distribution. The exact position of this "trough" is
rather variable though it generally occurs amongst animals between 40 - 50 mm
in length - the approximate size at which Modiolus in these waters matures.
Wiborg (1946) reports that growth in Norwegian mussels slows appreciably after
maturation. Gonad development is known to cause temporary cessation of growth
in other bivalves and somatic growth in Cerastoderma locally (unpub. data)
increases rapidly only after reproduction is complete. Unlike many bivalves
Modiolus does not mature until several years old. The reasons for this may
possibly be sought in terms of the pattern of predation. Although predators
are numerous on the mussel beds only the largest crabs and starfish can open
mussels over 50 mm (Roberts 1975). Selection therefore favours faster growing individuals since these will be available to the predators for a much
shorter period than slow growing animals. Selection for faster growth
ensures that a greater number of mussels reach the size range where longevity
is greatly enhanced. Once beyond this size, energy resources can be re-
directed towards reproduction rather than somatic growth. The bulk of the
Modiolus population may thus represent a whole series of overlapping year
groups each growing rather slowly. The variable peak of small mussels, on
the other hand, may consist of faster growing animals which are subject to
extremely heavy predation pressures. Our growth data suggest that many of the
mussels in Strangford Lough must be well over 20 years of age.

Differences in relative growth between the intertidal and subtidal populations
may reflect the different environmental conditions experienced by these
populations. Declining variability with age in subtidal animals suggest some
mechanism of compensatory growth such as that described for various other
bivalves. The more extreme conditions intertidally may account for the
higher variability exhibited by the Bangor population. Greater emphasis on
shell growth in intertidal populations has previously been demonstrated in
Mytilus. That the differences between the two Modiolus populations are not
more marked is probably a reflection of the more restricted intertidal distribution of this species. In this respect its allometry contrasts sharply
with those species having a more extended intertidal range.

CONCLUSIONS

Although M. modiolus is abundant and widely distributed in the sublittoral
waters of North Europe its behaviour and physiological limitations prevent
any extensive penetration of the littoral zone. Size frequency distributions
in those populations examined by the authors were typically bimodal and this
picture showed little change through the year. An explanation of this
apparently stable situation is sought in terms of local recruitment, growth
and mortality rates. Reproduction in the subtidal population showed little
evidence of any seasonal pattern and ripe mussels could be found throughout
the year. A more pronounced cycle was demonstrated in the intertidal
population where environmental factors are much less stable. Protein, lipid
and calorific values showed little seasonal variation in subtidal mussels
though glycogen levels did increase slightly during the summer. Many of the
larger mussels in our populations are probably well over 20 years of age and
growth rates estimated from disturbance rings and from observations of caged
individuals have proved to be relatively slow. Intertidal mussels are more
variable in growth and form and have proportionately heavier shells than

their subtidal conspecifics. The Modiolus beds in Strangford Lough support a rich and vibrant community of associated species; the structure of this community is briefly considered.

REFERENCES

Ansell, A.D. 1974 Seasonal changes in biochemical composition of the bivalve Nucula sulcata from the Clyde Sea area. Mar. Biol. 25, 101-108.

Brown, R.A., R. Seed & R.J. O'Connor 1976 A comparison of relative growth in Cerastoderma edule, Modiolus modiolus and Mytilus edulis (Mollusca: Bivalvia) J. Zool., Lond. 179, 297-315.

Coleman, N. 1973 Water loss from aerially exposed mussels. J. exp. mar. Biol. Ecol. 12, 145-155.

Coleman, N. & E.R. Trueman 1971 The effect of aerial exposure on the activity of the mussels Mytilus edulis L. and Modiolus modiolus (L.). J.exp.mar.Biol. Ecol. 7, 295-304.

Dare, P.J. & D.B. Edwards 1975 Seasonal changes in flesh weight and biochemical composition of mussels (Mytilus edulis L.) in the Conwy Estuary, N. Wales. J. exp. mar. Biol. Ecol. 18, 89-97.

Kuenzler, E.J. 1961a Structure and energy flow of a mussel population in a Georgia Salt marsh. Limnol. Oceanogr. 6, 191 - 204.

Kuenzler, E.J. 1961b Phosphorus budget of a mussel population. Limnol. Oceanogr 6, 400-415.

Lent, C.M. 1968 Air gaping by the ribbed mussel Modiolus demissus (Dillwyn): effects and adaptive significance. Biol.Bull.mar. biol.Lab., Woods Hole, 134, 60-73.

Lent, C.M. 1969 Adaptations of the ribbed mussel Modiolus demissus (Dillwyn) to the intertidal habitat. Am. Zool. 9, 283 - 292.

Read, K.R.H. & K.B. Cumming 1967 Thermal tolerance of the bivalve molluscs Modiolus modiolus L. and Brachidontes demissus Dillwyn. Comp. Biochem. Physiol. 22, 149-155.

Roberts, C.D. 1975 Investigations into a Modiolus modiolus (L) (Mollusca: Bivalvia) community in Strangford Lough, N.Ireland. Rep.Underwater Ass. 1, 27-49.

Rowell, T.W. 1967 Some aspects of the ecology, growth and reproduction of the horse-mussel Modiolus modiolus M.Sc. Thesis Queen's University Ontario,138pp

Schlieper, C., R. Kowalski & P. Erman 1958 Beitrag zur ökologisch - zellphysiologischen charakterisierung des borealen lamellibranchiers Modiolus modiolus L. Kieler Meeresforsch. 14, 3 - 10.

Schweinitz, E. & R.A. Lutz 1976 Larval development of the northern horse-mussel Modiolus modiolus (L.) including a comparison with the larvae of Mytilus edulis L. as an aid in plankton identification. Biol. Bull. mar. biol. Lab., Woods Hole, 150, 348 - 360.

Seed, R. & R.A. Brown 1975 The influence of reproductive cycle, growth, and mortality on population structure in Modiolus modiolus (L.), Cerastoderma edule (L.) and Mytilus edulis L. (Mollusca: Bivalvia) In - Proc. 9th Europ. mar. biol. Symp. 257-274. Aberdeen University Press.

Stanley, S.M. 1972 Functional morphology and evolution of byssally attached bivalve molluscs. J. Paleont. 46, 165-212.

Wiborg, K.F. 1946 Undersøkelser over oskjellet (Modiola modiolus (L.)) Rep. Norw. Fishery mar. Invest. 8, 1-85.

Winter, J. 1970 Filter feeding and food utilization in Arctica islandica L. and Modiolus modiolus L. at different food concentrations. In - Marine Food Chains 196-205, Oliver & Boyd.

DISTRIBUTION AND MAINTENANCE OF A *LANICE CONCHILEGA* ASSOCIATION IN THE WESER ESTUARY (FRG), WITH SPECIAL REFERENCE TO THE SUSPENSION—FEEDING BEHAVIOUR OF *LANICE CONCHILEGA*[1]

K.-J. Buhr and J. E. Winter

Institut für Meeresforschung, 285 Bremerhaven, Fed. Rep. of Germany

ABSTRACT

The distribution of the polychaete *Lanice conchilega* (Pallas) was investigated in 1976 at about 150 stations in the outer part of the Weser Estuary (FRG). Special attention was given to a well-defined area (Nautical Buoys 'L', 'K' and 'J') where, at least since 1963, very high population densities of *L. conchilega* have been maintained. Over the past 14 years, *L. conchilega* was the numerically dominant species in this area, associated with the passive suspension-feeding anthozoan *Sagartia troglodytes* Price. The interactions between these two dominant species are discussed with respect to food availability and competition for food. The density of *L. conchilega* in this area amounted to 20,000 individuals/m^2 (= 1090 g dry meat wt), and the density of *S. troglodytes* to 700 individuals/m^2 (= 108 g dry meat wt). The highest density of *S. troglodytes* (1010 individuals/m^2), however, was found in the adjacent sand-coral area which has been included in the present investigations.

These high population densities may be maintained by strong tidal currents (up to 1 m/sec) supplying a high energy input of (1) primary benthic production (diatoms) from large tidal flats surrounding 2/3 of the research area, and of (2) precipitated planktonic organisms from the open sea.

From the high population densities of *L. conchilega* observed (grazing area: 0.5 cm^2/animal), it is unlikely that *L. conchilega* engages solely in surface-deposit-feeding. Laboratory experiments revealed that in addition to surface-deposit-feeding, suspension-feeding plays a very important role in the nutrition of this polychaete. The suspension-feeding experiments showed that the amount of food ingested, expressed in percentages of body wt (dry tissue wt), increases from 3.9 to 35.7% with decreasing body size. Such values agree well with those obtained for highly specialized obligatory filter-feeders. Thus, the final explanation of the high population density is based on the high amounts of suspended matter and the ability of *L. conchilega* to utilize alternative sources of food.

---

[1]This work was made possible through a research grant from the "Deutsche Forschungsgemeinschaft" in connection with the program "Litoralforschung - Abwässer in Küstennähe".

INTRODUCTION

The polychaete *Lanice conchilega* (Pallas) is well-known to be a wide-spread and common species in coastal areas, especially of the Atlantic Ocean and the North Sea where it usually lives in depths of up to 50 m. It even penetrates into estuarine regions where it often experiences high turbidity and considerable amounts of pollutants. The distribution of *L. conchilega* in the German Bight is shown in Fig. 1.

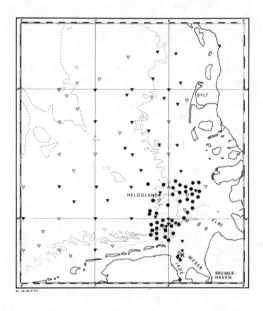

Fig. 1. Distribution of *Lanice conchilega* in the German Bight.
Circles: investigations during 1966/67 by Stripp (1969);
rhomboids: Dörjes (personal communications);
triangles: present investigations;
open triangles: no findings of *L. conchilega*.

The abundance of *L. conchilega* ranges from solitary living individuals to densities of several thousands/m^2. Such high densities, however, have been reported from near-shore areas only. Patches of high densities of this polychaete may disappear completely after short periods of time, or may be maintained for many years as it is described in this paper for a well-defined area in the outer part of the Weser Estuary (Fig. 1, Nautical Buoy 'K'). In this area a very high population density has been maintained for at least 14 years. During this time, *L. conchilega* has been the dominant species in this area, associated with the passive suspension-feeding anthozoan *Sagartia troglodytes* Price.

This area was investigated by Riemann-Zürneck (1969) in 1966/67, and re-investigated (present paper) 10 years later, in 1976, to evaluate the stability of the *Lanice conchilega* association. Quantitative data on abundance and biomass are calculated to obtain a rough estimation of the area's productivity. Abiotic and biotic factors have been measured to present, if possible, an explanation for the stability and high population density of this *Lanice conchilega* association. Furthermore, the feeding behaviour of *L. conchilega* is discussed in the present paper on the basis of laboratory experiments, carried out by Buhr (1976) to evaluate the filter-feeding capacity of the polychaete.

MATERIALS AND METHODS

Field Observations

Sampling procedure. Bottom samples were collected in 1976 at 150 stations within a well-defined area in the outer part of the Weser Estuary (FRG; Fig. 2) using a Van-Veen-type grab covering an area of 0.2 m^2. The weight of the Van-Veen grab (increased by lead) was about 100 kg to make sure that it penetrated deep enough into the bottom. For the calculation of abundance and biomass only grabs have been used which had penetrated into the bottom at least 25 cm, thus, avoiding an underestimation as described by Beukema (1974).

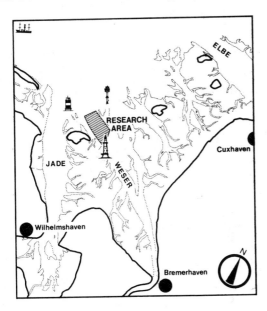

Fig. 2. Location of the research area in the Weser Estuary (FRG).

Determination of abundance and biomass. For the determination of the number of individuals of *Lanice conchilega* and *Sagartia troglodytes*, only the anterior parts (including the head) of the polychaetes and the oral disks of the anthozoans were counted. Three or 4 size groups of *L. conchilega* and of *S. troglodytes* were differentiated and counted separately. At least 20 undamaged individuals of each size group were individually weighed (60°C, 24 h) for the calculation of the biomass of each size group. The associated fauna was quantitatively handled in the same manner; results are not included in this paper.

Determination of abiotic and biotic factors. Factors such as temperature, salinity, current speed, direction of current, seston content, and the organic C content of the seston have been determined during the last 10 years. In order to evaluate temporal changes in all these factors, measurements were made at hourly intervals over complete tidal cycles in June and July, 1976. The organic C content of the seston was determined by wet oxidation with potassium dichromate, according to the method described by Dietrich and Höhnk (1961), and the organic material was then calculated using a factor of 1.72.

Laboratory Experiments

Determination of filtration rate. The determination of filtration rate (Fig. 3) was carried out using the automatic recording apparatus described by Winter (1973). The photoaquaria were modified (Fig. 4) according to Buhr (1976). By this apparatus the food concentration of the experimental medium is continuously monitored and automatically regulated by a photometer. Thus, the particle concentration is kept constant at a definite, predetermined food level by additions of small volumes of new algal suspension to replace those algal cells that have been removed by the filter-feeding activity of the animals. The calculation of the filtration rate is based on the volume of fresh algal suspension which has to be added per unit time to maintain the predetermined level of food concentration (for further details, see Winter 1973).

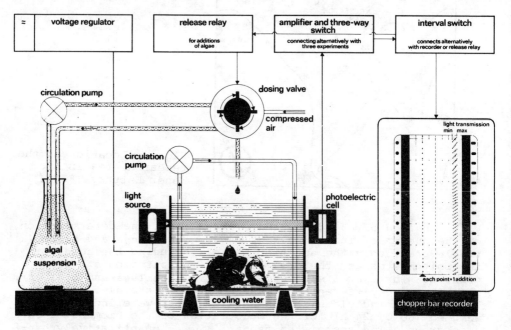

Fig. 3. Schematic diagram of the apparatus for measuring filtration rates used by Winter (1973).

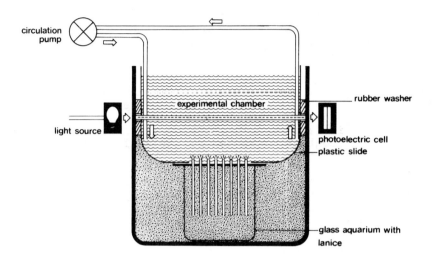

Fig. 4. Schematic diagram of a photoaquarium used for measurements of suspension-feeding in *Lanice conchilega* (Buhr 1976).

Experimental conditions. In all the suspension-feeding experiments, reported here, pure algal cultures of *Dunaliella marina* Butcher (7.5 µm x 5 µm) were used as food. The food concentration used in all experiments was 40×10^6 algal cells/l. The experiments were carried out at $12°C$, at a salinity of $27.5°/oo$, and over periods of at least 20 days. For further details, see Buhr (1976).

Experimental animals. Lanice conchilega was collected near Nautical Buoy 'K' in the research area described above. Polychaetes were placed into five experimental size classes on a tube-diameter basis which is significantly correlated with the dry meat wt of *L. conchilega*.
The dry meat weights of the experimental polychaetes ranged from 1.3 mg to 35.0 mg. For further details, see Buhr (1976).

RESULTS AND DISCUSSION

Field Observations

Abiotic and biotic factors influencing the research area. As the research area lies within the Weser Estuary, it still experiences the tidal cycle with a change of approximately 3 m in depth between low and high tide. Consequently, the direction of the current changes periodically with the tidal cycle. The current speed in this area may be as high as 1 m/sec. Two thirds of the research area (Fig. 5) are surrounded by tidal flats.

The temperature varies from $-0.9°$ to $18.4°C$ with season and the salinity from 25 to 33 $°/oo$ in relation to tide and season. The amount of suspended particulate matter (seston content) ranges from 5.6 mg to 77.5 mg dry wt/l with a corresponding organic content of 1.3 mg to 20.4 mg.

The measurements carried out at hourly intervals to evaluate temporal changes within the water column revealed that salinity and temperature in this area were more or less constant with tide and sample depth. The oxygen content and the amount of suspended particulate matter were also fairly constant with time, but the oxygen content decreased slightly with depth.

The high particulate loads in this region of the Weser Estuary, mentioned above, represent the main food source for the *Lanice conchilega* association in this area, chiefly composed of passive and active suspension-feeding organisms. The suspended particulate loads come from at least 3 different major sources:
- firstly, during ebb-tide, from the Weser River itself which has a particulate load as high as 1000 mg dry wt/l in the region of Bremerhaven,
- secondly, from the tidal flats surrounding 2/3 of the research area, and
- thirdly, during flood-tide, from the open sea.

The surrounding tidal flats are well-known as regions of high primary production, mainly of benthic diatoms. In this connection, especially the tidal flats situated to the west and northwest of the research area are of considerable importance, because of the wind which comes very often from these directions throughout the whole year. In addition, the large variety of larvae and zooplanktonic organisms coming from the open sea during flood tide, should not be underestimated as a food source of high nutritive value.

With respect to the amount of food available, the special topographic characteristics of the research area should be further elucidated. The main stream of the Weser River, indicated in Fig. 5 by the 10 m depth lines, widens from about 1 km to about 2 km within a distance of about 5 km. By this expansion the current speed is reduced, and consequently a good deal at least of the larger fraction of the suspended matter settles by sedimentation to the bottom, where it is partly caught by the dense network of the fringed ends of the *Lanice* tubes. Furthermore, the flood stream coming from the open sea, produces a high degree of turbulence by which bottom material is brought into suspension.

The different types of sediments found in the research area are specified in Fig. 5. The *Lanice* patch was found to be more or less restricted to areas of fine and medium sand, sometimes containing a high fraction of silt. Only a very limited number of individuals occurred in the north-western part of the research area which is characterized by large colonies of the sand-coral, *Sabellaria spinulosa* Leuck. The coarse sand areas were not colonized by *L. conchilega*.

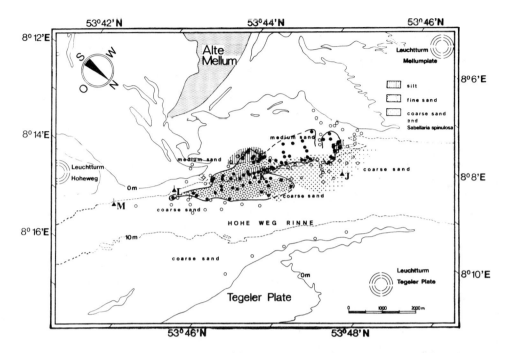

Fig. 5. Types of sediment in the research area. The distribution area of *Lanice conchilega* (solid circles) is surrounded by a solid line. Open circles: no findings of *L. conchilega*.

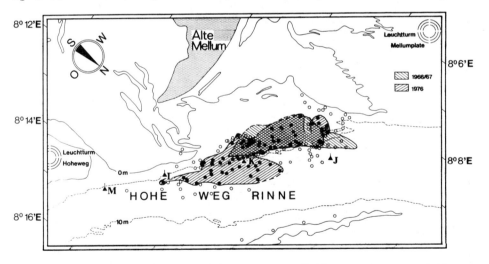

Fig. 6. Comparison between the extent of the *Lanice conchilega* association found in 1966/67 (Riemann-Zürneck 1969) and in 1976 (present work).

Extent of the Lanice conchilega association found in 1966 and 1976. In Fig. 6 a comparison is made between the extent of the *Lanice conchilega* association found in 1966/67 by Riemann-Zürneck (1969) and in 1976 during the present investigations. In general, the centre of the old *Lanice* patch still existed. While during later years the old *Lanice* patch had been only reduced seawards, a considerable colonization of the deeper regions had taken place. The overall impression obtained from the present investigations is a shift of the patch upstream, directed to the inner part of the Weser Estuary.

Abundance and biomass of Lanice conchilega in the research area. The population density of *Lanice conchilega* is represented in Fig. 7. The number of individuals/m^2 varied considerable from one station to another. The highest densities occured in depths from 10 to 20 m, in most cases within the recently colonized regions. In these depths, the number of individuals was mostly much higher than 10,000/m^2 and reached values of up to 20,000/m^2, or of 2 individuals/cm^2. An analysis of size frequency revealed that these very high densities were due to the recruitment of young individuals. As can be seen from Fig. 7, however, stations with a high fraction of young individuals were not restricted to the deeper regions only.

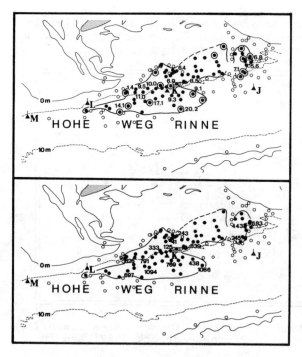

Fig. 7. *Lanice conchilega*. Number of individuals x $10^3/m^2$. Rings around solid circles indicate stations at which very small individuals predominated.

Fig. 8. *Lanice conchilega*. Biomass in g dry meat wt/m^2.

The biomass of L. conchilega is represented in Fig. 8. The highest values of biomass have been found in the deeper areas, inspite of the large fraction of very young individuals, mentioned above. At two stations the total biomass of L. conchilega amounted to more than 1 kg dry meat wt/m^2, corresponding to more than 5 kg fresh meat wt. These high values of biomass were mainly due to the proportion of older individuals which were found to have very high individual weights in this area.

Sagartia troglodytes, the second species in dominance in the research area. As pointed out already by Riemann-Zürneck (1969), at nearly all stations at which L. conchilega has been found, Sagartia troglodytes was the second most dominant species in the area. It lives burrowed in the sediment with its relatively short tentacles spread slightly above the sediment. The arrows in Fig. 9 symbolize sedimentation of detrital material which is brought by the tentacles to the oral disk as soon as the tentacles get in contact with the detrital material.

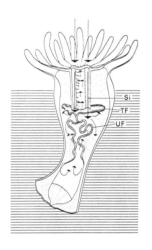

Fig. 9. Sagartia troglodytes. Schematic section (from Riemann-Zürneck 1969).

From investigations of the stomach content carried out by Riemann-Zürneck (1969), and from the present studies, there cannot be any question that detrital material plays a very important role in the nutrition of this anthozoan species. From the present investigations, we concluded that S. troglodytes feeds to a high degree (which has to be quantified by further experiments) on the faecal material expelled by L. conchilega, and that this faecal material used as food may be essential in explaining, at least partly, the obviously very successful association of S. troglodytes and L. conchilega.

Determinations of the calorific content (Buhr 1976) of the faecal material of L. conchilega (fed on Dunaliella marina) revealed that the mean calorific value of 1 mg ash-free dry wt of the faecal material was about 5.0 cal. From this it is obvious that the calorific content of the faeces of L. conchilega (on an ash-free dry wt basis) is of the same magnitude as the calorific content of unicellular algal species.

Abundance and biomass of Sagartia troglodytes in the research area. The distribution area of S. troglodytes corresponded more or less to that of L. conchilega. The population density (Fig. 10) reached a maximum of 700 individuals/m^2. The highest population density, however, was found in the area of the sandcorals (1010 individuals/m^2) where such high densities seem to be possible because of the very low number of L. conchilega, or the complete absence of this polychaete in this area. A determination of the total biomass of *S. troglodytes* showed that the highest values of biomass did not occur in the sand-coral area, in spite of the very high population density (Fig. 11).

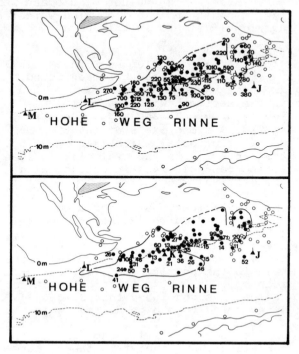

Fig. 10. *Sagartia troglodytes*. Number of individuals/m^2.

Fig. 11. *Sagartia troglodytes*. Biomass in g dry meat wt/m^2.

The relationships between the population density of *L. conchilega* and *S. troglodytes* as well as the effects of the population density of *L. conchilega* on the body size of *S. troglodytes* have been pointed out already by Riemann-Zürneck (1969). Her investigations showed that *S. troglodytes* prefers high population densities of *L. conchilega*, and that the total biomass of *S. troglodytes*/m^2 was highest even at the highest population densities of *L. conchilega*.

This clearly demonstrates that there is no competition for food between these two species, not even in areas of very high densities of *L. conchilega*. On the contrary, *S. troglodytes* seems to benefit from a high population density of *L. conchilega*, at least partly by the intake of faecal material expelled by the polychaete. In addition to this, by the fringed ends of the *Lanice* tubes a large fraction of the suspended particulate material is removed from suspension and may settle to some extent directly on the oral disks. Because of a reduction in current speed below the fringed ends and between the tubes, *S. troglodytes* may benefit from sedimentation of even a very fine fraction of suspended material.

Laboratory Experiments

From the high population densities observed, it is unlikely that *Lanice conchilega* engages solely in surface deposit-feeding. As the fringed ends of the tubes form an extremely dense network, only a very limited amount of detritus will reach the bottom. Consequently, food requirements have to be fulfilled by an alternative way of feeding. It is well-known from literature (summarized by Buhr 1976) that various invertebrates have developed additional suspension-feeding mechanisms. Thus, with respect to *L. conchilega* food uptake is performed not only from the bottom surface, but also from the fringed ends of the tubes and, as schematically shown in Fig. 12, from the surrounding water by the long, ciliated tentacles.

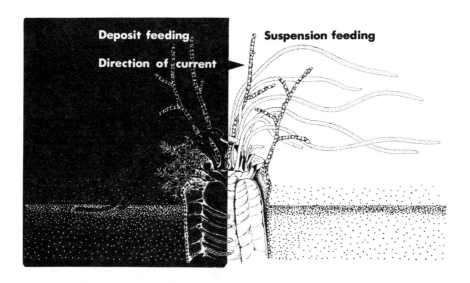

Fig. 12. *Lanice conchilega*. Schematic presentation of suspension-feeding and surface-deposit-feeding.

Laboratory experiments on the suspension-feeding capacity of *L. conchilega* (Buhr 1976; Fig. 13) had shown that up to 26 ml/h of the surrounding water could be filtered free of particles, corresponding to 1.35 mg algal dry wt/day, feeding at a constant food concentration of 40×10^6 *D. marina* cells/l. With increasing body size, the amount of food ingested increased, and expressed in percentages of body wt, decreased from 35.7 to 3.9% within the size range investigated. These values (Fig. 14) com-

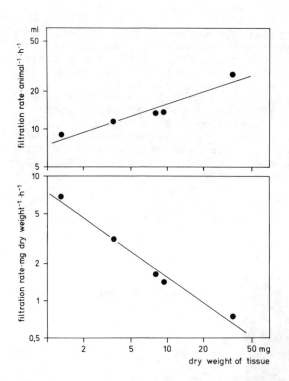

Fig. 13. *Lanice conchilega*. Filtration rates at 12°C in relation to body size (from Buhr 1976).

pared to those obtained for a roughly corresponding species-specific size range of a highly specialized obligatory filter-feeder, such as *Mytilus edulis* L. (Winter 1973), had given evidence that the classification of *L. conchilega* as an effective suspension-feeding organism is justified. Even the assimilation efficiencies calculated for *L. conchilega* (Buhr 1976) varying between 70.6 and 77.2% under the experimental conditions described above, were within the range typical for obligatory suspension-feeding organisms.

From all these data it is obvious that *L. conchilega* is capable of replacing surface-deposit-feeding by suspension-feeding. This ability of alternative feeding is, beyond doubt, of special importance, especially when the amount of food on the

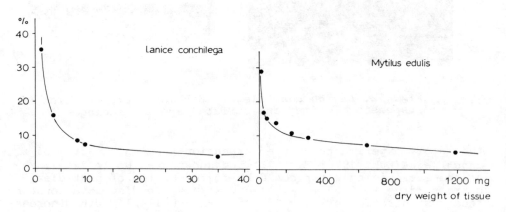

Fig. 14. Percentage of body weight ingested relative to body size in *Lanice conchilega* (Buhr 1976) and *Mytilus edulis* (Winter 1973).

bottom is limited as it is the case in areas of such high population densities as mentioned above. Thus, the final explanation of the high population density of *L. conchilega* is based on the considerable amounts of suspended matter in this special area and, on the other hand, on the ability of this polychaete to utilize alternatively different sources of food.

ACKNOWLEDGEMENTS

We would like to express our profound gratitude to Miss I. Antholz and Mr. M. Kern (students of the University of Bremen) for their inexhaustible help in collecting and sorting the vast number of bottom samples. We would like to extend our thanks to the laboratory assistants Mrs. E. Barwich, Mrs. M. Walter and Mrs. J. Schönfeld for carrying out the analyses of the water samples. Furthermore, we are also grateful to the crew of our research vessel 'Victor Hensen'.

REFERENCES

Beukema, J.J. 1974 The efficiency of the Van Veen grab compared with the Reineck box sampler. *J. Cons. int. Explor. Mer* 35, 319-327.

Buhr, K.-J. 1976 Suspension-feeding and assimilation efficiency in *Lanice conchilega* (Polychaeta). *Mar. Biol.* (In press).

Dietrich, R. and W. Höhnk. 1961 Studien zur Chemie ozeanischer Bodenproben. *Veröff. Inst. Meeresforsch. Bremerh.* 7, 15-35.

Riemann-Zürneck, K. 1969 *Sagartia troglodytes* (Anthozoa). Biologie und Morphologie einer schlickbewohnenden Aktinie. *Veröff. Inst. Meeresforsch. Bremerh.* 12, 169-230.

Stripp, K. 1969 Die Assoziationen des Benthos in der Helgoländer Bucht. *Veröff. Inst. Meeresforsch. Bremerh.* 12, 95-142.

Winter, J.E. 1973 The filtration rate of *Mytilus edulis* and its dependence on algal concentration, measured by a continuous automatic recording apparatus. *Mar. Biol.* 22, 317-328.

LE MACROBENTHOS DES FONDS MEUBLES DE LA MANCHE: DISTRIBUTION GENERALE ET ECOLOGIE

Louis Cabioch, Franck Gentil, René Glaçon et Christian Retière

Station Biologique, 29211 Roscoff, France.
Institut de Biologie Maritime et Régionale, 62930 Wimereux, France
Laboratoire maritime du Museum National d'Histoire Naturelle, 35800 Dinard, France

RESUME

Une association entre les laboratoires maritimes de Roscoff, Dinard et Wimereux a permis d'entreprendre, au cours des dernières années, une exploration d'ensemble du macrobenthos de la Manche. Les résultats de l'analyse de la distribution des principales espèces sont présentés sur l'exemple de l'épifaune sessile et les corrélations entre ces répartitions et les principaux paramètres écologiques sont commentées.

INTRODUCTION

La Manche appartient au plateau continental du Nord-Ouest de l'Europe. La profondeur n'y dépasse 100 m que très localement, dans la fosse Centrale, et les fonds constituent presque partout une plaine sédimentaire faiblement inclinée vers l'Ouest. On rencontre toutefois des substrats durs accidentés, essentiellement au voisinage des côtes françaises et anglaises ou des îles et des plateaux rocheux du Golfe normanno-breton.

La turbidité des eaux limite la pénétration de la lumière en profondeur, de telle sorte que les dernières Algues pluricellulaires disparaissent à partir de 45 m dans les eaux les plus claires de la région de Roscoff et les formations de Laminaires à partir de 25 m environ dans les mêmes conditions. Dans les baies des côtes bretonnes et encore plus en Manche orientale, cette dernière limite, qui marque le passage de l'étage infralittoral à l'étage circalittoral, se situe à une profondeur nettement plus faible. La signification et la distribution de ces limites en fonction de la pénétration quantitative et qualitative de la lumière ont été discutées par Cabioch (1968) et Boutler *et al.* (1974). Nous en retiendrons que la quasi-totalité de la plaine profonde de la Manche appartient à l'étage circalittoral; l'étage infralittoral ne s'étend que sur une bordure généralement étroite, allant de la partie inférieure de la zone des marées jusqu'à une profondeur de 25 m au plus.

Le second caractère essentiel de la Manche est le fait qu'elle est animée par des marées atteignant, par endroits, des amplitudes très élevées. L'alternance des émersions et des immersions crée, dans l'espace intertidal, une zonation bionomique, qui a été reconnue et décrite de longue date. En profondeur, les courants de marée déterminent dans une large mesure la nature et la distribution des sédiments. Aux forts courants, supérieurs à 2,5 noeuds environ en vive-eau moyenne, correspondent généralement des zones sédimentaires où les dépôts caillouto-graveleux résiduels et parfois le substrat rocheux demeurent dégagés. A partir de ces zones, les gradients d'hydrodynamisme décroissant se traduisent au niveau du substrat par des séquences sédimentaires de granulométrie également décroissante. Il résulte de ce fait et de la répartition des

vitesses des courants dans la Manche, que les sédiments fins ne se rencontrent
guère que dans les baies, là où les vitesses des courants de marée ne dépassent pas 1 à 1,5 noeuds en vive-eau moyenne. On trouvera dans les travaux de
Boillot (1964), Hommeril (1967), Ruellan et al. (1967), Larsonneur (1971),
Auffret et al. (1975), des descriptions régionales, des discussions et des références bibliographiques concernant ce phénomène essentiel dans la Manche,
que constitue l'existence de gradients édaphiques très accusés, hydrodynamiques et sédimentaires. A défaut d'une carte sédimentologique globale, la carte des vitesses des courants de marée (Fig. 3) donne une représentation approchée, assez bonne, de ces gradients.

Enfin, les eaux de la Manche présentent un gradient climatique illustré par
les cartes classiques des moyennes mensuelles de température et de salinité de
surface publiées par Lumby (1935), et dont Pingree et al. (1975) ont récemment
analysé l'origine en construisant un modèle mathématique explicatif. Les isothermes de février et d'août représentent les conditions extrêmes de température (Fig. 1). Disposées en chevrons, elles montrent, en hiver, une décroissance de la température d'Ouest en Est et de la région axiale vers les côtes,
mais au contraire, une croissance inverse en été, suivant les mêmes directions.
On passe ainsi depuis des eaux océaniques relativement sténothermes, à des
eaux de climat plus continental, eurythermes. Les courbes isohalines, moins
variables au cours de l'année, présentent la même disposition en chevrons, avec décroissance vers l'Est et vers les côtes qui, dans l'un et l'autre cas,
n'est pas sans relation avec la tendance générale à une dérive d'eau de l'Atlantique vers la Mer du Nord.

La Manche présente ainsi une structure climatique fortement zonée d'Ouest en
Est, se recoupant avec une succession de gradients édaphiques, locaux ou régionaux, d'orientations diverses.

Peu de mers ont fait l'objet de tant de récoltes zoologiques et botaniques,
mais les explorations générales et les travaux de synthèse sur l'ensemble de
la distribution de la faune et de la flore ont été bien moins nombreux. Les
travaux de Crisp et Southward (1958) en ce qui concerne le littoral, et de
Holme (1961, 1966) pour les principales espèces de l'endofaune et de l'épifaune vagile des fonds non exondables en donnent des éléments. On trouvera des
données complémentaires sur la distribution des Algues dans le travail de Dizerbo (1969). Les uns et les autres ont observé l'existence de limites faunistiques et floristiques souvent très nettes qui, dans l'ensemble se traduisent
par une réduction du nombre des espèces d'Ouest en Est.

MATERIEL ET METHODES

Depuis 1972, la Station Biologique de Roscoff a entrepris une exploration de
l'ensemble des peuplements benthiques de la Manche, en coopération avec le Laboratoire maritime du Muséum National d'Histoire Naturelle à Dinard, et l'Institut de Biologie Maritime et Régionale de Wimereux (Université de Lille).
1746 stations de dragage ont été effectuées par le "Pluteus II", du Golfe
normanno-breton au Pas-de-Calais. Il a en outre été procédé à 700 prélèvements
dans la région axiale de la Manche et le long des côtes Sud de l'Angleterre,
lors d'une campagne organisée en Septembre 1975 à bord de la "Thalassa", de
l'Institut Scientifique et Technique des Pêches Maritimes. Ces opérations ont
fait suite à la prospection des fonds du Nord de la Bretagne occidentale (Cabioch, 1968) et aux 330 stations effectuées par C. Retière avant 1972 dans la
partie méridionale du Golfe normanno-breton. Nous disposons ainsi actuellement
d'un échantillonnage couvrant la totalité de la Manche.

A chaque station, un volume de 30 dm^3 de sédiment, prélevé à la drague Rallier du Baty, a été tamisé et trié à bord. Un relevé faunistique provisoire est établi immédiatement, en pointant les espèces aisément reconnaissables. Nous disposons ainsi, moyennant éventuellement un contrôle des identifications au laboratoire, d'informations sur la distribution d'environ 300 espèces de l'endofaune et de l'épifaune dans l'ensemble de la Manche. A côté de l'utilisation de ces données immédiates, nous procédons progressivement à l'établissement des relevés définitifs des stations. Des descriptions régionales préliminaires de la distribution des peuplements ont ainsi été publiées, concernant les secteurs où le dépouillement des données est le plus avancé (Cabioch et Gentil 1975; Cabioch et Glaçon, 1975; Gentil, 1976; Retière, 1975). Le stockage informatique de ces données est effectué en collaboration avec le Centre Océanologique de Bretagne; leur collecte doit aboutir à la constitution d'un fichier "Manche", dans lequel sont introduits les résultats faunistiques, accompagnés d'informations aussi complètes que possible sur les caractéristiques des stations et les paramètres écologiques du milieu, notamment la granulométrie (Laboratoire de Géologie de l'Université de Caen).

Sans attendre l'achèvement de ce travail de longue haleine, qui devrait permettre d'aboutir à une analyse autoécologique et synécologique des peuplements de l'ensemble de la Manche, il nous est possible de dégager dès à présent certains traits essentiels de l'écologie benthique de cette mer, en nous appuyant sur la distribution des espèces immédiatement identifiables, ou de détermination contrôlable sans difficulté.

RESULTATS

Nous avons vu que les fonds non exondables de la Manche appartiennent, sur la majeure partie de leur étendue, à l'étage circalittoral. Ils sont le plus généralement recouverts, à ces profondeurs, par des dépôts grossiers variant, selon l'intensité hydrodynamique, depuis des fonds de blocs et de cailloux jusqu'à des sables grossiers plus ou moins hétérogènes. Les fonds grossiers circalittoraux forment ainsi un ensemble continu, depuis l'entrée de la Manche jusqu'au Pas-de-Calais. Les gradients hydrodynamiques et sédimentaires s'y développent généralement avec régularité. Opposés en cela à la mosaïque de conditions qui règnent au voisinage du littoral, ils offrent d'excellentes possibilités d'observation de la distribution d'un ensemble cohérent d'espèces animales, dans un cadre écologique relativement simple.

Nous tenterons par conséquent d'analyser la distribution qualitative des principales espèces de la faune des fonds grossiers circalittoraux de la Manche, telle qu'elle a été observée de 1972 à 1976, en nous limitant à l'épifaune sessile, sur laquelle les connaissances antérieures faisaient le plus défaut. Ces espèces, fixées sur le fond, constituent un ensemble de témoins permanents des conditions écologiques, entre lesquels ne règnent guère que des relations de compétition pour l'espace. Des cartes de répartition ont été établies pour 133 espèces, représentant un échantillonnage de l'épifaune sessile basé sur la facilité d'identification et qui n'est pas lié, par conséquent, aux affinités écologiques des espèces.

On constate, en comparant entre elles les cartes de distribution obtenues, qu'elles se classent pour la plupart sans difficulté, en des catégories qui reflètent de manière assez évidente les effets séparés ou conjugués des conditions climatiques et édaphiques très accusées de la Manche. On peut ainsi proposer, au moins provisoirement, la classification suivante parmi ces distributions.

Distributions Générales

Nous classerons dans cette catégorie les espèces que l'on rencontre dans toutes les régions de la Manche et qui sont, par conséquent, relativement indifférentes, en termes de présence-absence, aux gradients édaphiques et climatiques. Telles sont :

Ciocalypta penicillus Bowerbank
Stelligera stuposa (Montagu)
Myxilla rosacea (Lieberkühn)
Halecium halecinum (Linné)
Hydrallmania falcata (Linné)
Abietinaria abietina (Linné)
Sertularia cupressina (Linné)
Nemertesia antennina (Linné)
Nemertesia ramosa Lamarck
Sarcodictyon catenata Forbes
Alcyonium digitatum Linné
Epizoanthus couchi (Johnston)
Hormathia coronata (Gosse)
Sargartia elegans (Dalyell)
Disporella mamillata (Lagaaij)

Alcyonidium gelatinosum (Linné)
Schizomavella auriculata (Hassall)
Schizomavella linearis (Hassall)
Porella concinna (Busk)
Parasmittina trispinosa (Johnston)
Celleporaria pumicosa (Pallas)
Turbicellepora avicularis (Hincks)
Ascidia virginea Müller
Ascidiella scabra (Müller)
Dendrodoa grossularia (van Beneden)
Sabellaria spinulosa Leuckart
Potamilla reniformis (O.F. Müller)
Protula tubularia (Montagu)
Serpula vermicularis Linné

Distributions Climatiques

De nombreuses espèces, dont la présence n'est pas très dépendante des gradients édaphiques, sont communes dans l'entrée de la Manche, mais se raréfient et disparaissent plus ou moins rapidement vers l'Est. On peut tracer pour chacune d'entre elles la limite à partir de laquelle elle n'est plus qu'exceptionnellement rencontrée et l'on remarque que ces limites successives d'Ouest en Est (Fig. 2) présentent une disposition assez semblable à celle des courbes isothermes et isohalines de la Manche. Bien qu'il soit difficile d'établir des ensembles bien nets dans leur succession assez progressive, nous répartirons ces distributions en quelques catégories, qui constituent autant de groupes écologiques d'espèces.

Groupe 1. Espèces localisées très à l'Ouest, ne dépassant pas la région de Roscoff, et n'atteignant pas la région de Start Point (exemples : Fig. 2 A,B):

Diphasia alata Hincks
Diphasia pinaster (Ellis et Solander)
Antennella siliquosa (Hincks)

Polyplumaria flabellata Sars
Polyplumaria frutescens (Ellis & Sol.)
Porella compressa (Sowerby)

Les espèces suivantes sont préférantes des mêmes conditions :

Quasillina brevis (Bowerbank)
Cellaria salicornioides (Lamouroux)

Palmicellaria skenei (Ellis & Solander)

Groupe 2. Espèces occidentales ne dépassant pas la région de Roscoff, mais atteignant et dépassant un peu le méridien de Start Point (exemple : Fig 2 C):

Sertomma tamarisca (Linné)
Thuiaria articulata (Pallas)

Thecocarpus myriophyllum (Linné)

Diphasia pinnata (Pallas) n'occupe que la moitié méridionale de cette zone.

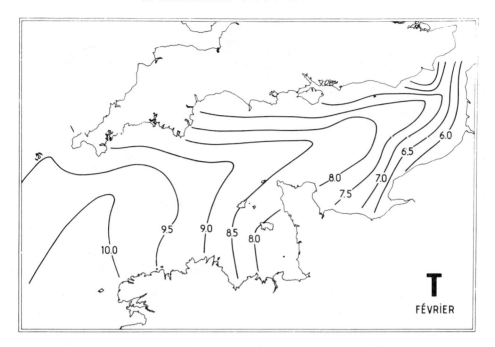
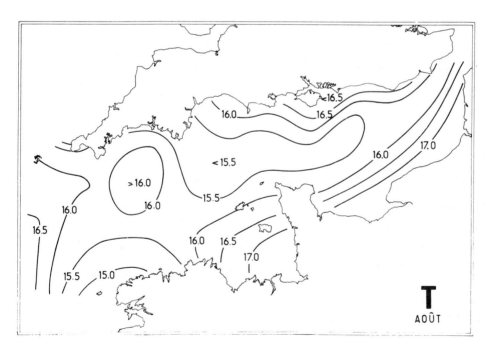

Fig. 1 : Distribution des températures de surface dans la Manche, d'après Lumby (1935).

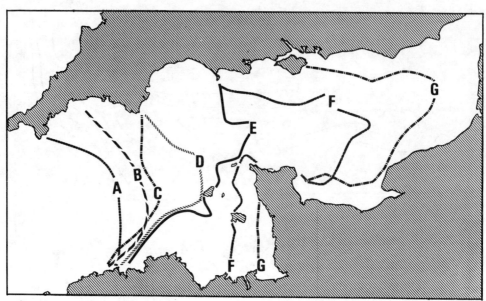

Fig. 2 - Distributions climatiques. Limites orientales successives, d'Ouest en Est de A : *Porella compressa*. B : *Diphasia pinaster*; C : *Thuiaria articulata*; D : *Lafoea dumosa*; E : *Caryophyllia smithi*; F : *Sertularella gayi*; G : *Rhynchozoon bispinosum*.

<u>Groupe 3</u>. Espèces occidentales ne dépassant guère la ligne Start Point - Guernesey - région de Roscoff (exemple : Fig. 2, D).

Lafoea dumosa Fleming
Antennella secundaria (Gmelin)
Plumularia catharina Johnston

Vermiliopsis infundibulum (Philippi)
Arca tetragona Poli

<u>Groupe 4</u>. Espèces occidentales ne dépassant guère la ligne Portland - Guernesey - région de Roscoff (exemple : Fig. 2 E).

Caryophyllia smithi Stokes

Monia squama (Gmelin)

Les espèces suivantes sont préférantes des mêmes conditions :

Cellaria salicornia (Pallas)
Cellaria sinuosa (Hassall)

Hydroides norvegica (Gunnerus)
Monia patelliformis (Linné)

Le Foraminifère sessile *Miniacina miniacea* (Pallas) n'occupe que la moitié méridionale de cette zone.

<u>Groupe 5</u>. Espèces pénétrant dans le Golfe normanno-breton et la Manche centrale, mais ne dépassant pas vers l'Est le niveau de la Baie de Seine (exemple : Fig. 2 F).

Sertularella gayi (Lamouroux)

Les espèces suivantes sont préférantes des mêmes conditions :

Polymastia agglutinans Ridley & Dendy
Diphasia attenuata Hincks
Omalosecosa ramulosa (Linné)

Anomia ephippium Linné
Scalpellum scalpellum (Linné)

Groupe 6. Espèces occupant la majeure partie de la Manche, à l'exception de son extrémité orientale et de la partie interne de la Baie de Seine (exemple: Fig. 2 G).

Rhynchozoon bispinosum (Johnston)

Préférante des mêmes conditions : *Verruca stroemia* (O.F. Müller).

Distributions Edaphiques

Il a été montré, dans la région de Roscoff, que la diversité qualitative de l'épifaune sessile décroît depuis les régions de fort hydrodynamisme vers les fonds plus ensablés correspondant à des courants plus faibles (Cabioch, 1968), l'essentiel de la réduction intervenant dans la direction du Nord-Ouest, entre les lignes correspondant à des vitesses de 2,5 à 2,1 noeuds en vive-eau moyenne. Le phénomène présente un caractère général et l'on peut distinguer un groupe d'espèces qui habitent, dans l'ensemble de la Manche, les fonds rocheux ou les nappes caillouto-graveleuses soumises à de fortes conditions de courant. Le Bryozoaire *Alcyonidium mytili* Dalyell en est un excellent exemple (Fig. 4 G) Les espèces suivantes appartiennent à ce groupe écologique :

Pachymatisma johnstonia (Bowerbank)
Tethya aurantium (Pallas)
Polymastia robusta Bowerbank
Hemimycale columella (Bowerbank)
Dysidea fragilis (Montagu)
Alcyonidium mytili Dalyell

Escharoides coccineus (Abildgaard)
Schizoporella unicornis (Johnston)
Chlamys varia (Linné)
Botryllus schlosseri (Pallas)
Polycarpa pomaria (Savigny)
Polycarpa violacea (Alder)

Les espèces suivantes sont préférantes des mêmes conditions :

Berenicea patina (Lamarck)

Ciona intestinalis (Linné)

Distributions Edapho-Climatiques :

Alors que les espèces précédentes sont répandues dans toutes les zones de forts courants, il en existe de nombreuses autres qui n'occupent que la partie la plus occidentale de ce même domaine édaphique, ou s'avancent plus ou moins loin dans la Manche. On observe ainsi dans cette direction, une série de limites faunistiques caractéristiques d'espèces des fonds caillouto-graveleux, qui manifestent une tolérance croissante vis-à-vis des conditions climatiques de plus en plus continentales vers l'Est. Mais il est important de noter qu'un petit nombre des espèces étudiées présentent des distributions inverses des précédentes. Nous qualifierons les premières d'espèces occidentales, les secondes d'espèces orientales et nous envisagerons en dernier lieu le cas de quelques espèces particulières qui paraissent limitées à la fois vers l'Est et vers l'Ouest.

Groupe occidental 1. Espèces ne dépassant pas vers l'Est la région de Roscoff (Exemples : Fig. 4 A,B).

Axinella infundibuliformis (Fleming)
Pseudaxinyssa digitata Cabioch

Diphasia delagei Billard
Hornera lichenoides (Linné)

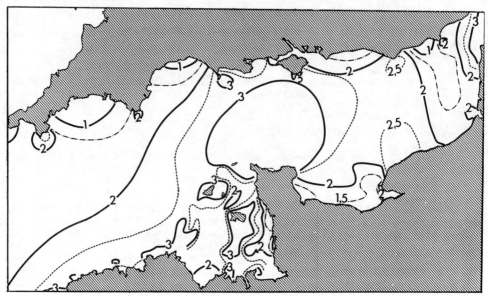

Fig. 3 - Vitesses maximales en noeuds, des courants de marée de surface, en vive-eau moyenne. Synthèse, d'après Sager (1963) et les ouvrages 550, 553 et 556 du S.H.O.M. (vitesses supérieures à 3 noeuds non représentées).

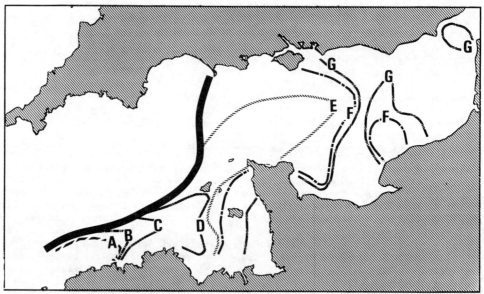

Fig. 4 - Distributions édapho-climatiques. Limites orientales successives, d'Ouest en Est de A : *Myxilla fimbriata;* B : *Diphasia delagei;* C : *Phakellia ventilabrum;* D : *Axinella agnata;* E : *Axinella dissimilis;* F : *Hymedesmia versicolor* . Distribution édaphique : G : *Alcyonidium mytili*. Trait épais = limite commune des distributions vers le N.-W., sur la nappe sédimentaire.

Myxilla fimbriata (Bowerbank)   *Sertella couchii* (Hincks)

Le Bryozoaire *Porella cervicornis* a été récolté dans la partie la plus occidentale de ce secteur, qui constitue sa limite septentrionale en Europe.

<u>Groupe occidental 2</u>. Espèces dépassant la région de Roscoff, mais sans atteindre l'entrée du Golfe normanno-breton (exemple : Fig. 4 C).

Craniella cranium (Müller)   *Desmacidon fruticosum* (Montagu)
Phakellia ventilabrum (Johnston)

Les espèces suivantes sont préférantes des mêmes conditions :

Axinella flustra Topsent   *Aglaophenia tubulifera* Hincks

<u>Groupe occidental 3</u>. Espèces ne dépassant guère l'entrée occidentale du Golfe normanno-breton (exemple : Fig. 4 D).

Axinella agnata Topsent   *Endectyon delaubenfelsi* Burton
Axinella damicornis (Esper)   *Endectyon teissieri* Cabioch
Axinella egregia (Ridley)   *Diazona violacea* Savigny
Phakellia rugosa (Bowerbank)

<u>Groupe occidental 4</u>. Espèces pénétrant plus ou moins largement dans la Manche centrale et dans une partie du Golfe normanno-breton, mais ne dépassant pas vers l'Est le niveau de la Baie de Seine (exemple : Fig. 4 E).

Axinella dissimilis (Bowerbank)   *Sertularia distans* (Lamouroux)
Adreus fascicularis (Bowerbank)   *Pista maculata* (Dalyell)
Eurypon major Sarà & Siribelli   *Distomus variolosus* Gaertner
Chelonaplysilla noevus (Carter)   *Stolonica socialis* Hartmeyer

La plupart de ces espèces se répandent largement entre le Cotentin et la région de l'Ile de Wight; certaines d'entre elles ont essentiellement été rencontrées du côté français de ce resserrement de la Manche. Ce sont : *Axinella dissimilis* (Fig. 4 E) et surtout *Adreus fascicularis* et *Stolonica socialis*. Enfin, d'autres espèces peuvent être classées comme préférantes du groupe occidental 4. Telles sont:

Haliclona viscosa (Topsent)   *Arca lactea* Linné
Corynactis viridis Allman   *Chlamys distorta* (da Costa)
Pentapora foliacea (Ellis & Solander)

<u>Groupe occidental 5</u>. Espèces pénétrant en Manche orientale jusqu'au Nord du Pays de Caux, mais n'atteignant pas, ou exceptionnellement, le Pas-de-Calais (Exemple : Fig. 4 F).

Halicnemia patera Bowerbank   *Sertularella ellisi* (Milne-Edwards)
Hymedesmia versicolor (Topsent)   *Polycarpa mamillaris* (Gaertner)
Iophonopsis nigricans (Bowerbank)   *Pyura microcosmus* (Savigny)
  Pyura tessellata (Forbes)

Les espèces suivantes sont préférantes de cette distribution :

Plumularia setacea (Linné)   *Microcosmus claudicans* (Savigny)

Groupe oriental. Le Bivalve boréo-arctique *Modiolus modiolus* (Linné) est le représentant le plus caractéristique de ce groupe. Assez commun sur les fonds caillouteux de la Manche orientale et centrale, il forme par endroits de véritables moulières. Ces formations denses, classiques dans les mers nordiques, ont été rencontrées en certains points du Pas-de-Calais, ainsi qu'au Nord et à l'Ouest de l'extrémité du Cotentin. L'espèce est présente dans le Golfe normanno-breton, mais ne le dépasse guère vers l'Ouest; sa récolte est très exceptionnelle dans la région de Roscoff. Seules quelques autres espèces présentent une distribution comparable, bien que moins nette; gagnant plus à l'Ouest que *Modiolus modiolus*, elles sont communes en général au Nord-Est de Roscoff, mais se raréfient vers l'entrée atlantique de la Manche. Telles sont :

Haliclona oculata (Pallas)
Flustra foliacea (Linné)
Balanus crenatus Bruguière
Ascidia mentula Müller
Botrylloides leachi (Savigny)

On peut également citer, en tant qu'espèce préférante des mêmes conditions, l'Actinie *Tealia felina* (Linné).

Groupe des espèces limitées à la fois vers l'Ouest et vers l'Est. La plupart des espèces de ce groupe se raréfient à la fois vers l'Ouest de la région de Roscoff et à l'Est de la Manche centrale ou du Pays de Caux. Elles occupent ainsi les fonds de forts courants dans la partie médiane de la Manche. Les espèces suivantes présentent une telle distribution :

Terpios fugax Duchassaing & Michelotti
Stelligera rigida (Montagu)
Amphisbetia operculata * (Linné)
*Modiolus adriaticus** Lamarck
*Modiolus barbatus** (Linné)
*Musculus discors** (Linné)

(Les espèces marquées d'un astérisque ne dépassent pas vers l'Est le niveau de la Baie de Seine, ou sont très peu communes au-delà).

Distributions Particulières :

Un nombre limité d'espèces parmi celles dont la répartition a été établie, échappent à la classification précédente. *Terebratulina caput-serpentis* (Linné) et *Caberea ellisii* (Fleming) occupent des aires circonscrites, au large des côtes nord-occidentales de la Bretagne (Cabioch, 1968); elles n'ont pas été retrouvées ailleurs. *Vesicularia spinosa* (Linné) présente une distribution fragmentée, dans la plupart des régions de la Manche.

DISCUSSION

Le classement écologique des principales espèces de l'épifaune sessile des fonds grossiers circalittoraux qui vient d'être proposé, aussi frappantes que paraissent les analogies entre les distributions comparées des espèces et des paramètres climatiques et édaphiques les mieux connus, ne doit être considéré que comme une interprétation provisoire, dans l'attente d'un traitement plus complet des données faunistiques et écologiques. Beaucoup reste encore à faire pour que soit acquise une connaissance d'ensemble des facteurs abiotiques et biotiques étendue à toute la Manche. Nous n'avons actuellement à notre disposition que les données générales concernant la profondeur, la température et la salinité en surface et la vitesse des courants de marée; nous connaissons d'autre part la nature du substrat en chacun des points de prélèvement. On remarquera toutefois que ce sont des paramètres généralement considérés comme essentiels. Certaines de ces données doivent néanmoins être utilisés avec précaution. Il est bien connu, par exemple, que les eaux de la Manche sont homothermes en hiver de la surface au fond, si bien que les courbes isothermes de surface représentent également les températures au niveau du substrat. Mais cette situation ne se maintient en été que dans la mesure où les courants de marée sont suffisamment forts pour empêcher l'établissement d'une stratification thermique. C'est le cas dans la majeure partie de la Manche, sauf dans son entrée nord-occidentale, où les températures en surface et au fond diffèrent sensiblement (Dietrich, 1950). Ces réserves étant faites, nous avons vu que l'effet de la profondeur intervient surtout par l'intermédiaire de la pénétration de la lumière, dont le rôle, essentiel dans la délimitation de l'étage infralittoral, s'efface rapidement dans l'étage circalittoral. De nettes similitudes apparaissent alors entre les distributions faunistiques et les conditions thermohalines d'une part, hydrodynamiques et sédimentaires d'autre part. On notera cependant, au sujet des conditions édaphiques, que des espèces limitées, sur la nappe sédimentaire, aux fonds cailloutteux, peuvent être rencontrées ailleurs, sur des pointements rocheux. Il conviendra en outre de faire la part, quand elle sera mieux connue, du rôle de la turbidité, causée notamment par les fortes actions hydrodynamiques. Il ne faudra pas non plus négliger l'influence du transport des larves par les courants périodiques ou généraux, ni celle de la pollution.

Les types de distribution que nous avons établis pour l'épifaune sessile rejoignent dans une large mesure les données publiées par Holme (1966), concernant l'endofaune et l'épifaune vagile. Dans l'un et l'autre cas, on distingue nettement, à côté de nombreuses espèces présentes dans toute la Manche, des espèces occidentales s'opposant dans leur distribution à des espèces orientales beaucoup moins nombreuses. Les distributions édapho-climatiques limitées vers l'Est et vers l'Ouest, que nous avons rencontrées, généralisent la notion d'espèce "sarnienne" introduite et commentée par Holme. Le bilan total se traduit par une forte diminution de la diversité de la faune benthique d'Ouest en Est dans la Manche, au fur et à mesure que l'on passe des conditions océaniques sténothermes vers des conditions climatiques de plus en plus continentales. Plusieurs limites d'espèces occidentales se concentrent vers la ligne des 80 m, surtout dans la région de Roscoff. Cette particularité avait justifié la caractérisation d'un horizon circalittoral côtier et d'un horizon circalittoral du large, ayant pour frontière commune cette convergence de limites faunistiques (Cabioch, 1968). Si les phénomènes, vus sur l'ensemble de la Manche, paraissent plus graduels, il n'en reste pas moins que la distinction entre un circalittoral côtier et un circalittoral du large présente un caractère général sur les plateaux continentaux du Nord-Ouest de l'Europe (Glémarec, 1973).

Nous avons voulu présenter un état de la Manche aussi instantané que le permettent les moyens logistiques : les données analysées correspondent à des échantillonnages réalisés entre 1972 et 1976, sauf dans la région de Roscoff, dont l'étude a été effectuée entre 1959 et 1968, mais avec cependant de nombreuses observations ultérieures, qui ont témoigné de la stabilité de la plupart des distributions. La comparaison de cet état actuel avec les données anciennes, concentrées pour la plupart au voisinage immédiat des laboratoires maritimes, est délicate, surtout dans le domaine de l'épifaune sessile, et il serait hasardeux d'en tirer des conclusions très assurées. L'exploration totale de la Manche montre en effet que l'on peut récolter exceptionnellement, loin de leur aire de présence normale, des exemplaires isolés d'espèces à distribution limitée. On constate néanmoins qu'il n'y a pas de divergence majeure entre les distributions climatiques que nous avons relevées et les données des inventaires faunistiques des laboratoires riverains de la Manche. La position très occidentale de certaines espèces, telles que *Porella compressa*, concorde bien, par exemple avec les observations de Crawshay (1912). On trouvera dans la publication de Holme (1966) une discussion sur quelques fluctuations constatées depuis le début du siècle dans l'abondance ou la distribution d'espèces de l'endofaune.

Enfin, l'interprétation des distributions des espèces benthiques en Manche prendra tout son relief à la lumière d'une connaissance de leur répartition biogéographique sur les fonds marins de l'Atlantique nord-oriental. Une compilation des données acquises sera nécessaire pour y parvenir; les difficultés de la systématique et le caractère lacunaire de beaucoup d'observations concernant l'épifaune sessile ne faciliteront pas la tâche. On peut toutefois, dès à présent, sur un certain nombre d'exemples bien connus, distinguer quelques articulations essentielles entre les distributions locales en Manche et la biogéographie générale :

a) Parmi les espèces de l'épifaune sessile présentant des distributions climatiques ou édapho-climatiques occidentales, on remarque :

      - des espèces largement répandues en Europe vers le Nord ou vers le Sud; nous les qualifierons, par rapport à la Manche, d'espèces océaniques. *Diphasia pinaster*, *Aglaophenia tubulifera*, *Caryophyllia smithi* en sont des exemples. Cette dernière espèce contourne l'Ecosse, pour ne pénétrer, comme en Manche, que d'une manière limitée en Mer du Nord (Wilson, 1975).

      - des espèces de la faune tempérée chaude (ou "lusitanienne") qui semblent atteindre dans la Manche une de leurs limites vers les mers boréales. Telles sont : *Porella cervicornis*, *Diphasia pinnata* ou *Vermiliopsis infundibulum*.

      - des espèces de la faune boréale ou boréo-arctique, qui ne supportent sans doute pas de fortes variations climatiques annuelles. Nous les qualifierons d'espèces boréo-océaniques. *Porella compressa* et *Phakellia rugosa* en sont des exemples.

b) Parmi les espèces présentant des distributions orientales, on note la fréquence d'espèces de la faune boréale ou boréo-arctique. Telles sont *Modiolus modiolus* et *Flustra foliacea*.

CONCLUSION

La Manche est l'une des voies de passage des régions tempérées chaudes aux régions boréales du plateau continental européen. Elle présente la particularité d'être marquée par un gradient climatique intense, d'Ouest en Est, conduisant depuis des eaux océaniques à faibles variations thermohalines annuelles, vers des eaux de caractère nettement plus continental.

Dans l'état actuel des connaissances, on peut proposer une interprétation provisoire de la distribution de la faune benthique de la Manche, sur l'exemple des peuplements des fonds grossiers circalittoraux. Au gradient climatique répond un gradient de peuplement : de nombreuses espèces sont limitées vers l'Est dans leur pénétration en Manche, tandis que d'autres, bien moins nombreuses, présentent une distribution orientale. Quelques espèces sont limitées à la fois vers l'Est et vers l'Ouest. Le bilan de ces distributions climatiques se traduit par une réduction importante, d'Ouest en Est, de la diversité de la faune benthique. Le groupe occidental comprend à la fois des espèces océaniques largement réparties en latitude, vers le Nord et vers le Sud par rapport à la Manche, des espèces de la faune tempérée chaude, mais aussi des espèces boréo-océaniques. Par contre, d'autres espèces boréales ou boréo-arctiques présentent une distribution orientale. Des gradients édaphiques très accusés recoupent le système climatique. Ils se traduisent par une réduction de la diversité de l'épifaune sessile depuis les fonds caillouteux, soumis à de forts courants, vers les fonds de plus en plus ensablés.

Carrefour biogéographique d'un grand intérêt, la Manche rassemble ainsi sur une étendue relativement restreinte, des séries de conditions écologiques plus largement étalées ou dispersées dans les autres mers épicontinentales du Nord-Ouest de l'Europe.

REMERCIEMENTS

La prospection benthique de la Manche a été effectuée dans le cadre d'une Recherche Coopérative sur Programme du Centre National de la Recherche Scientifique et avec l'aide de contrats du Centre National pour l'Exploitation des Océans, dans le cadre des études entreprises par cet organisme sur les effets éventuels de l'extraction des sables et graviers sur l'environnement marin. Les auteurs témoignent leur reconnaissance aux commandants et aux équipages du "Pluteus II" et de la "Thalassa" pour leur amicale collaboration lors des opérations à la mer. Ils remercient les autorités britanniques, qui ont bien voulu les autoriser à étendre leurs recherches dans les eaux territoriales de l'Angleterre et des Iles anglo-normandes.

REFERENCES

AUFFRET, J.P., HOMMERIL, P. et LARSONNEUR, C. 1975 La mer de la Manche, modèle de bassin sédimentaire épicontinental sous climat tempéré. *9è Congrès intern. Sédimentologie, Nice,* 25-31.

BOILLOT, G. 1964 Géologie de la Manche occidentale. *Ann. Inst. Océanogr. Monaco* 42, 1-220.

BOUTLER, J., CABIOCH, L. et GRALL, J.R. 1974 Quelques observations sur la pénétration de la lumière dans les eaux marines au voisinage de Roscoff et ses conséquences écologiques. *Bull. Soc. phyc. Fr.* 19, 129-140.

CABIOCH, L 1968 Contribution à la connaissance des peuplements benthiques de la Manche occidentale. *Cah. Biol. mar.* 9, 493-720.

CABIOCH, L. et GENTIL, F. 1975 Distribution des peuplements benthiques dans la partie orientale de la Baie de Seine. *C.R. Acad. Sci. Paris* 280, 571-574.

CABIOCH, L. et GLACON, R. 1975 Distribution des peuplements benthiques en Manche orientale de la Baie de Somme au Pas-de-Calais. *C.R. Acad. Sci. Paris* 280, 491-494.

CRAWSHAY, L.R. 1912 On the fauna of the outer western area of the English Channel. *J. mar. biol. Ass. U.K.* 9, 292-393.

CRISP, D.J. & SOUTHWARD, A.J. 1958 The distribution of intertidal organisms along the coasts of the English Channel. *J. mar. biol. Ass. U.K.* 37, 157-208.

DIETRICH, G. 1950 Die animale Jahresschwankungen des Warmeinhältnis im Englischen Kanal, ihre Ursachen und Auswirkungen. *Deutsche hydrogr. Z.* 3, 184-201.

DIZERBO, A.H. 1969 Les limites géographiques de quelques algues marines du Massif armoricain. *Proc. 6th internat. Seaweed Symp., Santiago de Compostela,* 141-149.

GENTIL, F. 1976 Distribution des peuplements benthiques en Baie de Seine. *Thèse 3è Cycle Océanogr. biol. Paris.*

GLEMAREC, M. 1973 The benthic communities of the European north Atlantic continental shelf. *Oceanogr. mar. Biol. Ann. Rev.* 11, 262-289.

HOLME, N.A. 1966 The bottom fauna of the English Channel. *J. mar. biol. Ass. U.K.* 41, 397-461.

HOLME, N.A. 1966 The bottom faune of the English Channel. II. *J. mar. biol. Ass. U.K.* 46, 401-493.

HOMMERIL, P. 1967 Etude de géologie marine concernant le littoral bas-normand et la zone prélittorale de l'archipel anglo-normand. *Thèse Doc., Caen.*

LARSONNEUR, C. 1971 Manche centrale et Baie de Seine : géologie du substratum et des dépôts meubles. *Thèse Doc., Caen.*

LUMBY, J.R. 1935 Salinity and temperature of the English Channel. *Fish. Invest.* (2) 14, 1-67.

PINGREE, R., PENNYCUICK, L. et BATTIN, G.A.W. 1975 A time-warying temperature model of mixing in the English Channel. *J. mar. biol. Ass. U.K.,* 55, 975-992.

RETIERE, C. 1975 Distribution des peuplements benthiques des fonds meubles du golfe normanno-breton. *C.R. Acad. Sci. Paris* 280, 697-699.

RUELLAN, F., BEIGBEDER, Y. et DAGORNE, A. 1967 Répartition des fonds sédimentaires dans la partie méridionale du golfe normanno-breton (au sud du parallèle de 48°46'48" - 54 G 20'). *C.R. Acad. Sci., Paris* 264, 1580-1583.

SAGER, G. 1963 *Atlas der Elemente des Tidenhubs und der Gezeitenströme für die Nordsee den Kanal und die Irische See.* Rostock.

SERVICE HYDROGRAPHIQUE ET OCEANOGRAPHIQUE DE LA MARINE Ouvrages n° 550, 553 et 556. Paris.

WILSON, J.B. 1975 The distribution of the coral *Caryophyllia smithii* S. & B. on the Scottish continental shelf. *J. mar. biol. Ass. U.K.* 55, 611-625.

COLONISATION ET DISTRIBUTION SPATIALE DES COPEPODES DANS DES LAGUNES SEMI-ARTIFICIELLES

Jacques Castel et Pierre Lasserre

Institut de Biologie Marine, Université de Bordeaux I, 33120 Arcachon, France

ABSTRACT

Benthic and periphytal copepods constitute a dominant group of species uniquely adapted to a very highly stressed system of semi-artificial lagoons, in the Arcachon basin, France. Highly productive, the pools are utilized for extensive acquaculture. These very shallow impoundments (0.4 - 2m depth) are subjected to varying salinity and they fall into the polyhaline, mesohaline and oligohaline subdivisions (Venice System). Monthly sampling, reveals low diversity - high density communities with great fluctuations in numerical abundance. In spring and summer (April-July), very high densities ($>10^6$ ind/m^2) have been found. Both benthic and periphytal copepods are much more abundant than planktonic forms. Maximum reproductive activity differs temporarily for the dominant species, so that they avoid competition for space. Furthermore, most of the successful species are very tolerant (examples are given for temperature and salinity), and they display a good resistance to the eutrophic conditions which prevail in late summer (August).

A different spatial pattern has been found for benthic deposits as compared to Chaetomorpha belts (bottom), Cladophora (air-water interface) and Ruppia and, therefore, gives some idea of the spatial heterogeneity of the habitat. Nineteen species of copepods inhabit benthic deposits of detritus, while twenty species are found in beds of filamentous algae, eleven of them being clearly ubiquitous. It is concluded that their success in colonizing stressed areas, at different levels of shallow lagoonal interfaces, depends notably on their large capability for ecophysiological adaptation allowing occupation of quite different ecological niches.

INTRODUCTION

Les caractéristiques adaptatives mises en évidence chez de nombreuses espèces meiofauniques littorales (revue in Lasserre, 1976 a), confèrent à ces communautés de micro-métazoaires des avantages compétitifs marqués pour coloniser les écotones estuariens, plus spécialement lagunaires (Lasserre, 1976 b). Récemment, Lasserre et al., 1976 ont mis en évidence dans les lagunes endiguées eutrophes du Bassin d'Arcachon, utilisées pour l'aquaculture, une compétition entre deux constituants importants de ces étangs mixohalins, la meiofaune benthique d'une part, les poissons mugilidés d'autre part, tous deux utilisateurs du tapis de débris végétaux et de ses productions associées (microflore benthique, bactéries). La très forte activité oxydative biologique de la couche détritique superficielle est due, pour une grande part, aux besoins respiratoires des communautés meiofauniques composées pour l'essentiel de copépodes et de nématodes en densités très élevées (Lasserre et al., 1976). Dans ces étangs peu profonds (40 cm à 2 m) et fondamentalement instables, les relations entre les interfaces air-eau-végétation-sédiment sont permanentes et intenses. Une étude de la distribution spatiale des copépodes et de ses variations saisonnières nous a parue exemplaire pour une analyse plus générale des interactions qui lient les différentes interfaces dans ces systèmes lagunaires.

MATERIEL ET METHODES

Les lagunes endiguées situées dans la partie est du Bassin d'Arcachon (les "réservoirs à poissons" de Certes), sont écologiquement bien définies (Amanieu, 1967; Lasserre et Gallis, 1975; Labourg, 1976). Elles sont envahies par une abondante végétation halophyle : Cladophora en surface, Chaetomorpha sur le fond, herbiers de Ruppia à rhizomes persistants (Fig. 5 et 8). Le sédiment réduit en profondeur est recouvert superficiellement par un tapis de détritus végétaux (épaisseur 1-2 cm) à microflore abondante. Certaines parties sont dépourvues de macrophytes mais présentent le même sédiment détritique d'une épaisseur de 1 à 3 cm (Fig. 5, Tableau 5). Trois stations appartenant à des réservoirs poly- et mésohalins (système de Venise) ont été suivies d'avril 1975 à mai 1976. Les régimes de température et de salinité sont résumés dans le Tableau 1.

TABLEAU 1 Variations saisonnières des températures et salinités

lagune		printemps	été	automne	hiver
poly-mésohaline	T	9-11	24-27	11-14	3-6
(D.d.l.)	S	14-25	30-38	25-30	14-22
poly-mésohaline	T	8-11	24-27	12-14	5-7
(D.d.c.)	S	16-25	28-33	22-30	16-22
meso-polyhaline	T	9-12	22-24	10-14	6-8
(D.d.c.)	S	16-18	20-26	20-25	16-20

T, température; S, salinité; D.d.l., domaine détritique libre; D.d.c., domaine détritique couvert.

Récolte et tri.

Le meiobenthos est récolté à l'aide d'un carottier de 2,2 cm de diamètre int., enfoncé de 5 cm dans le sédiment. La colonne d'eau (30-50 cm env.) surmontant le sédiment est éliminée par aspiration. Les animaux sont presque tous distribués dans le 1° cm de sédiment (détritus : Lasserre et al., 1976). On considère une surface de base de 4 cm^2 par prélèvement. Quatre prélèvements répartis sur une aire de 1 m^2 sont effectués deux fois par mois.

Les récoltes de meiofaune phytophile ont lieu chaque semaine dans les trois types de végétation. Dans chaque cas, on prélève 4 échantillons sur une surface de 1 m^2 présentant une couverture végétale homogène. Pour les algues filamenteuses, on utilise une pince à mâchoires concaves (Labourg, non publié). La surface prélevée est de 100 cm^2. Pour les herbiers (Ruppia), nous avons retenu le "phyto-isolateur" de Dejoux et Saint-Jean (1972). La surface prélevée est de 200 cm^2.

Les récoltes de zooplancton sont effectuées dans les parties les plus profondes (1 à 2 m) d'un réservoir poly-mésohalin. Le filet spécialement mis au point (Castel non publié), possède une embouchure rectangulaire (35 x 25 cm), plus large que haute, pour éviter d'échantillonner dans les macrophytes. Deux types de tissus filtants (Tripette et Renaud) sont utilisés : 120 µm pour la récolte d'adultes, 60 µm pour la récolte des stades larvaires. La longueur de la partie

filtrante est de 1 m. La durée du trait est de 5 min. sur 100 m, la tranche
d'eau filtrée est comprise entre - 25 et - 60 cm. Une pompe filtrante autonome a été également employée et donne des résultats comparables. Le volume d'eau
filtré est ramené au m^2 en considérant une profondeur moyenne de 1,5 m.

La meiofaune benthique et périphytale est comptée pour la totalité du prélèvement sauf dans le cas d'échantillons de périphyton très riches, qui font
l'objet d'un sous-échantillonnage comme pour le plancton. Le taux d'efficacité
dans l'extraction et les dénombrements est de 95 % au moins pour les copépodes
meiobenthiques et de 90 % pour les copépodes phytophiles. Les récoltes de
zooplancton sont sous-échantillonnées à trois reprises, le sous-échantillon
étant replacé dans la récolte après comptage. Le calcul de l'erreur au niveau
95 % est au plus égal à 30 %, pour un comptage de 100 individus (Frontier,
1972).

Taille et variabilité de l'échantillonnage.

Nous avons déterminé la taille minimum de l'échantillon et le nombre de prélèvements requis pour avoir une image représentative des communautés. Les récoltes faites dans ce but comprenaient :
- domaine détritique, 16 carottages de 4 cm^2 dans un carré de 1 m de côté;
- domaine phytal, niveau à Chaetomorpha et Cladophora, 10 prélèvements de
100 cm^2 dans un carré de 1 m de côté; niveau à Ruppia, 10 prélèvements de 200
cm^2 dans un carré de 1 m de côté;
- domaine pélagique, 5 récoltes de 6 m^3.

Les rapports de variance à la moyenne (s^2/m) sont compris entre 1,8 et 65. Ils
indiquent un mode de distribution binomial négatif, significativement différent
d'une distribution de Poisson. Ce mode de répartition en agrégat est signalé
chez de nombreux invertébrés pélagiques et benthiques (Taylor, 1971; Heip, 1975).
Comme le rapport s^2/m décroît linéairement avec la densité (Pielou, 1969),
en accord avec Heip (1975), nous pensons qu'il est illusoire de vouloir évaluer
un degré plus ou moins grand d'agrégation. Le calcul du paramètre binomial
négatif k (Pielou, 1969) et de la pente 1/k de la régression liant s^2/m à m
indiquent que la variabilité statistique et l'erreur relative (30 % au max.)
sont indépendantes de la densité, si cette dernière est suffisamment élevée.
En revanche, l'erreur relative augmentera pour de faibles densités (voir également Frontier, 1972 et Heip, 1975). L'erreur relative diminue vers une valeur
constante pour des densités supérieures à 30.

Pour des comparaisons de moyennes acceptables, les échantillons comptés doivent
donner des moyennes toujours supérieures à 20, 45 et 100 pour les copépodes
meiobenthiques, phytophiles et planctoniques. En pratique, 4 échantillons sont
généralement suffisants pour avoir une idée de l'évolution bi-mensuelle des
populations dominantes. Une normalisation de la distribution (variance indépendante de la moyenne) est obtenue par une transformation logarithmique des
données. Les intervalles de confiance au seuil 95 % ont été calculés pour les
espèces dominantes. Les coefficients figurés dans le Tableau 2, multipliés ou
divisés par les moyennes de N échantillons cumulés donnent des écarts à la
moyenne acceptables pour détecter des différences significatives entre les
milieux et les saisons. Pour une surface suffisamment importante (100 m^2 env.),
de règle dans les stations étudiées, nous considérons que l'agrégation est
suffisamment reproductible et dense pour permettre des comparaisons statistiques
saisonnières et une extrapolation des données à une unité de surface du m^2.

TABLEAU 2   Intervalles de confiance des m pour un seuil 95 %

copépodes du domaine détritique	0,686 m à 1,469 m
copépodes du domaine phytal	0,496 m à 2,017 m
copépodes du domaine pélagique	0,598 m à 1,672 m

RESULTATS
Liste faunistique

	distribution
Calanoides	
* Acartia bifilosa Giesbrecht, 1881	Pl
* Calanipeda aquaedulcis Kritschagin, 1873	Pl
Cyclopoides	
* Halicyclops neglectus Kiefer, 1935	U
* Cyclopina gracilis Claus, 1863	U
* Paracyclopina nana Smirnov, 1935	U
Harpacticoides	
* Canuella perplexa T. & A. Scott, 1893	D
Brianola stebleri (Monard, 1926)	D
* Ectinosoma melaniceps Boeck, 1864	U
* Tachidius discipes Giesbrecht, 1882	U
* Harpacticus littoralis Sars, 1910	U
Tisbe holothuriae Humes, 1957	U
Diarthrodes nobilis (Baird, 1845)	Ph
Dactylopodia tisboides (Claus, 1863)	Ph
Dactylopodia micronyx (Sars, 1905)	Ph
* Paradactylopodia sp.	Ph
Amphiascus sp.	Ph
* Amonardia normani (Brady, 1872)	U
Amphiascopsis sp.	Ph
* Bulbamphiascus inermis (Sewell, 1940)	D
Schizopera sp.	U
Metis ignea Philippi, 1843	Ph
Nitocra typica Boeck, 1864	Ph
* Mesochra lilljeborgi Boeck, 1864	U
* Enhydrosoma gariene Gurney, 1930	D
* Enhydrosoma caeni Raibaut, 1965	D
* Cletocamptus confluens (Schmeil, 1894)	D
* Nannopus palustris Brady, 1880	D
* Heterolaophonte strömi (Baird, 1837)	U
Heterolaophonte phycobates (Monard, 1935)	Ph
Onychocamptus sp.	D

On notera que 19 espèces sont distribuées dans le domaine détritique (D), 20 dans le domaine phytal (Ph), 11 d'entre elles sont ubiquistes (U). Les espèces marquées d'une astérisque sont dominantes à un moment de l'année pour l'un au moins des trois domaines : pélagique, phytal et détritique.

Fluctuations mensuelles au niveau du benthos

Les courbes de fluctuations numériques des trois stations envisagées montrent un maximum estival et un maximum automno-hivernal (Fig. 1 et 2). Les densités sont considérablement plus fortes dans les étangs à dominance polyhaline (maximum 1,6 10^6 ind/m^2 dans le domaine détritique libre: Fig. 2). Les faibles densités du mois de septembre résultent incontestablement de la crise dystrophique (août), on note également la diminution du % de femelles ovigères ou leur disparition (Fig. 1 et 2). Les populations se réinstallent à partir du mois d'octobre.

Les espèces les plus abondantes sont Clotocamptus confluens dans les étangs poly-mésohalins et Mesochra lilljeborgi dans les étangs méso-polyhalins (Fig. 1B et 2A). C. confluens est plus abondant dans les tapis de débris végétaux dépourvus de macrophytes. M. lilljeborgi (espèce à préférence mésohaline: voir Noodt, 1957 et Fig. 3) est surtout abondant en hiver et au printemps; la population estivale est également importante dans la station méso-polyhaline, la salinité se maintenant relativement basse (24-26 °/°°). Enfin, M. lilljeborgi est très largement dominant d'octobre à avril dans le domaine détritique libre de la station poly-mésohaline, alors qu'il est absent quand les détritus sont recouverts par les algues (Fig. 1B).

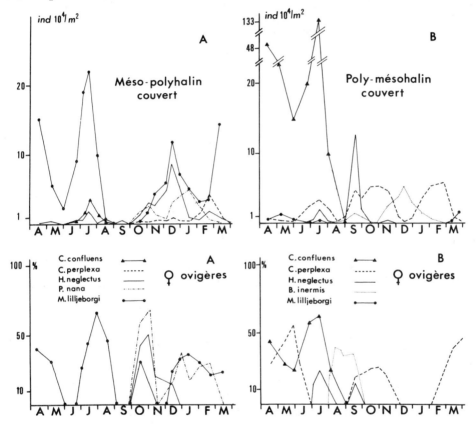

Fig. 1. Fluctuations mensuelles des principaux copépodes et % de femelles ovigères

A, station méso-polyhaline, domaine détritique couvert par Chaetomorpha
B, station poly-mésohaline, domaine détritique couvert par Chaetomorpha

La distribution de quelques espèces non dominantes est intéressante à considérer. Dans les réservoirs polyhalins, Enhydrosoma caeni et E. gariene sont présents en été (Fig. 2A) et Nannopus palustris au printemps, surtout au niveau du domaine détritique libre. Bulbamphiascus inermis (station poly-mésohaline) vit dans les détritus recouverts par Chaetomorpha, c'est une espèce estivale dans les réservoirs (Fig. 1B). Sa présence à la fin de l'automne et au début de l'hiver est accidentelle: on n'observe pas de femelles ovigères à ces périodes. B. inermis provient probablement de la zone intertidale où il est présent presque toute l'année (Lasserre et al., 1976). Halicyclops neglectus est une espèce pérenne dans toutes les stations, mais en densités très variables (Fig. 1 et 2). Ubiquiste dans les réservoirs, elle est habituellement considérée comme planctonique (Dussart, 1969). Cependant, les espèces du genre Halicyclops à affinité benthique ne sont pas rares : H. magniceps (Heip, 1973), H. rotundipes et H. incognitus (Herbst, 1962) entre autres. Nous avons constaté des différences morphologiques entre les individus de H. neglectus récoltés en pleine eau et les spécimens benthiques. Paracyclopina nana a une aire de répartition assez comparable, mais elle est beaucoup moins abondante. Canuella perplexa est une espèce pérenne qui peuple surtout les réservoirs poly-mésohalins (Fig. 1B et 2A). Sa reproduction a lieu principalement au printemps et en été dans ces milieux. Elle ne semble pas se reproduire dans la station méso-polyhaline.

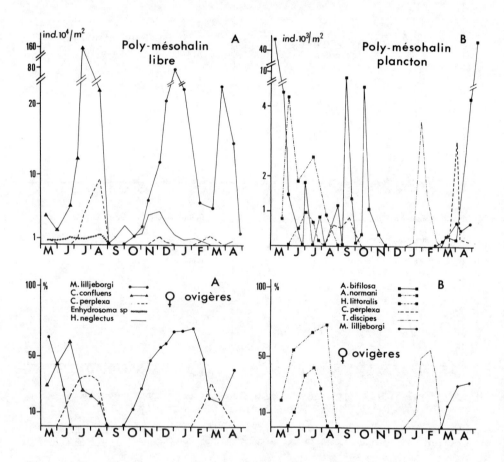

Fig. 2. Fluctuations mensuelles des principaux copépodes et % de femelles ovigères
A, station poly-mésohaline, domaine détritique libre
B, station poly-mésohaline, domaine pélagique

Périodes et potentiels de reproduction

Dans les trois station étudiées, les périodes de reproduction ne sont pas synchrones pour les espèces dominantes (Fig. 1 et 2). D'autre part, on constate des décalages, parfois importants ou des différences d'amplitude d'une station à l'autre. M. lilljeborgi présente deux périodes de reproduction dans la station poly-mésohaline (détritus libre) : une de 6 mois en automne-hiver et une de 3 mois au printemps (Fig. 2A). En revanche, l'espèce ne se reproduit pas dans la station poly-mésohaline à sédiment recouvert par les chaetomorphes (Fig. 1B). Dans les réservoirs méso-polyhalins, on note 4 périodes de reproduction de 2 à 3 mois correspondant aux saisons (Fig. 1A). C. perplexa se reproduit dans le domaine détritique couvert de février à juin et de septembre à novembre (station poly-mésohaline: Fig. 1B). Les période de reproduction sont décalées dans le domaine détritique libre: juillet-août et février-avril (Fig. 2A). Enfin, aucune femelle ovigère n'a été récoltée dans la station meso-polyhaline (Fig. 1A). Chaque espèce réagit donc différemment suivant le type de milieu où elle se trouve. Ce n'est pas seulement l'effet de la température qui induit la reproduction mais l'action conjuguée de plusieurs autres facteurs: la salinité, l'oxygène, la nourriture, le pH etc. (étude en cours).

Afin d'avoir une idée du potentiel de reproduction, nous avons calculé le taux d'accroissement intrinsèque de quelques espèces dans les conditions naturelles. Ce potentiel est estimé pendant la phase ascendante d'une population.

$$r = \frac{1}{t} \ln \frac{N_t}{N_0} \qquad \text{(Fenchel, 1968; Heip, 1972)}$$

t est le nombre de jours séparant les 2 récoltes; N_t et N_0 sont les nombres d'individus aux temps considérés.

Le temps nécessaire pour doubler la taille de la population est :

$$t = \frac{\ln 2}{r}$$

TABLEAU 3 Potentiel de reproduction (r) et temps nécessaire pour doubler la population (t) de Canuella perplexa et Mesochra lilljeborgi dans des conditions naturelles.

périodes	méso-polyhalin D.d.c.*				poly-mésohalin D.d.c.*		poly-mésohalin D.d.l.*			
	C. perplexa		M. lilljeborgi		C. perplexa		C. perplexa		M. lilljeborgi	
	r	t	r	t	r	t	r	t	r	t
printemps	---	---	---	---	---	---	---	---	0,074	9
été	---	---	0,053	13	0,075	9	0,074	9	---	---
automne	---	---	0,041	17	0,042	17	---	---	0,082	9
hiver	0,054	13	0,051	14	0,048	14	0,113	6	---	---

* D. d. c. : domaine détritique couvert, D. d. l. : domaine détritique libre.

Les valeurs du taux d'accroissement intrinsèque des populations de C. perplexa et de M. lilljeborgi sont comprises entre 0,041 et 0, 113/jour. Il faut 6 à 17 jours aux populations pour doubler leur taille (Tableau 3). L'influence d'un facteur particulier est difficile à déceler. Cependant, la moyenne élevée des températures estivales (25-27 °C en juin-juillet) exerce probablement une action favorisante sur la reproduction. Ce phénomène est net pour la station poly-mésohaline, dans le D.d.c. (Tableau 3) : C. perplexa double sa population en 14 et 17 jours en hiver et en automne, 9 jours seulement en été. D'autre part, C. perplexa et M. lilljeborgi se reproduisent plus rapidement dans le D.d.l. que dans le D.d.c. C'est également le cas de C. confluens (r = 0,044/j dans l'interface à Chaetomorpha, r = 0,073/j dans l'interface libre). Ces phénomènes sont à mettre en relation avec la moins forte eutrophisation des stations poly-mésohalines à détritus libre (étude en cours).

Aires thermohalines de reproduction et de tolérance

L'observation du moment d'apparition des premières femelles ovigères permet de déterminer avec une bonne précision la température et la salinité nécessaires à la reproduction des espèces dans le milieu naturel, en tenant compte des durées approximatives d'incubation (2 à 14 j d'après Heip, 1973). La température nécessaire à la reproduction varie avec la salinité chez C. perplexa et H. neglectus. En revanche, M. lilljeborgi est peu influencé dans sa reproduction par la salinité (au moins au-dessus de 20 °/$_{oo}$). D'une manière générale, à salinité égale, les températures nécessaires à la reproduction sont différentes suivant les espèces. Ces faits vont dans le sens de Heip (1973); cet auteur avance l'idée que des périodes de reproduction liées à des températures différentes seraient un moyen de limiter la compétition entre espèces. On peut observer également dans le Tableau 4 que plus la salinité est forte, plus la température permettant la reproduction est élevée.

TABLEAU 4 Température nécessaire à la reproduction de copépodes

espèces	méso-polyhal. D.d.c.*		poly-mésohalin D.d.c.*		poly-mésohalin D.d.l.*		Heip, 1973
	T	S	T	S	T	S	T
T. discipes	---	---	4-6	16	---	---	3,5-6,5
C. perplexa	---	---	8-10	22	8-10	16	9-12
	---	---	20-25	32	17-20	25	
M. lilljeborgi	9-10	20	---	---	13-16	28	10-12
	13-16	21	---	---	---	---	
	22-24	22	---	---	---	---	
H. neglectus	17-19	23	20-25	31	---	---	---
C. confluens	---	---	12-15	18	9-15	18	---
B. inermis	---	---	22-25	31	---	---	---

* Abréviations: voir Tableau 5. Températures (T) données à 2 et 14 j avant l'apparition des premières femelles ovigères; salinités moyennes (S) correspondantes.

Les Figures 3 et 4 indiquent les températures et les salinités où l'on a récolté les 12 espèces dominantes dans des réservoirs poly-, méso- et oligohalins (avril 1973 - mai 1976). Les polygones représentent les limites de température et de salinité pour lesquelles ont été observées des femelles ovigères (sauf pour Acartia bifilosa qui pond ses oeufs isolément). Ils déterminent les aires de reproduction de l'espèce. Les pointillés représentent les limites de tolérance (gros points: valeurs observées, petits points: valeurs extrapolées). Ces limites sont encore plus élevées en élevage (Raibaut, 1967, et résultats non publiés). Ces diagrammes donnent une idée de l'aptitude d'une espèce à coloniser des milieux à salinité variable.

Espèce limitée aux milieux oligo-mésohalins :

Tachidius discipes est la seule espèce de ce groupe à se reproduire uniquement dans ce type de réservoir oligo-mésohalin (Fig. 3).

Espèces colonisant les milieux oligo-, méso et polyhalins :

P. nana, H. neglectus et M. lilljeborgi se reproduisent dans les réservoirs méso- et polyhalins, même pour des températures basses (3 à 7°C). Ces espèces ont le plus grand domaine de tolérance (Fig. 3). Elles colonisent également les domaines pélagique et benthique.

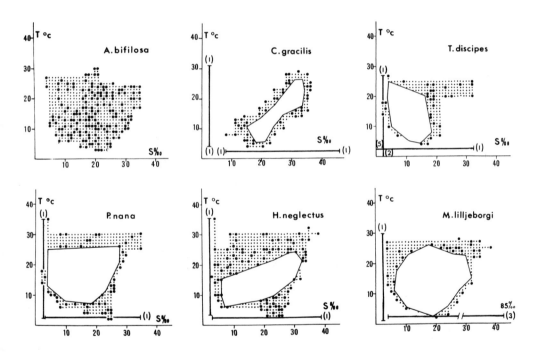

Fig. 3. Aires thermohalines naturelles

Présence de femelles ovigères à l'intérieur des polygones, limites de tolérance en pointillés. Traits parallèles aux coordonnées: limites expérimentales ou naturelles (1), données originales; (2), Lang, 1948; (3), Raibaut, 1967; (4), Raibaut, 1962; (5), Dussart, 1967.

Espèces présentes dans les milieux méso-, poly- et euhalins :

Cyclopina gracilis est la moins euryhaline du groupe. Haute température et basse salinité sont deux facteurs défavorables pour cette espèce qui prospère surtout en automne et à la fin de l'hiver (Fig. 3). La reproduction de Heterolaophonte strömi s'effectue au-dessus de 30 °/$_{oo}$, quelle que soit la température (9-30 °C). Cette espèce peut se reproduire à des salinités inférieures quand la température diminue (Fig. 4). D'origine marine, H. strömi peut coloniser tous les milieux polyhalins. Amonardia normani et Harpacticus littoralis ont à peu près le même domaine de tolérance. Ils prospèrent dans les réservoirs polyhalins et peuvent coloniser des milieux hypersalés (Fig. 4). C. perplexa est moins apte à coloniser ces milieux, mais son mode de reproduction particulier (nauplii pélagiques, adultes benthiques) doit l'avantager.

Espèces colonisant les milieux poly- et euhalins :

C. confluens est bien adapté au milieu polyhalin. Ses limites de tolérance sont beaucoup moins étendues que celles des espèces précédentes (Fig. 4). Enfin, B. inermis est un immigrant temporaire typique. Il provient du milieu marin (sa reproduction s'effectue entre 30 et 35 °/$_{oo}$, Fig. 4).

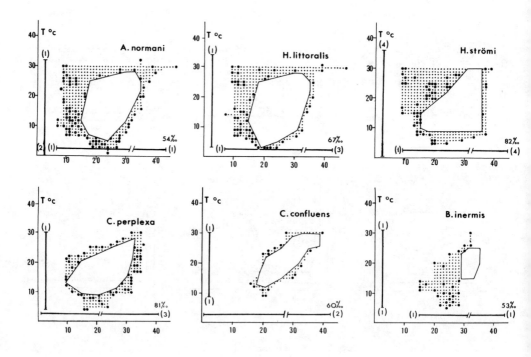

Fig. 4. Aires thermohalines naturelles
Même légende que la Figure 3

Distribution spatiale

Nous distinguerons suivant un plan vertical d'un réservoir type (Fig. 5) les domaines résumés dans le Tableau 5.

TABLEAU 5 Domaines et niveaux verticaux d'un réservoir type*

domaine pélagique (D.p.)	: eau libre sans macrophytes
domaine phytal (D.ph.)	: niveau supérieur à Cladophora (Cl) niveau moyen à Ruppia (R) niveau inférieur à Chaetomorpha (Ch)
domaine détritique (D.d.)	: couvert par Chaetomorpha (D.d.c.) libre (D.d.l.)

*Profondeur moyenne de 50 cm

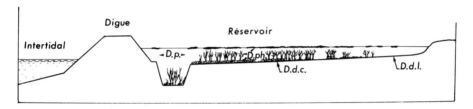

Fig. 5. Profil schématique d'une lagune semi-artificielle
(réservoir à poissons du Bassin d'Arcachon)

Les calanoïdes Acartia bifilosa et Calanipeda aquaedulcis sont les seules espèces véritablement planctoniques. Les harpacticoïdes C. perplexa (adultes) B. inermis, E. caeni, E. gariene, C. confluens et N. palustris sont des formes fouisseuses exclusivement benthiques. On les rencontre fréquemment dans les vases d'estuaires ou dans le sable vaseux (Lang, 1948; Bodin, 1976). Un groupe non négligeable d'espèces (10) colonise tous les domaines; leurs affinités pour un ou plusieurs niveaux sont difficiles à établir.

(1) Mouvements nyctéméraux.

Un échantillonnage régulier (toutes les 2 hr) dans les 3 domaines : pélagique, phytal et détritique, démontre l'existence de mouvements nyctéméraux. Certaines espèces vivant parmi les chaetomorphes auront tendance à remonter vers le niveau supérieur à Cladophora, au milieu de la journée (Amonardia normani et Cyclopina gracilis), tandis que d'autres tendront à s'enfoncer vers les chaetomorphes (Paradactylopodia sp. et Mesochra lilljeborgi).

Il est très probable que ces mouvements de faible amplitude ne correspondent pas à de véritables migrations (observées pour la macrofaune), ils ne dépassent jamais en intensité les variations quantitatives saisonnières. Les valeurs moyennes de l'évolution saisonnière des espèces représentent une image significative des modifications de structure des différentes communautés.

(2) Distribution saisonnière verticale.

Les Figures 6 et 7 donnent une idée de la répartition verticale des différentes espèces de copépodes ainsi que de leurs dominances dans les différents niveaux. Les trois communautés: pélagique, phytale et benthique présentent des fluctuations quantitatives saisonnières parallèles (Fig. 6 et 7). L'importance relative des trois communautés est seule inégale. Au printemps (mars-mai), la biomasse végétale augmente considérablement, permettant à la meiofaune phytophile de se développer (Fig. 6). Une bonne partie de ces espèces se retrouve au niveau benthique (D.d.). Les populations de copépodes benthiques du D.d.c. sont plus abondantes que celles du D.d.l. . Dans le domaine phytal, M. lilljeborgi et C. gracilis sont dominants au niveau des cladophorales; Paradactylopodia sp. est récolté surtout dans le niveau inférieur à Chaetomorpha. En revanche, H. littoralis et A. normani sont répartis dans tout le domaine phytal.

En été (Fig. 6), les copépodes meiobenthiques réussissent particulièrement bien dans le D.d.l., probablement grâce à l'absence de compétition entre meiobenthos et meiofaune phytophile. Amonardia normani, Cyclopina gracilis et Ectinosoma melaniceps sont particulièrement abondants dans les niveaux supérieurs et moyens, tandis que H. littoralis et H. strömi sont dominants au niveau des chaetomorphes. H. neglectus est diversement abondant dans le domaine phytal et le domaine détritique, sans que l'on puisse lui attribuer une préférence.

Toutes les communautés sont très touchées par la crise dystrophique estivale (août, Fig. 6). Les espèces persistant dans les algues et les Ruppia ne présentent plus d'affinité pour un milieu donné. Les espèces inféodées au tapis de débris végétaux (D.d.l.) se maintiennent cependant en nombre relativement élevé (Fig. 2 et 6).

A l'automne, on assiste à une reprise des populations phytophiles et benthiques, si ces dernières sont protégées par les chaetomorphes. Seul A. normani conserve sa répartition verticale, les autres espèces étant à peu près également distribuées dans le domaine phytal (Fig. 7).

D'une manière générale, les densités diminuent en hiver, la biomasse végétale déclinant, la meiofaune phytale a tendance à devenir benthique et à se développer dans les zones sans macrophytes (Fig. 7). Les véritables copépodes benthiques disparaissent quasiment au niveau du D.d.l., ils sont relativement nombreux dans le D.d.c. . En revanche, la faible salinité ambiante permet à M. lilljeborgi de se développer considérablement au niveau benthique et de se maintenir jusqu'au printemps. Cette espèce sera alors dominante parmi les Cladophora et on notera sa présence dans le plancton. Ces différentes colonisations restent inexpliquées. Enfin, C. gracilis est dominant dans les niveaux supérieurs et moyens du domaine phytal, alors que Paradactylopodia sp. et H. littoralis vivent plutôt dans les niveaux moyens et inférieurs (Ruppia et Chaetomorpha).

Les communautés phytophiles et détritiques représentées Figure 8 ne sont pas nettement séparées dans l'espace. Nous les considérons comme des populations liées aux interfaces macrophytes-détritus. Malgré des mouvements importants entre ces différentes interfaces, nous avons pu distinguer des communautés d'espèces préférentiellement distribuées dans les trois domaines: pélagique, phytal et benthique (Fig. 8).

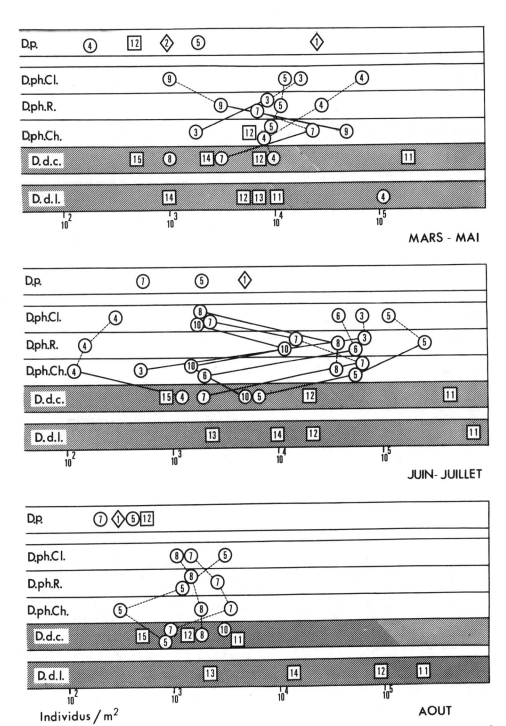

Fig. 6. Distribution verticale des copépodes (légende, voir Fig. 7)

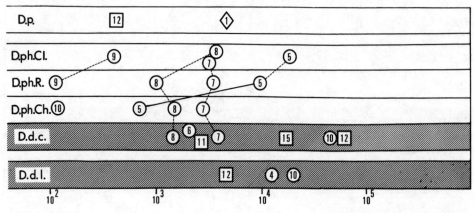

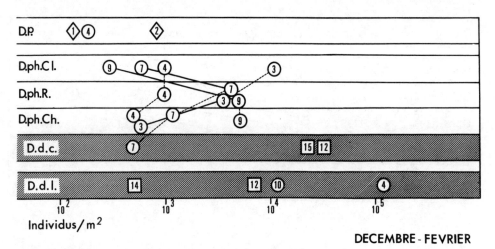

Fig. 7. Distribution verticale des copépodes dans une lagune poly-mésohaline

D.p., domaine pélagique; D.ph.Cl., domaine phytal à Cladophora; D.ph.R., domaine phytal à Ruppia; D.ph.Ch., domaine phytal à Chaetomorpha; D.d.c., domaine détritique couvert; D.d.l., domaine détritique libre.

Les nombres d'individus par m² sont portés sur une échelle logarithmique. Les numéros indiquent les espèces: (1), A. bifilosa; (2), T. discipes; (3), C. gracilis; (4), M. lilljeborgi; (5), A. normani; (6), E. melaniceps; (7), H. littoralis; (8), H. strömi; (9), Paradactylopodia sp.; (10), H. neglectus; (11), C. confluens; (12), C. perplexa; (13), N. palustris; (14), E. caeni et E. gariene; (15), B. inermis.

Losanges : espèces du D.p.; ronds : espèces du D.ph.; carrés : espèces du D.d. Traits pleins : différences significatives; pointillés : diff. non significat.

… 143

DISCUSSION

Les nématodes et les copépodes représentent environ 80 à 90 % des communautés meiobenthiques (Lasserre et al., 1976) et périphytales, dans les lagunes semi-artificielles du Bassin d'Arcachon. Les copépodes sont globalement 10 à 100 fois plus abondants dans les domaines benthique et phytal que dans le domaine pélagique. D'origine lacustre (Calanipeda aquaedulcis) ou marine (Acartia bifilosa), les copépodes zooplanctoniques ne colonisent que sporadiquement les lagunes poly-et mésohalines.

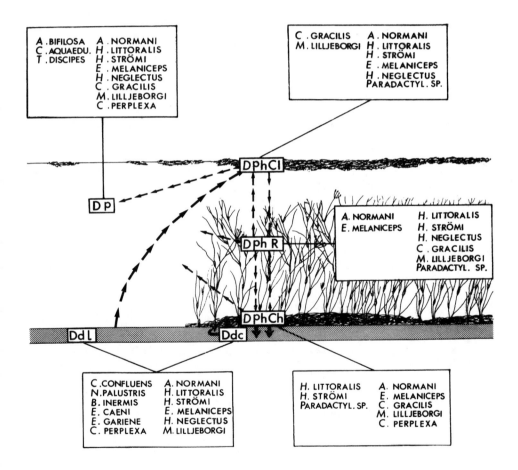

Fig. 8. Schéma récapitulatif de la distribution spatiale des copépodes dans une lagune poly-mésohaline. Les flèches montrent les affinités interfaciales. Mêmes abréviations que Fig. 7.

La Figure 8 résume la distribution spatiale des espèces dominantes ainsi que leurs affinités interfaciales. Ce schéma est établi à partir de l'analyse d'abondance saisonnière des espèces dominantes (Fig. 6 et 7). Nous distinguons trois domaines : pélagique, phytal et détritique. A chaque domaine est associé un groupement d'espèces qui colonisent préférentiellement un niveau (partie gauche de chaque rectangle, Fig. 8). Un groupe de 6 espèces est considéré comme benthique (domaine détritique : D.d.). Huit espèces, au moins, ont une préférence pour le domaine phytal (D.ph.), 2 espèces seulement (A. bifilosa et C. aquaedulcis) étant strictement planctoniques (D.p.). T. discipes, présent dans le plancton, colonise également le benthos mais surtout dans des aires intertidales ouvertes (Lasserre et al., 1976). Les espèces ubiquistes (partie droite des rectangles, Fig. 8) peuvent occuper des niches écologiques très différentes (pélagique, phytale, détritique). M. lilljeborgi vit dans les deux domaines phytal et détritique. Les 15 espèces dominantes possèdent une très vaste répartition géographique. Noodt (1957, 1970) montre que peu d'espèces des milieux saumâtres sont séparées dans l'espace. Il en va de même dans les lagunes semi-artificielles du Bassin d'Arcachon, mis à part le petit groupe d'espèces meiobenthiques (C. confluens, N. palustris, B. inermis, E. caeni, E. gariene, C. perplexa), pouvant essaimer dans l'interface détritus-Chaetomorpha (par exemple C. perplexa).

Les espèces du domaine phytal ont des aires thermohalines naturelles plus larges que les espèces benthiques (Fig. 3 et 4). Il semble que ces dernières soient plus sensibles aux conditions hivernales et qu'elles manifestent des préférences marquées pour un régime thermohalin estival (température et salinité élevées). D'autre part, les populations benthiques estivales persistent pendant les périodes de forte eutrophisation (voir également Lasserre et al., 1976), ce qui indiquerait une bonne adaptation écophysiologique de ces espèces à un milieu pauvrement oxygéné. Cette hypothèse a pu être vérifiée au laboratoire (non publié), elle est également explicitée pour d'autres copépodes par Vernberg et Coull (1975). Il est remarquable que Paraleptastacus spinicauda (T & A Scott), caractérisé comme un sténoxybionte, soit toujours absent des milieux eutrophisés (Lasserre et Renaud-Mornant, 1973).

Les espèces phytophiles, abondantes en hiver et en été (juin-juillet) sont moins adaptées aux basses salinités assorties de températures relativement élevées : périodes de février (7°C) à avril (15°C, Fig. 3 et 4 et Castel, 1976). D'autre part, les déficits périodiques en oxygène (crise dystrophique) entraînent une disparition des populations phytophiles. Les espèces ubiquistes (M. lilljeborgi, H. neglectus, P. nana) ont les aires thermohalines les plus larges, mais elles résistent assez mal à des milieux pauvrement oxygénés, tout en maintenant, cependant, un petit stock de femelles ovigères (Fig. 1 et 2).

Les potentiels de reproduction sont généralement élevés (r = 0,041 à 0,113/j) et s'apparentent à ceux de la littérature (Heip, 1972; Heip et Smol, 1976; Volkmann-Rocco et Battaglia, 1972). Coull et Vernberg (1975) ont souligné qu'une reproduction continue des copépodes meiobenthiques était observée plutôt exceptionnellement, même chez les espèces à large répartition géographique. Dans la plupart des cas, la reproduction est cyclique. Nos observations vont dans le même sens. Tous les copépodes distribués dans les lagunes semi-artificielles du Bassin d'Arcachon ont une ou plusieurs périodes de reproduction. Celles-ci ne sont pas les mêmes suivant les profils mixohalins des réservoirs. Le décalage des cycles dans le temps est souvent assorti de différences dans la durée et le nombre des périodes reproductrices, phénomène également noté par Heip (1973). Ces faits tendent à démontrer une absence de compétition entre espèces dominantes et une influence directe du milieu sur leurs cycles reproducteurs.

REMERCIEMENTS

Ce travail a bénéficié du concours financier du Centre National pour l'Exploitation des Océans (CNEXO), contrat "Ecotron" n° 76/5311 à P.L. Nous remercions Brigitte Volkmann (C.N.R., Venise) et Philippe Bodin (C.N.R.S., Brest) d'avoir bien voulu confirmer certaines déterminations. Notre reconnaissance va également à M. Jean Boisseau et Mme Jeanne Renaud-Mornant pour leurs conseils.

REFERENCES

Amanieu, M. 1967 Introduction à l'étude écologique des réservoirs à poissons de la région d'Arcachon. Vie Milieu 18 (2B), 79-94.

Bodin, P. 1976 Les copépodes harpacticoïdes (Crustacea) des côtes charentaises (Atlantique). Données écologiques et biologiques sur les principales espèces. Bull. Mus. Nat. Hist. Nat. Paris (3° sér.) 363, 1-45.

Castel, J. 1976 Développement larvaire et biologie de Harpacticus littoralis Sars, 1910 (Copepode, Harpacticoïde) dans les étangs saumâtres de la région d'Arcachon. Cah. Biol. Mar. 17, 195-212.

Coull, B.C. et Vernberg W.B. 1975 Reproductive periodicity of meiobenthic copepods: seasonal or continuous ? Mar. Biol. 32, 289-293.

Dejoux, C. et Saint-Jean, L. 1972 Etude des communautés d'Invertébrés d'herbiers du lac Tchad. Recherches préliminaires. Cah. ORSTOM, sér. Hydrobiol. 6, 67-83.

Dussart, B. 1967 Les Copépodes des eaux continentales. I. Calanoïdes et Harpacticoïdes. Boubée, Paris, 500 pp.

Dussart, B. 1969 Les Copépodes des eaux continentales. II. Cyclopoïdes et Biologie. Boubée, Paris, 292 pp.

Fenchel. T. 1968 The ecology of marine microbenthos. III. The reproductive potential of ciliates. Ophelia 5, 123-136.

Frontier, S. 1972 Calcul de l'erreur sur un comptage de zooplancton. J. exp. mar. biol. ecol. 8, 121-132.

Heip, C. 1972 The reproductive potential of copepods in brackish water. Mar. Biol. 12, 219-221.

Heip, C. 1973 Partitioning of a brackish water habitat by copepod species. Hydrobiologia 41, 189-198

Heip, C. 1975 On the significance of aggregation in some benthic marine invertebrates. Proc. 9th Europ. Mar. Biol. Symp. (H. Barnes, ed), pp. 527-538, Aberdeen University Press, Aberdeen.

Heip, C. et Smol, N. 1976 Influence of temperature on the reproductive potential of two brackish water harpacticoids (Crustacea: Copepoda). Mar. Biol. 35, 327-334.

Herbst, H.V. 1962 Marine Cyclopoida Gnathostoma (Copepoda) von der Bretagne-Küste als Kommensalen von Polychaeten. Crustaceana 4, 191-206.

Labourg, P.J. 1976 Les réservoirs à poissons du Bassin d'Arcachon et l'élevage extensif de poissons euryhalins (muges, anguilles, bars, daurades). La Pisciculture française 45, 35-52.

Lasserre, P. 1976 a Metabolic activities of benthic microfauna and meiofauna. Recent advances and review of suitable methods of analysis. In: The Benthic Boundary Layer (I.N. McCave, ed), pp. 65-142. Plenum, New York.

Lasserre, P. 1976 b Osmoregulatory responses to estuarine conditions : chronic osmotic stress and competition. In: Estuarine Research (M.L. Wiley, ed.), Academic Press, New York (sous presse).

Lasserre, P. et Gallis, J.L. 1975 Osmoregulation and differential penetration of two grey mullets, Chelon labrosus (Risso) and Liza ramada (Risso) in estuarine fish ponds. Aquaculture 5, 323-344.

Lasserre, P. et Renaud-Mornant, J. 1973 Resistance and respiratory physiology of intertidal meiofauna to oxygen deficiency. Neth. J. Sea Res. 7, 290-302.

Lasserre, P., Renaud-Mornant, J. et Castel, J. 1976 Metabolic activities of meiofaunal communities in a semi-enclosed lagoon. Possibilities of trophic competition between meiofauna and mugilid fish. Proc. 10th Europ. Sympos. Mar. Biol. vol. 2 (G. Persoone et E. Jaspers, eds), pp. 393-414. Universa Press, Wetteren, Belgium.

Lang, K. 1948 Monographie der Harpacticiden. I, II. Hakan Ohlsson, Lund, 1682 p

Noodt, W. 1957 Zur Ökologie der Harpacticoidea (Crust. Cop.) des Eulitorals der Deutschen Meeresküste und der angrenzenden Brackgewässer. Zeitschr. Morphol. Ökol. 46, 149-242.

Noodt, 1970 Zur Ökologie der Copepoda Harpacticoidea des Küstengebietes von Tvärmine (Finnland). Acta Zool. Fennica 128, 1-35.

Pielou, E.C. 1969 An introduction to mathematical ecology. John Wiley, New York, 286 pp.

Raibaut, A. 1962 Les harpacticoïdes (Copepoda) de l'étang des Eaux Blanches et la Crique de l'Angle. Naturalia Monspeliensa, sér. Zool. 3, 87-99.

Raibaut, A. 1967 Recherches écologiques sur le copépodes harpacticoïdes des étangs côtiers et des eaux saumâtres temporaires du Languedoc et de Camargue. Bull. Soc. Zool. France 92, 557-572.

Taylor, L.R. 1961 Aggregation, variance and the mean. Nature 189, 732-735.

Vernberg, W.B. et Coull, B.C. 1975 Multiple factor-effects of environmental parameters on the physiology, ecology and distribution of some marine meiofauna. Cah. Biol. Mar. 16, 721-732.

Volkmann-Rocco, B. et Battaglia, B. 1972 A new case of sibling species in the genus Tisbe (Copepoda, Harpacticoida). In : Fifth Europ. Mar. Biol. Symp. (Battaglia B., ed), pp. 67-80. Piccin, Padova.

RECRUTEMENT ET SUCCESSION DU BENTHOS ROCHEUX SUBLITTORAL

Annie Castric

Laboratoire de Biologie Marine - 29110 Concarneau
Université de Bretagne Occidentale - Laboratoire d'Océanographie Biologique -
29283 Brest Cédex - France

ABSTRACT

In order to study recruitment and species succession in the sublittoral rocky benthos, periodic observations were made on new substrates immersed in the sea. Work was carried out over a period of 26 months, in the vicinity of Glenan Island, 9 miles from the coast. Experimental plates, presenting 3 kinds of slope, were positioned at a depth of 13 meters.
The study focused on groups of sessile fauna and flora whose identification is possible even in the young stages. More than 187 animal and 48 plant species were screened. A calendar of species settlement, related to seasonal thermal variations has been established for 2 annual cycles.

Where species succession is concerned, the various types of "behaviour" differentiated between annual, pioneering and climactic species. Synecological synthesis shows that a state of "qualitative climax" was reached at the end of the 26 months of the study. This foreshadows the true climax, which takes longer to develop.

INTRODUCTION

Après avoir décrit les épibioses des fonds rocheux infra- et circalittoraux observés en plongée aux Glénan - Sud-Finistère (Castric-Fey, 1973, Castric-Fey, 1974), il s'est avéré intéressant de comprendre comment s'édifient de telles épibioses. Les facteurs essentiels de ce milieu sont lumière, agitation et turbidité. Dans le problème de la colonisation, à ces derniers s'ajoute le facteur temps sous ses 2 formes : durée absolue et qualité thermique du temps d'immersion. Seul cet aspect est étudié ici. La technique des plaques-tests immergées est une des meilleures solutions pour le but proposé. Elle permet en outre de préciser les exigences de quelques espèces vis à vis des facteurs essentiels. L'expérience réalisée aux Glénan a eu lieu en mer ouverte, dans l'infralittoral et porte sur 26 mois.

MATERIEL ET METHODES

L'archipel de Glénan, situé à 9 milles de la côte, dans le Sud de Concarneau (47°43' N, 3°60' W) ne subit que peu ou pas d'influence terrestre. Les facteurs physiques les plus importants y sont l'hydrodynamisme, la turbidité incluant la lumière, et la température (variant annuellement de 8° à 17°5).

Les plaques-tests sont constituées d'ardoise non polie, de 30 x 20 x 1 cm. Elles sont amarrées sur des châssis présentant 3 inclinaisons : verticale, oblique et horizontale. L'examen recto-verso des plaques procure 6 surfaces différentes. Les chassis, convenablement lestés, reposent sur fond rocheux, vers 13-15 m. Immersion et collecte des plaques ont été réalisées en plongée, à raison d'une fois par mois. Examens *"in vivo"* et photographies sont effectués au laboratoire.

Protocole expérimental : les opérations comprennent 3 parties :
- expérience a : examen de la séquence temporelle de colonisation à partir d'une date donnée. Immersion d'un stock de plaques dont quelques-unes sont relevées chaque mois, et dont la dernière a 26 mois d'immersion.
- expérience b : examen de l'influence de la saison d'immersion sur les séquences temporelles = immersion de 3 autres stocks de plaques aux 3 autres saisons.
- expérience c : examen des époques de fixation des diverses espèces : immersion chaque mois d'une série de 6 plaques "mensuelles", c'est-à-dire ne restant qu'un mois dans l'eau.

RESULTATS

L'influence des facteurs écologiques intervenant dans le recrutement des espèces est étudiée, mais seuls sont exposés ici les résultats concernant la dynamique de la succession.
Etant donné la durée de l'étude - 26 mois - la succession n'est qu'appréhendée par l'isolement de groupes écologiques et par la définition pour certaines espèces de tendances climaciques.
Il est procédé à l'analyse de l'abondance de chaque espèce, sur l'ensemble des plaques de même inclinaison mises en jeu dans les expériences a) b) et c). L'espèce est donc étudiée en fonction du temps selon ses deux composantes : durée absolue et qualité thermique, grâce à l'abaque Fig. 1.

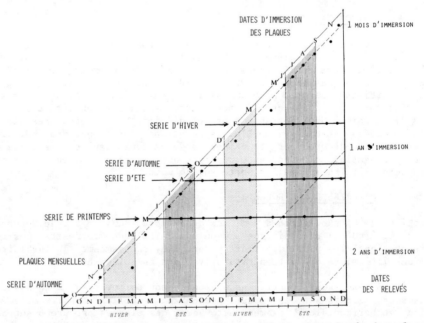

Fig. 1. Abaque avec en abscisse les dates de relevés des plaques et en ordonnée leurs dates d'immersion.

Sur cette abaque, l'abondance de l'espèce est représentée par des trames de densité croissante. Les diagrammes ainsi obtenus peuvent se lire dans les 3 sens : 1) sens vertical : évènements survenant à un instant donné sur les différentes séries de plaques. 2) sens horizontal : devenir des espèces une fois fixées. 3) axe oblique : comparaison entre des plaques ayant séjourné le même temps dans l'eau, mais à des époques différentes (influence de la qualité thermique du temps).
Selon leur structure - disposition des minima et maxima - des diagrammes des 30 espèces choisies se partagent en 5 grands types résumés dans le Tableau I.

Dans la dynamique de la succession, si l'on s'en tient aux populations animales, 4 strates sont à distinguer :
- Strate Crustacée se divisant en 3 :
1= strate des Bryozoaires "en croûte mince" (à petites loges, faible calcification). 2= strate des Bryozoaires "en croûte épaisse" (grosses loges, forte calcification). 3= strate Crustacée molle (Botryllidés, Didemnidés, *Alcyonidium mytili* (Dalyell).
- Strate dressée basse, comportant Aeteidés, Ctenostomes rampants, Ectoproctes petits Hydraires gymnoblastiques. Son importance volumétrique est négligeable.
- Strate muscinale : Crisidés, Scrupariidés, Scrupocellariidés, *Clytia spp. Campanulina sp.*
- Strate herbacée : *Obelia dichotoma* (Linné), *Bougainvillia ramosa* (Van Beneden).
D'autre part, les individus isolés sont susceptibles de former d'importantes populations (Anomies, Balanes, Serpulidés) ainsi que les colonies de taille finie (*Lichenopora* et *Tubulipora*). Individus isolés et colonies de taille finie s'opposent par leur mode de croissance aux éléments formant les strates qui sont tous à croissance indéfinie.

Dynamique de la succession

Elle est décrite ici sur une seule inclinaison - les faces inférieures - où le peuplement est dominé par le développement des *Pomatoceros* : *P. triqueter* (Linné) essentiellement et *P. lamarcki* (Quatrefages). Cette dynamique est illustrée par la Fig. 2.

Série de plaques immergées à l'automne.

Très peu de *Pomatoceros* s'installent en fin d'été, et la strate crustacée bryozoologique peut s'établir sous la forme d'espèces pionnières ou semi-pionnières : *Lichenopora radiata* (Audouin), *Tubulipora plumosa* (W. Thomson), *Tubulipora phalangea* (Couch) et d'espèces en croûte mince , *Chorizopora brongniarti* (Audouin), *Microporella ciliata* (Pallas) ... La concurrence apparaît au printemps avec l'apparition des *Balanus crenatus* (Brugnière). La présence des Balanes est éphémère et un nouveau recrutement de *Lichenopora* est possible, puis les *Anomia* apparaissent ainsi que les *Pomatoceros* qui croissent et se maintiennent au moins 1,5 an. Les espèces à caractère pionnier disparaissent et seront absentes la deuxième année. Une 2ème strate de Chilostomes en croûte mince se maintiendra malgré la forte concurrence des *Pomatoceros*. Au 2ème printemps, un nouveau naissain de *Balanus crenatus* s'installe, suivi des *Anomia* et des *Pomatoceros*. Au cours du 2ème été apparaissent les Chilostomes en croûte épaisse, à forte vitalité et tendance perennante : *Escharoides coccinea* (Abildgaard) *Umbonula ovicellata* (Hastings), *Schizomavella hastata* (Hincks) et qui peuvent - ainsi que les Didemnidés - étouffer les *Pomatoceros* et toute portion de population sous-jacente. La strate muscinale est assez dense au bout de 26 mois ; les Crisidés et Scrupocellariidés dominent avec deux groupes d'espèces principales, à comportement différent : les *Scrupocellaria ssp.*, espèces saisonnières, se surimposant au peuplement préexistant quelque soit

TYPE 1 Espèce se fixant indifféremment sur tous les substrats

Espèce à vie brève disparaît avant un nouveau recrutement	L'espèce ne disparaît pas avant un nouveau recrutement	L'espèce évite toutefois les plaques neuves

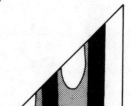

- 1 mois
- 1 an
- 2 ans

Balanus crenatus
Anomia ephippium typica

Microporella ciliata
Scrupocellaria reptans
Scrupocellaria scruposa
Hydroides norvegica

Anomia ephippium acul.
Protula tubularia
Serpula vermicularis
Verruca stroemia
Schizomavella h., Umbonula o., Caberea b., Crisia ramosa

TYPE 2
Espèce choisissant un certain degré de colonisation. Influence de la date d'immersion.

TYPE 3
Pionnière ne se fixant que sur des substrats vierges (ou légèrement colonisées lors de naissains abondants)

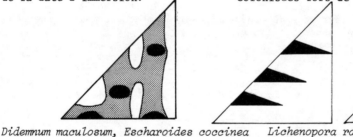

Didemnum maculosum, Escharoides coccinea
Crisia cf. eburnea, Celleporina hassali

Lichenopora radiata, Chorizopora brongniarti, Celleporella hyalina Alcyonidium mytili

TYPE 4
Espèce semi pionnière, évite les plaques vierges ainsi que celles trop colonisées

TYPE 5
Espèce pérennante en abondance sur des plaques immergées depuis quelque temps, se maintient si l'immersion est réalisée à un moment favorable

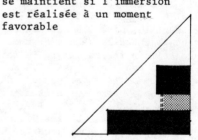

Tubulipora plumosa
Tubulipora phalangea

Aetea spp., Pomatoceros triqueter
Sabellaria spinulosa

TABLEAU

l'âge de la plaque ; les Crisiidés et *Caberea boryi* (Audouin) montrant une tendance climacique. Cette strate très réduite en hiver ne devient abondante qu'au premier été, car ce sont toutes des espèces estivales.

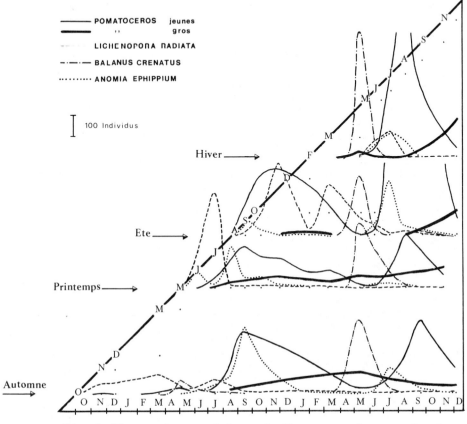

Fig. 2. Fluctuations numériques de diverses espèces sur les 4 séries de plaques (faces inférieures).

Série de printemps
L'espèce pionnière *Lichenopora radiata* se fixe massivement suivie d'*Anomia*. Puis ce sont les *Pomatoceros* qui bénéficient de températures élevées et croissent rapidement. Le succès des *Lichenopora* s'explique par la non-conccurence (absence de *Balanus crenatus* et l'immersion des plaques à une date propice. Par contre, les *Lichenopora* sont étouffés dès leur fixation par *Alcyonidium mytili* s'installant à la même période. Les plaques évoluent ensuite de la même façon que pour la série d'automne.

Série d'été
La colonisation débute massivement par les *Pomatoceros* et par quelques *Anomia*, *Lichenopora* leur succède, leur stock étant rechargé par un nouvel afflux de larves en mars. Les fortes densités de gros *Pomatoceros* ne sont présentes qu'au bout d'un an contrairement à la série de printemps, ce qui peut s'expliquer par une croissance ralentie durant l'hiver, les températures anormalement élevées de l'automne 71 provoquant une accélération soudaine de la

croissance. L'évolution ultérieure est la même que celle décrite précédemment
Série d'hiver
Elle est très courte. La colonisation débute par l'invasion de *Balanus crenatus* et de *Pomatoceros lamarcki*, celui-ci grossit très rapidement puis disparaît, créant ainsi sur le graphique, le petit pic de mai. Puis arrivent *Lichenopora Anomia* et *Pomatoceros triqueter*, ces derniers croissant rapidement grâce à un été-automne très chaud.

Au cours de leur immersion, les plaques subissent divers évènements qui se traduisent par un certain nombre de pics plus ou moins prononcés. La hauteur des pics résulte de la combinaison de 3 facteurs : a) meilleure réussite de l'espèce d'une année sur l'autre, b) préférence des espèces pour un certain degré de colonisation du substrat, c) immersion des plaques par rapport aux époques de fixation. L'étalement des pics dépend de la faculté de prospérer de l'espèce une fois installée. Interviennent ici des phénomènes de croissance (liés à la température, c'est-à-dire à la qualité thermique du temps d'immersion) et de concurrence interspécifique. Ces pics surgissent toujours dans le même ordre, autrement dit toutes les séries sont passées par les mêmes évènements, sauf celle de printemps qui a subi en plus la colonisation des *Alcyonidium mytili*. Si la séquence des évènements reste la même pour toutes les séries suivant leur date d'immersion, certaines étapes disparaissent, mais par la suite les 4 séries de plaques évoluent jusqu'à se ressembler fortement. Cette étape de convergence se caractérise par un peuplement à gros Serpuliens et une strate muscinale importante. Elle survient entre 16 et 18 mois selon les séries.

Sur les plaques des 24 et 26 mois, le peuplement est toujours à base de *Pomatoceros triqueter*. Dans les interstices dominent les Chilostomes à croûte épaisse, les Crisiidés et Scrupocellariidés. On note aussi quelques Démosponges, Calcisponges, Didemnidés et l'apparition de quelques *Caryophyllia sp*. A côté du développement de ces formes climaciques, les espèces qui ont dominé dans les premiers mois se raréfient : *Anomia, Verruca, Balanus, Hydroides, Serpula, Protula, Tubulipora, Lichenopora, Chorizopora* ... Pour certaines, le recrutement va s'atténuer d'année en année, pour les autres les fixations seront fortes mais peu d'entre elles seront destinées à évoluer, la place disponible se faisant de plus en plus rare et la compétition de plus en plus sévère. C'est ainsi que les Bryozoaires rigides et les Didemnidés surmontent les Serpuliens - seuls les plus gros survivent en érigeant leur tube - et que les Spongiaires ennoient les bases de Crisiidés et les jeunes organismes.

DISCUSSION

Ces formations de Serpuliens n'ont pas d'équivalent sur les parois rocheuses, et l'état climacique, du moins pour les faces inférieures et verticales, n'est pas atteint. Cependant, la majorité des espèces présentes sur le rocher le sont aussi sur les plaques, mais avec des tailles et des proportions différentes. Cet état préclimacique atteint en 2 ans que nous appelons "Climax qualitatif" correspond au stade où presque toutes les espèces de la roche sont représentées, mais où elles ne sont pas dans leurs rapports d'abondance, de taille et de structure de population observés dans les peuplements de la roche en place. Les organismes n'ont pas eu le temps d'atteindre leur plein développement et les interrelations ne sont qu'ébauchées, le laps de temps nécessaire à l'apparition du climax étant celui de la durée de génération des espèces quantitativement les plus importantes, à croissance et reproduction les plus lentes. L'évolution vers le climax se fera selon la voie indiquée plus haut :

la concurrence deviendra de plus en plus forte, les formes annuelles vont être peu à peu supplantées par des espèces à développement plus lent, mais pérennantes ou à forte potentialité de régénération comme Spongiaires et Bryozoaires. La prédation elle-même continue à favoriser l'apparition du climax, en faisant disparaître certains individus isolés au profit d'organismes coloniaux. Comme le fait remarquer Stebbing (1973), il est significatif que, dans les peuplements sessiles, les animaux coloniaux ou à bourgeonnement prédominent sur les types isolés, les peuplements sont ainsi beaucoup plus stables.

Ce peuplement à Serpuliens est à rapprocher de celui des biotopes tels que champs de petits blocs, milieux à salinité faiblement variable, valves de *Pecten*, bancs d'huîtres, coquilles mortes, carapaces de Crustacés vivants, tous ces milieux ont en commun des conditions physiques instables ne permettant pas à la succession de se poursuivre plus avant. C'est une étape par laquelle passent généralement tous les peuplements de substrat nouvellement immergé, comme le remarquait Simon-Papyn (1963) en parlant de *Pomatoceros triqueter* "qui caractérise habituellement le stade juvénile des peuplements rocheux".

CONCLUSION

Les étapes de la succession sont donc les suivantes :
- la phase de recrutement dépendant de la période d'immersion. Les larves sont présentes dans le plancton et sont susceptibles de se fixer à la dite période. Certaines se fixeront mais n'évolueront pas.
- la phase d'installation, s'étalant de 2 à 11 mois, au terme de laquelle 75 % des espèces sont présentes sur les collecteurs (Bellan-Santini, 1970), ce qui est réalisé au Glénan également vers 11 mois. Si le nombre des espèces ne croît plus que très lentement à partir de cette période, les séries de plaques immergées à des saisons différentes ne convergent en une physionomie commune qu'à partir de 16 mois. Tout en subissant les mêmes séquences d'évènements, elles prennent des allures différentes par la combinaison des facteurs physiques et biologiques.
- la phase de convergence, de 11 à 16 mois, pendant laquelle toutes les séries tendent à se rassembler, quelque soit leur point de départ.
Ces trois phases correspondraient, pour la première, au stade initial de Huvé (1970) et au stade de dominance pour les deux autres.
- la quatrième phase de 16 à 26 mois correspondrait au stade de prédestination de Huvé. Si le climax qualitatif est atteint vers 24-26 mois, l'équilibre quantitatif est loin de l'être. Le stade de prédestination n'est donc pas achevé au sens de Huvé, le stade de maturation du même auteur étant à fortiori hors de portée de notre étude.

Les trois premières étapes sont sous le contrôle étroit des facteurs physiques, principalement de la température. La deuxième dépend de la saison d'immersion comme l'a montré Persoone (1971). Les organismes en aménageant progressivement leur milieu (modification de l'hydrodynamisme, de l'éclairement, envasement des surfaces, production de mucus et de sources de nourriture nouvelles, multiplication des surfaces de fixation ...) créent à divers niveaux de stratification, des conditions microclimatiques nouvelles, favorables - voire indispensables - à de nouvelles espèces caractérisant l'état climacique. Les interrelations biologiques prennent peu à peu, le relais des facteurs physiques. Le milieu devient de moins en moins "physically controled" - pour les 3 premiers stades - et de plus en plus "biologically accomodated" (Sanders, 1968) pour les 2 derniers. La marche de la succession correspond à la prédominance successive de 2 catégories de facteurs, désignés par Pérès (1971) comme "facteurs temporels" : "ceux qui affectent le milieu physicochimique... et ceux inhérents aux êtres vivants en eux mêmes".

REMERCIEMENTS

Ce travail a été réalisé grâce au support financier du Centre National pour l'Exploitation des Océans (Contrats n° 73-755 et 74-1035), avec l'aide du navire "Armorique" et l'assistance technique de Mme Quiguer, Mlle Le Liart, M. Castric. Nous tenons à remercier MM. Glémarec et Chassé pour la révision du manuscrit, ainsi que Mme Thiriot et Mlle Lafargue pour la détermination des Spirorbes et Ascidies.

REFERENCES

Bellan-Santini, D. 1970 Salissures biologiques de substrats vierges artificiels immergés en eau pure, durant 26 mois, dans la région de Marseille (Méditerranée nord-occidentale). Etude qualitative.
Tethys, 2 (2), 335-356.

Castric-Fey, A. et al, 1973 Etagement des algues et des invertébrés sessiles dans l'archipel de Glénan. Définition biologique des horizons bathymétriques.
Helgolander wiss. Meeresunters.24, 490-509.

Castric-Fey, A. 1974 Les peuplements sessiles du benthos rocheux de l'archipel de Glénan, Sud-Bretagne. Ecologie descriptive et expérimentale.
Thèse Dr. Sc. Nat. Fac. Sci. Paris, 333 p.

Huvé, P. 1970 Recherches sur la genèse de quelques peuplements algaux marins de la roche littorale dans la région de Marseille.
Thèse Dr. Sc. Nat. Fac. Sci. Paris, 480 p.

Pérès, J.M. 1971 Considérations sur la dynamique des communautés benthiques.
Thalassia Jugoslavica, 7 (1), 247-277.

Persoone, G. 1971 Ecology of fouling on submerged surfaces in a polluted harbour.
Vie et Milieu, 22, 613-636.

Sanders, H.L. 1968 Marine benthic diversity : a comparative study .
American Naturalist, 102 (925), 243-282.

Simon-Papyn , L. 1965 Installation expérimentale du benthos sessile des petits substrats durs de l'étage circalittoral en Méditerranée
Recl.Trav. Stn. mar. Endoume, 39 (55), 53-94.

Stebbing, A.R. 1973 Competition for space between the epiphytes of *Fucus serratus*.
J.M.B.A. 53, 247-261.

ANNUAL MACROFAUNA PRODUCTION OF A SOFT BOTTOM IN THE NORTHERN BALTIC PROPER

Hans Cederwall

Department of Zoology and the Askö Laboratory, University of Stockholm, Box 6801, S-113 86 Stockholm, Sweden

ABSTRACT

Quantitative samples were collected on 15 separate occasions over two years at a single station (46 m deep) in the northern Baltic proper.

Production was estimated for the 5 most important species. Pontoporeia affinis (Lindstr.) produced 3.2 g dry weight equalling 71 kJ m^{-2} yr^{-1}. Pontoporeia femorata (Kröyer) produced 3.0 g dry weight (68 kJ), Harmothoe sarsi (Malmgren) 0.2 g dry weight (5 kJ), Halicryptus spinulosus (v. Sieb.) 0.2 g dry weight (4 kJ) and Macoma balthica (L.) 0.1 g shell-free dry weight (2 kJ) per m^2 and year.

Other species occasionally found in very low numbers are assumed to contribute only negligibly to the production.

The sum of macrofauna production for this bottom was estimated at 6.8 g dry weight or 150 kJ per m^2 per year.

INTRODUCTION

In the northern Baltic the Pontoporeia-dominated soft bottom community covers vast areas of the deeper bottoms and plays an important role in the ecosystem. Its macrofauna serves as important fish food for species like herring, cod, fourhorn sculpin and eel-pout (Aneer 1975). The present study was thus a natural part of the Askö Laboratory research project "Dynamics and energy flow in the Baltic ecosystem".

Sampling was carried out at a station situated at 46 m depth near Asenskallen in the Askö-Landsort area, northern Baltic proper (Lat 58° 46.5' Long 17° 41.6'). The station is one of ten standard monitoring stations sampled by the Askö Laboratory for the Swedish Environment Protection Board. The temperature of the bottom water at a nearby station (38 m deep) averaged 4.3°C, (range 1-10°C, N=62). The loss on ignition (500°C) of the upper 6cm of sediment was 3.6% of the dry weight (Nov. 1971).

MATERIAL AND METHODS

Quantitative samples
Quantitative samples were collected 15 times during two years (March 1972 - February 1974, see Fig. 1). A 0.1 m^2 van Veen grab, and sieves of 1 mm and 0.5 mm mesh were used. The sieving residues were preserved in 4% Formaldehyde solution buffered with Hexamethylene tetramine (40 g l^{-1}) and stained with Rose Bengal. The two sieve fractions were treated separately.

After not less than 3 months storage the animals were picked out. All animals in the 1 mm fractions were measured, except the Pontoporeia which were subsampled using a device similar to that described by Elmgren (1973). The whole 0.5 mm fractions were subsampled and one eighth was used. Only those animals which were normally found in the 1 mm net as adults were picked out from the 0.5 mm fraction. The limits between the different generations were deduced from the size distributions.

Regressions
Several times during the investigation period, fresh material was collected at the sampling station with a dredge. The fresh animals were measured under a stereomicroscope, and divided into size classes of 0.5 mm (for Harmothoe sarsi 0.25 mm). The aim was to obtain 10 individuals of each size class, which of course was not always possible. All animals in the same size class were weighed together, fresh and after 14 days of drying at 60°C. The regression values were determined with a computer program which fits a curve $y = a\, x^b$ to values $(x_i; y_y)$, that is, it determines the optimal values for a and b in the least square sense in the y direction. The a and b values obtained are thus a better fit than values obtained by ordinary linear regression analysis applied on logarithmic x^- and y^- values.

Fecundity
The number of eggs per female was investigated for Pontoporeia affinis and P. femorata. The lengths of the females were measured as above. Regression analysis of the values was made as above (Table 1).

TABLE 1 Length-egg number regressions ($e = a\, l^b$) for the Pontoporeia spp. Values for a, b, the correlation coefficient r and the number of females in March 1973.

Species	a	b	r	n
Pontoporeia affinis	.117	2.68	.78	25
Pontoporeia femorata	.046	2.84	.66	18

Biomass
For Pontoporeia affinis, P. femorata and Harmothoe sarsi the size distributions and the size-weight regressions were used to calculate the biomass for each year class and sample. For the Pontoporeia species, the biomass of generation 0 and 0' when hatching was determined by drying (60°C) and weighing. For Pontoporeia affinis a value of 16.3 µg per egg was obtained (360 eggs weighed) and for P. femorata 26.9 µg (92 eggs). With the length - egg number regression equation, the length distribution of the females (at March 1st 1972 and February 22nd 1973) and the dry weight of one egg, the initial biomasses were computed.
For Macoma balthica and Halicryptus spinulosus biomass was calculated for each sample using the measured parameter figures and regressions to dry weight.

Production
For Pontoporeia affinis, P. femorata and Harmothoe sarsi, production was calculated according to the growth increment method as described by Crisp (1971) pp. 207-226).

For Macoma balthica and Halicryptus spinulosus the mean biomass for each year was calculated and annual turnover ratios from literature were used to estimate production. For Macoma the P/B value of 0.37 found by Ostrowski (in print) was applied and for Halicryptus the "minimum production" value of P/B = 0.9 estimated by Arntz (1971).

TABLE 2 Average annual macrofauna (0.5 mm) abundance, biomass or standing crop ($\bar{B}$), production (P) per m^2, and calculated turnover ratios (P/$\bar{B}$).

Species	Year	Abundance ind.	$\bar{B}$ g dr wt	P g dr wt	P kJ	P/$\bar{B}$
Pontoporeia affinis	72/73	4176.7	1.70	4.42	99.0	2.60
	73/74	3028.0	1.62	1.92	42.9	1.19
	Mean	3602.4	1.66	3.17	71.0	1.90
Pontoporeia femorata	72/73	3448.3	2.22	4.63	103.6	2.08
	73/74	3069.6	1.84	1.44	32.2	0.78
	Mean	3258.9	2.03	3.03	67.8	1.43
Harmothoe sarsi	72/73	397.0	0.12	0.41	8.2	3.36
	73/74	359.0	0.09	0.05	1.0	0.61
	Mean	378.0	0.11	0.23	4.6	1.99
Halicryptus spinulosus	72/73	47.1	0.38	0.35	7.4	x
	73/74	88.7	0.13	0.12	2.5	x
	Mean	67.9	0.26	0.23	4.4	x
Macoma balthica	72/73	56.1	0.31	0.11	2.0	x
	73/74	48.1	0.20	0.07	1.2	x
	Mean	52.1	0.26	0.09	1.6	x
Others	72/73	10.5	x	x	x	x
	73/74	7.9	x	x	x	x
	Mean	9.2	x	x	x	x
Sum (for biomass and production excl. others)	72/73	8135.7	4.73	9.92	220.2	2.10
	73/74	6601.3	3.88	3.60	79.8	0.93
	Mean	7368.5	4.31	6.75	149.3	1.57

^{x}Not calculated.

RESULTS AND DISCUSSION

On the investigated station Pontoporeia affinis was the most abundant species, closely followed by Pontoporeia femorata. Less important species were Harmothoe sarsi, Halicryptus spinulous and Macoma balthica (Table 2). Other species occassionally appearing in the samples are assumed to contribute only negligibly to the production.

Pontoporeia affinis and Pontoporeia femorata

A detailed report on the production of Pontoporeia affinis has already been given (Cederwall, in print), but as the life cycles of Pontoporeia femorata and P. affinis are very alike they will be dealt with together.

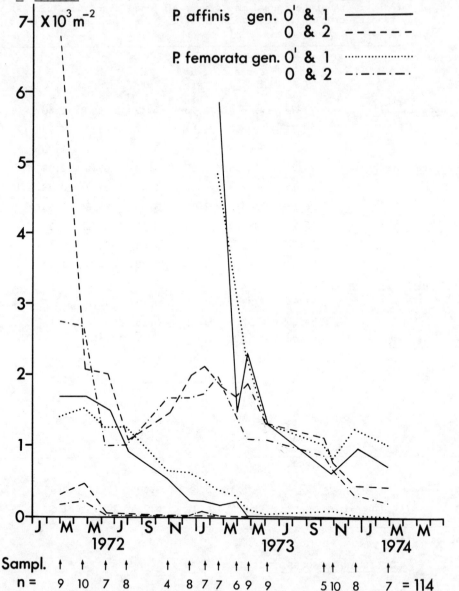

Fig. 1. Variations in abundance of the different generations of Pontoporeia affinis and P. femorata. Arrows at the bottom show sampling occasions, figures number of samples.

At this depth the two Pontoporeia species generally have a two year life cycle (cf. Segerstråle 1950). A small number of animals live more than two years, some may breed during the summer (Segerstråle 1967), but a few live to be three years before they propagate and die (Fig. 1). As can be seen in Fig. 1 the mortality is greatest during the first months of a generations life. Mortality is also high during and just after the propagation period. In the autumn, during the copulation period, the animals live a more semi-pelagic life, and naturally predation rises.

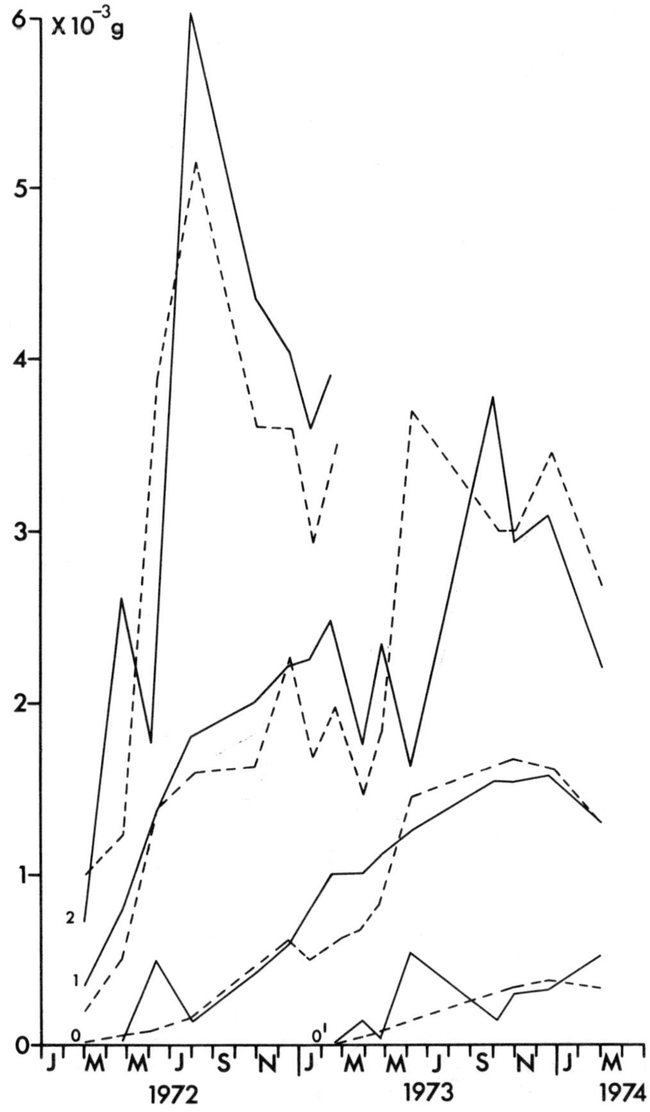

The growth of the different generations is shown in Fig. 2. Growth is almost entirely confined to spring, with a cessation or even a decrease during the rest of the year. A similar situation was found by Johnson & Brinkhurst in Lake Ontario (1971).

There are some peculiarities in the curves. In gen.0 and 0' the number of survivors sometimes rises above the earlier values. This is probably due to sampling variability but there may also have been some immigration.

The main difference between the two Pontoporeia spp. is the lower fecundity and mortality of P. femorata (Table 1, Fig. 1).

For Pontoporeia affinis the production during the first year (Table 2) was more than twice as high as during the second. A mean production of 3,17 g dry weight per m^2 and year was calculated. Using a calorific value of 5346 cal per g dry weight (Cederwall, in print)

Fig. 2. Variations in individual mean weight (g dry weight) of the different generations of Pontoporeia affinis (interrupted line) and P. femorata (continuous line).

this corresponds to 71 kJ per m^2 and year (1 J = 0.24 cal.).

The calculated annual P/B is 1.90 giving a daily P/B of 0.0052, which is exactly the same value as Grez (1951) found for Pontoporeia affinis with a two year cycle. It also agrees very well with the annual P/B of about 1.95 found by Johnson and Brinkhurst (1971).

For Pontoporeia femorata the production during the first year (Table 2) was more than three times as high as during the second. The mean production was 3.03 g dry weight m^{-2} yr^{-1}. Using the same calorific value as for P. affinis this equals 68 kJ m^{-2} yr^{-1}. The estimated annual P/B of 1.43 agrees closely with Ostrowski (in print).

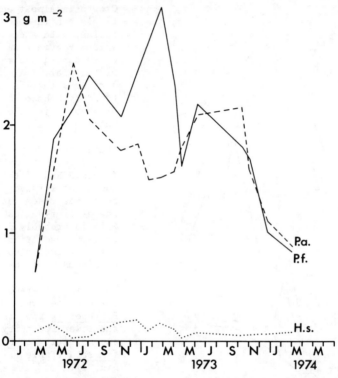

For both species the mean biomasses are nearly the same in both years (Table 2). It may then seem a bit odd that the production is so much higher the first year than the second. The reason is obvious in Fig. 3 however. After one year the biomass level had increased due to a build up over the year. During the following year however the mortality was higher and at the end of that year a level more like the initial one of two years earlier was reached. Obviously the mean value for the two years is a more reliable value.

Fig. 3. Variations in the mean biomasses (g dry weight m^{-2}) of Pontoporeia affinis (interrupted line), P. femorata (continuous line) and Harmothoe sarsi (dotted line).

Harmothoe sarsi

On this bottom Harmothoe sarsi also generally live for two years before they propagate and die and just as for the Pontoporeia spp. a small part live more than two years (cf. also Sarvala 1971, in print) (Fig. 4).

Since the smallest sieve used had a 0.5 mm mesh size net, not only the planktonic stages but also the smallest settled specimens are excluded from the calculations. Therefore the production values given here are underestimates.

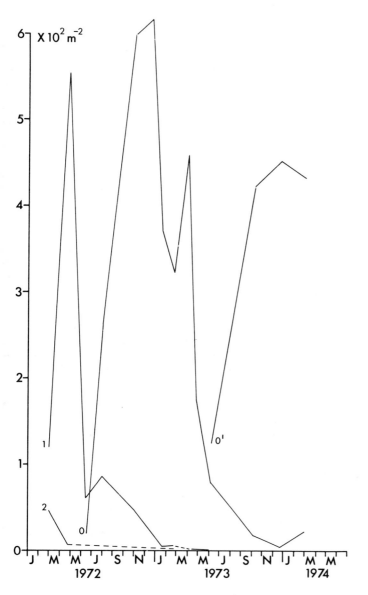

Fig. 4. Variations in abundance of the different generations of *Harmothoe sarsi*.

During the first year after settling the growth is remarkably slow (Fig. 5), as already pointed out by Sarvala (1971, in print).

The production was 0.23 g dry weight m^{-2} yr^{-1}, (using the value of 4789 cal g^{-1} dry weight given by Thayer et al., 1973, for polychaetes).

Halicryptus spinulosus

For *Halicryptus* as well as for *Macoma balthica* the production estimates are less exact, due to the different method of calculation employed.

Halicryptus was assumed to produce 0.23 g dry weight m^{-2} yr^{-1} equalling 3.9 kJ (energy content 23.4 kJ g^{-1} ash free dry weight, 90% estimated to be ash-free dry weight, Ankar & Elmgren 1976).

Macoma balthica

For *Macoma*, production was estimated at 0.9 g dry weight m^{-2} yr^{-1} including shell. 90% of dry weight is assumed to be shell weight (Ankar pers. comm.). This gives a production of roughly 0.1 g dry weight m^{-2} yr^{-1} equalling 1.6 kJ (energy content 4252 cal g^{-1} dry weight, Thayer et al. 1973).

Sum of production

The sum of macrofauna production on this bottom was 6.8 g dry weight corresponding to about 150 kJ m^{-2} yr^{-1} (Table 2).

This is only on third of the value estimated for the Kiel Bight (Arntz 1971, Arntz & Brunswig 1975) which is not unreasonable since the Kiel Bight is more productive and much shallower.

The value found here is however also lower than the value of 225 kJ m^{-2} yr^{-1} estimated by Ankar & Elmgren (1976) for the Askö-Landsort area. The reason for this is partly the greater depth of this station.

If we assume that Harmothoe and Halicryptus, which are the predators of this system are living mainly off the macrofauna production and consuming 5 times their own production they will utilize 2.? g dry weight m^{-2} yr^{-} That leaves a maximum fish food basis of 4.5 g dry weight m^{-2} yr^{-1}. A realistic estimate is that between 3 and 4 g dr weight m^{-2} yr^{-1} is utilized as food for the fish.

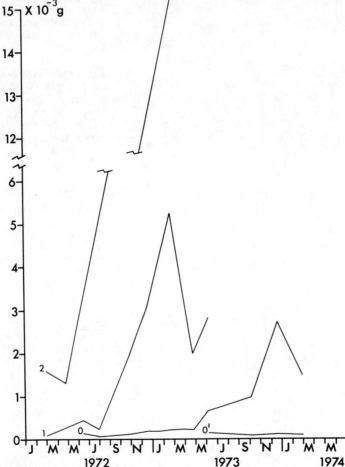

Fig. 5. Variations in individual mean weight (g dry weight) of the different generations of Harmothoe sarsi.

Meiofauna was collected at this station twice during this investigation and the biomass was 6.6±1.9 (May 1972) and 7.4 ± 1.1 (Oct. - Nov. 1973) g wet weight m^{-2}, giving us a mean of 7 g (Elmgren, pers. comm.). We can assume that around 10% of the wet weight is ash free dry weight (Elmgren, pres. comm.) and that the turnover ratio is 4.6 (Ankar & Elmgren 1976). This would give a meiofauna production of 3.2 g dry weight equalling 75 kJ m^{-2} yr^{-1} energy content 23.4 kJ g^{-1} ash free dry weight, Ankar & Elmgren 1976).

To sum up, this bottom is estimated to produce 225 kJ per m^2 annually, ciliates, bacteria and fish excluded.

The most striking feature of all the production estimates presented, is the much higher production during the first year. For the Pontoporeia spp. this i largely due to a much higher growth rate during 1972, as shown in Fig. 2. Available temperatures data from 35 - 60 m depth show a slightly lower (about 1°C) mean temperature this year than during 1973. This is not a likely explanation for the differences in production. On the other hand the phytoplankton spring bloom was much larger in 1972, with higher peaks of both ^{14}C primary production (72: 1,140, 73: 730 mg C m^{-2} day $^{-1}$) and algal biomass (72: 38, 73; 16 g wet weight m^{-2}), and a longer duration (72: about 11/2, 73: 1 month) (Hobro & Nyqvist, pers. comm.). Since almost 40% of the phytoplankton spring bloom sediments to the bottom in the Askö area (Hobro, Larsson & Wulff, in print), higher food availability seems to explain the higher production during the first year. This supports Wiederholms (1973) opinion, that food is a more important regulating factor for the growth rate of Pontoporeia affinis than temperature.

CONCLUSIONS

At the soft bottom station studied, annual production of the five most important species was; Pontoporeia affinis 71 kJ, Pontoporeia femorata 68 kJ, Harmothoe sarsi 4.6 kJ, Halicryptus spinulosus 3.9 kJ and Macoma balthica 1.6 kJ per m^2 and year. This adds up to an annual macrofauna production of 150 kJ per m^2, out of which around 80 kJ could be utilized as fish food.

ACKNOWLEDGEMENTS

I want to express my gratitude to Kerstin Rigneus and Elsa Linnerstam for sorting the material, to Rangnar Elmgren for correcting the manuscript and to Birgit Mayrhofer for making the drawings.

REFERENCES

Aneer, G. 1975 Composition of the Food of the Baltic Herring (Clupea harengus L.), Fourhorn sculpin (Myxocephalus quadricornis L.) and Eel-pout (Zoarces viviparus L.) from deep soft bottom trawling in the Askö-Landsort area during two consecutive years. Merentutkimuslait. Julk./ Havaforskningsinst. Skr. 239, 146-154.

Ankar, S. and R. Elmgren 1976 The Benthic Macro- and Meiofauna of the Askö-Landsort area (Northern Baltic Proper), A Stratified Sampling Survey. Contr. Askö Lab. 11.

Arntz, W.E. 1971 Biomasse and Produktion des Makrobenthos in den tieferen Teilen der Kieler Bucht im Jahr 1968. Kieler Meeresforsch. 25 (1)36-72.

Arntz, W.E. and D. Brunswig 1975 An approach to estimating the production of macrobenthos and demersal fish in a western Baltic Abra abra community. Merentutkimuslait. Julk./Havsforskningsinst. Skr. 239, 195-205.

Cederwall, H. (in print) Production of Pontoporeia affinis in the northern Baltic proper. To be published in Prace Morskiego Instytutu Rybackiego.

Crisp, D.J. 1971 Energy Flow Measurements. In Holme & McIntyre (ed.): Methods for the Study of Marine Benthos, I.B.P. Handbook No 16, Blackwell, Oxford & Edinburgh, 197-279.

Elmgren, R. 1973 Methods for sampling sublittoral softbottom meiofauna. Oikos (suppl. 15), 112-120.

Green, R.H. 1971 Lipid and Calorific Contents of the Relict Amphipod Pontoporeia affinis in Cayuga Lake, New York. J. Fish. Res. Bd. Canada 28, 776 - 777.

Greze, V.N. 1951 (The production of Pontoporeia affinis and a method for its determination). Trudy Vses. Gidrobiol. Obshch. 3, 33 - 43.

Hobro, R., U. Larsson and F. Wulff (in print) Dynamics of a phytoplankton spring bloom in a coastal area of the northern Baltic proper. To be published in Prace Morskiego Instytutu Rybacki.

Johnson, M.G. & R.O. Brinkhurst 1971 Production of Benthic Macroinvertebrates of Bay of Quinte and Lake Ontario. J. Fish. Res. Bd. Canada 28, 1699-1714

Ostrowski, J. (in print) Production of Pontoporeia femorata and Macoma baltica in the Gdansk Bay. To be published in Prace Morskiego Instytutu Rybackie;

Sarvala, J. 1971 Ecology of Harmothoe sarsi (Malmgren) (Polychaeta, Polymoidae) in the northern Baltic area. Ann. Zool. Fenn. 8, 231-309.

Sarvala, J. (in print) Production of Harmothoe sarsi (Polychaeta) at a soft bottom locality near Tvärminne, southern Finland. To be published in Prace Morskiego Instytutu Rybackiego.

Segerstrale, S.G. 1950 The amphipods on the coast of Finland - some facts and problems. Soc. Sci. Fenn. Comm. Biol. 10 (14), 1-28.

Segerstrale, S.G. 1967 Observations of summer-breeding in populations of the glacial relict Pontoporeia affinis (Lindstr.)(Crustacea Amphipoda), living at greater depths in the Baltic Sea, with notes on the reproduction of P. femorata (Kröyer). J. exp. már. Biol. Ecol. 1, 55-64.

Thayer, G.W., W.E. Schaaf, J.W. Angelovic and M.W. LaCroix 1973 Caloric measurements of some estuarine organisms. Fish. Bull. 71 (1), 289-296.

Wiederholm, T. 1973 On the life cycle of Pontoporeia affinis (Crustacea Amphipoda) Lake Malaren. ZOON 1, 147-151.

EVOLUTION DANS LE TEMPS DES PEUPLEMENTS DES SABLES ENVASES EN BAIE DE CONCARNEAU (BRETAGNE)

Pierre Chardy et Michel Glémarec

Centre Océanologique de Bretagne - 29273 Brest Cédex
Université de Bretagne Occidentale - Laboratoire d'Océanographie
Biologique - 29283 Brest Cédex - France

ABSTRACT

Qualitative and quantitative studies were carried out at three localities at different times of the year, from 1970 to 1974. Multivariate analysis, using inertia techniques, elucidated the major features of the principal sources of variation in time and space. The first main variation, in spite of the similarity of edaphic factors, was related to the spatial distribution of the samples among the stations. Replicate sampling procedures facilitated identification of the heterogeneous components of the environment. To estimate the importance of the time fluctuations at each of the three stations, an attempt was made to locate them along a climatic gradient. Time evolution is governed by two phenomena: annual cycle and climactic trend. Species diversity analysis gives an interpretation of this evolution.

INTRODUCTION

En baie de Concarneau, les entités de peuplement sont bien connues, ainsi que les facteurs écologiques qui les gouvernent. Le présent travail a pour but de rechercher une éventuelle synchronisation des fluctuations dans le temps des peuplements caractéristiques des sables envasés de la baie de Concarneau.

L'analyse s'effectue en deux temps :
- étude globale des facteurs "temps" et "espace" sur la composition qualitative et quantitative des peuplements. Importance relative de ces deux sources de variations ;
- étude particulière du facteur temps : mise en évidence des variations saisonnières et pluriannuelles, communes aux trois stations (principal objectif de l'étude).

MATERIEL ET METHODES

Récolte des données

Durant cinq années, de 1970 à 1974, 3 stations ont été régulièrement échantillonnées, 4 fois par an à intervalles réguliers, à raison de 4 à 8 bennes Smith-Mac Intyre par station ; ce qui représente 331 prélèvements. Les trois stations appartiennent à la même unité édaphique, les sables envasés définis par le type de peuplement à *"Amphiura"*. Elles s'étagent le long du gradient climatique (Glémarec, 1973).

<u>Station Beg-Meil</u> : 10 mètres, étage infralittoral, à l'abri des houles dominantes, communauté à *Acrocnida brachiata - Clymene oerstedii*.
<u>Station Baie de Concarneau</u> : 28 mètres, étage circalittoral côtier, hydrodynamiquement stable, communauté à *Amphiura filiformis - Lumbrinereis gracilis*.

Station Mousterlin : 18 mètres, station intermédiaire entre l'infralittoral et le circalittoral côtier. Soumise aux houles de sud-est fréquentes. Peuplement également intermédiaire.

Entre ces 3 stations, l'analyse granulométrique globale ne montre pas de variation significative du taux de pélites. Ce paramètre mesuré au niveau du premier centimètre de surface, permet d'appréhender le caractère de stabilité de la couche superficielle : Boucher (1975), montre ainsi que la station de Beg-Meil est plus stable et légèrement plus envasée que les deux autres. Les 23 espèces les mieux représentées sont retenues pour l'analyse.

METHODE D'ANALYSE DES DONNEES

L'outil mathématique utilisé est l'analyse d'inertie, permettant de visualiser globalement les ressemblances faunistiques entre prélèvements ou groupes de prélèvements caractérisés par une date et une position dans l'espace. Une analyse des diverses variantes possibles a déjà été proposée par Chardy, Glémarec & Laurec (1976). Cependant, il est utile de préciser la signification des options retenues pour cette étude.

Etude globale des variations spatio-temporelles

Afin de dégager une structure qui ne soit pas uniquement due aux fluctuations des espèces à forte densité (telle qu'*Amphiura filiformis*), les effectifs des espèces sont réduits et centrés. Le centrage a la propriété de faire disparaître le décalage systématique de densité entre deux espèces. L'intérêt de l'analyse consiste à prendre en compte les espèces dont les effectifs varient de la même façon ; les différences d'abondance importent peu. La réduction permet de pondérer le rôle des espèces dont la densité peu fluctuer largement d'un prélèvement à l'autre. Il faut noter par ailleurs, que ces options permettent de prendre en compte les différences de densité entre prélèvements. Afin d'éliminer les biais dus aux phénomènes de surdispersion (particulièrement sensibles dans le cas de l'échantillonnage par benne) les données ont subies une transformation logarithmique. La distance retenue est une distance euclidienne, qui nous ramène au cas de l'analyse d'inertie d'une matrice de corrélation entre espèces (Analyse en composantes principales, mode R).

Etude particulière des variations temporelles

Le but de cette analyse consiste à rechercher les axes factoriels qui traduisent uniquement les variations dans le temps, en maintenant fixées les variations dans l'espace (on reconnaît le principe de l'analyse de variance où la variation totale est décomposée en sources de variations élémentaires). Cette variante, de l'analyse précédente consiste à centrer chaque variable, non pas par la moyenne générale des observations, mais par la moyenne partielle des observations de chaque station (on procède de même pour la réduction des effectifs). Le centrage des données par "blocs" (déjà utilisé par Laurec et Le Gall, 1974) permet d'éliminer tout décalage systématique de densité des espèces entre les trois stations et de privilégier ainsi les variations dans le temps. Effectuer séparément une analyse d'inertie sur chacune des trois stations aboutirait à trois structures issues de trois systèmes d'axes différents qu'il serait impossible de relier les uns aux autres, l'analyse perdant ainsi toute unité.

Diversité

La diversité spécifique est estimée par l'indice de Shannon ($H_{(s)} = -\sum p_i \log_2 p_i$; $p_i = \frac{n_i}{N}$ = fréquence relative de l'espèce i) qui présente l'avantage de n'être subordonné à aucune hypothèse préalable sur la distribution des espèces et des individus. Afin d'obtenir une meilleure estimation des fréquences relatives de chaque espèce, et par la même, d'éviter les biais dûs à la distribution contagieuse des individus, l'indice de diversité est calculé, non pas sur les effectifs de chaque benne, mais sur les effectifs cumulés des bennes effectués au même endroit à la même date.

RESULTATS

Analyse globale des variations spatio-temporelles

La répartition des barycentres saisonniers (barycentres des prélèvements effectués à la même date) de chaque station (Fig. 1) dans le plan I-II de l'analyse d'inertie met en évidence l'hétérogénéité de la composition faunistique entre les 3 stations. Les 3 unités de peuplement sont bien séparées selon l'axe I (24,1 % de la variance totale) qui représente essentiellement l'axe climatique. La station Mousterlin apparaît bien intermédiaire entre les deux autres. L'axe II (11 % d'inertie) discrimine vers son pôle positif les prélèvements effectués à Mousterlin, caractérisés par des conditions écologiques instables dues à un fort hydrodynamisme. La dispersion des points dans l'espace factoriel témoigne de cette instabilité.

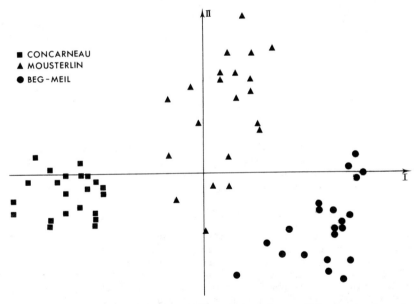

Fig. 1. Analyse globale - Répartition des prélèvements dans le plan I-II de l'analyse d'inertie - Chaque point représente le barycentre des prélèvements effectués au cours de chaque sortie à chaque station de 1970 à 1974.

Il n'est pas possible de repérer les cycles saisonniers au sein de chaque station. Néanmoins, la projection des barycentres annuels (Fig. 2) dans le même système d'axes I-II de l'analyse, suggère l'existence de variations pluriannuelles non négligeables. Une évolution régulière apparait de 1970 à 1972, à l'intérieur de chaque station ; les projections des années 73 et 74 indiquent un retour vers une situation initiale en baie de Concarneau, tandis qu'à Beg-Meil et surtout à Mousterlin, l'évolution se poursuit vers une nouvelle composition faunistique. Les axes I et II traduisent essentiellement les effets spatiaux et la structure temporelle recherchée s'en trouve masquée. Pour lever cette hypothèque, l'étude des variations dans le temps est abordée par une analyse ayant les propriétés d'éliminer les variations inter-stations, pour ne considérer globalement que les variations intra-stations.

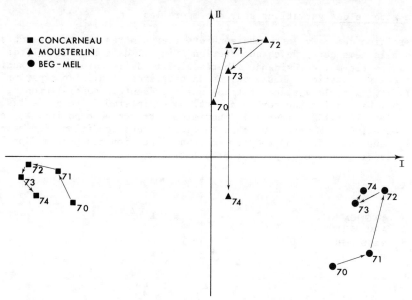

Fig. 2. Analyse globale - Répartition des barycentres des prélèvements effectués chaque année à chaque station dans le plan I-II de l'analyse d'inertie - L'espace de référence est le même que celui de la figure 1 ; les symboles également.

Etude particulière des variations temporelles

Les axes d'inertie extraits par cette analyse (Fig. 3) traduisent uniquement les variations temporelles de l'ensemble faunistique étudié. Pour simplifier la représentation graphique les barycentres sont groupés par station, pour 1970-71 et 72 d'une part, pour 73 et 74 d'autre part. On retrouve en effet, cette évolution en deux phases signalée plus haut :
- de 1970 à 1972, évolution quasi-parallèle de la composition faunistique des stations, Fig. 3, a, b, c ;
- 1973 et 1974, évolution divergente, désynchronisation des fluctuations du peuplement aux 3 stations, Fig. 3, d, e, f.

A la station baie de Concarneau, le phénomène est le plus net ; les modifications de la composition faunistique décrivent un cycle saisonnier distinct

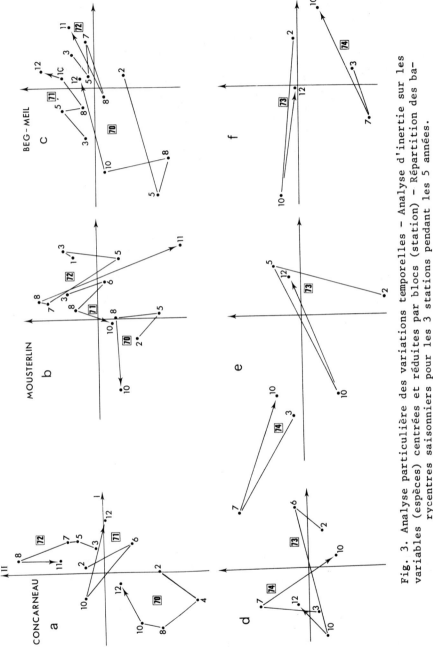

Fig. 3. Analyse particulière des variations temporelles - Analyse d'inertie sur les variables (espèces) centrées et réduites par blocs (station) - Répartition des barycentres saisonniers pour les 3 stations pendant les 5 années.

chaque année. On constate néanmoins qu'il n'y a pas reproductibilité des variations saisonnières de chaque cycle annuel. Ces 3 cycles montrent une dérive évidente de la composition faunistique ; cette même dérive existe à Mousterlin et à Beg-Meil, tout en étant moins nette.

Durant cette période, il y a augmentation du nombre des effectifs du peuplement, enrichissement en espèces et élévation de l'indice de diversité spécifique (Fig. 4). Cette phase apparaît comme une période d'occupation de plus en plus poussée du milieu, faisant certainement suite à des évènements survenus avant 1970. La population d'*Amphiura filiformis* étudiée particulièrement passe de 400 / m à 700 / m^2 à la station baie de Concarneau ; de 250 / m^2 à 500 / m^2 à la station Mousterlin. Parallèlement, des populations de Polychètes et de Bivalves s'accroissent régulièrement aux stations. A la station Beg-Meil, les *Amphiura filiformis* absentes en 1970 apparaissent ensuite sans toutefois dominer. Ce coefficient de diversité y atteint 3,5 , ce qui est très élevé.

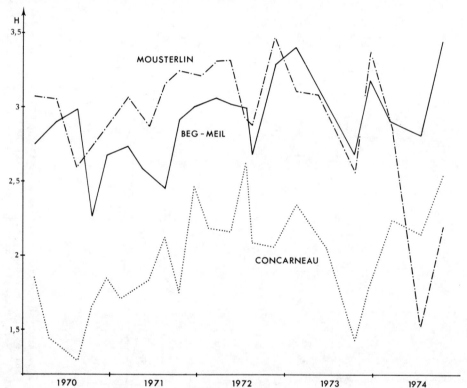

Fig. 4. Variations temporelles du coefficient de diversité H aux 3 stations.

La deuxième phase de cette évolution (Fig. 3, d, e, f) montre en 1973 un retour aux conditions faunistiques rencontrées précédemment en 1970 et 1971, ceci aux 3 stations. Inversement, en 1974, les peuplements de Mousterlin et de Beg-Meil évoluent dans une direction opposée traduisant des différences faunistiques marquées par rapport aux conditions initiales. En baie de Concarneau la situation semble se stabiliser.

On retrouve dans l'étude des variations temporelles comme dans celles des variations spatiales, le caractère éminemment instable de la station de Mousterlin. Les fortes fluctuations de l'indice de diversité en 1973 et 74, en cette station, confirment ce caractère.

DISCUSSION

Les phénomènes évolutifs sont plus accusés à la station baie de Concarneau, où les valeurs de l'indice de diversité sont les plus faibles; le peuplement étant dominé par un nombre très restreint d'espèces abondantes, telles qu' *Amphiura filiformis*, les fluctuations de la composition faunistique sont très liées à la dynamique de ces populations. Ainsi le printemps de 1973, marqué par des conditions climatiques catastrophiques, a exercé un effet négatif sur le recrutement des *Amphiura filiformis* et vraisemblablement sur d'autres espèces qui l'accompagnent dans cette période de colonisation.

Le point de rupture entre les deux phases semble lié à un phénomène climatique. Ses effets ont été d'autant plus accusés, qu'il est survenu à un moment où la compétition pour la place disponible était la plus forte. Passé ce point de rupture, les relations d'équilibre entre espèces sont modifiées, le peuplement semble désorganisé, chaque station voit son peuplement évoluer de façon particulière.

Sur l'ensemble de la baie de Concarneau, on note au niveau de plusieurs communautés, un envasement et un enrichissement croissant. On peut donc se demander si l'accident survenu dans l'évolution des sables envasés n'est que passager et si l'évolution avec enrichissement constant reprendra son cours. La deuxième hypothèse consiste à considérer ce point de rupture entre les deux phases, comme tout à fait naturel, impliquant l'existence de cycles pluriannuels de l'ordre de 5 à 6 ans, appelés à se répéter. Avant de pouvoir se prononcer sur ces deux hypothèses et sur ce que peut être l'état climax au niveau de chaque peuplement, il serait souhaitable de disposer de séries temporelles plus longues.

CONCLUSION

L'étude des fluctuations faunistiques sur 5 années au niveau d'un même ensemble édaphique fournit une série de données particulièrement propices à l'analyse des variations temporelles. Ces variations sont marquées par l'hétérogénéité spatiale inter-stations gouvernée principalement par le facteur climatique. L'étude particulière des variations temporelles permet de décomposer cette période de 5 années en deux phases :
- une phase à évolution faunistique parallèle aux 3 stations, phase de colonisation caractérisée par une augmentation de la densité, de la diversité et de la richesse faunistique ;
- une phase de désorganisation où les relations d'équilibre entre espèces sont modifiées et où les stations évoluent chacune de façon particulière.

REMERCIEMENTS

Ce travail est réalisé avec l'aide financière et matérielle du Centre National pour l'Exploitation des Océans.

REFERENCES

Boucher, D. 1975 Production primaire saisonnière du microphytobenthos des sables envasés en baie de Concarneau.
Thèse 3è Cycle, Université de Brest, 113 p.

Chardy, P., Glémarec, M. & Laurec, A. 1976 Application of inertia methods to benthic marine ecology : practical implications of the basic options.
Estuar. Coastel Mar. Sc. 4, 179-205.

Glémarec, M. 1973 The benthic communities of the european north atlantic continental shelf.
Oceanogr. Mar. Biol. Ann. Rev. 11, 263-289.

Laurec, A. & Le Gall, J.Y 1974 Application des méthodes d'analyse multivariable à l'étude d'une pêcherie plurispécifique : la Pêcherie Palangrière thonière en Atlantique.
Comm. Int. pour la Conservation des Thonidés de l'Atlantique, Vol. III.

DISTRIBUTION OF BENTHIC PHYTO- AND ZOOCOENOSES ALONG A LIGHT GRADIENT IN A SUPERFICIAL MARINE CAVE

Francesco Cinelli,* Eugenio Fresi,* Lucia Mazzella*
Maurizio Pansini,** Roberto Pronzato** and Armin Svoboda***

*Reparto di Ecologia Marina della Stazione Zoologica di Napoli, Punta S. Pieto, Ischia Porto (Napoli), Italy.
**Instituto di Zoologia dell'Universitá, Via Balbi 5, Genova, Italy
***Lehrstuhl für spezielle zoologie, Ruhr - Universität Bochum, Postfach 2148, D-463 Buchum Querenburg, Fed. Rep. of Germany

ABSTRACT

Superficial marine caves not only afford a strong light gradient, but also a relative homogeneity of water movement and related ecological variables. In such a situation one can reasonably presume to be dealing with a simplified light gradient which is the principal factor responsible for the benthic community zonation observed along the cave.

This hypothesis was tested in a superficial marine cave, the Grotta del Mago (Island of Ischia). Descriptive and structural analysis showed that the response of the community to light changes is mostly indirect, as is mediated by the floristic component, and largely obscured both by a positive water movement gradient and the unfavourable substratum in the inner section of the cave.

INTRODUCTION

One of the best research opportunities offered by marine caves is the possibility of studying the biological effects of a strong light gradient in an easily accessible environment. This is particularly true for the so-called superficial caves, defined by Riedl (1966) as 'Grotten', consisting of blind ended cavigies, partly submerged by the sea. These biotopes, in fact, not only afford a light gradient as a function of their length, but also a relative uniformity of water movement. As has been stated by Biedl (op. cit.), in the typical superficial caves there is no significant reduction of hydrodynamic forces, except in the innermost parts. Such a situation should ensure throughout the biotope a relative uniformity of all those ecological variables which are largely affected by water exchange rates. One can therefor reasonably presume to be dealing with a simplified or 'pure' light gradient, which is the principal factor responsible for the benthic community zonation observed along the cave.

In order to test this hypothesis, a superficial marine cave with the required conditions was selected. The Grotta del Mago (Wizard's Cave) was found to be the best place for the experiment amongst the several caverns existing along the coast of the Gulf of Naples.

THE BIOTOPE

The Grotta del Mago opens on the East coast of the Island of Ischia (Fig. 1) in rock of volcanic origin. According to Rittmann (1930), it was formed by wave erosion on softer extrusive material, embedded in a solid thrachytic matrix.

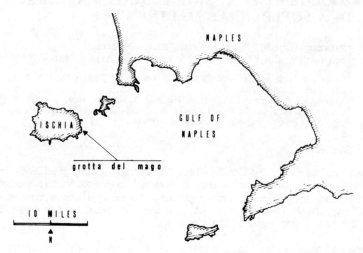

Fig. 1. Location of the Grotta del Mago in the Gulf of Naples.

It consists of a funnel-shaped outer chamber connected by a long narrow passage to an inner chamber, at the end of which there is a small beach (Fig. 2). The total length is 110m. The bottom steadily rises to the surface from a depth of 6m at the entrance and is composed of irregular boulders and pebbles interspersed in a finer sediment.

The cave is exposed to eastwinds, the frequency of which does not exceed 20%, westerlies being the dominant winds of the region (Düing, 1965). For this reason the East coast of Ischia can be regarded as relatively sheltered. Nevertheless there are indications that strong water movement occurs throughout the biotope, especially in the narrow corridor, which is clearly eroded near the bottom. There is also biological evidence of strong hydrodynamic forces. In the Porifera, for instance, encrusting species are largely dominant over erect ones, and flattening occurs in species which are normally massive and erect (e.g. <u>Agelas oroides</u> (Schmidt), <u>Acanthella acuta</u> Schmidt, <u>Hippospongia communis</u> Lam. and <u>Spongia officinalis</u> L.)

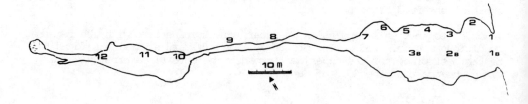

Fig. 2. Schematic map of the Grotta del Mago.
Numbers indicate the location of sampling sites.

A complete light gradient exists in the Grotta del Mago (Fig. 3). Both transmitted light and albedo curves steeply decrease, being roughly parallel up to 31m from the entrance. Transmitted light cannot be detected at a distance of 61m whilst albedo is still measurable 10m further on (71m). Reflected and diffused light therefore play a prime role in the illumination of the innermost portion of the cave.

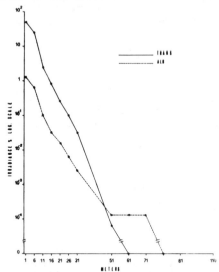

Fig. 3. Light gradient in the Grotta del Mago. Figures are given as % of the total radiation reaching the sea surface at an unshaded station outside the cave.

METHODS

Biological quantitative sampling was designed to maintain all the conditions along the light gradient as uniform as possible. Twelve samples were taken along the northern wall of the cave, at 5 meter-intervals up to 31m from the entrance (St. 7), and at 10 meter-intervals thereafter. A depth of 1.5m was selected to fit the bottom profile of the cave. Three further samples were collected from the bottom at 1, 11, and 21m from the entrance, at a depth of 6m. They were established as reference points for water movement effects. These stations were in fact rather different from the wall stations because they lay well below the 'first critical depth', defined by Riedl (1971) as the change from translating to oscillating water movement.

A sampling area of 400 cm^2 (20 x 20 cm quadrats) was used; a size which should be greater than or correspond to the 'minimal area' for macrophytobenthos in the poorest Mediterranean hard bottom communities (the problem of the 'minimal area' in marine environment is extensively discussed by Boudouresque, 1974). For the purposes of this work, it was assumed that the phytocoenosis 'minimal area' is equivalent to that of zoocoenosis.

After the sampling area had been photographed, each sample was collected by chiseling out scales of substratum until the bare rock was reached. The material was preserved and subsequently examined in the laboratory. Individuals of identified species were fully counted. Per cent cover values were determined for algae and sedentary animals (Boudouresque, 1970). Microphytes were recorded on a presence/absence criterion.

The structural analysis of the community was carried out by "factor analysis of correspondences" (Cordier, 1965) – the mathematics of which are extensively discussed in Benzécri et al. (1973). More recently, Lauro & Mongelluzzo (1976) and Chardy et al. (1976) have discussed this method and its various applications.

The significance of the extracted axes (or factors) has been tested by a 'simulation' method, as described by Lauro & Mongelluzzo (op. cit.).

The analysis was performed on the single taxocoenosis of the community, both with abundance values and with presence/absence scores. It was seen that the outcoming ordination models are similar and therefore it was decided to use the presence/absence criterion, especially for the analysis of the entire community. This overcame the problem of variates commensurability.

Calculations were obtained by the UNIVAC 1106 computer of the Centro di calcolo della Facoltá di Scienze, University of Naples, using the STACOR programme written by N.C. Lauro and M.P. Ponticelli.

DESCRIPTIVE ANALYSIS

The quantitative sampling yielded 324 species, of which 38 were Microphytes, 83 Macrophytes, 32 Porifera, 105 Polychaetes, 11 Crustaceans, 2 Sipunculids, 35 Molluscs, 5 Echinoderms and 13 Polyzoans. The species of other taxa present in the samples have not yet been identified. An additional 53 species were collected or observed during the qualitative survey of the cave.

Pattern of Zonation

The wall stations. The most obvious aspect of the community changes along the wall is the massive decrease both of the substratum cover and the species richness. Cover values are near 100% in the outer chamber and range between 10 and 40% in the connecting passage and the inner chamber. The number of species (Table 1) shows a similar trend, the mean value being 51 species per sample in the outer chamber and of 14 in the inner.

TABLE 1 Species distribution in the samples

Samples	1	1B	2	2B	3	3B	4	5	6	7	8	9	10	11	12
MICROPHYTA	20	19	21	22	5	19	9	3	1	2	2	5	0	3	1
MACROPHYTA	24	40	19	46	8	9	3	1	1	3	0	1	0	0	0
PORIFERA	12	3	10	2	7	1	5	2	3	5	3	5	5	4	1
POLYCHAETA	38	48	19	28	26	43	8	10	16	9	4	7	4	11	6
SIPUNCULIDA	1	1	1	1	1	1	0	0	0	0	0	0	0	0	0
MOLLUSCA	3	19	6	11	6	10	3	4	4	2	5	2	0	1	1
CRUSTACEA	8	7	3	4	2	4	1	0	0	1	0	0	0	0	0
POLYZOA	1	0	8	0	2	3	2	0	0	1	0	0	0	0	0
ECHINODERMATA	3	2	1	2	2	4	1	1	0	0	0	0	0	0	0
TOTAL	110	139	88	116	59	94	32	21	25	23	14	20	9	19	9

This reduction affects all the components of the community and is very evident in the inner sections of the cave, where biological settlement shows a marked patchiness. Clumps occur especially on juttings, and in cavities and crevices of the substratum which elsewhere appear bare and smooth.

As far as the single taxa are concerned, a few comments should be made.

Flora. The strongest reduction occurs at St. 3, where cover is reduced to 1/10 and the number of species to 1/4 of St. 1. Macroflora, with the exception of the Cyanophyte Lyngbia gracilis Rab. disappears completely beyond St. 7, but cover is already negligible at St. 5. Hildebrandtia prototypus Nardo seems to be the last algal element of some importance. Microfloristic elements such as the diatoms Amphora richardiana Cholnoky (St. 9), Cocconeis scutellum Ehr. (St. 8), Grammatofora marina (Lyngb.) (St. 11, 12), and Rhabdonema adriaticum Kutzing (Sta. 11) occur also at the innermost stations.

Sponges. The boring species (Clionids) are lacking, except Cliona viridis (Schmidt) which is restricted to the outer chamber. Lithistids and pharetronids, in particular Petrobiona massiliana Vacelet & Lévi, are also absent.

Coelenterates. A great number of coelenterates, reported in other caves (e.g. Riedl, 1959; Abel, 1959), are missing. Stony coral fauna is particularly poor. No diurnal rhythmical activity occurs in species living beyond 45m from the entrance. Astroides calycularis (Pall.) settled in this zone always shows open tentacles.

Polychaetes. The number of species exceeds that reported in other caves (e.g. Banse, 1959). Sedentaria represent 1/3 of the total, but serpulids are not as numerous as in other caves, especially in the dark region.

Crustacea. The most conspicuous species, Balanus perforatus (L.) does not extend beyond St. 7. This is also true of isopods. Amphipods and tanaids are represented, though scarce, in the innermost samples.

Molluscs. The majority of the species are restricted to the illuminated chamber. Only bivalves such as Amygdalum phaseolinum (Phil.), Musculus costulatus (Risso), and Anomia ephippium (L.) settle in the dark portion, where no gastropods are represented.

Other groups. Polyzoans, sipunculids and echinoderms are not present at St. 8 - 12. In the inner chamber however, small populations of Arbacia lixula (L.) and Paracentrotus lividus (Lam.), feed on fragments of Posidonia oceanica (L.) and algae brought in by the surf.

The bottom stations. Bottom stations are characterized by high algal dominance and generally richer in species than the corresponding wall stations (Table 1). This is true for all the taxa, with the exception of Porifera, the specific richness of which is higher on the wall.

As far as the changes along the bottom transect are concerned, it can be stated that there is a decrease in cover values and species numbers from St. 1B to 3B. The macroflora is the major cause of this phenomenon.

Distance from the mouth of the cave determines, within certain limits, the evolution of the community as described above. Assuming relative hydrodynamic uniformity, the variations in the community should be regarded both as indirect and direct responses to the light gradient.

Actually, there is no biological evidence for a water movement gradient along the cave. Species known as fine indicators of high water movement, such as Halocordyle disticha (Goldf.) and Aglaophenia octodonta (Stechow)(Riedl, 1959), do not show any clear zonation, but are distributed throughout the cave.

The flora shows a clear light-dependent zonation as can be seen from changes of cover values and species number. A first critical point can be located between St. 2 and 3 on the wall, where illumination decreases by one order of magnitude, passing from 25% to 2.5% of the total incident radiation. Another critical point is situated between St. 7 and 8, where the relative illumination decreases by three orders of magnitude, from 0.03% to 0.00006%. This can be assumed to be the threshold beyond which photosynthetic macrophytes cannot survive.

As far as the animal components of the community are concerned, a direct response to the light gradient can be detected in few instances. Species such as the sponge Chondrilla nucula (Schmidt) and the scleractinian Cladocora cespitosa L., are restricted to the most illuminated portion of the cave (1 - 10m from the entrance), coinciding with the needs of their symbiotic algae. A few others show a clear preference for the dim and the dark portions of the cave, e.g. the sponges Timea fasciata Topsent, T. unistellata (Bow.), Terpios fugax Duch. & Mich., Geodia cydonium (Jameson), the coelenterate Astroides calycularis (Abel, 1959), the mysid Hemimysis speluncola Ledoyer (Wittmann, pers. comm.), the brotulid fish Oligopus ater (Risso). All these are well known cave dwellers (Riedl, 1966).

Nevertheless, the light gradient can be considered indirectly responsible for the animal community zonation, through the influence of the floristic component on the quality of the substratum. It is clearly seen that in areas outside the algal zone, and excepting some irregular surfaces, the bare rock does not facilitate settlement by the larvae of sessile species. Additionally the hardness of the substratum, even prevents the establishment of endolithic forms (cyanophytes, sponges, molluscs) which would cause a biological 'conditioning' of the substratum itself and create a favourable situation for epilithic species. The combination of strong waver movement and substratum quality can be considered as the main cause of the patchiness and the poverty of the settlements in the dark parts of the cave.

On the contrary, though the substratum in the outer chamber is of the same type as in the inner sections, the light is sufficient for the establishment of algal species, particularly the crustose corallines which are well adapted to strong water movement. These form a thin layer of soft 'secondary' substratum in which endolithic species can bore, this creating settlement surfaces for the larvae and spores of sessile organisms. Finally, the latter provide favourable microenvironments for the vagile fauna.

As was assumed in the beginning of this paper, a water movement gradient explains the differences between the wall and the bottom stations. At the bottom stations, stiller water allows for the growth of a high number of algal species, the most conspicuous of which (e.g. Halopteris filicina Kütz., Cladostephus verticillatus Ag. and Phyllophora nervosa Grev.) form erect and dense populations. This fact, in combination with the horizontal

substratum, favours the accumulation of fine sediment and organic detri-
on the bottom where shelter and better food supply explains the richness of
the vagile fauna, all groups of which are well represented.

STRUCTURAL ANALYSIS

For brevity, only the results of the structural analysis of the entire
biocoenosis are discussed here.

It must first be stated that the analysis yeilded low variance percentages
for the first three axes: FI = 12,07%; FII = 10,66%; FIII = 9.64%. This
is often the case when factorial analysis of correspondences is performed on
dichotomic variates and does not affect the results (Céhessat, 1976).

Ordination Pattern

"Station-points" are ordered in the factorial space (Fig. 4) as follows:
- St. 1B, 2B, 3B, are isolated in the IV Quadrant, where they form a well
defined 'galaxy' (Gl);
- St. 1 to 7 lie in the II Quadrant, forming another galaxy (G2) in which
two sub-galaxies can be recognized: St. 1, 2, 3, 7 and St. 4, 5, 6, the
latter being very compact;
- St. 8 - 12 are scattered in the I Quadrant.

"Species-points" are clustered in the vicinity of Gl and G2 where several
'multiple points' (sites of coinciding coordinates) are seen. In the I
quadrant no clusters can be identified.

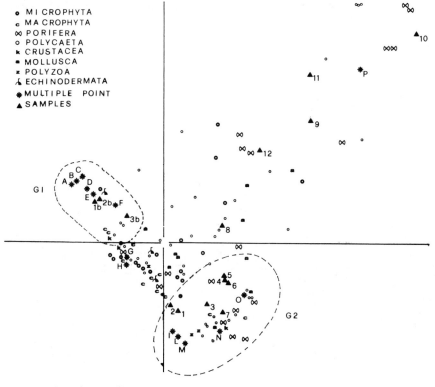

Fig. 4. Ordination pattern.

Axes Interpretation

It is essential, before interpreting the axes, to remember that they do not represent ecological factors in the classical sense, but rather a combination of highly correlated environmental parameters.

Factor I. As the bottom stations are opposed to all the others in the space of FI, it seems reasonable to interpret this axis as water movement and correlated variables (sedimentation, food transport, etc.). If this is so, there should be a positive hydrodynamic gradient from the entrance to the end of the cave, where the outer stations are less exposed than the inner ones. This is in contrast to the already mentioned opinion of Riedl (see above) and to the assumption of hydrodynamic uniformity put forward earlier.
A more detailed study of the hydrological regime of the cave in relation to its topography is therefore required. It could be inferred here that a reduction of water movement occurs in the outer section because of its funnel shape. On the contrary, the 'Venturi effect' in the corridor, the rise of the bottom and the presence of the terminal beach could amplify the water movement counteracting the expected 'cul de sac effect'.

Factor II. This second axis might be interpreted as the light factor. Indeed excepting the anomalous position of St. 7 and the bottom stations, the ordination seems to be rather consistent with this hypothesis. Three discontinuities are seen in the space of FII: between ST. 1, 2, 3 and 4, 5, 6; between St. 6 and 8; and between St. 8 and the remaining inner stations. The first might represent the boundary value between photophilic and sciaphilic biotopes. The most evident of these discontinuities (between St. 6 and 8) seems to coincide with the critical value of illumination for macrophytobenthos

The location of St. 7 and bottom stations can be a matter of speculation only. Bottom stations are less well illuminated than the corresponding ones on the wall, but this fact should be compensated for. at least partially, by their horizontal situation. The assumption could be made that the dense algal population (at least for St. 1B and 2B) favours bottom-settlement by a great number of sciaphilic species. This should largely contribute to the 'drift' of the bottom station-points in the 'dim' part of the factor.

Biocoenotic Characterization

It can be seen, from the arrangment of species-points, that two opposite clusters exist in association with G1 and G2 respectively. In G1 there are 5 remarkable 'multiple points' (A, B, C, D, F) indicating a strong similarity of the component species as far as the ecological needs are concerned. On the basis of the previous interpretation, this cluster should be composed of sciaphilic species, favouring calm biotopes. Actually it contains algae such as Peyssonnelia squamaria Decaisne, Phyllophora nervosa, Pseudolithophyllum expansum Lemoine, Rodriguezella strafforellii Schmitz, Ceramium bertholdi Funk, Botryocladia boergesenii Feld., Acrosorium venulosum Kylin, Heterosiphonia würdemannii Falkbrg., Hypoglossum woodwardii Kützing, the gastropod Mathilda retusa Brugnone, the bivalve Striarca lactea (L.) and the polychaetes Eusyllis lamelligera Marion & Bobretzky, Pionosyllis pulligera (Krohn), Sclerocheilis minutus Grube, Potamilla torelli Malmgren, which, according to various authors (Boudouresque, 1970, 1971; Bombace, 1970; Pérés & Picard, 1964; Bellan, 1964; Cognetti, 1957), belong to deep infralittoral and circalittoral biotopes. Species such as Sphaerosyllis hystrix Claparéde, _Exogone verugera_ (Claparéde), _Notomastus latericeus_ Sars, _Caulleriella bioculata_ (Keferstein), _Thelepus concinnatus_ (Fabricius) and _Lumbrinereis gracilis_ (Ehlers) can be considered as incursives from soft bottoms.

In G2 (see especially 'miltiple points' I, L, M, N) the community is composed partly of more photophilic elements, such as _Amphiroa rigida_ Lamour, _Cladophora coelotrix_ Kützing, _Chondrilla nucula_, _Clathrina contorta_ (Bow.), _Leuconia aspera_ (Schmidt), mixed with sciaphilic species, e.g. _Gelidium pectinatum_ Montagne, _Ceramium codii_ Mazoyer and _C. gracillimum_ Mazoyer, all tolerating strong water movement. Species inhabiting microcavities as well as boring elements are present here, too.

The weaker cluster existing between G1 and G2, is composed of species having a greater ecological spectrum. They seem to form a sort of 'continuum' between the two opposite biocoenotical 'noda' of G1 and G2.

As has been previously seen, no cluster can be identified in the first Quadrant, indicating that a real biocoenotic complex does not exist in the dark section of the cave. The ordination suggests the idea of an impoverished 'facies', a sort of an embryo of the 'biocoenose des grottes obscures' described by Pérés & Picard (1964), of which it contains a few elements capable of standing the strong water movement and the unfavourable substratum.

Notwithstanding the strong bias introduced by the quality of the substratum, the indication that the majority of sessile animal species adapted to the darkness or dim light are also specialized for low water movement appears very interesting. This had been inferred by Harmelin (1969) and by Zibrowius (in press) and would explain the absence from the Grotta del Mago of extremely sciaphilic and rheophobic species such as lithistid and pharetronid Porifera (see also Vacelet, 1964).

CONCLUSIONS

In the light of the information obtained both from the descriptive study and the structural analysis of the community, the following conclusions can be drawn:

-- the assumption of hydrodynamic uniformity of the biotope has probably to be rejected, and a positive water movement gradient should be assumed;

-- the effects of the light gradient are biased or obscured both by the water movement gradient and the quality of the substratum;

-- the response of the community to light changes is mostly indirect, as it is mediated by the algal component, especially as far as the biological 'conditioning' of the microenvironment is concerned;

-- sciaphilic and spelaeophilic species seem to react negatively to strong water movement;

-- two coenotic complexes can be distinguished in the outer section of the cave, the first composed of sciaphilic species requiring low water movement, the second both photophilic and sciaphilic elements tolerating strong water movement with a progressive reduction of the algal component;

— no association can be identified in the inner portion of the cave, where the settlement appears as an extremely impoverished 'facies' of the dark cave biocoenosis.

REFERENCES

Abel, E. 1959 Zur Kenntnis der marinen Höhlenfauna under besonderer Berücksichtigung der Anthozoa. Pubbl. Staz. Zool. Napoli (Suppl.) 30, 1-94.

Banse, K. 1959 Ueber die Polychaeten. Besiedlung einiger submariner Höhlen. Pubbl. Staz. Zool. Napoli (Suppl.) 30, 417-469.

Bellan, G. 1964 Contribution à l'étude systematique, bionomique et ecologique des Annélides Polychètes de la Méditerraneé. Rec. Trav. St. Mar. Endoume (34-49), 1-372.

Benzécri, J.P. et Al. 1973 L'Analyse des données. I. La Taxinomie. II. L'Analyse des Correspondences. Bunod, Paris.

Bombace, G. 1970 Notizie sulla Malacofauna e sulla Ittiofauna del Coralligeno di falesia. Unioncamere Palermo, Quaderni di ricerca e Sperimentazione 14, 1-77.

Boudouresque, C.F. 1970 Recherches de bionomie analytique, structurale et expérimentale sur les peuplements benthiques de Méditerranée Occidentale (fraction algale). Thèses CNRS, Paris.

— 1971 Recherches de bionomie analytique, structural et expérimentale sur les peuplements benthiques de Méditerranée Occidentale (fraction algale). La soustrate sciaphile des peuplements des grandes Cystoseira de mode battu. Bull. Mus. Hist. Nat. Marseille 31, 141-151.

— 1974 Aire minima et peuplements algaux marins. Soc. Phycol. de France, Bull. N° 19, 141-157.

Céhessat, R. 1976 Exercises commentés de Statistique et Informatique Appliquées Dunod, Paris.

Chardy, P. et Al. 1976 Application of Inertia Methods to Benthic Marine Ecology: Practical Implications of the Basic Options. Estuarine and Coastal Marine Science 4, 179-205.

Cognetti, G. 1957 I Sillidi del Golfo di Napoli. Pubbl. Staz. Zool. Napoli, 30, 1-100.

Cordier, B. 1965 L'Analyse des Correspondences. Théses. Fac. Sci. Rennes.

Düing, W. 1965. Strömungsverhältnisse im Golf von Neapel. Pubbl. Staz. Zool. Napoli 34, 256-316.

Harmelin, J.G. 1969 Bryozoaires des Grottes Submarines Obscures de la Region Marseillese. Faunistique et écologie. Tethys 1 (3), 793-806.

Lauro, N.C. & R. Mongelluzzo 1976 Metodologie statistiche per la analisi dei sistemi. Atti Simp. 'Ingegneria dei Sistemi', Roma, 11-12 Dic. 1975. Associazione Elettrotecnica ed Elettronica Italiana, 1-18.

Péres, J.M. & J. Picard 1964 Nouveau manuel de bionomie benthique de la Me Mediterranée. Rec. Trav. St. Mar. Endoume (31-47), 1-137.

Riedl, R. 1959 Die Hydroiden des Golfes von Neapel und ihr Anteil an der Fauna unterseeischer Höhlen. Pubbl. Staz. Zool. Napoli (Suppl.) 30, 589-755.

-- 1966 Biologie der Meereshohlen. Parey, Hamburg.

-- 1971 Water movement. General aspects of water movement. Animals. In O. Kinne Marine Ecology, Vol. 1, pt. 2, 1123-1156.

Rittmann, A. 1930 Geologie der Insel Ischia. Reimer, Berlin.

Vacelet, J. 1964 Etude monographique de l'Eponge calcaire Pharétronide de Méditerranee Petrobiona massiliana Vacelet & Levi. Les Pharétronides actuelles et fossiles. Theses Fac. Sci. Univ. Aix-Marseille.

Zibrowius, H. (in press) Les Scléractinaires des grottes sousmarines en Méditerranée et dans l'Atlantique Nord-Oriental.

BIONOMIE BENTHIQUE DU PLATEAU CONTINENTAL DES ILES KERGUELEN. 8. VARIATIONS SPATIALES ET TEMPORELLES DANS LE PEUPLEMENT DES VASES A SPICULES

Daniel Desbruyères* et Alain Guille**

*Laboratoire de Zoologie-Vers, Muséum National d'Histoire Naturelle, 43 rue Cuvier, 75005 Paris (France)
**Laboratoire de Biologie des Invertébrés Marins, Muséum National d'Histoire Naturelle, 55 rue de Buffon, 75005 Paris (France)

ABSTRACT

The changes through space and time in the faunistic composition of two taxa (Polychaeta, Echinodermata) were studied during 1974-1975 at a station in Morbihan Bay (Kerguelen isles).

Inertia analysis was employed in this study using such factors as Euclidian distance and Spearman's rank correlation coefficient. Seasonal variations are obscured by the spatial heterogeneity of populations, by the presence of a dominant species, Amphicteis gunneri (50% of density of Polychaeta), and by the seasonal introduction of species from neighbouring biotopes.

INTRODUCTION

L'étude des écosystèmes marins antarctiques et subantarctiques revêt un grand intérêt par l'importance des concentrations présentes en matière vivante (plancton, poissons, benthos animal et végétal) liées, au plan physique, aux mouvements des masses d'eau au niveau des fronts hydrologiques (convergence et divergence antarctiques)et aux importants apports concomitants en sels nutritifs. Pourtant aucune étude de biocénotique analogue à celles réalisées il y a plus d'un demi-siècle dans l'Océan Glacial Arctique, n'avait été menée jusqu'à ces dernières années marquées, notamment, par l'important mémoire d'Arnaud (1974), axée essentiellement sur le benthos côtier de la Terre Adélie.

Les îles Kerguelen offrent à cet égard des possibilités particulières de contribution à l'économie marine subantarctique par la présence d'un plateau continental bien développé. Ces îles sont situées à la limite de la convergence antarctique, baignées par des eaux froides dont l'amplitude thermique annuelle ne dépasse pas 6°C. Les biomasses benthiques, suivant nos premiers résultats limités au golfe du Morbihan, sont exceptionnellement élevées (moyenne : 45.87 g/m^2, poids sec décalcifié), nettement supérieures à celles signalées dans des conditions similaires dans l'Océan Glacial Arctique, et de dix à quarante fois plus importantes que celles connues des mers tempérées et tropicales (Desbruyères et Guille, 1973). Un programme de recherches benthiques est mené à Kerguelen depuis 1972 (Guille et Soyer, 1976), programme caractérisé par deux directions complémentaires : l'une synécologique, inventaire numérique et pondéral de la faune, reconnaissance, délimitation, dynamique des communautés présentes, l'autre autécologique, mise en évidence des adaptations écophysiologiques à l'environnement subantarctique au niveau reproduction, croissance, longévité, éthologie alimentaire, métabolisme de quelques benthontes "leaders". L'étude présentée ici s'intègre danc dans la première direction de ce programme.

Les vases à spicules d'éponges caractérisent la plus grande partie des fonds du golfe du Morbihan, véritable mer intérieure de l'archipel de Kerguelen ; ils constituent également un des principaux types sédimentaires du plateau continental de l'Antarctique. Nous avons limité l'étude de l'évolution spatiale et temporelle de la faune de ce biotope, pour des raisons matérielles, à deux taxons : les Annélides Polychètes et les Echinodermes. Ce choix arbitraire, malgré l'importance de l'un des deux taxons, les Annélides Polychètes, largement dominant qualitativement et quantitativement dans la communauté, ne donnera donc qu'une image incomplète de la dynamique de celle-ci.

PHYSIOGRAPHIE

Géomorphologie et Sédimentologie

L'archipel des îles Kerguelen est situé par 50° de latitude sud et 70° de longitude est, environ. Il est l'un des sommets émergés de la dorsale Kerguelen-Gaussberg qui sépare, dans l'Océan Austral, le bassin atlantico-indien du bassin antarctico-indien. L'archipel, d'une superficie de 7000 km^2, groupe environ 300 îles et îlots. Le Golfe du Morbihan, dont la superficie est d'environ 700 km^2, est situé à l'est de l'île principale (Grande Terre) et s'ouvre sur l'Océan Austral par la Passe Royale dont la largeur et la profondeur atteignent respectivement 12 km et 45 m. Le Golfe présente deux parties distinctes : la région nord-est, bien dégagée, a une profondeur moyenne de 50 m mais est entaillée de trois fosses dont la profondeur dépasse 100 m ; la région sud-ouest est au contraire parsemée de nombreuses îles et se prolonge par des fjords étroits dont la longueur peut dépasser 50 km.

Les vases à spicules d'éponges sont localisées uniquement dans la région nord-est du golfe du Morbihan et en occupent la plus grande partie à partir de 30 m de profondeur jusqu'au fond des fosses qui l'entaillent (Fig. 1). Ces vases sont constituées de sablons et de poudres (80% des particules sont inférieures à 40 µ) dont la fraction siliceuse insoluble, très importante, provient de frustules de diatomées et de spicules d'éponges des genres <u>Tetilla</u> et <u>Cinarchyra</u>. Ces spicules peuvent former de véritables feutrages sur le fond, représentant parfois plus de 3.5 kg/m^2 en poids sec lavé ; elles permettent la stabilisation des vases fines, favorisent la fixation des benthontes sessiles, jouant un rôle biocénotique important par la formation de microbiotopes. Transportées par la houle et les courants, elles proviennent du talus continental ou de plus lointaines distances et se déposent à l'intérieur du golfe, véritable bassin de sédimentation.

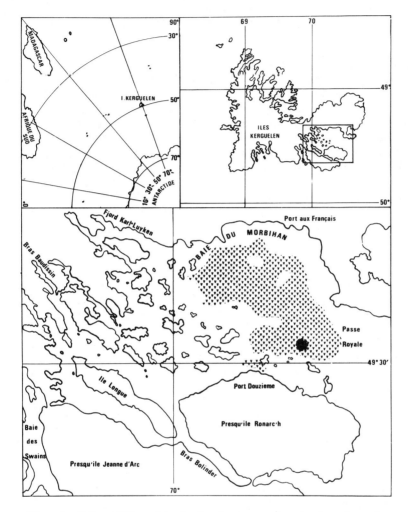

Fig. 1. Répartition des fonds de vases à spicules dans le Golfe du Morbihan (îles Kerguelen). Localisation de la station d'étude.

Hydrologie

Le caractère de bassin à seuil de dilution du golfe du Morbihan modèle son régime hydrologique. C'est le cas des bassins situés à des latitudes où les apports d'eau par ruissellement et précipitations sont supérieurs aux quantités enlevées par évaporation. Cette structure rend compte des brusques variations hydrologiques erratiques qui explique la profonde originalité de ce golfe dans le milieu subantarctique.

En hiver, de juin à octobre, l'homothermie est de règle sur la verticale au-dessus des fonds de vases à spicules, et la température est de l'ordre de 2°C. Cette dernière est au contraire maximale à la fin de l'été, en février-mars, atteignant ou dépassant légèrement 8° C. A cette époque une thermocline plus ou moins marquée est présente.

La salinité a une évolution plus complexe. Au printemps et en été ses variations sont faibles, comprises entre 32.80 et 33.50 ‰. Les précipitations automnales, abondantes, provoquent une baisse importante de la salinité (31.50 ‰) brusquement compensées par des apports océaniques liés aux tempêtes de vecteur sud. Celles-ci poussent les eaux extérieures de l'Océan Austral dans le golfe ; ces eaux plus denses, une fois le seuil de la Passe Royale franchi, s'enfoncent en profondeur puis se mélangent progressivement. La salinité au voisinage du fond augmente alors brutalement (33.10 ‰), conjointement aux valeurs de l'oxygène dissous et des silicates. En hiver, la salinité se stabilise progressivement autour de 33 ‰. L'amplitude de ces variations erratiques diminue vers le fond du golfe mais est maximale au niveau des vases à spicules.

BIONOMIE

Comparativement aux autres peuplements du golfe du Morbihan, la macrofaune des vases à spicules d'éponges est caractérisée par les valeurs les plus faibles de l'indice de diversité de Shannon-Wienner (Desbruyères, 1976). La situation du biotope exposé à l'amplitude maximale des variations erratiques de la salinité constitue sans doute un des facteurs limitant la diversification ; il est de même de la nature du substrat, notamment pour les espèces fouisseuses.

La densité spécifique moyenne est, à 50 m de profondeur, de 33 espèces/m^2 dont 42 % d'Annélides Polychètes, de 24 % de Crustacés, suivis des Mollusques (15 %) et des Echinodermes (9 %). Les Annélides Polychètes sont également largement dominants numériquement puisqu'ils participent pour 47 % à la densité moyenne (1800 ind./m^2) alors que les Echinodermes ne représentent que 5.8 %. Parmi les Crustacés (18 %), les isopodes Arcturidae, les Amphipodes Lysianassidae, Aoridae, Lilljeborgiidae et Corophiidae sont très abondants. Les Ascidies (15 %) sont aussi très denses, des Molgulidae et des Pyuridae pouvant atteindre la densité de 200 ind./m^2 dans certaines stations. Les mollusques ne participent que pour 2.7 % à la densité numérique moyenne avec notamment un Struthiolariidae (Perissodonta mirabilis) et deux Nuculanidae (Yoldia isonota et Yoldia subaequilateralis). Il n'en est pas de même pondéralement où, au contraire, les Mollusques représentent plus de 51 % de la biomasse moyenne (29.88 g/m^2, en poids sec décalcifié), suivis des Annélides Polychètes (25.5 %), des Crustacés (11 %), des Echinodermes (6.2 %) et de divers groupes et particulièrement les Ascidies (5.6 %).

Au plan biocénotique, les vases à spicules d'éponges sont caractérisées par la présence de l'Ampharetidae Amphicteis gunneri, espèce la plus abondante du peuplement, pouvant atteindre la densité de 1540 ind./m^2. La localisation de la station d'étude de l'évolution du peuplement a été choisie dans la partie du biotope où Amphicteis gunneri présente une densité maximale. Cette espèce cosmopolite, décrite des côtes de Norvège, a une très large distribution latitudinale et bathymétrique. La comparaison entre les exemplaires de Kerguelen et ceux des côtes de France montre l'existence d'un nanisme des formes sub-antarctiques. La distribution de cette espèce dans les vases à spicules est en aggrégats comme c'est le cas pour toutes les espèces abondantes du peuplement. Sa reproduction a lieu au printemps et en été, révélée par l'existence de deux cohortes de jeunes ; sa croissance est faible marquée par une importante perte de poids hivernale (Desbruyères, 1977).

Aux côtés d'Amphicteis gunneri et parmi les espèces des deux taxons dont nous avons suivi l'évolution, deux autres Annélides Polychètes et un Echinoderme sont abondants et constants des vases à spicules (Desbruyères, 1977; Guille, 1976) : l'Orbiniidae Haploscoloplos kerguelensis (densité maximale 580 ex./m^2) et le Lumbrineridae Lumbrineris magalhaensis (densité maximale 210 ex./m^2), enfin l'Amphiuridae Amphiura eugeniae (densité maximale 81 ex./m^2).

ECHANTILLONNAGE

La station d'étude est située au sud de la fosse de l'Océanographie, par 51 m de profondeur (Fig. 1). Toutes les opérations de prélèvements ont été effectuées dans un rayon de 0,1 Mille nautique autour d'un signal sur mouillage permettant une localisation précise, à l'aide d'une benne Smith-Mc Intyre de 0.1 m^2 d'ouverture. La faune a été triée sur un tamis de 1mm de vide de maille. Les prélèvements ont été réalisés, au nombre de 3 à 10 suivant les mois, en juin, novembre et décembre 1974, et en avril, mai, juillet, août et septembre 1975. L'étude d'un cycle synécologique nécessite un rythme très régulier de l'échantillonnage, cependant les conditions logistiques et météorologiques à Kerguelen ont rendu celui-ci particulièrement insuffisant pendant la période estivale (décembre/février).

METHODES D'ETUDE

Les méthodes d'inertie, dont l'emploi se généralise en bionomie benthique, permettent de résumer les relations entre les "points-espèces" ou les "points-prélèvements" en ajustant un ensemble de points munis de masses et de distances par un sous-ensemble de dimensions réduites (Chardy et al., 1976). La recherche des axes d'inertie du nuage de points permet une hiérarchisation de l'influence des groupes de facteurs écologiques qui leurs sont liés. Deux méthodes d'analyse complémentaires ont été utilisées :
a) l'analyse générale (Lebart et Fenelon, 1971) permet une représentation euclidienne dans un espace multidimensionnel de points ayant tous la même masse.

La distance choisie est la distance euclidienne :

$$d^2(i_1, i_2) = \sum_{j=1}^{n} \left[X(i_1, j) - X(i_2, j)\right]^2$$

où X est le nombre d'individus de l'espèce j dans le prélèvement i, les effectifs ayant subi la transformation $y = \log(x + 1)$.

Afin de conserver l'ensemble des axes de variation, l'origine n'est pas déplacée au centre du nuage de points (les observations ne sont pas centrées). Dans cette analyse, une parfaite symétrie existe entre les variables (espèces) et les observations (prélèvements) permettant de superposer les deux représentations et constituant ainsi un outil très puissant. Comme Chardy et al (1976) l'ont souligné, l'analyse générale est bien adaptée pour l'étude de l'économie de systèmes faunistiquement homogènes: c'est une approche fonctionnelle, puisqu'elle tient directement compte des effectifs des espèces.

b) la seconde méthode est une variation de la première par l'emploi comme distance dans l'analyse générale du <u>coefficient de corrélation de rang de Spearman</u>, les prélèvements étant rangés selon les espèces (matrice "R") :

$$\mu = 1 - \left[\frac{6 \sum_{i=1}^{n} D_i^2}{n^3 - n}\right] \quad \text{où } D_i^2 = \left[R(j_1, i) - R(j_2, i)\right]^2$$

Dans cette analyse, chaque observation (espèce) a la même contribution absolue à l'inertie de l'analyse; R étant le rang du prélèvement j pour l'espèce i. Compte-tenu du nombre important d'absences dans la matrice des données, il existe un très grand nombre d'exaequo de rang élevé ayant donc un poids important dans la distance entre prélèvements. De ce fait, cette analyse met en évidence une structure liée aux espèces rares et constitue une approche indicatrice à dominance qualitative.

39 prélèvements et 43 espèces (cf annexe) sont entrés dans la matrice de chaque analyse; les espèces présentes dans un seul prélèvement ont été éliminées.

RESULTATS

Approche Fonctionnelle. (Fig. 2)

Le premier axe (non figuré) donne l'importance de la stabilité relative du phénomène dans le temps sur le plan quantitatif. Il représente 78.7 % de l'inertie totale et est dû aux trois espèces constantes dans les prélèvements : <u>Amphicteis gunneri</u>, <u>Haploscoloplos kerguelensis</u> et <u>Lumbrineris magalhaensis</u>. Les effectifs de ces trois espèces subissent des variations qui ne sont pas interprétables dans le temps mais qui sont dues à l'hétérogénéité spatiale très forte des différentes populations. Ainsi l'évolution des populations d'<u>Amphicteis gunneri</u> montre une surdispersion des classes d'âge dans l'espace au cours de la même période (Desbruyères, 1977). L'effectif global de cette espèce ne peut donc traduire l'influence de la mortalité et du recrutement ; les variations de cet effectif ne peuvent induire dans le temps un cycle synécologique. Or cette espèce représente 50.66 % de l'effectif total et a une contribution absolue de 33 % à l'inertie de l'analyse.

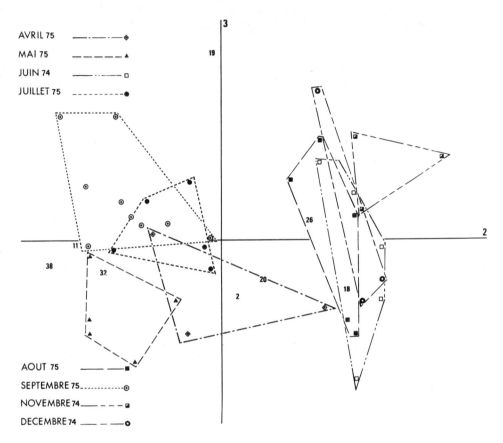

Fig. 2. Analyse générale (distance euclidienne), représentation dans le plan des axes 2 et 3 (codification des espèces : cf annexe).

Une fois extraite la variabilité spatiale dues aux espèces abondantes, la dispersion reste importante à l'intérieur des échantillons mensuels, en liaison sans doute avec une microdistribution des populations. La tendance générale mise en évidence selon l'axe 2 (5.3 % de l'inertie totale) fait apparaitre une modification faunistique globale entre les années 1974 et 1975 bienque les prélèvements d'août 1975 montrent un retour passager aux conditions de l'année précédente.

Les espèces qui contribuent le plus à l'inertie de l'axe 2 sont Sphaerosyllis perspicax (32), avec une contribution de 24.01 % à l'inertie de l'axe, Chaetozone setosa (11) (16.36 %), Lumbrineris magalhaensis (18) (13.18 %),

Trichobranchus glacialis (38) (12. 12 %) et Pherusa kerguelarum (26) (6. 20 %).
Ce sont des espèces d'effectifs moyens qui ne représentent au total que 14. 5 %
du peuplement. Le retour passager en août 1975 aux conditions de 1974 est dû
à une poussée quantitative de Pherusa kerguelarum et une quasi disparition
des Chaetozone setosa et des Trichobranchus glacialis.

Selon l'axe 3 (2. 9 % de la variance) Maldane sarsi (19) est responsable de
la distribution des barycentres d'avril 1975 à septembre 1975, mettant en
évidence un enrichissement en Maldane sarsi au cours de la même période.

Approche Indicatrice. (Fig. 3)

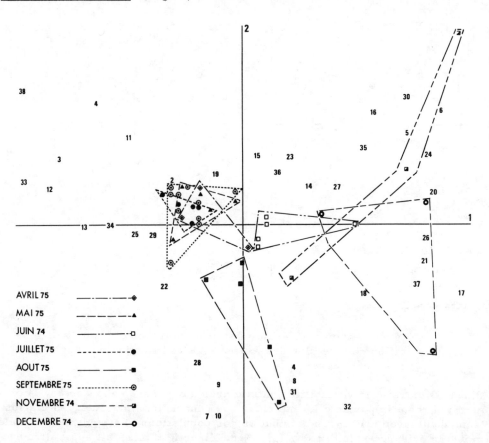

Fig. 3. Analyse générale (coefficient de corrélation de rang
de Spearman), représentation dans le plan des axes 1 et
2 (codification des espèces : cf annexe).

Le premier axe n'est plus trivial et représente 16. 6 % de l'inertie totale.
Les prélèvements effectués en 1974 et août 1975 ont une dispersion très accusée dans le plan 1-2 alors que l'homogénéité faunistique des autres échan-

tillons mensuels est remarquable. Les prélèvements d'avril, mai, juillet, et septembre 1975 sont caractérisés par la présence d'espèces limicoles (Trichobranchus glacialis, Sphaerosyllis perspicax, Spiophanes tcherniai, Eulalia sp., Amphicteis gunneri, Aglaophamus ornatus etc...) à l'exception de Haploscoloplos kerguelensis qui est constante dans le biotope mais cependant préférante des milieux sablo-vaseux.

L'axe 2 représente 12 % de la variance. Les échantillons prélevés en août 1975 sont séparés vers le pôle négatif des autres prélèvements mensuels de la même année par la présence d'espèces peu fréquentes, caractéristiques de substrats hétérogènes voisins, en particulier organogènes, (Amphitrite kerguelensis (4), Anobothrus sp. (8), Scoloplos marginatus (31) et Platynereis magalhaensis (28)). Le cas de cette dernière espèce a été plus particulièrement approfondi (Desbruyères, 1977). Platynereis magalhaensis est un Nereidae tubicole qui vit à la face inférieure des galets et des blocs marquant la limite bathymétrique inférieure des herbiers à Macrocystis pyrifera. Au cours de l'hiver se produit une reproduction à l'état atoque, les adultes gagnent les frondes des Macrocystis où les femelles pondent puis meurent. Les mâles entretiennent les pontes situées dans des tubes à la surface des frondes. Sous l'action des très violentes tempêtes de cette période ces frondes sont arrachées entrainant la dissémination des juvéniles et des mâles vers le biotope des vases à spicules. Un autre cas intéressant est celui des Amphiuridae juvéniles (7) explicable par le recrutement hivernal des espèces Amphiura eugeniae et Amphiura angularis dans le peuplement. De même Sphaerosyllis perspicax est un Exogoninae germmophore à reproduction hivernale (août).

L'axe 2 est fortement affecté par l'hétérogénéité spatiale des prélèvements de novembre et décembre 1974 et d'août 1975.

CONCLUSION

Les deux analyses permettent des approches complémentaires dans la mise en évidence des variations faunistiques du peuplement. Elles montrent l'existence de deux ensembles faunistiques disjoints dans le temps : 1°) les prélèvements de 1974, 2°) les prélèvements de l'année 1975 à l'exception de celui du mois d'août. Ce dernier se regroupe avec les prélèvements de 1974 dans le cas de l'approche fonctionnelle et s'isole dans celui de l'approche indicatrice. Les différences entre les deux ensembles se manifestent, au niveau quantitatif, par des poussées temporaires des effectifs d'espèces moyennement représentées et non exclusives des vases à spicules telles Lumbrineris magalhaensis et Pherusa kerguelarum. Au niveau qualitatif, ces différences s'expliquent par la présence accidentelle d'espèces, liée à un mode de reproduction (formes d'essaimage et formes juvéniles), ainsi que par l'apparition de formes juvéniles indéterminées mais d'espèces constantes du peuplement.

Les deux analyses montrent également une très forte variabilité spatiale du peuplement (70 % de la variabilité quantitative) pouvant être attribuée à deux causes, la répartition en mosaïque des populations, la faible diversité du peuplement. La première de ces causes est mise en évidence par l'étude de

photographies sous-marines du biotope de la station étudiée (Fig. 4. A. B.).
Elles montrent l'existence d'îlots de faune de substrats durs (Ascidies, Spongiaires, Actiniaires) (Fig. 4. B.). La fixation de ces benthontes sessiles est
liée à la formation et à l'accumulation de sortes de nodules de spicules d'éponges cimentés par de la vase fine, sous l'action des courants. De même
Soyer et de Bovée (1976) ont souligné la surdispersion des populations de la
méiofaune dans le même biotope. La seconde de ces causes, la faible diversité du peuplement, donne une très forte importance quantitative à l'espèce
dominante, Amphicteis gunneri, dont l'étude de la dynamique de sa population au cours de la période considérée a montré également une surdispersion
tant au niveau des effectifs totaux qu'à ceux des différentes classes d'âge.

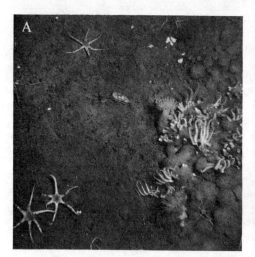 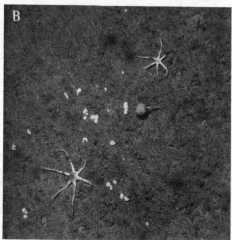

Fig. 4. Aspects des fonds de vases à spicules à la station d'étude
(51 m de profondeur). A - Accumulation de nodules de spicules d'éponges formant un substrat dur sur lequel sont fixés des Spongiaires (? Isodictya kerguelenensis (Ridley et
Dendy), des Actinies (? Glyphoperidium bursa Roule), des
Ophiurides (Ophiacantha vivipara Ljungman).; B - Aspect
général (présence de l'Ophiuride Ophionotus hexactis et de
Calcisponges).

Si les deux analyses, à partir de la distance euclidienne et du coefficient de
corrélation de rang de Spearman, dégagent les mêmes résultats principaux,
l'approche indicatrice indique que les variations saisonnières au sein des vases à spicules sont essentiellement liées à l'intrusion d'espèces non caractéristiques du biotope à certaines périodes de l'année, phénomène en relation
avec l'histoire des communautés voisines. Les variations saisonnières propres au peuplement ne sont pas mises en évidence, masquées par l'hétérogénéité spatiale de celui-ci, sans que l'on puisse en conclure quant à leur
amplitude. L'évolution du peuplement apparait sous forme d'une dérive synécologique dans le temps au lieu d'un cycle.

D'une manière plus générale, de précédents résultats sur la dynamique de population et la croissance - faible - de trois espèces de Polychètes des vases à spicules (Desbruyères, 1977) laissent supposer que le benthos sub-antarctique, malgré sa grande richesse pondérale, présente un turn-over faible, essentiellement dû au recrutement. Cette hypothèse s'accorde avec les particularités de la faune de ces eaux froides australes, le nanisme de nombreuses espèces, ou au contraire, le gigantisme allié à une grande longévité.

ANNEXE

Codification des espèces retenues dans les deux analyses.

1. Abatus cordatus (Verrill)
2. Aglaophamus ornatus Hartman
3. Amphicteis gunneri Sars
4. Amphitrite kerguelensis Mc Intosh
5. Amphiura angularis Lyman
6. Amphiura eugeniae Ljungman
7. Amphiuridae juvéniles
8. Anobothrus sp.
9. Brada mammillata Grube
10. Capitella capitata (Fabricius)
11. Chaetozone setosa Malmgren
12. Maldanidae indét.
13. Eulalia sp.
14. Eumolpadia violacea (Studer)
15. Flabelligera pennigera Ehlers
16. Haploscoloplos kerguelensis (Mc Intosh)
17. Harmothoë gourdoni Gravier
18. Lumbrineris magalhaensis Kinberg
19. Maldane sarsi Malmgren
20. Neanthes kerguelenensis (Mc Intosh)
21. Notomastus latericeus Sars
22. Ophiacantha vivipara Ljungman
23. Ophionotus hexactis (Smith)
24. Amphiophiura brevispina (Smith)
25. Ophiuridae juvéniles
26. Pherusa kerguelarum (Grube)
27. Phyllocomus crocea Grube
28. Platynereis magalhaensis Kinberg
29. Polycirrus kerguelensis (Mc Intosh)
30. Potamilla antarctica (Kinberg)
31. Scoloplos marginatus (Ehlers)
32. Sphaerosyllis perspicax Ehlers
33. Spiophanes tcherniai Fauvel
34. Spionidae indét.
35. Stereoderma laevigata (Verrill)
36. Terebellides stroemi kerguelensis Mc Intosh
37. Thelepus setosus (Quatrefages)

38 Trichobranchus glacialis Malmgren
39 Cirratulidae indét.
40 Euchone pallida Ehlers
41 Exogone verrugera (Claparède)
42 Ammotrypane sp.
43 Hermadion magalhaensis Kinberg

REMERCIEMENTS

Cette étude a été réalisée et financée dans le cadre du programme Benthos-Ker de la Direction des Laboratoires Scientifiques des Terres Australes et Antarctiques Françaises. Les calculs ont été effectués sur l'ordinateur CII 10070 du Centre Océanologique de Bretagne. P. Chardy nous a aidé dans le traitement mathématique des données.

REFERENCES BIBLIOGRAPHIQUES

Arnaud, P.M. 1974 Contribution à la bionomie benthique des régions antarctiques et sub-antarctiques.
Téthys 6 (3), 465-656.
Chardy, P., Desbruyères, D., et A. Laurec 1976 Bionomie benthique du plateau continental de l'Archipel des îles Kerguelen. Macrofaune. 4. Analyse multivariable des taxocénoses annélidiennes du Golfe du Morbihan.
Com. natl. Fr. Rech. antarct. 39, 97-105.
Desbruyères, D. 1976 Bionomie benthique du plateau continental de l'Archipel de Kerguelen. Macrofaune. 2. Diversité des peuplements annélidiens benthiques dans un système fjordique adjacent au Golfe du Morbihan.
Proc. Thd. SCAR Symp. Antarct. Biol. (Washington), Gulf Publ. Comp., Houston (sous presse).
Desbruyères, D. 1977 Bionomie benthique du plateau continental des îles Kerguelen. Macrofaune. 6. Evolution des populations de trois espèces d'Annélides Polychètes en milieu sub-antarctique.
Com. natl. Fr. Rech. antarct. 42 (sous presse),
Desbruyères, D. et A. Guille 1973 La faune benthique de l'Archipel de Kerguelen. Premières données quantitatives.
C.R. Hebd. Séanc. Acad. Sci., Paris, sér. D 276, 633-636.
Guille, A. 1976 Benthic bionomy of the continental shelf of the Kerguelen Islands. Macrofauna. 1. Echinoderms of the Morbihan gulf. Quantitative data.
Proc. Thd. SCAR Symp. Antarct. Biol. (Washington), Gulf. Publ. Comp., Houston (sous presse).
Guille, A. et J. Soyer 1976 Prospections bionomiques du plateau continental des îles Kerguelen. Golfe du Morbihan et Golfe des Baleiniers.
Com. natl. Fr. Rech. antarct. 39, 49-82.
Lebart, L. et J.P. Fenelon 1971 Statistique et informatique appliquées.
Dunod, Paris, 425 p.
Soyer, J. et F. de Bovée 1976 Premières données sur les densités en méiofaune des substrats meubles du Golfe du Morbihan (archipel de Kerguelen).
Proc. Thd. SCAR Symp. Antarct. Biol. (Washington), Gulf. Publ. Comp., Houston (sous presse).

SOME OBSERVATIONS ON THE RELATIVE ABUNDANCE OF SPECIES IN A BENTHIC COMMUNITY

R. A. Eagle and P. A. Hardiman

MAFF, Fisheries Laboratory, Burnham-on-Crouch, Essex, UK

ABSTRACT

The macrobenthos retained on a 0.5 mm sieve was sampled at 14 stations in Lyme Bay (western English Channel). Thirteen stations were occupied by an Echinocardium cordatum/Amphiura filiformis community, which objective analysis showed had a uniform species composition throughout the area. Numerically the fauna was dominated by Chaetozone setosa and Magelona filiformis. Other small polychaetes, bivalves and crustacea were also abundant. The data from the 13 sites were pooled, giving a total sample area of 2.6 m^2, so that an examination of the community structure could be made.

On comparing the data to some proposed species' relative abundance models, it appeared that four abundance classes of species were present. Although of no apparent biological significance, this observation will be of interest to ecologists searching for a unifying theory for species' relative abundances. Perhaps it will be eventually shown that there is no such unity underlying community structure.

Over 80% of the animals present in the community were detritivores feeding at the sediment/water interface. Though the lack of precise information on the type of food utilized by each benthic species was highlighted, it appeared that there was competition between many species at this layer. A possible factor involved in this apparently unbalanced competitive situation is the magnitude and period of hydrographic fluctuations in the western Channel, relative to the response time of the community, such that a climax community has not been attained.

INTRODUCTION

The purpose of this paper is to emphasize how difficult it can be to fit some community data into general theories of diversity. Furthermore, we will point out that we may not yet have sufficient knowledge about the biology of many benthic species to assess the role of competition in maintaining benthic community composition. Consequently we have been unable to arrive at any constructive conclusions. Our implication is that there may be no theory capable of describing the composition of every community, although some of the interacting regulatory processes (e.g. competition) can be modelled.

The data we present here were collected as part of an ongoing programme of the Ministry of Agriculture, Fisheries and Food to monitor the effects on the benthos of waste disposal at sea around England and Wales. The area under investigation was in the west of Lyme Bay (Fig. 1) where sewage sludge from Exeter is dumped. Water depths ranged from 30-45 m and the sediment was a very fine muddy sand with clay.

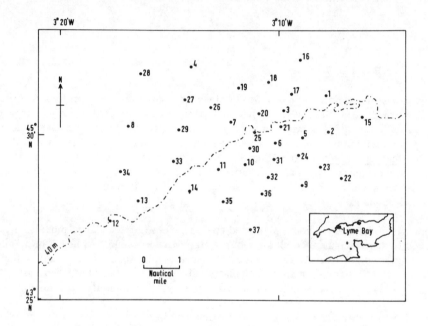

Fig. 1. Sampling positions and location of Lyme Bay.

We ascertained that the benthic fauna showed no detectable effect of sludge dumping. The results allowed an investigation into the structure of the community to be made because it became apparent that the bottom community was very uniform and that we had in effect collected from it a single large sample. Because we did not collect the data to test any preconceived hypothesis, we believe that it is unbiased.

METHODS

During October 1975, 37 stations in Lyme Bay (Fig. 1) were sampled from RV CORELLA using a 0.1 m^2 Day grab. Two samples were collected at each station for faunal analysis and sieved through a 2 mm sieve. At 14 of these stations the fauna was also removed on 1 mm and 0.5 mm sieves. The animals were identified to species, as far as possible, and nomenclature follows the Plymouth Marine Fauna

TABLE 1 Rank and abundance of important species in the community and species unique to sample 4 (0.5 mm sieve size)

Rank	Specimen	Numbers/m^2	% frequency of occurrence in 0.2 m^2 samples	Coefficient of variation: S.D./$\bar{x}$ density
1	Chaetozone setosa	948	100	0.29
2	Magelona filiformis	638	100	0.37
3	Ampharete grubei	281	100	0.42
4	Corbula gibba	275	100	0.56
5	Nemertini	234	100	0.43
6	Exogone hebes	199	100	0.83
7	Abra alba	154	100	0.46
8	Pectinaria koreni	143	100	0.51
9	Eudorella truncatulata	141	100	0.39
10	Owenia fusiformis	127	100	0.61
11	Amphiura filiformis	104	100	1.56
12	Cythereis sp.	100	100	0.83
13	Ampelisca tenuicornis	92	100	0.82
14	Nematoda	92	100	0.77
15	Nephthys hombergi	65	100	0.64
16	Glycera rouxii	64	100	0.31
17	Aricidea jeffreysii	63	100	0.56
18	Phaxas pellucidus	63	100	0.56
19	Perioculodes longimanus	60	100	0.53
20	Mysella bidentata	57	100	0.77
21	Scoloplos armiger	56	92	0.62
22	Cylichna cylindracea	55	100	0.36
23	Diplocirrus glaucus (?)	48	92	0.89
24	Turritella communis	48	100	0.63
25	Spio filicornis	37	100	0.65
26	Pectinaria auricoma	35	100	0.87
27	Notomastus filiformis	30	85	1.18
28	Abra nitida	29	77	1.02
29	Pariambus typicus	27	85	0.76
30	Diastylis bradyi	24	100	0.45
31	Melinna palmata	22	92	1.0
32	Phoronis sp.	22	92	0.88
43	Montacuta ferruginosa	16	62	1.69
55	Leptosynapta inhaerens	7	62	1.30
60	Echinocardium cordatum	5	69	0.90

St 4: Antennularia antennata (large colony); Doto coronata (46/0.1 m^2); Melita palmata (7/0.1 m^2); Anthozoa (4/0.1 m^2); Stenothoe marina (3/0.1 m^2); Nereis sp., Macropodia rostrata, Porcellana longicornis, Chlamys opercularis (all at 1/0.1 m).

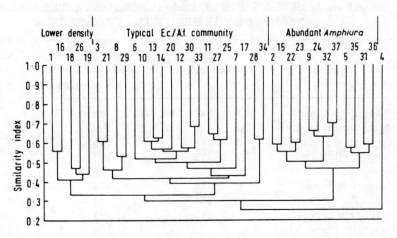

Fig. 2. Dendogram of station similarity (2 mm sieve).

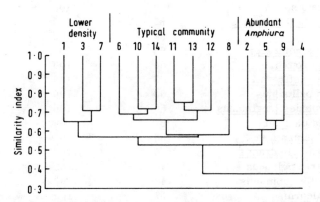

Fig. 3. Dendogram of station similarity (1 mm sieve).

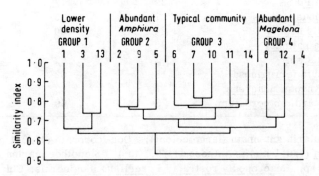

Fig. 4. Dendogram of station similarity (0.5 mm sieve).

(Marine Biological Association 1957). The species abundance data were analysed by the group average procedure of classification (Lance and Williams 1966) using the Bray Curtis index of similarity (Bray and Curtis 1957). This produces a hierarchy in which stations with a similar faunal content are grouped together.

RESULTS

It was apparent when the samples were collected initially that the community corresponded to Petersen's Echinocardium cordatum/Amphiura filiformis community (Petersen 1918). Apart from these two type species, Turritella communis, Phaxas (= Cultellus) pellucidus and Corbula gibba were also obvious. However, it was not until closer examination of the samples that it was realized that smaller polychaetes, bivalves and crustaceans were vastly superior in numbers. Table 1 is a list of the more important species with their average abundances. The preponderance of polychaetes in the upper ranks can be seen, with Chaetozone setosa and Magelona filiformis dominating. Echinocardium was ranked 60th with a density of $5/m^2$ and occurred in only 69% of the $0.2\ m^2$ samples. The total number of animals found was $4685/m^2$, which compared broadly to densities collected by several other workers in coastal muds and muddy sand regions that were sheltered from wave disturbance.

The results of classification analysis seemed to show faunal variations within the community. Figure 2 is the hierarchy for the > 2 mm data from all 37 stations and we could identify distinguishing faunal features for the three groups differentiated at about the 0.4 similarity level. One group had a higher density of Amphiura filiformis than the 'typical' community, whilst the total density of animals was lower in the other group. A similar pattern of three groups and an isolated site was distinguished when the fauna larger than 1 mm, collected at 14 stations, was considered (Fig. 3). At the 0.5 mm level of sieving (Fig. 4) four groups of sites were evident and differentiated at the 0.65-0.70 similarity level. The fourth cluster was attributable to the relatively high density of Magelona at these sites.

Station 4 was isolated at all sieving levels and had an average of only 0.5 similarity with the other sites in the analysis of the 0.5 mm data. Its unique fauna was dominated by a large colony of the hydroid Antennularia antennata which supported its own epifaunal association including the nudibranch Doto coronata and several crustaceans.

We have no measure of the statistical significance of the variations within the community (i.e. omitting station 4) but in the final column of Table 1 the coefficient of variation for each species is presented. This index (standard deviation ÷ mean density) indicates the spatial evenness in the abundance of the species; numbers much less than 1 being indicative of a uniform distribution. Of the common species only Amphiura had a coefficient of variation greater than 1. Figure 5 shows that this was due to a bimodal frequency/log abundance distribution, with significantly higher densities at the three stations in the classification cluster. Magelona densities had a very narrow range as did the total number of animals per sample. This implied that the other three clusters were abstractions from observed continua which were delineated artificially by the divisive classificatory technique used.

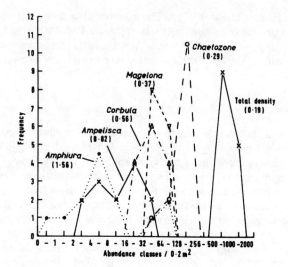

Fig. 5. Frequency of abundance classes of selected species and of total density in the 0.2 m² samples (coefficient of variation $S.D./\bar{x}$ in brackets).

We therefore felt justified in treating the 13 samples as coming from the same community, with the only reservation being that <u>Amphiura</u> was aggregated in three stations. In these 13 samples, covering 2.6 m² of sea bed, 117 different species were identified. The rate of addition of new species with increasing sample size is shown in Fig. 6. Only about half of the total observed species were collected in any one 0.2 m² sample. Extrapolation of this curve is not very precise but it would suggest that there were only about another 10-15 species in the community which would have been collected if 4.0-4.5 m² had been sampled. We believe, therefore, that we have very nearly fully sampled a benthic community at the 13 stations.

DISCUSSION

Having postulated that the community was uniformly distributed and that it was fully represented by the 13 samples, we scrutinized the data in two ways. Our first approach was to compare the observed species' relative abundances to some proposed models. It has been shown that these models are of minimal value in interpreting ecological data because it cannot be determined what biological, or even statistical, factors control the observed distributions (Cohen 1968). However, our results did show some interesting points.

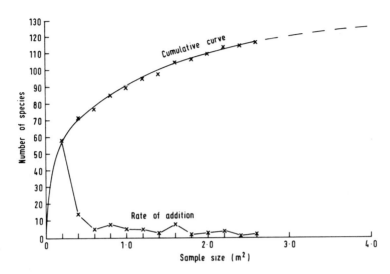

Fig. 6. Increase in species collected against increase in sample size.

Figure 7 is the logged abundance against rank curve and it bears some resemblance to MacArthur's broken stick model (MacArthur 1960). However, one peculiar characteristic is that several species frequently have similar abundances so that the curve has a stepped appearance.

The second curves of interest were based on the log normal distribution of Preston (1948). Figure 8 shows the frequency of species in log abundance classes for stations 2, 7, 9 and 12. The feature of interest here was the bimodality; the modes being at about 8-16 and at 1 with a third, intermediate mode sometimes present. This was also evident in the summed community data (Fig. 9), though we have not been able to decide whether there is one principal mode here or several peaks. Alternative interpretations based on a log normal prediction have been sketched in. Both the single and triple normal curves give a good fit to the data. A third option with two normal curves did not fit well despite the two apparent modes. Which of the alternatives is correct we are not sure, though it is interesting to note that at the abundance values where the four suggested curves overlap there is a slight change of gradient on the abundance-rank curve (Fig. 7).

We have been unable to determine the biological reasons why the species fall into at least two, and probably four, orders of abundances in such a homogeneous community but this observation must surely be of interest to any ecologists attempting to identify a unifying theory for species' relative abundances.

Our second approach to the data was from a biological standpoint in an attempt to identify some relationships between the animals. The benthos is peculiar in that it

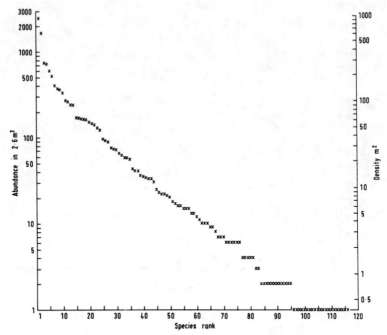

Fig. 7a. Species rank v. abundance.

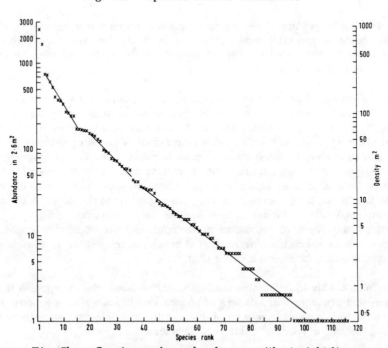

Fig. 7b. Species rank v. abundance; with straight lines superimposed.

has very few trophic levels, the majority of animals in many benthic assemblages being of the same trophic level - the detritivores. Some carnivores are encountered but most of the predatory species are large epifauna invertebrates and fish which are not collected in grab samples, although they do form part of the integrated benthic community.

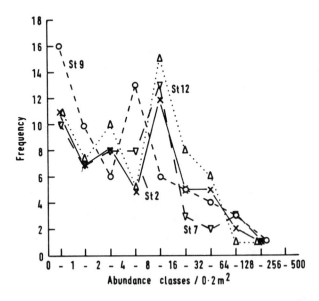

Fig. 8. Frequency of species' abundances (by log classes) in stations 2, 7, 9, 12.

The detritivore trophic level has traditionally been subdivided on the method of feeding rather than on the basis of the type of food taken. We found the suspension/deposit feeding classification unsuitable to represent the Echinocardium/Amphiura community in Lyme Bay because more than 80% of the animals were apparently feeding at the sediment/water interface (e.g. Chaetozone, Magelona, spinoids and several other polychaetes, the molluscs Corbula, Abra spp., Phaxas, Mysella and Turritella, the amphipods and cumaceans, and Phoronis and Amphiura). Only Amphiura, and perhaps larger Phoronis, would appear to have the potential to collect food from more than 1 cm above the sediment surface while true deposit feeding infauna, such as Pectinaria, Scoloplos and nemertines, made up only about 15% of the community and its main food source may have been the faecal pellets of the interface feeders.

At the interface layer the animals probably make no clear distinction between material in suspension and material recently deposited, and furthermore there is probably a continual flux of food particles between the two states. Thus, for the most part the polychaetes present use a pair of palps or tentacles to collect food from the sediment surface and from suspension; the molluscs filter out suspended material from just over the sea bed, except Abra which sucks up the sediment surface layer; while the crustaceans stir up the surface layer and filter the resulting suspension.

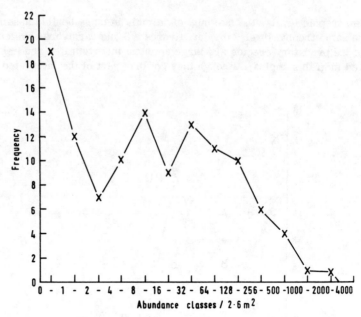

Fig. 9a. Frequency of species' abundances (by log classes) in 2.6 m^2.

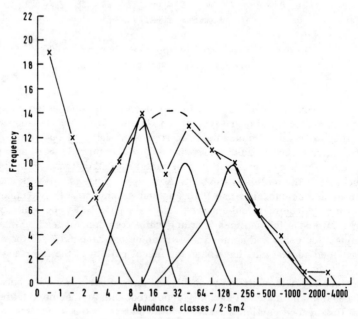

Fig. 9b. Frequency of species' abundances (by log classes) in 2.6 m^2; with alternative interpretations.

Unfortunately, our knowledge of benthos feeding does not go much further than mechanisms. Little specific information is available on the type or size of food ingested by the interface feeders, though the scope for specialization along these lines by the small animals concerned would seem to be limited. In addition, we might guess that because of the fluctuations in the type and input rate of food, detritus-eating species living in a coastal marine environment would not benefit by becoming very specialized in their feeding habits.

We are therefore left with the hypothesis that within the described community there is competition between many species for the same food resource. Whether or not this is a stable association is uncertain though the lack of spatial patchiness in the community indicated that it might be, at least in the short term.

We do not feel justified in attempting to discuss the maintenance of this apparently unbalanced competitive association without more information on feeding selectivity. Data on the density of larval settlement and on the parent populations if the larvae are not indigenous to Lyme Bay, and on the effects of fish predation would also be required. However, it may be that the variations in hydrographic conditions in the western English Channel, which have been observed to cause dramatic changes in the pelagic fauna (Russell et al. 1971), together with the fact that it may take many generations to rectify competitive unbalance (Miller 1969), are controlling factors.

The fauna which we investigated was mainly composed of detritivores feeding at the sediment/water interface. Despite this relatively simple situation the pattern of species' relative abundances was complex and we could not determine the factors regulating the community's diversity. Our opinion is that if this simple community evades interpretation by general theories then the understanding of fully integrated communities of producers, herbivores, carnivores and detritivores will not be assisted by generalized hypotheses.

REFERENCES

Bray, J.R. and J.T. Curtis 1957 An ordination of the upland forest communities of Southern Wisconsin.
Ecol. Monogr. 27, 325, 349

Cohen, J.E. 1968 Alternative derivations of a species-abundance relation.
Amer. Nat. 102, 165-172

Lance, G.N. and W.T. Williams 1967 A general theory of classificatory sorting strategies.
Comp. J. 9, 373-380

MacArthur, R.H. 1960 On the relative abundance of species.
Amer. Nat. 94, 25-36

Marine Biological Association 1957 Plymouth Marine Fauna, 3rd Edition.
MBA, Plymouth, 457 p.

Miller, R. S. 1969 Competition and species diversity. In: **Diversity and Stability in Ecological Systems**, eds G. M. Woodwell and H. H. Smith. Brookhaven Symposium in Biology, No. 22.

Petersen, C. G. J. 1918 The sea bottom and its production of fish food. Rep. Dan. Biol. Sta. 25, 1-62.

Preston, F. W. 1948 The commonness, and rarity, of species. Ecology 29, 254-283.

Russell, F. S., A. J. Southward, G. T. Boalch and E. I. Butler 1971 Changes in biological conditions in the English Channel off Plymouth during the last half century. Nature, Lond. 234, 468-470.

THE POLYCHAETE *Eulalia viridis* (O.F. Müller) AS AN ELEMENT IN THE ENERGY DYNAMICS OF INTERTIDAL MUSSEL CLUMPS

R. H. Emson

Zoology Department, King's College, Strand, London WC2R 2LS

ABSTRACT

Eulalia viridis has a scavenging role on exposed mussel dominated rocky shores. It is relatively abundant and preliminary observations suggest that the worm consumes a high proportion of the sedentary animals which die within the system.

INTRODUCTION

The importance of scavenging organisms in energy flow through marine intertidal communities is relatively little known. In Pembrokeshire and elsewhere the mid-intertidal areas of exposed shores are dominated by *Mytilus edulis* (L). The standing crop of these mussels and the other major component of the fauna at mid-tide level, *Balanus balanoides* (L), is often of the order of $350g/m^2$ dry tissue weight and this represents a large continually available food resource for animals able to take advantage of it. Such animals may be divided into predators which actually kill, and scavengers which utilize the tissues of dead or damaged animals. Predator pressure appears to be relatively low at the sites studied, although the effects of birds and fish are unknown. On the other hand the phyllodocid polychaete *Eulalia viridis* (O.F. Müller) is apparently abundant. This species has recently been shown, Emson (in press), to behave principally as a scavenger, at least in mussel-barnacle associations. Thus the animal was found to feed upon dead or damaged animals, of any kind, found in the mussel-barnacle association. Neither in the field nor in experimental situations in the laboratory were attacks on living undamaged animals observed. This paper attempts to show the relative importance of *E. viridis* in the passage of energy through this community sub-web.

AREA OF STUDY AND METHODS

The field work was carried out on the South Coast of Dyfed (Pembrokeshire) at Broadhaven (O.S. reference sheet 138/151 978937). The distribution and abundance of *E. viridis* and the other fauna was determined at 0.615 m (= 2 foot) height intervals along a belt transect running from E.L.W.S.T. upwards.

All the fauna, found within four separate 0.0625 m^2 quadrats, was identified and counted. Biomass estimates of the important species were made by drying samples to constant weight at $50^{\circ}C$ and the mean tissue weights of *E. viridis* and mean size and tissue weights of *M. edulis* from mid-tide level populations were determined. Respiration rates of *E. viridis* in water were measured at constant temperatures, in a continuous flow respirometer using the Beckman oxygen electrode to detect oxygen uptake. Animals placed in such an apparatus remain inactive under conditions of gentle flow, but perform irrigatory movements when flow is discontinued. Respiration in air was measured by means of a Scholander type respirometer (Turner and Batten, 1976). Animals placed in the respiration chambers of this apparatus generally stay inactive after initial exploratory movements, and only animals which did so were monitored for oxygen uptake. All experiments were carried out within one week of collecting the animals. Calorific values for food materials consumed by E. viridis were determined using a Phillipson type micro-bomb calorimeter.

RESULTS

Distribution and Abundance of Fauna. The results of the belt transect survey are summarised in Table 1 and clearly show that the shore is dominated by M. edulis and barnacles. The mussels are for the most part associated in clumps and much of the remaining fauna lives in the interstices between the mussels with the exception of the barnacles and limpets which occupy the intervening spaces between mussel clumps. It can be seen that the clumps harbour a number of species but few are found in abundance and fewer still are abundant, large and present over an extensive vertical height range.

TABLE 1. Fauna of transect at Broadhaven (Abundance in numbers per m^2)

Species	Height of sample site above E.L.W.S.T. (m)								
	0.3	0.9	1.5	2.1	2.7	3.4	4.0	4.6	5.2	5.8
B. balanoides	0	514	P	16000	P	27000	P	25000	P	39000
M. edulis	P	2450	P	4000	P	9800	P	10400	P	1500
B. perforatus	P	6032	P	1632	P	0	0	0	0	0
E. viridis	0	0	96	140	255	350	160	270	100	0
Gammarids	30	16	128	300	400	144	112	16	48	0
T. lapillus	16	48	208	224	112	80	32	32	32	0
C. maenas*	16	48	208	80	80	48	48	0	16	0

Species listed below were found in small numbers

Patella vulgata   Actinia equina   Anurida maritima
Cirratulus cirratus   Hiatella arctica   Lepidonotus clava
Nereis pelagica   Lasaea rubra   Idotea baltica
Amphipholis squamata   Ophiothrix fragilis   Idotea pelagica
Aphrosylus celtiber   Nemertines (P)   Nematodes (P)

(P = present but not counted)   * includes some Cancer pagarus

Only a small number of species, therefore, contribute significantly to the biomass. The species falling into this category include E. viridis, Marinogammarus sp, Thais lapillus (L), the two species of crab and possibly nemertines whose abundance was not determined because of the difficulty of extracting them. Of these, the known predatory species T. lapillus and the two crab species are the least abundant and their potential importance is diminished by the fact that the specimens found were small, since they must fit into spaces between small mussels. The five largest *T. lapillus* had a mean length of 15.62 mm and the largest crabs had a carapace width of 16 mm. *E. viridis* is the most abundant and the largest of the remaining species. The average of the mean dry weights of samples of *E. viridis* taken from mid-tide level (the zone of greatest abundance) in June, September, and December, is 30.5 mg. Since there are 350 worms/m^2 at this level, this represents a biomass of over 10 gms dry wt/m^2 and only *M. edulis* exceeds this. Evidently *E. viridis* is the major component of the fauna associated with mussel clumps. Since *E. viridis* is a scavenger and only *M. edulis* and *B. balanoides* are large and abundant throughout the range of *E. viridis* the latter species must depend principally upon the death of mussels and barnacles for food. The importance of *E. viridis* in terms of its role in energy flow in the community will depend upon its calorific requirements.

Respiration of *E. viridis*. The results shown in Table 2 reveal that respiration rate is consistent over the size range of animals used and that respiration in air is very much less than that in water.

TABLE 2. Oxygen uptake in *E. viridis*.

Respiration in air (mg O_2/gm dry wt/hr)

   10°C $\bar{x}$ 0.178 ± 0.0134 (2 x S.E.) n = 9   Weight range 5-48 mg

   15°C $\bar{x}$ 0.324 ± 0.055 (2 x S.E.) n = 6   Weight range 13-51 mg

Respiration in water (mg O_2/gm dry wt/hr)

   10°C $\bar{x}$ 0.796 ± 0.18   (2 x S.E.) n = 5   Weight range 11-29 mg

   15°C $\bar{x}$ 1.54 ± 0.371 (2 x S.E.) n = 5   Weight range 16-36 mg

Estimation of calorific requirements of *E. viridis*. For the sake of simplicity, calculation has been done for mid-tide level. In order to obtain calorific requirements from respiration data the oxycalorific coefficient is required. Assuming *E. viridis* respires tissue components in the same proportions as they occur in the food, then, from the data of Dara and Edwards (1975) for *M. edulis*, an oxycalorific coefficient value of 3.245 cals/mg O_2 is obtained. Animals at mid-tide level respire in air and water for equal periods so the average of respiration rates at mean sea water temperature (12.5°C) will give a close approximation to real requirements. This figure is 0.709 mg/gm dry wt. 350 worms weigh 10.67 gms thus O_2 uptake/day = 0.709 x 10.67 x 24 = 181.56 mg/day and calorific requirement/day = 181.56 x 3.245 cals = 589.0 cals/day. To obtain an accurate figure for the food required to supply this need the assimilation efficiency is required. The small size of the worms and the fact that faeces must be liquid, since no solid faeces have been seen, prevented the determination of this factor. However, since the animal is feeding on animal tissue, often semi-fluid in character, a high assimilation efficiency is likely and a figure (80%) of the same order as those for *Neanthes (Nereis) virens*, (Sars) obtained by Kay and Brafield (1973) and Tenore and Gonoplan (1974), is assumed. Adjustment of the calorific requirement, having taken this into account, gives a figure of 736 cals/day for respiration. Kay and Brafield (1973) also report that in *Neanthes virens* respiration accounts for only 22.5% of the energy intake and it is obvious that the actual energy required by *E. viridis* will be considerably more than the value calculated above.

Availability of food for *E. viridis*. At mid-tide level mean size of mussels was in March 14.5 mm and in August 17.4 mm, with corresponding dry tissue weights of 65 mg and 340 mg respectively. Thus mussel biomass is about 65gm/m^2 in March and 350 gm/m^2 in August. Calorific value of mussel tissue was found to be 5800 cals/gm ash free dry weight and, as ash forms 10% of mussel tissue, an animal of 207 mg dry weight (average of two values above) contains 1203 cals. Clearly therefore, if all tissues of a mussel are available to the worms, the death of one mussel per m^2 per day would provide almost double the calories required for the respiration of 350 quiescent worms. Seed (1969) has shown that on exposed shores, where predation is low, mortality is between 10 and 30% of the population per year. As the population of mussels is 10,000 per m^2 at mid-tide level, it can be calculated that between 3 and 8 mussels die per m^2 per day. Barnacles provide an additional source of food material. There are 26,000 *B. balanoides* per m^2 at mid-tide and the data of Connell (1961) indicates a mortality of 20-50% where predation is low. Thus between 14.2 and 35.6 die per m^2 per day. Barnacles have been found to have approximately 1 mg dry weight tissue of which 15% is ash and, as barnacle tissue has a calorific value of 5675 cals/g ash free dry weight, each barnacle contains 4.82 cals. Barnacle mortality would therefore supply the respiratory requirements of 32 to 80 worms.

DISCUSSION

E. viridis is shown to be a major component of the fauna of mussel clumps and is the only component of the mussel-clump infauna potentially capable of processing a significant proportion of the available tissue. Exact data on the proportion of available tissue actually utilized by the worm population is not to hand at present. In particular two crucial items of information are missing *i.e.*, (a) the proportion of the animals calorific intake devoted to growth, reproduction etc., and (b) accurate figures for the mortality rate of mussels at this site.

If however, *E. viridis* resembles *Neanthes virens* and less than 25% of absorbed energy is used in respiration (Kay and Brafield 1973), then it can be predicted (Table 3) that the tissue of at least 30% of mussels dying will be required to supply the needs of *E. viridis* at its zone of maximum abundance.

Only detailed further study will reveal the actual level of uptake but even without such study the importance of the species in the energy flow through mussel clumps is clearly indicated. One further point re-inforces this view. It has been assumed throughout that all tissues are available to *E. viridis* and this may not be so.

Although mid-tide level was chosen for calculations, the wide vertical range of the species and the numbers in which it is found indicate that it has a significant role over much of the intertidal of exposed shores. It seems

TABLE 3.   The relationship between actual and predicted requirements
of E. viridis and availability of food

Energy requirement of 350 worms (cals)		Energy available	Proportion used (%)
Quiescent respiration	736	3609 cals	20
Total uptake 4 x Q.R.	2944	(at 10% mortality)	81
Quiescent respiration	736	9624 cals	7
Total uptake 4 x Q.R.	2944	(at 30% mortality)	30

probable that E. viridis achieves its high level of abundance in mussel clumps because its body form makes it well suited to the habitat, and because other scavengers, with the possible exception of nemertines are not.   Certainly E. viridis is replaced by N. pelagica at low level on these shores (see Table 1) and on less exposed shores, where other known scavengers are found, E. viridis is less abundant.   It still remains to be ascertained whether the greater part of the energy bound up in the tissues of E. viridis leaves the system as reproductive products, passes to predatory species, perhaps nemertines, or is retained within the E. viridis population by cannibalization of defunct worms.   Clearly a study of the population dynamics of this species would be very worthwhile.

ACKNOWLEDGEMENTS

I thank Bob Faller-Fritsch for his generous assistance with this work.

REFERENCES

Connell, J.H. 1961.   Effects of competition, predation by Thais lapillus and other factors on natural populations of the barnacle Balanus balanoides.   Ecol. Monogr. 31, 61-104.

Dare, P.J. and Edwards D.B. 1975.   Seasonal changes in flesh weight and biochemical composition of mussels (Mytilus edulis L.) in the Conway Estuary, North Wales.   J. exp. mar. biol. ecol. 18, 89-97.

Emson, R.H. 1977.   J. mar. biol. Ass. U.K. 57,1. (in press).

Kay, D.G. and Brafield, A.E. 1973.   The energy relations of the polychaete Neanthes virens (Sars).   J. anim. Ecol. 42, 3, 673-693.

Seed, R. 1969.   The ecology of Mytilus edulis (Lamellibranchiata) on exposed rocky shores.   II.   Growth and Mortality.   Oecologia (Berlin) 3, 317-350.

Tenore, K.R. and Gonoplan, U.K. 1974.   Feeding efficiencies of the polychaete Nereis virens Cultured on Hard-Clam tissue and Oyster detritus.

J. Fish. Res. Bd. Can. 31, 1675-78

Turner, B.D. and Batten G.L. 1976. Robust and sensitive microrespirometers for class use. Lab. Pract, 25, 9, 585-6.

A DIVING SURVEY OF STRANGFORD LOUGH: THE BENTHIC COMMUNITIES AND THEIR RELATION TO SUBSTRATE — A PRELIMINARY ACCOUNT

David G. Erwin

Ulster Museum, Botanic Gardens, Belfast BT9 5AB

ABSTRACT

The diving team of the Ulster Museum Botany and Zoology Department has carried out a survey of the benthic communities of Strangford Lough, Northern Ireland, a marine lough of approximately 150sq Kms, with a range of water movement between zero and 350cms/sec. Because of the range of water movement a wide variety of substrate exists grading from bed rock to silt. Benthic communities have been recorded and shown to be related to substrate/water movement variations. Ten communities, some exhibiting different facies have been identified in depths from ten to fifty metres.

INTRODUCTION

Before the current work, Strangford Lough, Co Down, (Fig. 1) was known to have a rich and varied benthic fauna. (Williams 1954, unpublished records held at the Queen's University Marine Station, and short notes by various authors). Yet no systematic survey of the area had been carried out. This was almost certainly due to the varied nature of topography and substrate necessitating a multiplicity of equipment for remote studies. The only practicable way in which all areas of the lough, except the very deepest, could be surveyed was by the use of scuba techniques. In 1972 the Ulster Museum diving team set out to answer the apparently simple question:- What is found, where in the lough? A survey method was evolved to maximise underwater work and to obtain useful data from a large area in a reasonable time. This initial survey which involved more than 450 underwater hours has now been completed. Results presented in this paper can only be a broad outline of the information obtained. Selected results illustrating the main trends are all that is possible in the allocated space and time. A mathematical treatment and a detailed descriptive and faunistic work will appear separately.

METHODS

The method utilised involved 'flying' behind a towing boat on a 'sledge' or 'flying machine' whilst giving a continuous description of the fauna and substrate. Photographic records and detailed analyses were made of the communities and substrate at more than 250 sites. (Details of method - Erwin 1976).

The fauna recorded was all macrofauna, mostly epifaunal with visible infauna. Many infaunal species show traces on the surface indicating their presence e.g. (fig. 9).

Sediment samples were dried and weighed, wet sieved through a 0.0625mm sieve to remove the silt/clay fraction, dried again and weighed.

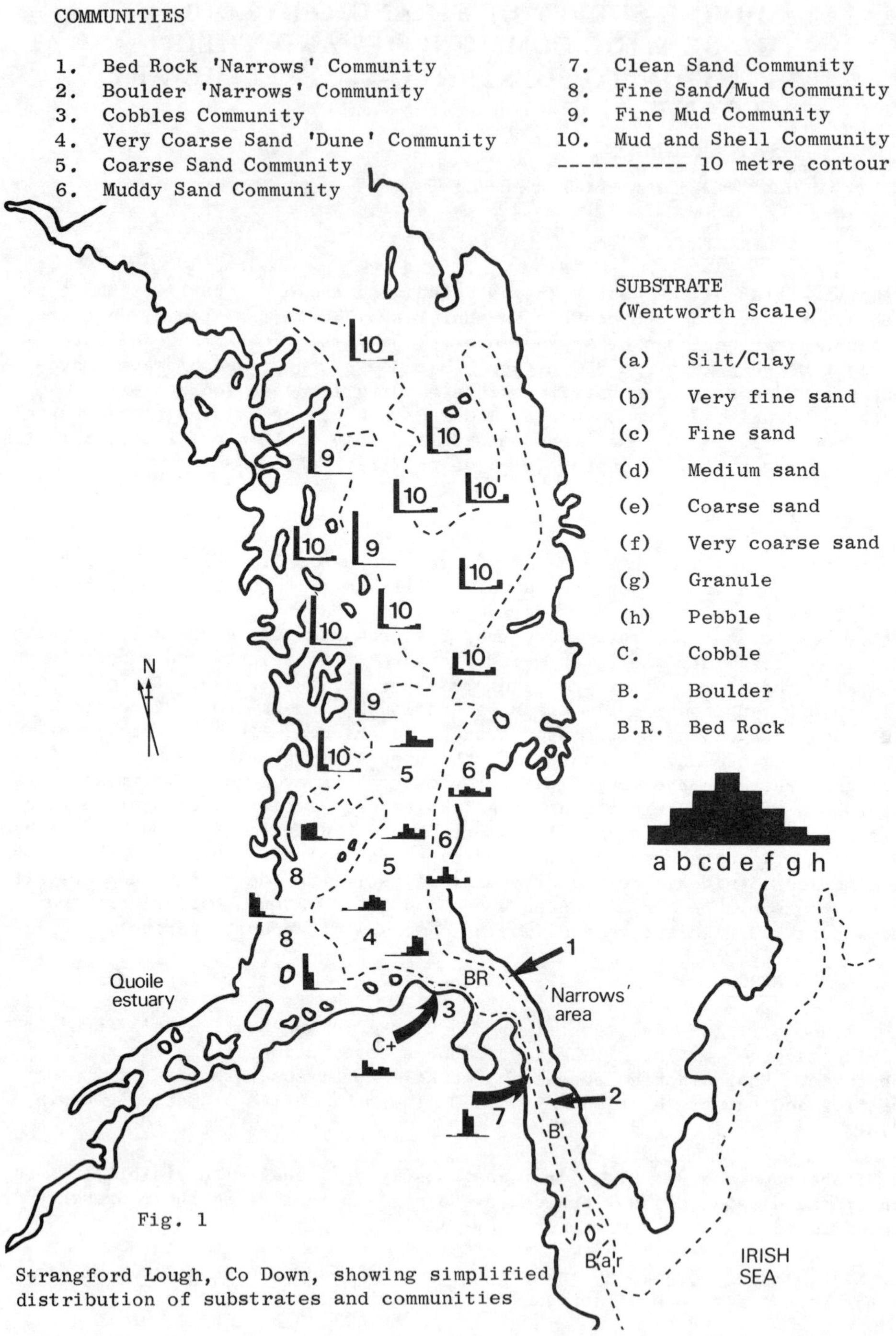
COMMUNITIES
1. Bed Rock 'Narrows' Community
2. Boulder 'Narrows' Community
3. Cobbles Community
4. Very Coarse Sand 'Dune' Community
5. Coarse Sand Community
6. Muddy Sand Community
7. Clean Sand Community
8. Fine Sand/Mud Community
9. Fine Mud Community
10. Mud and Shell Community
----------- 10 metre contour
SUBSTRATE
(Wentworth Scale)
(a) Silt/Clay
(b) Very fine sand
(c) Fine sand
(d) Medium sand
(e) Coarse sand
(f) Very coarse sand
(g) Granule
(h) Pebble
C. Cobble
B. Boulder
B.R. Bed Rock
a b c d e f g h
N
Quoile estuary
BR
'Narrows' area
C+
B
Bar
IRISH SEA

Fig. 1

Strangford Lough, Co Down, showing simplified distribution of substrates and communities

The coarser fractions were then dry-sieved through a nest of sieves separating the standard Wentworth size classes. (Wentworth 1922).

RESULTS

The topography and tidal streams of Strangford Lough have been outlined by Boyd (1973a, 1973b). A summary only will be given here. Strangford Lough is a land protected, entirely marine lough (32‰ -34‰ salinity), 31Kms long and 4 to 6Kms wide, connected to the Irish Sea by a narrow channel known as the 'Strangford Narrows'. Numerous partially submerged drumlins, and banks which barely reach sea level known locally as 'pladdies', are found in the main body of the lough. Depth is extremely variable to a maximum of 60m. Major tidal water movements occur through the 'Narrows' area. Rate of water movement varies from 350cms/sec in the fastest part of the 'Narrows' to virtually nil. Because the lough is almost entirely land protected by surrounding drumlins and because of the short fetch available for the production of waves there is little or no evidence of wave action below a depth of 10m. Substrates are therefore laid out on a gradient of tidal water movement in depths greater than 10m, a pattern only disturbed by biotic factors. Recognisably different communities follow the substrate series with a precision which repeats itself many times. Where a particular substrate type exists in different areas of the lough (indicating a particular water movement regime) the same community of animals will be present. It is not suggested that the substrates are distributed on a stepped system but rather that one merges into the next along the gradient. In some circumstances where there is a large change of rate of water movement in a short distance, the substrates will follow this pattern and change rapidly. When all the sediment analyses carried out are viewed on a chart it becomes evident where the current gradients are, where areas of similar current exist and where there are areas of mixing. Neither do the communities simply start or stop at particular points. Most commonly one community merges into another with a mixed area between.

THE SUBSTRATES AND THEIR ASSOCIATED COMMUNITIES

Only the major 'indicator species' of each community will be mentioned. Full species lists of the communities will appear later.

1. Bed Rock 'Narrows' Community

This occurs in the 'Narrows' areas of fastest water movement where the substrate is 100% bed rock. It is typified by massive encrusting forms exhibiting 100% cover of the rock surface. Typical species are Alcyonium digitatum (L.) (fig. 2), Pachymatisma johnstonia (Bowerbank) and the massive form of Cliona celata Grant. The community has been observed on near vertical faces in depths down to 48m below which it appears to continue unchecked.

2. Boulder 'Narrows' Community

Where the water movement is slightly reduced the substrate is composed of boulders of various sizes. Species diversity is greater than on the bed rock, typified by almost 100% cover of Tubularia indivisa L. (fig. 3) underlain by various encrusting sponges. Large branching hydroids (e.g. Sertularia argentea L.) and bryozoans are also much in evidence together with many mobile species. Particularly numerous is the caprellid Caprella linearis (L.) and this is the only community in the lough where Macropipus puber (L.) is common.

Fig. 2 Alcyonium digitatum

Fig. 3 Tubularia indivisa with associated fauna

Fig. 4 Actinothoë sphyrodeta

Fig. 5 'Tentacles' of Neopentadactyla mixta

Fig. 6 Ophiothrix fragilis

Fig. 7 Pecten maximus

Fig. 2-13 all represent 13cms X 9cms of benthos

Two more widespread species which are found in this community are Echinus esculentus L., which has been found commonly well below the photic zone apparently grazing on sponge and on Tubularia, and the butterfish Pholis gunnellus (L.) which is abundant in this situation at depths down to at least 40m.

3. Cobbles Community

In several bays off the 'Narrows' tidal flow, cobbles are found covering 50% to 80% of the bottom. Between them is a coarse sand probably accounted for by a reduction in water movement produced by the cobbles. The most obvious 'indicator species' in these areas is Actinothoë sphyrodeta (Gosse) (fig. 4). Some cobbles are limestone and are heavily bored by the piddock Hiatella arctica (L.), and have a complex cryptic fauna typified by the ophiuroid Ophiopholis aculeata (L.). The animals associated with the sand between the cobbles are closely related to community 6 and will be described there.

One very strange fairly extensive cobble and pebble area exists in the 'Bar' area where the 'Narrows' meet the open sea. It is a virtual desert of well rounded cobbles and pebbles totally devoid of life. Occasional bed rock outcrops occur with typical bed rock 'Narrows' community on the flat surfaces and typical Boulder Tubularia community in recessed crevices. The 'Bar' is an area of great turbulence due to the tide and wind interference, and this together with its relative shallowness probably induces great mobility of the cobbles.

4. Very Coarse Sand 'Dune' Community

At either end of the 'Narrows' where the current begins to disperse over a wider area, very coarse sand 'Dunes' are set up running at right angles to the line of the 'Narrows'. They vary from 7-9m apart and 0.5m high to 3-5m apart and 1m high. The community is characterised by the holothurian Neopentadactyla mixta (Östergren) (fig. 5), the crab Atelecyclus rotundatus (Olivi) and the bivalve Glycymeris glycymeris (L.).

5. Coarse Sand Community

This is found in extensive areas of the lower lough and is typified by extremely dense aggregations of brittle stars, the great majority being Ophiothrix fragilis (Abildgarrd) (fig. 6) but with Ophiocomina nigra (Abildgarrd) also present. At the fringes of the community, when Ophiothrix fragilis ceases to be present, often on a sharp line, Ophiocomina nigra regularly continues for some distance at approximately the same concentration in numbers as in the dense Ophiothrix bed.

6. Muddy Sand Community

This substrate is fairly extensive in relatively shallow water (down to 20m) on the lower east side of the lough and is found in small areas on the west side where a suitable current regime exists. The animal component is quite variable but is best represented as a commercial Pecten maximus (L.) bed (fig. 7). Other species much in evidence are Marthasterias glacialis (L.), Solaster papposus (L.), Asterias rubens L. (although this is present in other communities), Echinus esculentus on occasional rocks and many species of hydroids with their associated faunas.

The association of the coarse sand between the cobbles of community 3 approximates to this community. Most of the species are those of community 6 but in much reduced numbers.

Fig. 8 'Arms' of Amphiura sp.

Fig. 9 Siphons of Cyprina islandica

Fig. 10 Virgularia mirabilis

Fig. 11 Nephrops norvegicus in mud burrow

Fig. 12 Modiolus 'clump' with associated fauna

Fig. 13 Ascidiella aspersa

An interesting point is that in the case of Pecten maximus only very young specimens have been found in 3 suggesting that primary settlement may take place in faster water areas than where the species finally settles.

7. Clean Sand Community
Clean sand communities are rare in the lough and tend to be in small pockets where they do exist. They are found in open shallow bays (5-15m), usually just to the side of fast water areas. The substrate is very mobile and apart from gobies and flatfish the scarce macrofauna is all infaunal, and is probably best represented by species of Ensis.

8. Fine Sand/Mud Community
This substrate is found mainly in the area where the lough approaches the Quoile estuary in depths from 10-15m, but where there is even less wave action than normal. The community seems to correlate almost completely with the 'Amphiura filiformis - Amphiura chiajei community' of Thorson (1957), originally described by Petersen (1913, 1918). It appears to be present in two facies dependant on the coarseness of the substrate. In one small area where the substrate is mainly fine sand Amphiura filiformis (O. F. Müller) and Amphiura chiajei Forbes (fig. 8) exist in large numbers along with Cyprina islandica (L.) (fig. 9) in fairly high numbers. In the majority of the range of the community, where the substrate is finer, the two species of Amphiura continue but Cyprina islandica is not present. However the pennatulid Virgularia mirabilis O. F. Müller (fig. 10) is very common and Aporrhais pes-pelecani (L.) and Turritella communis Risso are present in quite high numbers. These latter are not present in the coarser Cyprina facies.

An interesting feature of this community is the mode of life of two of its members which seems to have been misrepresented in the past. A. pes-pelecani seems to be almost entirely epifaunal and a large element of the T. communis population is also epifaunal. The two species of Amphiura live semi-infaunally as suggested by Thorson (1950).

9. Fine Mud Community
This substrate seems to be confined largely to the north of the lough. The most obvious member of the community is Nephrops norvegicus (L.) (fig. 11) inhabiting burrows in the very soft mud. Other common species are Goneplax rhomboides (L.) and Aphrodite aculeata L. One very interesting species which appears with this community is the anemone Hormathia coronata (Gosse) as a pure white form. This is not, as first appears, living simply in the mud but is always attached to large shell fragments. Until now this anemone was not known to exist further north than Plymouth and is almost certainly Lusitanian. It is oviviviparous and a chance colonisation into suitable conditions could have produced the existing high population levels. Isolation in the lough could explain the occurrence of the pure white form.

10. Mud and Shell Community
Mud with a coarse shell fraction is widespread in the northern half of the lough, both in the main channel and between islands and 'pladdies'. The shell fraction permits settlement onto the mud and colonisation by species which would not normally do so. These may themselves modify the substrate and permit other species to colonise. By far the most widespread occurrence of this is based on Modiolus modiolus (L.) (fig. 12) which occurs in extensive beds in a wide range of depths down to 40m and possibly deeper. Modiolus are always found attached to shell and around this epicentre a clump of

Modiolus builds, all connected by byssus threads. This 'clump' provides an artificial biotic hard substrate in an area where it would not normally exist and many other species are to be found associated with it. Some of the most obvious are Chlamys varia (L.), Ascidiella aspersa (O. F. Müller), various serpulids and sponges and the predator/scavengers Asterias rubens and Buccinum undatum L. Between the 'clumps' a special fauna also exists best indicated by holothurians Thyone fusus (O. F. Müller) and Thyonidium commune (Forbes) and the bivalve Chlamys opercularis (L.). More detailed direct observations on this community have been published by Roberts (1975) on work intimately connected with the present survey.

A further variation on the colonisation of shell in mud occurs mainly in relatively shallow water, where the shell is colonised by Ascidiella aspersa (fig. 13) with no Modiolus modiolus. The relationship of this 'Ascidiella facies' to the 'Modiolus facies' is open to conjecture.

DISCUSSION

All natural processes of erosion tend to make sediment finer and given enough time for stabilisation of the system, hard or coarse substrates will only exist where they are held in hydrodynamic balance by water movement, except where the substrate is relict or where sediment is being transported from elsewhere. Most natural systems are thus assumed to be in hydrodynamic balance. Generally, the gross water movement in an area is produced by two types of energy input - wave action and tidal action. In low tidal current areas hard or coarse substrates will exist only due to wave action. In low wave action areas hard or coarse substrates will exist only due to tidal current. The latter is the case in Strangford Lough where, apart from the 'Bar' area, the sediments below 10m are laid out in hydrodynamic balance only with tidal current. Thus sediments and associated communities which would normally only be found on extensive offshore areas are found in relatively discrete shallow areas accessible to the diver.

It is possible that the pattern described represents an evolution of community along the substrate gradient towards a 'climax' in the fines. On this basis it would seem that the climax community of Strangford Lough is the mud Nephrops/Goneplax community. However as this shows a reduction in diversity from coarser sediments it seems more likely that the final state, where Modiolus colonises shell in mud and brings in a much greater diversity is the true climax. Once established this community will provide its own shell for colonisation by mortality, thus producing a strong, diverse self-perpetuating system. It is in fact the most extensive community in Strangford Lough.

The great majority of community studies previously undertaken have depended on remote quasiquantitative sampling and the building of a community concept through statistical analyses. Petersen (1911) employed hard-hat divers with some success because he was not satisfied with the results of his sampling methods. Parker (1976 p252) refering to remote sampling and sensing methods together with their subsequent complex analyses states "In fact there is no computer now, or likely to be in the future, which can integrate and correlate with the speed of the human brain the hundreds of variables noted with a single human glance". This statement could not be improved upon in relation to benthic communities. The current work has shown that a diver moving over the benthos in the course of a survey can recognise a community without necessarily seeing all its components. Only when the main communities in an area have been located it is necessary, if required, to go back to sites to carry out full quantitative work. Even here the diver is at a great advantage over the remote sampler.

CONCLUSIONS

1. Ten Benthic communities, some exhibiting different facies have been identified in Strangford Lough in depths between 10m and 50m.

2. The communities in Strangford Lough can be closely correlated to substrate/water movement.

3. A diving biologist can see and recognise communities as such without the need to carry out statistical analyses.

ACKNOWLEDGEMENTS

I thank Messrs. C. Lyle and B. Picton, colleagues in the Ulster Museum, and Mr C. Roberts, who have been deeply involved in the work leading to this paper, often in extremely arduous conditions. I also extend thanks to Messrs. R. Brachi, T. Bruton, E. Ferguson, M. Lyle and D. Michael. A special debt of gratitude is due to Dr. P. J. S. Boaden and the staff of the Queen's University Marine Station who have generously facilitated us on many occasions.

REFERENCES

Boyd, R.J. 1973 A survey of the plankton of Strangford Lough, Co. Down.
    Proc. R. Ir. Acad.   73B, 231-267

Boyd, R.J. 1973 The relation of the plankton to the physical chemical and biological features of Strangford Lough, Co. Down.
    Proc. R. Ir. Acad.   73B, 317-353

Erwin, D.G. 1976 A cheap scuba technique for epifaunal surveying using a small boat.
    Rep. Underwater Ass. 2 (N.S.), (in press)

Parker, R.H. 1975 The Study of Benthic Communities, a model and a review.
    Elsevier, Amsterdam

Peterson, C.G. 1911 The possibility of cambating the noxious animals in the fishery of the Limfjord, especially Buccinum undatum and Nassa reticulata.
    Kobenhavn Ber. Biol. Stat.   19, (20)

Petersen, C.G. 1913 Valuation of the Sea II. The animal communities of the sea bottom and their importance for marine zoogeography.
    Rep. Dan. Biol. Stat.   25, (62)

Petersen C.G. 1918 The sea bottom and its production of fish food. A survey of the work done in connection with the valuation of Danish waters from 1883-1917
    Rep. Dan. Biol. Stat.   25, (62)

Roberts, C. D. 1975 Investigations into a *Modiolus modiolus* (L.) community in Strangford Lough, Northern Ireland.
Rep. Underwater Ass. 1 (N.S.), 27-49

Thorson, G. 1950 Reproductive and Larval behaviour of marine bottom invertebrates.
Biol. Reviews 25, 1-45

Thorson, G. 1957 Bottom Communities.
In J. W. Hedgpeth (Ed.) Treatise on Marine Ecology and Paleoecology.
Mem. Geol. Soc. Am. 67 (1), 461-534

Wentworth, C. K. 1922 A scale of grade and class terms for clastic sediments.
J. Geol. 30, 377-392

Williams, G. 1954 Fauna of Strangford Lough and neighbouring coasts.
Proc. R. Ir. Acad. 56 B, 29-133

REPRODUCTIVE STRATEGIES OF THE WINKLE *Littorina rudis* IN RELATION TO POPULATION DYNAMICS AND SIZE STRUCTURE

R. J. Faller-Fritsch

*Zoology Department, University of London King's College,
London WC2R 2LS, England
Now at: Biology Department, University of Exeter, Devon, England*

ABSTRACT

Studies have been made on the reproductive output and size at maturation in populations of L. rudis experiencing differences in exposure and substrate type. In exposure, large numbers of small embryos are produced per female per annum, whereas in shelter embryo production involves fewer numbers but greater individual size. It is suggested that these differences constitute adaptive responses to particular features of each type of habitat. In exposure, recruitment involves colonization of suitable small crevices by the small juvenile winkles. Selection therefore favours the production of numerous offspring by each female. Populations of L. rudis on unstable boulder shores in shelter may be susceptible to very high rates of mortality among the smallest juveniles, caused by crushing, burial and dessication. In this situation large embryos, hence large size at birth, may be advantageous.

Size at maturation is related to population structure in different conditions. Exposed populations tend to consist of small individuals which mature at a small size, thereby maximizing their total reproductive effort. Maturation is relatively delayed in shelter, perhaps because energy is devoted to somatic growth, allowing individuals to outgrow the size range in which high mortality rates occur

INTRODUCTION

The viviparous winkle L. rudis (Maton) occupies a wide range of marine and estuarine habitats throughout Europe. It is extremely variable in all aspects of its biology, and the results presented in this paper form part of a study of the ecology and underlying causes of variation in this species. Variation, to some extent reflecting the occupancy of many habitat types, in conjunction with low rates of gene flow resulting from viviparity and low mobility, has been responsible for taxonomic difficulties in this and related species. The form rudis, regarded by Dautzenberg & Fischer (1912) and James (1968) as a variety or subspecies of L. saxatilis (Olivi), has been accorded specific status by Heller (1975), whose nomenclature is used in this paper.

As aspect of variability of particular importance in L. rudis is that of population size structure, which varies considerably between habitats, though often remaining relatively stable in a particular place. In general, exposed populations consist of small individuals (Ballantine, 1961; James, 1968), and within the size range encountered smaller shells often predominate (Table 1). Emson & Faller-Fritsch (1976) have shown experimentally that the lack of large individuals in exposure may be due to a scarcity of suitable crevices upon which they show a dependence. A contrasting situation prevails in sheltered conditions, where a greater size range of individuals is found, and where the larger size classes usually assume a greater relative abundance. The size structure of several populations of L. rudis in a range of conditions is summarized in Table 1.

Any attempt to relate observed population structure and dynamics to features of the habitats occupied must include an account of reproductive performances in these habitats. The results presented here concern the effects of certain habitat features, notably the degree of exposure in relation to the type of substrate, upon some aspects of reproduction in three L. rudis populations.

TABLE 1   SIZE STRUCTURE IN SOME L. rudis POPULATIONS

Population		Size class (mm) and frequency (%)				Mean
		2+	6+	10+	14+	
BEDROCK						
1. Greenala	(Dyfed, S. Wales)	73.0	25.2	1.7	-	5.0
2. Newhaven	(Sussex)	79.2	17.2	3.6	-	4.9
3. West Angle	(Dyfed, S. Wales)	47.4	52.6	-	-	5.6
4. Clevedon	(Somerset)	75.9	20.4	3.8	-	5.2
BOULDERS						
5. Whitstable	(Kent)	11.9	88.2	-	-	6.8
6. Greenhithe	(")	33.0	56.9	10.0	-	7.2
7. Allhallows	(")	7.6	30.8	51.9	9.6	10.4
8. Landshipping	(Dyfed, S. Wales)	52.9	34.8	11.9	0.3	5.7
9. West Angle		11.2	32.0	27.6	29.3	10.8

Summary of size structure in some populations of L. rudis from bedrock and boulder substrates. Population 1; exposed, 2 & 3; semi-exposed, 4 & 5; sheltered, 6 - 9; very sheltered.

MATERIALS AND METHODS

Of the three sites at which reproduction has been principally studied, one is semi-exposed and the others very sheltered. The semi-exposed site, at Newhaven on the Sussex coast of England, has been described by Emson & Faller-Fritsch (1976). The substrate consists of large, stable chalk boulders on which L. rudis occurs at a density of approximately 150 m^{-2}. The size structure of this population (Table 1) shows a preponderance of small animals. Their abundance corresponds to the availability of empty barnacle shells upon which they are reliant for shelter, and the scarcity of larger crevices results in the presence of few large individuals.

The second population, at Greenhithe on the south bank of the Thames estuary 30 km. east of London, is subject to markedly sheltered estuarine conditions. The winkles occupy an unstable boulder and pebble substrate with densities reaching 1400 m^{-2}, reflecting an abundance of food. An important feature of the dynamics of this population is the very high mortality among small juveniles, caused by crushing, burial and dessication (Faller-Fritsch, unpubl.) The population size structure (Table 1) reveals an abundance of animals in the middle and larger size classes.

The third population, at Landshipping on the Cleddau estuary in Dyfed, South Wales, occupies a sheltered shore composed of boulders, many of which are bound in the substrate and are therefore stable. Densities are approximately 100 m^{-2}, and the size structure of the population is intermediate between those at Newhaven and Greenhithe (Table 1).

Reproductive activity in each population extends throughout the year, with a maximum in May and June, followed by a more or less conspicuous lull (Faller-Fritsch, unpubl.). The study, carried out between 1972 and 1975, included two reproductive years at Newhaven and Landshipping, and three at Greenhithe. At approximately monthly intervals, collections of standard (9-11mm. shell length) animals were made. This size range was chosen so that the results would be comparable with those of Berry (1961). Counts were made of the numbers of embryos present in the brood pouches of the females, to establish estimates of monthly and annual brood pouch contents for each population. The latter estimate, in conjunction with the gestation period in L. rudis of approximately 65 days (Berry, 1956), has been used to estimate annual embryo output by the standard females.

It was discovered during the embryo counting that the size of embryos in the stage immediately prior to hatching varies consistently between populations. The mean diameter and dry weight of batches of embryos were therefore established during the counting proceedure.

The other aspect of reproduction studied concerns the size at maturation in these and other populations. Maturation in male winkles, accompanied by the presence of a fully developed penis, is more readily determined than that in females. Size at maturation for each population has therefore been taken as the smallest size at which functional males are evident.

RESULTS

Annual reproductive output by standard females in the three populations is outlined in Table 2, which also contains data recalculated from Berry (1961) concerning a sheltered population at Whitstable, Kent. It will be seen that when expressed numerically, embryo production is highest at Newhaven and lowest at Greenhithe, the other populations being intermediate. However, mean embryo size and dry weight show an opposite trend, Greenhithe embryos being significantly larger than those from all other populations. Data expressing annual embryo output as dry weight indicate that the large size of the Greenhithe embryos counterbalances the effects of low numbers, this population having the greatest annual reproductive output on a dry weight basis. Table 2 therefore suggests that reproductive output in L. rudis is not adequately represented merely in terms of embryo numbers. In particular there are significant differences in embryo size, and an apparent negative correlation between the numbers of embryos produced and their size.

Data concerning size at maturation in the 7 populations for which this has been established are shown in Table 3. There is a close correspondence between size at maturation and population size structure, as represented by the mean and maximum shell lengths found in each population. This suggests that in general, exposed populations of L. rudis consisting of small individuals, become mature at a significantly smaller size than do those in shelter.

TABLE 2   REPRODUCTIVE OUTPUT OF L. rudis

	Embryos			Total output per female per annum (mg. dry Wt.)
	Nos. per female per annum	Mean Size (mm)	Mean Wt. (mg. dry Wt 100)	
1. Newhaven	1089	0.521	1.84	20.0
2. Whitstable	814	0.569	2.51	20.5
3. Landshipping	736	0.598	3.28	24.1
4. Greenhithe	674	0.661	5.04	34.0

Some aspects of reproductive output in standard (9-11mm) L. rudis from four populations. Whitstable data partly recalculated from Berry (1961).

TABLE 3   MATURATION AND SIZE STRUCTURE IN L. rudis

	Shell length (mm)		
	Mean	Max.	at Maturation
BEDROCK			
1. Newhaven	3.7	6.0	3.0 - 4.0
2. Watchet (Somerset)	3.8	8.0	2.0 - 3.0
3. Newhaven	4.9	13.0	4.0 - 5.0
BOULDERS			
4. Newhaven	11.7	17.0	7.0 - 8.0
5. Watchet	10.5	17.0	7.0 - 8.0
6. Greenhithe	7.2	13.0	6.0 - 7.0
7. Landshipping	5.7	14.0	6.0 - 7.0

Relationship between size structure and size at maturation in seven populations of L. rudis from bedrock and boulder substrates. Populations 1 and 2 from harbour walls, 3 is the Newhaven population referred to in text.

DISCUSSION

Previous studies on reproduction in L. rudis (Thorson, 1946; Berry, 1961; Muus, 1967; all as L. saxatilis) indicate appreciable variations in fecundity between different habitats. Insofar as fecundity reflects the quality and quantity of available food, as well as the conditions for its utilization, such variations are to be expected. The high reproductive output of Greenhithe females, measured as dry weight (Table 2), may indicate favorable conditions on this shore, since food is visibly abundant throughout the year, and the substrate is usually sufficiently wet to allow feeding during the emersion period (Faller-Fritsch, unpubl.).

It is however clear from Table 2 that the expenditure of energy per embryo, i.e. mean embryo dry weight, also shows variation between populations. Thus a given reproductive energy expenditure may involve large numbers of small embryos (e.g. Newhaven), or fewer larger embryos (e.g. Greenhithe). Preliminary investigations of L. rudis in many parts of Britain support the view that these differences are general in their occurrence, larger embryos being characterisitc of sheltered populations. A satisfactory account of these differences must relate them to prevailing conditions which may influence the observed strategy in a particular situation.

There is indeed evidence that differences in reproductive output and embryo size are correlated with habitat features important in their effects on the dynamics of the Newhaven and Greenhithe populations. At the former site, where juvenile winkles are largely reliant on empty barnacle shells for shelter, success in recruitment must partially depend upon the degree of colonization of these suitable microhabitats. A reproductive strategy involving production of numerous juveniles may therefore be adaptive here. In contrast, both the causes of mortality and the size range of the most susceptible winkles are quite different in sheltered boulder shore populations, exemplified by that at Greenhithe. Very high mortality rates, caused principally by the unstable substrate, affect the young. Mortality rates decrease with increasing shell length, and are low among large individuals. The production of large embryos may be a response to this mortality regime, and is apparently achieved at the expense of embryo numbers (Table 2). The size of the embryos in a particular population may therefore be considered to reflect the relative advantages of large numbers, or large size, according to local conditions.

Size-specific mortality may also be responsible for the observed variations in size at maturation which have been recorded (Table 3). The lack of large individuals in exposed populations, e.g. those from harbour walls in Table 3, must necessarily increase the reproductive role of small individuals. It may be envisaged that selection in such populations would favour maturation at a small size, thereby increasing the time period over which a reproductive contribution may be made. In sheltered habitats large individuals, with their greater reproductive capacity, are often abundant (Table 1), and the role of small animals would therefore be reduced. Nevertheless, early maturation would still be expected to confer some selective advantage, unless accompanied by a reduced expectancy of life and so of reproductive contribution. It has already been mentioned that the main incidence of mortality at Greenhithe is among small individuals, and this is certainly true of the mechanical damage resulting from movements of the unstable boulders. In other populations, though not at Greenhithe, small shells are also more susceptible to predation by crabs (Heller, 1976). An adaptive strategy in these conditions may involve the diversion of energy resources into somatic growth, the rapid attainment of a size at which mortality rates are reduced, and then the commencement of reproduction. The size at which populations of L. rudis become mature is, on this argument, dependent upon a balance between the advantages of maturation while small, and those in certain environments of rapidly attaining a relatively large size. Further work, in particular some comparative growth studies, will be required in the testing of this hypothesis.

It is evident that the factors controlling size at maturation in L. rudis may alter in their effects over short distances in some localities (Table 3). Many other attributes of this species are similarly variable, especially shell coloration (Fischer-Piette et. al., 1963, as L. saxatilis). From what is known concerning the determinants of shell colour in other molluscs, it is clear that this must be genetically controled in L. rudis. It is possible that the features studied here, namely embryo size and size at maturation, may also be under direct genetic influence, though there is at present no evidence relevant to this suggestion. There can however be no doubt that the ability of L. rudis to adapt on a very local scale to differing environmental conditions is in large part responsible for the widespread distribution of this species.

ACKNOWLEDGEMENTS

I would like to thank Professor Don R. Arthur of University of London King's College, for the use of facilities in his Department; and Dr. R.H. Emson for his advice at all stages of the work, which was carried out during the tenure of grant from the Natural Environmental Research Council.

REFERENCES

Ballantine, W.J., 1961. A biologically defined exposure scale for the comparative description of rocky shores. Fld. Stud., Vol. 1, pp. 1-19.

Berry, A.J., 1956. Some factors affecting the distribution of Littorina saxatilis (Olivi). Ph.D. Thesis, University of London.

Berry, A.J., 1961. Some factors affecting the distribution of Littorina saxatilis (Olivi). J. Anim. Ecol., Vol. 30, pp. 27-45.

Dautzenberg, P., & H. Fischer, 1912. Mollusques provenant des campagnes de l'Hirondelle et de la Princesse-Alice dans les Mers du Nord. Resultats des campagnes "Prince Albert de Monaco", Tome 37, pp. 187-201.

Emson, R.H., & R.J. Faller-Fritsch, 1976. An experimental investigation into the effect of crevice availability on abundance and size structure in a population of Littorina rudis (Maton). J. exp. mar. Biol. Ecol., Vol. 23, pp. 285-297.

Fischer-Piette, E., J.M. Gaillard et B.L. James, 1963. Etudes sur les variations de Littorina saxatilis. V. Deux cas de variabilite extreme. Cah. Biol. Mar., Vol. 4, pp. 1-22.

Heller, J., 1975. The taxonomy of some British Littorina species, with notes on their reproduction. (Mollusca: Prosobranchia). Zool. J. Linn. Soc., Vol. 56, pp. 131-151.

Heller, J., 1976. The effects of exposure and predation on the shell of two British winkles. J. Zool., Vol. 179, pp. 201-214.

James, B.L., 1968. The characters and distribution of the subspecies and varieties of Littorina saxatilis (Olivi, 1792) in Britain. Cah. Biol. Mar., Vol. 9, pp. 143-165.

Muus, B.J., 1967. The fauna of Danish estuaries and lagoons. Medd. Danm. Fisk. Havundersog., Vol. 9, pp. 1-316.

Thorson, G., 1946. Reproductive and larval development of Danish marine bottom living invertebrates, with special reference to the planktonic larvae in the Sound (Oresund). Medd. Komm. Havundersog., ser. Plankton, Vol. 4, pp. 1-523.

STRUCTURAL FEATURES OF A NORTH ADRIATIC BENTHIC COMMUNITY*

Kurt Fedra

*I. Zoologisches Institut, Universität Wien, Lehrkanzel für Meeresbiologie
A-1090 Wien, Währingerstrasse 17/VI, Austria*

ABSTRACT

The main structural characters of a macro-epibenthic community in the Gulf of Triest are described. The description is based on direct observation by means of a TV- and photo camera sled as well as on divers' observations and diver taken quantitative samples. 380 sampling units of 0.25 m^2 each from 49 positions along 80 km of TV-transect, and 147 samples of 1 m^2 each taken during a two year period from the center of the community are evaluated together with the observations, TV-recordings, and bottom photographs. The community was found to be dominated by the brittle star *Ophiothrix quinquemaculata* (D.Ch.), which contributes 28% to the community biomass (measured as wet weight) or 55% of the number of individuals of the 88 taxa considered. Species abundance relations are examined, using different importance values such as number of individuals, wet weight, caloric values or respiratory activity. Functional groupings were compared, and the community was found to be dominated by suspension feeders, representing 89% of the total biomass and of the main species and groupings are examined. Aggregation was found to be the dominant pattern, and the structure of the observed multi-species clumps is analysed with regard to its ecological function.

INTRODUCTION

Following Odum (1971), a biotic community is any assemblage of populations living in a prescribed area or physical habitat. Study and analysis of such biotic communities require, as a first step, an adequate census against a background of as much data on the environment as possible. Intracommunity classification, from the viewpoint of the concept of ecological dominance (Odum 1971),

---

*This paper is part of a joint program between the Lehrkanzel für Meeresbiologie, University of Vienna, and the Marine Biological Station Portoroz, University of Ljubljana.

provides the basis for a model of the functional structure of communities, leading to the analysis of coupled metabolic transformations and cycles. Primary classification should be based on trophic or other functional levels. Within these levels species or species groups are then ranked according to their ecological importance, i.e. the degree to which they control or determine the communities' features such as the energy flow.

Several measures of ecological importance as e.g. number of individuals, area occupied, biomass measures or metabolic rates have been used and discussed. Numbers of individuals, although frequently used for various approaches and numerical methods are only of ecological significance within comparable functional groups and within limited size classes, and if distinct individuals can be distinguished. Often enough this will not be the case when dealing with marine epibenthic communities; sizes of the macrofauna species alone range over several orders of magnitude, and many species such as most sponges, ascidians, or coelenterates will not allow reasonable counting. In addition to these problems, energy-convertible measures of biomass and metabolic activity provide more useful estimates from the ecological point of view.

Delimitation and a tentative classification of the epibenthic macrofauna community, dealt with below, are described and discussed by Fedra et al. (in press). The community, named according to its biomass dominants the *Ophiothrix - Reniera - Microcosmus* community, occupies large areas in the Gulf of Triest, the northernmost part of the Adriatic Sea. Based on references in Riedl (1961) and Czihak (1959), who record the occurrence of a similar bottom fauna about 50 kms south of the area investigated by Fedra et al., the authors believe that the area they scanned forms only the north-eastern edge of the *Ophiothrix - Reniera - Microcosmus* community. The range of this community probably reaches far to the south, occupying considerable areas of the North Adriatic Sea. The community's environment can be characterized as a shallow bay (average depth about 20 m) with muddy to sandy sediments with a conspicious detric component, classified as "fonds sablo-détritiques plus ou moins envasés (DC-E)" (Gamulin-Brida 1974). Besides sedimentation, the high density of seston (average 1.2 mg dry wt per liter; Stirn, personal communication) and a more or less continuous current (in the 0.1 m/sec range) seem to be the most important parameters for the community.

MATERIALS AND METHODS

12 transects, with a combined total length of 80 km, were laid during September and November 1974 in the North Adriatic Sea along the coast of SR Slovenia between 45°29' N and 45°36' N and east of 13°25' E (Fig. 1). 380 sampling units of 0.25 m^2 each were evaluated. Methods applied and materials evaluated are described by Fedra et al. (in press). From May 1973 to May 1975 another 147 1 m^2 samples from the center of the community, namely positions 38, 39, and 40 were collected in nine sampling series. They were preserved in formaldehyde, sorted to species level, and the numbers and weights of all species were determined. After shaking the organisms to remove adherent water the wet weight of the pre-

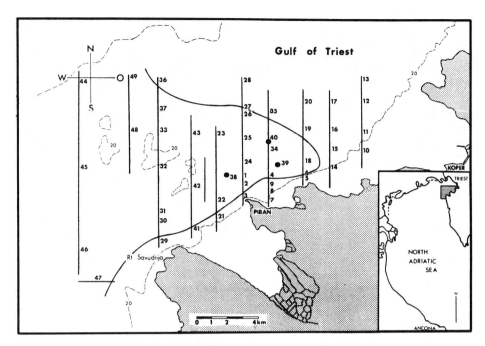

Fig. 1 Area under investigation in the North Adriatic Sea (redrawn after Fedra et al. (in press). Straight lines indicate TV-transects, numbers indicate sampling positions. Dots in the center of the community (38, 39, 40) show positions of long-term investigations and sampling program. Curved solid line marks community border towards north, east, and south.

served material was taken to an accuracy of 0.01 g. Hermit crabs, molluscs, and sedentary polychaetes were removed from their shells or tubes. The material was dried in a drying oven at 80° C for a minimum of 24 hours and echinoderms and crustaceans were decalcified in 10% hydrochloric acid. Dry weight and decalcified dry wt were determined to an accuracy of 0.001 g. Biomass conversion factors were calculated. Caloric equivalents of the main species and species groups were determined by means of dry combustion (adiabatic macrobomb). For species not included in these measurements, literature values (Cummins 1967), or estimates derived from values of comparable species were used. Respiration rates were determined in the lab by means of bell jars and temperature compensated polarographic electrodes (YSI 5419) with continuous recording on chart recorders. The experiments were performed at 15° C, the mean temperature of the community's habitat. More than 300 experiments with 33 species were carried out. Calculations were performed according to Weber (1967), Pielou (1969), and Elliot (1971). Self designed FORTRAN programs were used, utilizing the CYBER 73 of the Univeristy of Vienna. Values in parentheses indicate 95% confidence limits of the means.

RESULTS

According to the wide-range observations of the benthos by means of the camera sled the community area was found to be chracterized by a high standing crop of the macro-epibenthos. A distinct numerical and weight dominance of the brittle star *Ophiothrix quinquemaculata*, followed by *Reniera* ssp., and *Microcosmus* ssp., could be observed. The sampling data from 29 positions, considered to lie within the community area, showed the following relations: total biomass (measured as wet eight) was evaluated at $370(\pm 73)$ g/m^2. *Ophiothrix quinquemaculata* yielded $102(\pm 14)g/m^2$, corresponding to 28% of the total community biomass. *Reniera* ssp. accounted for $74(\pm 28)g/m^2$ or 20%, and *Microcosmus* ssp. contributed $61(\pm 24)$ g/m^2 or 16% to the total community biomass (Fedra et al., in press).

The species abundance relations of sampling series G (November 1974, at the time of the wide-range sampling program) were analysed in detail. Number of individuals, wet weight, biomass expressed as calories, and respiration activity are used and compared as importance values. Table 1 presents a list of the 50 most important species, ranked according to their contribution to the community respiration.

If we assume respiration to be the most appropriate of the four chosen measures to elucidate the degree to which species determine the community's energy budget, the following conclusions can be drawn: Numbers of individuals clearly underestimate larger species such as sponges or ascidians or bigger motile predators. On the other hand, small but abundant species are slightly overestimated, although their metabolic rates may well exceed those of the larger species. Wet weight and caloric values were found to be in good agreement with each other. Only minor differences were found in the order of ranks. In contrast to the numbers of individuals, both biomass measures tend to overestimate the importance of the larger species (with comparatively low metabolic rates) and conversely to underestimate the smaller species (with comparatively higher metabolic rates).

Importance-value curves (Whittaker 1965) for numbers of individuals, biomass (expressed as calories as well as g wet weight) and respiration rates were drawn, plotting the percentage of the respective importance value of each species vs species sequence. The curves for numbers of individuals and biomass (calories) are presented in Fig. 2. A histogram of number of species per octave vs importance value percents (in octaves) is included in the graph.

The curves for both biomass measures as well as for the respiration activity show the same sigmoid shape or a bell-shaped histogram. A lognormal distribution of the importance values is therefore suggested (Preston 1948). Extent of the niche hypervolumes of the single species may therefore be determined by a number of independent variables with different impact on different species. According to Whittaker (1970) the importance values of homogeneous samples, from communities rich in number of species, approach

TABLE 1 Species abundance relations.

Rank	Taxon	$mgO_2/m^2 d$	%	N/m^2	gwW/m^2	cal/m^2
1	Ophiothrix quinquemaculata	165.78	60.11	103.00	98.67	24546
2	Reniera ssp.	41.97	15.22	0.88	107.97	25743
3	Ophiura lacertosa	6.61	2.40	4.18	4.35	530
4	Microcosmus vulgaris	5.96	2.16	2.00	32.04	7745
5	Arca noae	5.49	1.99	1.35	5.45	3932
6	Microcosmus polymorphis	4.33	1.57	0.53	18.82	4550
7	Paguristes oculatus	3.23	1.17	1.47	1.20	475
8	Polychaeta sedentaria	3.18	1.15	2.65	0.44	321
9	Gastrochaena dubia	3.16	1.15	10.82	1.91	1201
10	Lima inflata	2.86	1.04	0.24	2.35	1265
11	Rhodymenia corallicola	2.67	0.97	0.18	2.85	521
12	Pisidia longicornis	2.12	0.77	20.88	0.98	220
13	Polychaeta errantia	2.03	0.74	3.59	0.34	244
14	Holothuria tubulosa	2.03	0.74	0.29	14.75	5000
15	Psammechinus microtuberculatus	1.93	0.70	2.00	5.33	492
16	Pilumnus hirtellus	1.91	0.69	1.00	0.92	437
17	Tethya aurantium	1.79	0.65	0.35	11.59	3019
18	Phallusia mammilata	1.40	0.51	0.06	4.53	726
19	Polycarpa gracilis	1.20	0.44	0.94	5.02	1700
20	Geodia cydonium	1.08	0.39	0.18	10.00	1600
21	Pyura dura	1.07	0.39	0.12	3.46	700
22	Cucumaria planci	1.00	0.36	0.65	8.04	2491
23	Arca lactea	1.00	0.36	4.47	0.52	394
24	Modiolus barbatus	0.95	0.35	1.47	1.53	880
25	Nemertini	0.85	0.31	0.35	0.14	85
26	Ophiura albida	0.80	0.29	4.65	0.67	102
27	Halichondria panicea	0.66	0.24	0.18	6.11	918
28	Ascidia mentula	0.63	0.23	0.06	2.02	191
29	Hippospongia communis	0.61	0.22	0.12	5.69	740
30	Ascidiella aspersa	0.61	0.22	0.18	1.98	183
31	Holothuria forskåli	0.59	0.21	0.12	4.21	1425
32	Murex trunculus	0.56	0.20	0.24	0.77	787
33	Polycarpa pomaria	0.56	0.20	0.53	2.35	750
34	Hemimycale columella	0.51	0.19	0.12	4.75	870
35	Chlamys varius	0.51	0.19	0.82	0.29	114
36	Brachyura (indet.)	0.31	0.11	0.12	0.15	40
37	Ircinia sp.	0.26	0.10	0.06	2.44	352
38	Dysidea sp.	0.24	0.09	0.18	2.26	340
39	Saxicava arctica	0.23	0.09	1.53	0.13	67
40	Axinella sp.	0.23	0.09	0.12	2.17	199
41	Epizoanthus arenaceus	0.22	0.08	0.06	0.11	56
42	Chaetopterus variopedatus	0.21	0.07	0.53	0.68	288
43	Hydroidea	0.20	0.07	0.29	0.17	43
44	Eurynome aspera	0.17	0.06	0.12	0.08	34
45	Botryllus schlosseri	0.15	0.05	0.06	1.75	55
46	Crella rosea	0.14	0.05	0.18	1.32	198
47	Tedania anhelans	0.14	0.05	0.06	1.29	233
48	Hippolytidae	0.13	0.05	3.59	0.05	13
49	Distoma adriaticum	0.10	0.04	0.06	1.16	190
50	Anomia ephippium	0.09	0.04	0.35	0.06	27

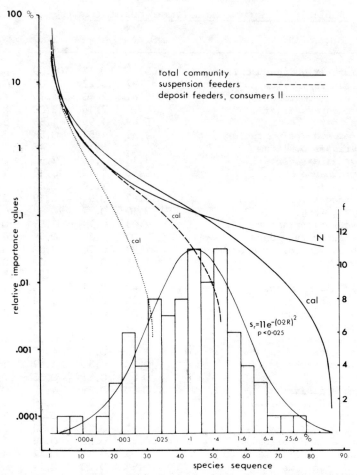

Fig. 2 Importance value curves for calories and numbers of individuals. Sampling series G (November 1974). Histogram for caloric values of total community, frequency (species per octave) vs relative importance values (octaves). S_r: species per octave, R: distance from modal octave.

lognormal curves. The curve drawn for the number of individuals as importance value shows a different shape. Rare species (by number) seemingly dominate the sample. The resulting histogram approaches a logarithmic series. As the rare species vary widely in size, they are partly of medium abundance when described in terms of biomass or metabolic activity. But one could assume that this curve also will approach the lognormal with increasing sampling size. A few very rare species would then be included in the sample.

For the analysis of the trophic structure and a comparison of functional compartments, the species were grouped according to the following divisions in trophic levels and subdivisions of feeding-groups:

A   primary producers
B   consumers I (secondary producers)
        active suspension feeders
        passive suspension feeders
    suspension feeders
    deposit feeders
C   mixed level (deposit feeders/carnivores)
D   consumers II (motile predators of higher levels)

The following relations were found: the community is predominated by consumers I, which account for 95.6% of the community biomass and 92.5% of the community respiration. 93.2% of the biomass of the consumer I compartment is represented by suspension feeders, corresponding to 89% of the community biomass and 91.5% of the community respiration. Active filter feeders, composed of sponges, ascidians, and mussels, comprise about two thirds of the suspension feeders compartment when measured in terms of biomass, but only one third according to individual-numbers and respiration. The passive filter feeders are predominated by *Ophiothrix quinquemaculata* (24.9% of the community biomass, 60.1% of the community respiration), followed by the dendrochirote holothurian *Cucumaria planci*. The passive filter feeders conversely account for one third of the community biomass and two thirds of the community respiration.

Deposit feeders, mainly represented by *Holothuria tubulosa* and *H. forskali* comprise 6.5% of the community biomass (in cal) but only 1% of the total respiration. Primary producers, composed exclusively of benthic macroalgae, are only of negligible importance. They contribute a mere 0.5% to the standing crop and the same proportion to the community respiration.

The mixed trophic level feeds on the pelagic import (namely the detritus component)as well as on smaller consumers I and most probably especially on their newly settled larvae (Thorson 1966). Mainly composed of smaller brittle stars (*Ophiura albida*, *Amphiura chiajei*), hermit crabs (*Paguristes oculatus*), and the sea urchin *Psammechinus microtuberculatus*, this compartment accounts for 1.2% of the biomass and 2.5% of the total respiration.

Higher trophic levels, combined in the consumer II compartment, are represented by *Ophiura lacertosa*, the crab *Pilumnus hirtellus*, and various snails, especially *Murex trunculus* and *M. brandaris*. Bigger, but rather rare species of this compartment are the sea stars *Astropecten* ssp.,*Anseropoda placenta*, and *Marthasterias glacialis*, and the sea urchin *Echinus melo*. Consumers II contribute 2.6% of the total biomass and 4.4% of the community respiration. The importance of this compartment is most probably underestimated, as benthic fishes (mainly *Gobiidae* and *Blenniidae*) are quite abundant but not included in the samples due to the sampling technique applied. The values discussed above are summarized in Table 2.

For the compartments described above, the numbers of individuals were found to provide a better estimate of the respective contributions to the community's energy flow than the applied biomass measures.

TABLE 2 Trophic levels and feeding groups.

Compartment	N/m^2	%	cal/m^2	%	mgO_2/m^2d	%
Primary producers	0.29	0.16	529	0.54	1.38	0.50
active susp.f.	51.18	27.21	60202	61.08	80.76	29.42
passive susp.f.	107.12	57.79	27617	28.02	170.50	62.12
Suspension feeders	158.30	85.40	87819	89.11	251.26	91.54
Deposit feeders	0.71	0.38	6430	6.52	2.68	0.98
Consumers I	159.01	85.79	94249	95.63	253.94	92.52
Mixed level	10.88	5.87	1179	1.20	6.98	2.54
Consumers II	15.35	7.20	2598	2.64	12.17	4.43
Community Total	185.53	100.00	98555	100.00	274.47	100.00

The importance values within the single compartments show the same pattern of distribution as for the whole community. Fig. 2 shows these relations (Consumer I compartment in detail).

The spatial dispersion of the total biomass - resulting from the sum of the component species populations - was obviously clumped or aggregated. In addition to the observed overdispersion in the distribution of most of the component species, a distinct interspecific tendency for aggregation was also obvious. Mathematical models for contagious frequency distributions were fitted to the sampling data distribution of the overall biomass (Fedra et al., in press). The theory of the resulting negative binomial series fits well with the assumption of a complex pattern of intra- and interspecific aggregation.

To test this assumption, the frequency distributions of the main species were analysed. The sampling value distribution of the dominant species were examined (220 sampling units of 0.25 m^2 from the whole community area). Numbers of individuals and average biomass values were used for *Ophiothrix quinquemaculata*, *Reniera* ssp., and *Microcosmus* ssp. respectively. For the brittle stars a Neyman A distribution turned out to provide the best fit ($p < 0.05$) whereas negative binomial series were fitted for the ascidians ($p < 0.01$) and sponges ($p < 0.05$). These models would indicate that the clumps of all three species are distributed at random, with the numbers of individuals or biomass units distributed independently in a Poisson series.

The counts from the center of the community (147 samples of 1 m^2) were examined in detail for 27 species or species groups and the number of species per sampling unit as well as the number of individuals. Only four out of the 27 examined species and groups showed random distributions. Agreement with a Poisson series as the appropriate model was accepted ($p < 0.05$) for the distributions of *Murex trunculus*, *M. brandaris*, *Holothuria tubulosa*, and *Chaetopterus variopedatus*. All other investigated species and groups showed positive contagion. Nine different indices of dispersion were calculated and found to be in good agreement with each other, due to the comparatively high number of sampling units evaluated, namely 147. Using the ratio of mean crowding (Lloyd 1967) to mean density as a measure of patchiness, the following rank of order was obtained: highest patchiness indices, ranging from 4.7 to 2.0 were obtained for *Pisidia longicornis, Arca noae, Modiolus barbatus, Microcosmus polymorphis, Polycarpa* ssp., and *Psammechinus microtuberculatus*. The lowest values among the species showing contaqion were calculated for *Ophiura lacertosa, Paguristes oculatus,* and *Ophiothrix quinquemaculata*. For the latter species the low value of patchiness may partly be due to the fact that the applied measure is sensitive to sample size. Clumps of the brittle stars are found in the 0.25 m^2 scale or smaller, and due to the high density, the 1 m^2 sample size tends to underestimate the degree of contagion. Nevertheless, the respective patchiness index, calculated from the 0.25 m^2 sampling units for the overall community area was evaluated at 1.41 as compared to 1.28 for the 1 m^2 sample size.

Comparing functional units, sedentary suspension feeders show a higher degree of contagion than the more mobile brittle stars or *Cucumaria planci*. Motile predators, carrion feeders, and the representatives of the intermediate trophic level are found in an intermediate position, whereas deposit feeders and large motile predators show only minute contagion or none at all. As a result from these patterns, the number of individuals per sampling unit also showed a contagious frequency distribution.

Besides the intraspecific aggregation, interspecific patterns were also obvious. The observed multi-species clumps (Fig. 3) are based on and around hard elements in the sediment, mostly mollusc shells and shell particles. Mussels such as *Arca noae, A. lactea, Modiolus barbatus,* and *Gastrochaena dubia*, calcareous algae, and sedentary polychaetes with calcareous tubes form elements of a secondary hard bottom on such structures. They are used as a substrate by numerous species of sponges, ascidians, bryozoa, as well as algae and hydrozoans. These aggregations then in turn serve as a substrate for *Cucumaria planci, Antedon mediterranea,* and especially *Ophiothrix quinquemaculata*. In addition, some of the more motile species such as *Pisidia longicornis* and benthic fishes are concentrated on and around such multi-species aggregations.

The number of species per sampling unit shows a contagious frequency distribution, i.e. the species are not independently distributed but tend to attract each other. To test this hypothesis, pairs of species were tested for association. Chi-square tests for 2 x 2 tables were applied, and the coefficient of Cole (1949) and Yule's indices Q and V (1912) were calculated and compared.

For the total community area (0.25 m^2 sampling units) the association between the three dominant taxa, namely *Ophiothrix quinquemaculata, Reniera* ssp. and *Microcosmus* ssp. was tested. The hypothesis of independency was rejected at the 0.01 probability level for *Reniera/Microcosmus*. For *Ophiothrix/Reniera* a respective probability of less than 0.3 was calculated, the value for *Ophiothrix/Microcosmus* was less than 0.5. The same relations could be deduced from the calculated indices. The correlation coefficient r, cal-

Fig. 3 Multi-species aggregation with *Ophiothrix quinquemaculata*, *Cucumaria planci*, sponges, hydrozoans, and anemone.

culated from the transformed (log(x+1)) biomass values, exceeded the tabulated values for independence at the 0.001 probability level for *Ophiothrix/Reniera*, and at the 0.05 level for *Ophiothrix/Microcosmus* and *Reniera/Microcosmus*. The samples from the center of the community were analysed for the 20 most important species and groups, frequently found in the aggregations. Several groups of species were found in close association: the ascidians *Microcosmus* ssp. and *Polycarpa* ssp. as well as several sponges are associated with the mussels *Arca noae* and *Modiolus barbatus*. Soft sponges, the synascidians *Didemnum* ssp., and *Lima inflata* were found together in another type of clump. The algae *Rhodymenia corallicola*, hydrozoans, and *Pilumnus hirtellus* obviously prefer the soft type of aggregation, whereas *Cucumaria planci* is rather found with the hard type with the ascidians. *Pisidia longicornis*, although found with both types of aggregations, seems to share its substrate requirements with *Pilumnus hirtellus* and *Lima inflata*. The calcareous siphons of *Gastrochaena dubia*, frequently found on the basis of both types of clumps, do not provide a preferred substrate for any species but the small bivalve *Arca lactea*; altogether 35 significant ($p < 0.05$) departures from the expectations under the hypothesis of independency were observed out of the 190 tested pairs. The same conclusions were drawn from the calculated indices and the correlation coefficient.

DISCUSSION

The relative importance of the investigated species, feeding groups and trophic levels was found to be different when measured with different importance values. Differences between standing crop and energy flow contributions were found to be mainly determined by the average size of the members of a compartment. It is well known that the ratio of respiration per biomass increases with decreasing size of the organisms. Average size of the members of a compartment can be defined as the ratio of its biomass to the number of individuals. As the ratio biomass to numbers is an inverse proportion of the ratio respiration to biomass, biomass to numbers show a distribution similar to biomass to respiration. In fact, numbers of individuals times biomass measurements provided a good approximation to the rank of order obtained from direct respiration measurements (cf. Pamatmat 1966, Banse et al. 1971).

Aggregation is a well known and frequently observed pattern in ecology. The observed differences in the degree of aggregation between the investigated species and species groups were mainly found to be related to their modes of feeding. Sessile filter feeders show a higher degree of contagion than the motile suspension feeders. Predators and carrion feeders can be found in aggregation around their prey, whereas deposit feeders show no contagion at all.

The random distribution of the multi-species clumps, as concluded from the frequency distributions of the main species, was confirmed by the detailed measurements of Wurzian (1977). The spatial dispersion of the clumps depends mainly on the availability of an appropriate substrate for the species to settle. Bigger shell particles, mainly from *Arca noae* and *Modiolus barbatus*, seem to be the preferred substrate, but also motile hard substrates, namely the housings of the hermit crab *Paguristes oculatus* are of considerable importance (Stachowitsch 1977). Besides size and texture of the substrate, sedimentation is also a determining factor. As the availability and distribution of appropriate substrate is obviously determined by several independent variables, the hypothesis of a random distribution might easily be accepted. The distribution of the species and individuals within the clumps is controlled by several ecological as well as behavioural factors. Once started, the generation of the multi-species clumps is mainly dependant on the substrate-relationships and patterns of growth of the sessile species involved. The observed size limit of 500 g to a maximum of about 1000 g per aggregation can easily be understood, if one takes into consideration that the center of the aggregation needs food supply and oxygen. If the support is no longer sufficient as a consequence of increasing size, the innermost species will die and its decay might possibly cause the disintegration or decay of the whole aggregation.

The distributions of the motile species on such aggregations are also determined by several factors. *Ophiothrix quinquemaculata* settles almost exclusively on several species of sponges. Juveniles are found only on the aggregations. Adults can also be found directly on the sediment, but they also prefer elevated substrates. Acrophilic behaviour was observed in *Ophiothrix quinquemaculata, Antedon mediterranea,* and *Cucumaria planci*. It must instead be interpreted as rheophilic behaviour, for all three passive suspension feeders show a dinstinct reaction to current direction and velocity. Any elevated substrate increases the exposure to the currents; water movement shows a remakable decrease in the immediate bottom layer. Tigmotactic behaviour is responsible for the relation of several other motile species to the aggregations. *Pisidia longicornis*, *Galathea* ssp., *Macropodia longirostris*, *Ethusa mascarone*, *Eurynome aspera*, and *Pilumnus hirtellus* are frequently found in, on, and around the multi-species clumps. Shelter- as well as food requirements are involved. The specific interactions between *Pilumnus hirtellus* and *Ophiothrix quinquemaculata* on their mutual substrate are described by Wurzian (1977).

The formation and maintenance of the observed complex biogenic structures on the originally flat and uniform sediment bottom is concluded to be of vital importance for the above described high-biomass community. The water body available for exploitation to the dominating passive suspension feeders is increased. The elevated substrates allow the passive filter feeders and especially the brittle stars to reach the first 50 cm above the bottom instead of only 15 cm when sitting on the sediment. This corresponds to a more than threefold increase of the available resources. In addition, the complex micro-pattern of currents, caused by the various biogenic structures, may well favour the mixing of the bottom layer and the sedimentation of organic detritus. The latter will increase the import for the deposit feeders. All these mechanisms tend to increase the efficiency of the community in transforming energy from the pelagos to the benthos (Ölscher and Fedra 1977). The increasing community ingestion will favour the growth and increase the density of the observed structures, resulting again in increasing community filter efficiency. The exponential growth, indicated by this positive feedback loop, will be cut down by several limiting factors. It seems likely, that a certain range of density and complexity of the multi-species aggregations will provide optimal conditions for the community, as the patchiness of the structures is an important condition for their impact.

ACKNOWLEDGEMENTS

This study was carried out within the frame of projects No. 2084 and 2758 of the "Fonds zur Förderung der wissenschaftlichen Forschung, Österreich" with additional support from grant No. 1940. The author's thanks are due to Yugoslavian authorities for research permits and their hospitality, and to director and staff of the Marine Biological Station, Portoroz, for providing essential facilities. The author feels indepted to his colleagues R. Machan, E. Ölscher, M. Stachowitsch, and S. Wurzian for their essential help during the field work and sampling program, and especially to C. Scherübel, who in addition classified the ascidians and determined the caloric equivalents.

REFERENCES

Banse, K., F.H. Nichols, and D.R. May 1971. Oxygen consumption of the seabed III. -On the role of the macrofauna at three stations. Vie Milieu (Suppl.) 22, 31-52.

Blacker, R.W. and P.M.J. Woodhead 1965. A towed underwater camera. J. mar. biol. Ass. U.K. 45, 593-597.

Brun, E. 1969. Aggregation of *Ophiothrix fragilis* (Abildgaard) (Echinodermata: Ophiuroidea). Nytt. Mag. Zool. 17, 153-160.

Cabioch, L. 1967. Resultats obtenus par l'emploi de la photographie sous-marine sur les fonds du large de Roscoff. Helgoländer wiss. Meeresunters. 15, 361-370.

Cole, L.C. 1949. The measurement of interspecific association. Ecology 30, 411-424.

Cummins, K.W. 1967. Calorific equivalents for studies in ecological energetics. Pymatuning Lab. Ecol., Univ. Pittsburgh. 52pp.

Czihak, G. 1959. Vorkommen und Lebersweise der *Ophiothrix quinquemaculata* in der nördlichen Adria dei Rovinj. Thalassia jugosl. 1(7), 19-27.

Elliot, J.M. 1971. Some methods for the statistical analysis of samples of benthic invertebrates. Freshwater Biological Association Scientific publication No. 25, 148pp.

Fedra, K., E.M. Olscher, C. Scherübel, M. Stachowitsch and R.S. Wurzian, on the ecology of a North Adriatic benthic community: Distribution, standing crop and composition of the macrobenthos. Mar. Biol. (in press).

Gamulin-Brida, H. 1974. Biocenoses benthiques de la Mer Adriatique. Acta adriat. 15 (9), 1-102.

Guille, A. 1964. Contribution a l'etude de la systematique et de l'ecologie d'*Ophiothrix quinquemaculata* (D.Ch.) Vie Milieu 15, 243-308.

Lloyd, M. 1967. Mean crowding. J. Anim. Ecol. 36, 1-30.

McIntyre, A.D., 1956. The use of trawl, grab, and camera in estimating marine benthos. J. mar. biol. Ass. U.K 35, 419-429.

Odum, E.P., 1971. Fundamentals of ecology. Saunders, Phila. 574 pp.

Olscher, E.M. and K. Fedra, 1977. On the ecology of a suspension feeding benthic community: Filter efficiency and behaviour. In <u>Proceedings of the 11th European Symposium on Marine Biology</u> ed. Keegan, B.F., P. O Ceidigh and P.J.S. Boaden.

Pamatmat, M.M. 1966. The ecology and metabolism of a benthic community on an intertidal sandflat (False Bay, San Juan Island, Wash.) Ph.D. Thesis, Univ. Washington, Seattle. 243 pp.

Peres, J.M. et J. Picard 1964. Nouveau manuel de bionomie benthique de la Mer Mediterranee. <u>Recl. Trav. Stn. mar. Endoume</u>, 31(47), 1-137.

Pielou, E.C. 1969. <u>An introduction to mathematical ecology.</u> Wiley, New York, 286 pp.

Preston, F.W. 1948. The commonness and rarity of species. <u>Ecology</u>, 29, 254-283.

Riedl, R. 1961. Etudes des fonds vaseaux de l'Adriatique. Methodes et resultats. <u>Recl. Trav. Stn. mar. Endoume</u> 23, 161-169.

Stachowitsch, M. 1977. The hermit crab microbiocoenosis - the role of mobile secondary hard bottom elements in a North Adriatic benthic community. In <u>Proceedings of the 11th European Symposium on Marine Biology</u> ed. Keegan, B.F., P. o Ceidigh and P.J.S. Boaden.

Thorson, G. 1966. Some factors influencing the recruitment and establishment of marine benthic communities. <u>Netherlands J. Sea Res</u>. 3, 267-293.

Vevers, H.G. 1952. A photographic survey of certain areas of the sea floor near Plymouth. <u>J. mar. biol. Ass. U.K.</u> 31, 215-221.

Warner, G.F. 1971. On the ecology of a dense bed of the brittle star *Ophiothrix fragilis*. <u>J. mar. biol. Ass. U.K.</u> 51, 267-282.

Weber, E. 1967. <u>Grundriss der biologischen Statistik</u>, 674 pp. Fischer Stut.

Whittaker, R.H. 1965. Dominance and diversity in land plant communities. <u>Science,</u> N.Y. 147, 250.259.

Whittaker, R.H. 1970. <u>Communities and Ecosystems</u>. 162 pp. Macmillan, Lond.

STRUCTURE OF THE ABYSSAL MACROBENTHIC COMMUNITY IN THE ROCKALL TROUGH

John D. Gage

Scottish Marine Biological Association, Dunstaffnage Marine Research Laboratory, P.O. Box 3, Oban, Argyll, Scotland PA34 4AD

ABSTRACT

A sampling study of the abyssal and bathyal benthos of the Rockall Trough area was initiated by the S.M.B.A. in 1973. This paper reports on a preliminary analysis of macrobenthic community structure from six large quantitative samples taken from RRS Challenger in 1975 with a $0.25m^2$ spade corer from a depth of 2875m in the southern Rockall Trough. The cores were taken within an area measuring 2 x 1 nautical miles between $55°03 - 04'N$ and $12°02 - 06'W$.

Total non-foraminiferal wet-weight biomass averaged $3.67g. m^{-2}$ from 1853 macrofaunal specimens.m^{-2}, disregarding the weights of the occasional very large specimens. These data exceed the range of published values obtained from comparable depths and positions, relative to the continental land mass, in the northwestern Atlantic. Moreover, the relative proportions of the macrofaunal groups represented in the box core samples more closely resembled samples obtained from depths greater, rather than lesser, than 4000m in the N W Atlantic and the central N Pacific.

The Rockall Trough fauna clearly displays a much smaller average size than benthos from shallow water, with many macrofaunal taxa being of meiofaunal dimensions. Meiofaunal taxa were also well represented, particularly nematodes and foraminiferans. Agglutinating foraminiferans were numerically most numerous, but biomass estimates were frustrated owing to difficulty in identifying living plasma.

The identities of the species collected and the proportional representation showed little variation between cores. Expected species diversities of the polychaete/bivalve fraction showed macrofaunal species diversity to be high, though not as high as at abyssal depths in the central N Pacific or bathyal depths on the Californian continental borderland.

Between-core differences in expected species diversity were computed to fall within the limits of multinomial sampling error. This did not support any hypothesis of large-scale patchiness in faunal dispersion that might enhance diversity estimates from towed samplers such as the epibenthic sledge.

INTRODUCTION

Despite the recent spectacular increase in effort, particularly in America, on the deep-sea ecosystem, sampling of the deep-sea macrobenthos continues predominantly to employ gears such as the epibenthic sled (Hessler & Sanders 1967) that are towed for often quite long distances over the ocean floor. Consequently information on any possible fine scale spatial inhomogeneity is lost and because of the sampling bias to which such gear is subject, quantitative data on the density and community structure of the fauna of this

MATERIAL AND METHODS

In order to obtain reliable sample estimates of community parameters where the degree of spatial homogeneity of the fauna is unknown, it is desirable to take as many replicate samples as possible. The present study is of a series of six successful samples (out of a total of 10 lowerings) using a modified form of the $0.25m^2$ United States Naval Electronics Laboratory (USNEL) spade or box corer within an area of 1 x 2 nautical miles in the southern Rockall Trough (Fig. 2), in a depth of 2875m. The cores were obtained on cruise 12B/75 of RRS Challenger in September, 1975.

Sample positions (Table 1) were fixed with satellite navigator.

TABLE 1   Positions of box-core samples (Stations 46 - 51)

Station	Latitude	Longitude
46	55°03.7'N	12°06'W
47	55°03.5'N	12°03.5'W
48	55°03.9'N	12°03.9'W
49	55°03.4'N	12°05.3'W
50	55°04.1'N	12°02.6'W
51	55°03.3'N	12°02.7'W

The box corer used in September, 1975 deviated slightly from the description of Hessler & Jumars as follows: a) a hinged hook rather than a friction release freed the spade arm on bottom contact of the gear and b) spring-loaded hinged steel flaps over large unscreened vents were mechanically restrained open in order to allow free flow of water through the sample box during lowering and hence a much reduced bow-wave effect on approach to the sediment surface; the flaps were allowed to close on release of the spade arm. Winch speeds and other operational details were as detailed by Hessler & Jumars (1974).

The corer was monitored acoustically using an IOS 'H'-type pinger.

The sediment surface of the recovered cores, after draining off the supernatant water, appeared to be only slightly disturbed. This usually only amounted to small surface cracks and fissures probably resulting from manipulating the heavy gear (0.7 tonnes in air empty) back on to the deck. There were usually burrow openings visible together with small ophiuroids and, on one core, a pocket of what appeared to be faecal pellets. The core depth ranged from roughly 17 - 30cm; this variation being caused in part by a slight tilting of the surface of the core. This was likely caused by the spade arm exerting a progressive lifting moment as it described its turning arc through the sediment.

The cores revealed a very marked vertical stratification in the sediment. Below a relatively soft, light brown superficial layer of calcareous pelagic ooze there was a sharp demarcation with an underlying, much harder material consisting of a very stiff greenish clay.

On this occasion it was not possible to remove the full sample box from the sampler. Consequently, the core could not be sectioned or dissected before washing.

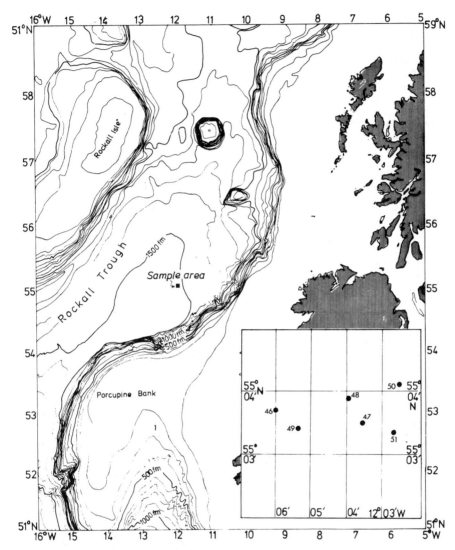

Fig. 1 Location of sample area within southern Rockall Trough. Inset map shows positions of separate core samples. Bathymetry is given in fathoms; contours drawn at 100 fm intervals.

still little-known biotope are still very few.

The present study reports on an analysis of the standing crop and community of the abyssal fauna of one small deep-sea area lying about 90 nautical miles west of the Irish mainland.

The entire core was then carefully washed through a 420μm square mesh sieve, using the elutination technique of Sanders, Hessler & Hampson (1965). Contamination of the sample with pelagic animals (especially copepods) that were present in the ship's fire hose supply used for sample washing was reduced by means of a 250μm sintered woven wire mesh industrial filter interposed in the system. The washed sample was then fixed in buffered formalin and transferred a few days later to a preservative sorting fluid buffered to pH 7.5 consisting of 2.5% formalin, 1% propylene phenoxytol, 10% propylene glycol and 86.5% deionized water. Rose Bengale was used to facilitate separation of living fauna from detrital material and to more easily separate living specimens from dead shells of animals such as molluscs, ostracods and foraminiferans.

Sorting aimed at a total extraction of the living fauna and a subsequent sorting to species level. A considerable effort was made to separate all fauna to the constituent species whether they could be identified or not. In fact, to date only a fraction of the material has been examined by taxonomic specialists. Consequently, the accuracy of sorting species will have varied with the different groups because the criteria separating species are more easily discernible, to a taxonomic generalist, in some groups than others. It is to be hoped that any undersorting will have been balanced by oversorting into species so that sample estimates of community structure based on the species abundance data obtained will not seriously be in error.

Biomass was estimated from the total fauna extracted from the samples before sorting to species. The organisms were gently blotted as a mass and air-dried on filter paper for 2 minutes before weighing. Probably because of the propylene glycol content of the preservative used, the fauna suffered little or no damage through dehydration as a result of this treatment.

BOTTOM ENVIRONMENT

Soundings taken both in the course of cruise 12B/75 and on previous cruises have suggested that the bottom topography in the sample area is flat. Observations of bottom temperature and salinity (P R Barnett personal communication) and dissolved oxygen (Ellett & Martin 1973) indicate stable conditions (2.7 - 2.8°C, 35.0% S, and ca. 6 ml.l^{-1} dissolved O_2). Seabed photographs (Fig. 2) taken by Dr Barnett in the area of the present sampling show elongated sediment tails and slight scour marks on opposing sides of small scale features such as projecting worm tubes. The unidirectional deflection of what appears to be protruding polychaete tubes in line with the sediment tails and scour marks suggests bottom currents of several centimetres per second. Recent indications (Fig. 2B) from the deflection of short string 'streamers' attached, with a compass, to a crosswire within the camera's field of view suggest that the bottom current flow may be intermittent. No clear directionality in flow has as yet been detected from such records over the relatively short (ca. 1 min) duration of bottom contact of the camera.

It is of interest to note that Heezen & Hollister (1964) consider that traction velocities necessary for transport of deep sea sediments probably range from 4 - 60 cm/sec. Geological evidence presented by Jones, Ewing, Ewing & Eittreim (1970) and Roberts (1975) suggests that a southgoing current may be an established feature at the floor of the Rockall Trough, although their same evidence suggests that its route is mainly constrained along the western margin of the Trough. We have as yet no information on the possible presence of a tidal component.

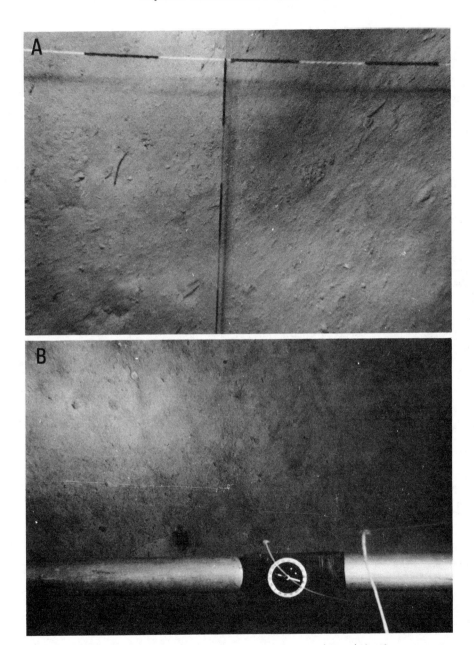

Fig. 2 The bottom at the sample area. <u>A</u> shows a view with the camera looking straight down onto the sediment surface. The crosswires visible are marked at 10cm intervals. <u>B</u> shows another overhead view. The limp state of the string 'streamers' visible do not suggest any appreciable bottom current. The compass indicates that N. is towards the left hand margin of the photograph.

Table 2 Faunal composition and standing crop in six 0.25m² box cores (listed as 'Stations 46 - 51') from the southern Rockall Trough

Macrofaunal taxa	Station 46 N	S	47 N	S	48 N	S	49 N	S	50 N	S	51 N	S
Porifera	✓	1	✓	2	✓	2	✓	2	✓	1	✓	1
Cnidaria	✓	2	✓	4	✓	3	✓	3	✓	3	✓	3
Polychaeta	261	42	291	62	252	52	294	56	284	46	118	36
Sipuncula	-	-	3	2	4	4	9	4	3	3	4	3
Nemertina	12	2	25	5	14	5	34	3	5	3	-	-
Pogonophora	-	-	1	1	-	-	-	-	-	-	-	-
Tanaidacea	44	15	53	12	45	18	40	15	38	13	19	8
Isopoda	25	10	13	8	20	13	26	13	19	8	8	5
Amphipoda	8	4	13	8	18	6	16	7	17	5	5	4
Cumacea	1	1	2	2	1	1	3	3	1	1	1	1
Other Crustacea	-	-	-	-	-	-	1	1	-	-	-	-
Aplacophora	8	5	3	2	5	2	3	2	3	1	2	1
Bivalvia	37	9	31	10	64	10	53	11	32	9	33	9
Gastropoda	3	2	3	2	3	3	9	3	1	1	2	2
Scaphopoda	13	4	11	3	33	3	23	4	5	3	4	3
Ophiuroidea	6	2	2	1	32	1	12	1	-	-	12	1
Asteroidea	-	-	-	-	2	1	1	1	1	1	-	-
Echinoidea	-	-	-	-	2	2	1	1	1	1	-	-
Holothuroidea	-	-	-	-	1	1	-	-	1	1	-	-
Ectoprocta	-	-	-	-	-	-	-	-	1	✓	1	✓	1
Ascidiacea	-	-	-	-	-	-	1	1	-	-	-	-
Total $\underline{N}$	431		451		496		526		412		208	
Total $\underline{N}.m^{-2}$	1724		1804		1984		2104		1648		832	
Total $\underline{S}$		99		124		127		132		101		78
Total $\underline{S}.m^{-2}$		396		496		508		528		404		312
Meiofaunal taxa												
Foraminifera	-		-		-		22		-		-	
Nematoda	301		200		198		231		221		127	
Copepoda	11	7	4	3	10	3	8	5	6	4	1	1
Ostracoda	-	-	3	2	5	1	3	1	12	2	5	2
Total $\underline{N}$	312		207		213		242		339		133	
Total W/W (g) excluding forams	1.18		0.64		0.82		0.92		1.54		0.40	
Total W/W Foraminifera	NM		NM		0.13		0.28		0.27		0.53	

Macrofaunal taxa	Mean N*	Mean S*	%N*	%S*	Total S*
Porifera		1.6		1.4	
Cnidaria		3.6		2.6	
Polychaeta	276.4	51.6	59.6	44.2	105
Sipuncula	3.8	0.4	1.1	2.2	
Nemertina	18	3.6	3.9	3.1	
Pogonophora	0.2	0.2		0.2	
Tanaidacea	44	14.6	9.5	12.5	
Isopoda	20.6	10.4	4.5	8.9	
Amphipoda	14.4	6.0	3.0	5.1	
Cumacea	1.6	1.6	0.4	1.4	
Other Crustacea	0.2	0.2		0.2	
Aplacophora	4.4	2.4	0.9	2.0	
Bivalvia	43.4	9.8	9.3	8.4	18
Gastropoda	2.4	2.2	0.9	1.9	
Scaphopoda	17	3.4	3.7	2.9	
Ophiuroidea	10.4	1.0	2.2	0.8	
Asteroidea		0.6	0.2	0.5	
Echinoidea	0.8	0.8	0.2	0.7	
Holothuroidea	0.4	0.4	0.1	0.3	
Ectoprocta		0.4		0.5	
Ascidiacea	0.2	0.2	0.04	0.2	
Total N	463.2				
Total $N.m^{-2}$	1852.8				
Total S		116.6			
Total $S.m^{-2}$		466.4			
Meiofaunal taxa					
Foraminifera	✓				
Nematoda	230				
Copepoda	7.8		4.4		
Ostracoda	4.6		1.2		
Total N	262.6				

Mean W/W(g) excluding forams = 0.92   (≡ 3.67 g m^{-2})

N = number of individuals, S = number of species, %N and %S = percentage of total N or total S.   ✓ designates a positive occurrence of taxa where species could not be enumerated as individuals. NM = not measured. W/W = wet weight. Owing to taxonomic difficulty the total number of species occurring through the sample series was enumerated only for polychaetes and bivalves.

*excluding station 51

RESULTS AND DISCUSSION

Standing Crop

The total numbers of the individuals of the various taxonomic groupings, together with total wet-weight values per core are given in Table 2. Agglutinating forms of the Foraminifera were numerically most numerous in the samples. However, because of difficulty in many cases in identifying living plasma with any great degree of certainty, their biomass was measured separately. The values given should be regarded as approximate until a more certain separation of living from dead tests is achieved.

Although not numerous, there were present also several small clusters of anastomosing tubules tentatively identified as belonging to the new rhizopod subclass Xenophyophoria (Tendal 1972, Hessler 1974).

Published data on deep sea standing crop are summarised in Zenkevitch, Filatova, Balyaev, Lukyanova & Suetova (1971), Menzies, George & Rowe (1973), Rowe, Polloni & Horner (1974) and Thiel (1975). Although much higher values than those of the present investigation have been recorded (e.g. Kusnetzov 1960) from the N Atlantic, these have come from depths less than 1500m. However, it is difficult to make close comparisons with other published data because of (a) differences in sampling gear and screen size and (b) because most of the values given tend to be derived from single samples, the sample variance is unknown.

In general, it seems that standing crop is high compared to data obtained from similar depths elsewhere. For example, the values of 3.67 g.m^{-2} non-foraminiferal wet-weight biomass and 1853 macrofaunal individuals m^{-2} that were obtained in the present study are considerably more than the largest values obtained by Rowe et al.(1974) at comparable depths on the continental rise of N America. However, the samples of Rowe et al. were taken with a deep-sea anchor dredge (Sanders, Hessler & Hampson 1965) or 0.5m^2 van Veen grab. The sampling efficiency in deep water of the van Veen compared to box corers is unknown; however, Smith & Howard (1972) and Beukema (1974) claim the 0.06m^2 USNEL and Reineck box corers to be considerably more efficient than grabs on shallow hard sand. The anchor dredge may be less efficient than either grab or box corer. In the sampler comparisons of Dickinson & Carey (1975) and Gage (1975) the anchor-box dredge (which is essentially similar to the Sanders anchor dredge) was found to yield samples defficient in vagile Crustacea (mostly amphipods) and bivalves compared to grab samples. It is also of interest to compare the results (Table 3) obtained by the present anchor-box dredge samples on positions very close to those of the present study. Assuming a 10-cm sampling depth, the fauna extracted give an estimate of non-foraminiferal standing crop at least half that from the present large box cores. Sanders, Hessler & Hampson (1965), however, give values, derived from their anchor-dredge samples of the range of standing crops, in terms of numbers of individuals to be expected along their Gay-Head Bermuda transect, that are considerably in excess of those given by Rowe et al. (1974) from the same area and over a comparable depth range (2500 - 2870m), but which are still less than the values obtained in the present study.

TABLE 3 Wet-weight biomass of total fauna excluding Foraminifera from two anchor-box dredge hauls (Stations 44 and 45) at positions close to box core samples

Station	44	45
Date	6 Sept. 1975	
Position	55°03.5'N 12°01.5'W	55°03.8'N 12°03'W
Depth (m)	2875	2875
Vol. sediment (1)	ca. 50	ca. 50
Biomass (g)	1.03	1.21
Biomass.m^{-2} (approx.)	2.06	2.42

Unfortunately, no other wet-weight biomass data from box core samples are known to the present author, although Hessler & Jumars (1974) and Jumars (1975, 1976) give data on faunal densities in the central N Pacific and Californian continental borderland, respectively. These authors estimate macrofaunal abundance (numbers of individuals) from samples taken with a $0.25m^2$ USNEL box corer similar to that used in the present study. Although a 297μm mesh sieve was used by Hessler & Jumars rather than the 420μm of the present study, values exceeding their mean value (from ten samples) by more than 16 times, were obtained in the present study. On the other hand, Jumars (1976) gives values from 5 box cores of macrofaunal density that are quite similar to those of the present study. Hessler & Jumars (1974) relate the low benthic standing crop in the central N Pacific to the unproductive (100 mg $C.m^{-2}$ day^{-1}, Koblents-Mishke 1965) nature of the euphotic zone. Considerably higher standing crops would be expected both in the Californian continental borderland and the Rockall Trough since both are situated in or near areas of regional upwelling with consequently relatively high (250 - 500 mg $C.m^{-2}$ day^{-1}, Koblents-Mishke, Volkovinskii, & Kabanova (1968) surface productivity.

Average size of fauna

As reported by other workers (see Thiel 1975), the average size of individual specimens of taxa normally considered as macrofaunal (see McIntyre 1969) seems minute compared to shallow water specimens. For example, the total macrofaunal standing crop of the soft mud in Loch Creran, Argyllshire, Scotland, was estimated from four $0.1m^2$ van Veen grab hauls at 169.8 $g.m^{-2}$ from 4360 individual specimens, giving a biomass/number of individuals ratio of 0.039, an order of magnitude difference when compared to the ratio of 0.002 for the six abyssal box cores.

It was also apparent that many individuals of taxa normally considered meiofaunal did not seem to follow this trend. For example many of the Nematoda which were numerically very conspicuous in the box cores were at least an order of magnitude larger than many of the traditionally macrofaunal specimens such as isopods. However, it remains likely that most of the Nematoda component was lost in washing through the 420 micron sieve used. Many of the Foraminifera were also of a relatively large size. Indeed the putative Xenophyophoria specimens were several millimetres in diameter.

Faunal composition and its proportional representation in the cores

The identities of the species collected and their proportional representation (Table 2) showed little variation between cores. Similarities are apparent when the percentage representation of the major taxonomic groups is compared (Table 4) with comparable data of Hessler & Jumars (1974) from the central N Pacific gyre. Hessler & Jumars compare their values with data obtained previously from the Gay Head-Bermuda transect in the N W Atlantic (Sanders, Hessler & Hampson 1965), but because these faunal data are derived from anchor-dredge samples there must remain some doubt of their accuracy as community estimates where the more vagile elements are concerned. However, it does seem safe to conclude that the present samples more closely resemble the values (summarised in Hessler & Jumars (1974), Table 4) from samples taken deeper rather than shallower than 4000m depth along the Gay Head-Bermuda transect. Because, however, the present taxon percentages, even if not showing a statistically significant difference from those from the abyssal central N Pacific, do in fact show most resemblance to the faunal composition reported by Sanders et al (1965) on the lower continental slope and abyssal rise (depth ranges 823 - 2086m and 2500 - 2870m, respectively).

TABLE 4 Faunal composition of southern Rockall Trough compared to central N Pacific.

Taxon	Pacific	Rockall
Polychaeta	55.1	59.6
Oligochaeta	2.1	-
Sipuncula	0.4	1.1
Nemertina) Pogonophora)	-	3.9
Tanaidacea	18.4	9.5
Isopoda	6.0	4.5
Amphipoda	-	3.0
Cumacea	-	0.4
Other crustacea	-	0.1
Aplacophora	0.4	0.9
Bivalvia	7.1	9.3
Gastropoda	0.4	0.9
Scaphopoda	2.5	3.7
Ophiuroidea	0.7	2.2
Echinoidea	-	0.2
Asteroidea) Holothuroidea)	0.4	0.2
Bryozoa	2.0	0.1
Brachiopoda	0.7	-
Ascidiacea	1.1	0.1

Values are taxon percentage contributions to the mean total number of macrofaunal individuals in the sample units of each study. Percentages whose differences are not significant at the 95% confidence level are jointly underlined.

As found by other workers using relatively fine screens for sample washing, the polychaetes comprise the single most abundant group of fauna. These have been sorted into at least 31 families and 105 species. The best represented were, in order of relative abundance, the Spionidae (13 species) followed by the Opheliidae (12), the Cirratulidae (9) and the Paraonidae (6). The second most abundant group were the Crustacea which were also represented by numerous species. In general, the relative proportions of the dominant crustacean subgroups, the tanaids, isopods, amphipods and cumaceans, fell within the range given by Sanders et al. (1965) for the abyssal rise of the N W Atlantic.

Unlike the other faunal components, the numerical proportion of protobranchs (80.6%) to eulamellibranchs (19.4%) showed most resemblance to samples from the abyssal plain (Sargasso Sea) area of the Gay Head-Bermuda transect. These results from the Rockall Trough support the contention that while the standing crop of the deep-sea depends mainly on local hydrographic conditions that control trophic input, the faunal composition at higher taxon level may be roughly predicted on the basis of depth.

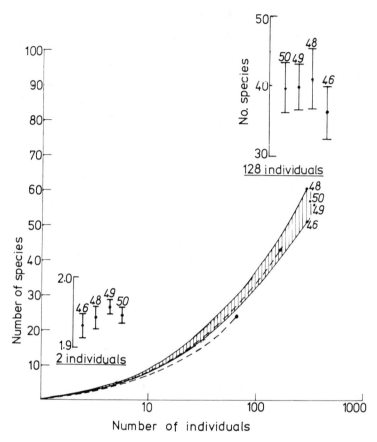

Fig. 3 Expected species diversity for four box-core replicates (Stations 46, 48, 49 and 50). All fall within shaded area of curve. Sampling errors for 2 and 128 individuals are given as vertical bars each two standard deviations above and below the estimated number of species. The two broken lines represent curves of expected species diversity from two anchor-box dredge samples taken from the box-core sampling area.

Faunal diversity

Although data on deep-sea benthic diversity are few, it is of considerable interest to compare values obtained from different areas, particularly in view of (a) the somewhat unexpectedly high deep-sea diversities reported in the last 15 years and (b) the current controversy (Slobodkin & Sanders 1969, Dayton & Hessler 1972, Grassle & Sanders 1973, and Jumars 1975, 1976) on the factors controlling benthos diversity.

Faunal diversity in the present investigation is expressed graphically (Fig. 3) an interpolation of the number of species to be expected amongst smaller numbers of individuals drawn randomly, without replacement, from the total sample. It is being better recognised by benthic workers and others that such a description of species diversity is to be preferred to the many single point estimator diversity measures, which have unsatisfactory sampling properties especially when the sample size is small (Sanders 1968, Hurlbert 1971, Smith & Grassle in press). The expected species curves indicate a slightly lower benthic diversity in the Rockall Trough compared to the central N Pacific (Hessler & Jumars 1974) and the San Diego Trough (Jumars 1975). Expected species curves in the two studies slightly exceed those for the Rockall Trough despite their being derived from the polychaete data only.

Curves generated (Fig. 3) from two anchor-box dredge samples (420μm sieved) from positions close to or within the box-core sampling area closely resemble those from the box cores. This may indicate that at least for the polychaete/bivalve fraction results from anchor dredging will not seriously underestimate species diversity.

Smith & Grassle, who should be consulted for a theoretical understanding of expected species diversity, have developed an unbiased estimator of sampling variance based on a multinomial sampling model. The estimated variance may itself be partioned into that due to multinomial sampling error and that caused by spatial variability over the total sampling area. For the present samples, Fig. 3 shows that for each core the 95% confidence limits for expected species diversity are generally all less than the total range of values for the cores.

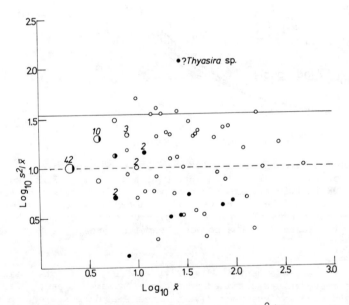

Fig. 4 Logarithm of 'coefficient of dispersion' ($s^2/\bar{x}$) plotted against logarithm of mean number of individuals ($\bar{x}$) per core for all polychaete (open circles) and bivalve (filled circles) species encountered in box-core samples. Numerals denote number of species in excess of one giving identical plots. Upper horizontal line denotes upper 95% confidence limit for Poisson distribution while dotted line denotes the Poisson expectation where $s^2 = \bar{x}$.

As has been pointed out by Jumars (1976) the resolution of spatial scales of deep-sea benthic diversity will likely clarify the issues involved concerning the possible causative mechanisms for the high diversity observed, and also help to determine whether the diversities recovered by towed samplers such as deep-sea trawls or dredges are artifactually enhanced by traversing a mosaic of patches, each at its individual and less diverse stage of succession following a previous disturbance event. Results from other box-core sampling (Hessler & Jumars 1974, and Jumars 1975, 1976) have likewise shown low sample variance. Jumars' (1975) results indicated that the characteristic scale of these controlling processes would be that approaching the size of single macrofaunal individuals, and hence not detectable between whole large box-core samples taken at the scale of the present sampling. Analysis of individual between-core dispersions (Fig. 4) from the present sample series supports such a conclusion: for only a modest number (11 out of 105 species) did the values of the 'coefficient of dispersion' $s^2/\bar{x}$ exceed the upper 95% confidence level of the Poisson expectation. This result may be compared to a shallow west Scottish sealoch where 41 - 56% of species showed a significant aggregation at a scale of 0.1km (Gage 1973).

ACKNOWLEDGEMENTS

The author is grateful to Vera Dunlop, Kareen MacLeod and Margaret Pearson for their help in sorting these samples, and to Dr Peter Barnett for allowing me to use his bottom photographs, and to Dr Fred Grassle for allowing me to use his computer program for calculating confidence limits of expected species diversity.

REFERENCES

Beukema, J.J. 1974 The efficiency of the Van-Veen grab compared with the Reineck box sampler. J. Cons. perm. int. Explor. Mer 35, 319-327

Dayton, P.K. and R.R. Hessler 1972 Role of biological disturbance in maintaining diversity in the deep sea. Deep-Sea Res. 19, 199-208.

Dickinson, J.J. and A.G. Carey 1975 A comparison of two benthic infaunal samplers. Limnol. Oceanogr.

Ellett, D.J. and J.H.A. Martin 1973 The physical and chemical oceanography of the Rockall Channel. Deep-Sea Res. 20, 585-625

Gage, J.D. 1975 A comparison of the deep-sea epibenthic sledge and anchor-box dredge samplers with the Van-Veen grab and hand coring by diver. Deep-Sea Res. 22, 693-702

Gage, J. and A.D. Geekie 1973 Community structure of the benthos in Scottish sea-lochs. III Further studies on patchiness. Mar. Biol. 20, 89-100

Grassle, J.F. and H.L. Sanders 1973 Life histories and the role of disturbance. Deep-Sea Res. 20, 643-659

Heezen, B.C. and C. Hollister 1964 Deep-sea current evidence from abyssal sediments. Mar. Geol. 1, 141-174

Hessler, R.R. 1974 The structure of deep benthic communities from central oceanic waters. in The Biology of the Oceanic Pacific C. Miller, editor Proc. 33rd Ann. Biol. Colloq., Oregon State University.

Hessler, R.R. and P.A. Jumars 1974 Abyssal community analysis from replicate box cores in the central North Pacific. Deep-Sea Res. 21, 185-209.

Hurlbert, S.H. 1971 The nonconcept of species diversity: a critique and alternative parameters. Ecology 52, 577-586.

Jones, E.J.W., Ewing, M., Ewing, J.I. and S. Eittreim 1970 Influences of Norwegian Sea overflow water on sedimentation in the northern North Atlantic and Labrador Sea. J. Geophys. Res. 75, 1655-1680

Jumars, P.A. 1975 Environmental grain and polycahete species diversity in a bathyal benthic community. Mar. Biol. 30, 253-266

Jumars, P.A. 1976 Deep-Sea Species diversity: does it have a characteristic scale? J. mar. Res. 34, 217-246.

Kusnetsov, A. 1960 Quantitative distribution of the seafloor fauna of the Atlantic Ocean. Dakl Akad. Nauk SSSR 130, 1345-1348 (in Russian)

Koblents-Mishke, O.I. 1965 Primary production in the Pacific. Oceanology 5, 104-116 (in Russian)

Koblents-Mishke, O.I., Volkovinskii, V.V. and Y.G. Kabanova 1968 Distribution and magnitude of the primary production of the Oceans. Sbornik Nauchne-Tekhnicheskoi Informatsii VNIRO, No. 5 (in Russian)

McIntyre, A.D. 1969 Ecology of marine meiobenthos. Biol. Rev. 44, 245-290

Menzies, R.J., George, R.Y. and G.T. Rowe 1973 Abyssal Environment and Ecology of the World Oceans Wiley, New York

Roberts, D.G. 1975 Marine geology of the Rockall Plateau and Trough. Phil. Trans. R Soc. A 278, 447-509

Rowe, G.T., Polloni, P.T. and S.G. Horner 1974 Benthic biomass estimates from the northwestern Atlantic Ocean and the northern Gulf of Mexico. Deep-Sea Res. 21, 641-650

Sanders, H.L. 1968 Marine benthic diversity: a comparative study An. nat 102, 243-282.

Sanders, H.L., Hessler, R.R. and G.R. Hampson 1965 An introduction to the study of deep-sea benthic faunal assemblages along the Gay Head-Bermuda transect. Deep-Sea Res. 12, 845-867

Slobodkin, L.B. and H.L. Sanders 1969 On the contribution of environmental predictability to species diversity. Brookhaven Symp. Biol. 22, 82-93

Smith, W.K. and J.F. Grassle Sampling properties of a family of diversity measures. Biometrics (in press)

Tendal, O.S. 1972 A monograph of the Xenophyophoria (Rhizopoda, Protozoa) Galathea Rep. 12, 7-99

Thiel, H. 1975 The size structure of the deep-sea benthos Int. Revue ges Hydrobiol. Hydrogr. 60, 575-606

Zenkevitch, L.A., Filatova, Z.A., Belyaev, G.M., Lukyanova, T.S. and L.A. Suetova 1971 Quantitative distribution of geobenthos in the world ocean. Byull. Mesk. Obshchest. Ispyt. Prir. (Otd. Biol.) 76, 27-33

ECOLOGY OF THE POGONOPHORE, *Siboglinum fiordicum* WEBB, IN A SHALLOW-WATER FJORD COMMUNITY

J. David George

British Museum (Natural History), Cromwell Road, London, SW7 5BD, U.K.

ABSTRACT

Most pogonophores known to science occur on the continental slopes and in the basins and trenches of the oceans. In the Norwegian fjords, however, Siboglinum fiordicum Webb is found at depths of as little as 30m. Thus it was possible to carry out an ecological investigation of this species and its associated fauna using SCUBA diving techniques, and to obtain living pogonophores in good condition for laboratory-based studies on their behaviour.

Samples taken at two stations from depths of 30m and 35m on a muddy-sand plain in Fanafjorden revealed that S. fiordicum occurred at densities of over $200/m^2$ at the deeper station and approximately $60/m^2$ at 30m. The significantly lower number at the shallower station was partly attributable to the coarser composition of the substratum and partly to the lower organic content. Diver observations revealed that the pogonophore tubes were sometimes aggregated, due possibly to the sediment-working activities of other macrofauna components.

Polychaetes were the dominant infaunal group both in number of species and in number of individuals with the suspension/indirect deposit feeders, Myriochele heeri Malmgren and Owenia fusiformis delle Chiaje, reaching densities of over $1000/m^2$. The molluscs were well represented in number of species but none were as abundant as the direct deposit feeding holothurian, Leptosynapta decaria (Ostergren) which reached densities of over $300/m^2$. Apart from the presence of S. fiordicum the general macrofauna composition was typical of North Atlantic shallow water muddy-sand communities and unlike that of the deep water basins of the Norwegian fjords.

No direct evidence was found to link predatory feeders with S. fiordicum, laboratory experiments testing response to sudden changes in light intensity and to macerated tissue extracts of possible infaunal predators also proving inconclusive. However, pogonophores in good physical condition responded to tactile stimuli applied at their anterior ends by withdrawing further into their tubes.

Observation of the reproductive behaviour of S. fiordicum prior to fertilisation showed that the eggs of a female are brooded in that part of the tube protruding from the sediment and that an effective flow of water is maintained passed the eggs. Males were observed to extend the anterior parts of their bodies from the tubes sufficiently to expose the genital papillae from which spermatophores were expelled into the water.

Late larvae of S. fiordicum showed a preference to orientate with their nerve cords in a ventral position, implying a closer relationship with the protostomian branch of the animal kingdom than with the echinoderm/chordate group to which the pogonophores have been assumed by many to belong.

INTRODUCTION

Morphologically the small phylum of tube-dwelling marine invertebrates called the Pogonophora is quite well-known. However, there is comparatively little information available on their ecology and only a few observations have been made on living worms due to the difficulties of obtaining specimens in good condition from the continental slopes, basins and trenches of the oceans, where the majority of species dwell. In the Norwegian fjords, however, one species, Siboglinum fiordicum Webb occurs at depths of as little as 30m. The species is thus within the range of the diving biologist and provides him with a unique opportunity to observe the worms in situ (George 1975) and to collect specimens in good condition for laboratory-based studies (e.g. Southward and Southward 1970).

This paper describes the results of an ecological investigation of S. fiordicum and its associated fauna using SCUBA diving techniques and of observations on the reproductive behaviour of adults prior to fertilisation. Additionally in an attempt to shed some new light on the vexed question of the phylogenetic position of the Pogonophora in the animal kingdom, observations have been made on the dorso-ventral orientation of living adults and larvae of S. fiordicum.

MATERIALS AND METHODS

Site of Field Investigations

Angel and Angel (1967) and Southward and Southward (1970) reported S. fiordicum from a depth of 30-35m in Fanafjorden, south of Bergen, Norway. It was decided therefore, to select a location in the shallows of Fanafjorden for the ecological investigation. The site chosen was a gently sloping muddy-sand plain at a depth of 30-35m north of the small island of Herøy (approx. 60° 14' 40" N, 5° 16' 30" E). On the landward side of the plain, rocks inclined steeply up to the shoreline, whilst seawards the bottom sloped away to depths in excess of 150m.

Field Sampling Methods

All investigations were carried out by a team of two divers. Two sampling stations were selected, one at 30m (Station I) and the other at 35m (Station II) approximately 60m seaward of Station I. Three random samples each with a surface area of 20 x 20cm, were dug to a depth of 10cm at both the Stations. The sediment was transferred in polythene bags to a launch anchored over the site and thence to the University of Bergen biological laboratory at Espegrend. The sediment was then washed through a 1mm mesh sieve, and the sieve contents fixed in 4% seawater formalin for later examination.

Three samples were taken at each Station for analysis of particle size. After handsorting to remove macrobenthic organisms these samples were retained with a little local seawater in air-tight jars until required for analysis. The sediment was washed with seawater through a series of sieves which fitted the Wentworth scale reasonably from 4–0.0625mm. The various grades were washed with distilled water, dried at 110°C, and the weight of dried material expressed as a percentage of the total dry weight of the sample.

Collecting Living Worms for Behavioural Studies

Living specimens of S. fiordicum were collected either by sieving from fresh sediment samples taken at Station II or by the diver carefully pulling individual tubes from the deposit with forceps, having first fanned away as much sediment as possible from around them. Sediment samples and worms were transferred in vacuum flasks to the laboratory less than 45 min by boat from the sampling site, where they were maintained at 6-7°C until required.

Behavioural Studies

Response to tissue extracts of possible infaunal predators. Specimens of Nereis zonata Malmgren, Nephtys caeca (Fabricius) and N. longosetosa Oersted were ground up in a little seawater and a few drops of the resulting fluid extracts introduced into dishes containing S. fiordicum which had been left undisturbed for 12 hrs.

Response to sudden decrease in light intensity. S. fiordicum specimens in their tubes were retained undisturbed for 12 hrs in artificial light of approximately the same intensity as that existing at 35m depth at the time of sampling. Measurements of light intensity were made using a photographic light meter with an incident light attachment. A sudden decrease in light intensity was effected by inserting an opaque screen between the light source and the worms.

Response to physical disturbance. The anterior ends of tubes containing worms that had been left undisturbed for 12 hrs were tapped sharply with a small implement to simulate the attack of a predator.

Dorso-ventral orientation. In the first set of experiments, adults of S. fiordicum retained within their tubes were placed in a horizontal position in order to observe the orientation that they would adopt within their tubes. In a second series of tests adults were removed from their tubes to allow observation of any inclination to take up a fixed orientation in relation to the substratum. In a third group of experiments, designed to test the orientation preferences of settling larvae, both early and late larvae were allowed to swim freely in a small dish, the bottom of which was covered with sediment.

RESULTS AND DISCUSSION

Field Investigations

The species recovered from the bottom samples are listed in Table I along with the number of individuals occurring in each sample. The list is somewhat incomplete since many of the meiobenthic organisms present in the sediment were washed through the 1mm mesh sieve used for sorting.

TABLE 1 Number of infaunal individuals in 400 cm^2 samples at Stations I and II on the muddy-sand plain of Fanafjorden on 20 July 1973

Species	\multicolumn{3}{c}{Station I}			\multicolumn{3}{c}{Station II}			Feeding niche
	1	2	3	1	2	3	
Siboglinum fiordicum Webb	1	3	3	9	12	6	DOM
Typosyllis armillaris (Müller)	0	0	1	1	1	0	P
Nereis zonata Malmgren	1	0	0	0	0	0	P
Nephtys caeca (Fabricius)	6	4	3	4	7	2	P
Nephtys longosetosa Oersted	0	1	1	0	0	0	P
Glycera alba (Müller)	3	7	4	4	3	8	P
Glycinde nordmanni (Malmgren)	2	0	3	2	2	1	P
Scoloplos armiger (Müller)	3	3	6	5	4	1	DD
Prionospio malmgreni Claparède	10	6	6	4	9	10	S, ID
Spio filicornis (Müller)	5	7	3	7	5	6	S, ID
Chaetopterus variopedatus (Renier)	1	0	0	0	0	0	S
Chaetozone setosa Malmgren	2	8	7	5	5	4	ID
Cirratulus cirratus (Müller)	0	1	0	2	0	0	ID
Diplocirrus glaucus (Malmgren)	1	0	3	0	1	0	DD
Ammotrypane aulogaster Rathke	1	7	2	2	4	6	DD
Arenicola marina (Linnaeus)	1	0	0	0	0	0	DD, S
Euclymene droebachiensis (Sars)	3	3	4	1	6	2	DD
Myriochele heeri Malmgren	50	41	36	39	49	63	S, ID
Owenia fusiformis delle Chiaje	65	73	44	67	50	42	S, ID
Amphictene auricoma (Müller)	0	0	2	1	0	0	DD
Sabellides octocirrata (Sars)	1	1	0	0	0	1	ID
Jasmineira caudata Langerhans	10	4	7	6	6	5	S, ID
Natica catena de Costa	0	1	0	0	0	0	P
Natica fusca Blainville	1	0	1	0	0	1	P
Turritella communis Risso	1	0	0	0	0	0	S, ID
Arctica islandica (Linnaeus)	0	2	0	1	1	0	S, ID
Loripes lucinalis (Lamark)	0	1	4	2	6	1	S, ID
Timoclea ovata (Pennant)	1	2	0	0	0	0	S, ID
Thyasira flexuosa (Montagu)	4	7	5	8	2	4	S, ID
Antalis entalis (Linnaeus)	3	6	2	7	3	5	ID
Ampelisca assimilis Boeck	10	2	14	6	7	9	ID, S
Ampelisca eschrichti Krøyer	5	13	2	3	0	8	ID, S
Ophiura affinis Lütken	0	0	1	0	0	1	P
Echinocardium cordatum Pennant	2	0	0	1	1	0	DD
Labidoplax buski (McIntosh)	3	1	0	0	1	2	DD
Leptosynapta decaria (Ostergren)	15	16	8	12	10	10	DD

DD, direct deposit feeder; DOM, feeds on dissolved organic matter; ID, indirect deposit feeder; P, predator; S, suspension feeder.

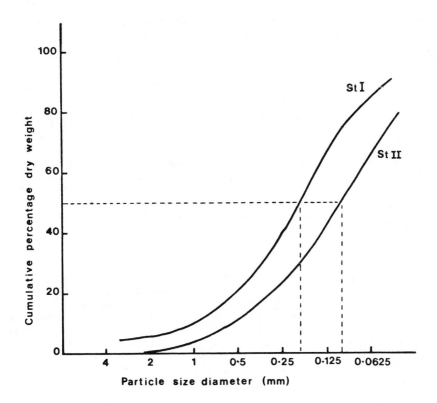

Fig. 1. Cumulative curves constructed from the results of the analysis of particle size at Stations I and II.

Distribution and abundance of S. fiordicum. The pogonophore was present at both stations. Approximately 65% of S. fiordicum tubes recovered from the sediment were empty, nevertheless at Station I the living specimens occurred at a mean density of $58/m^2$ and at Station II at a density in excess of $200/m^2$. The significant difference in population density between the two Stations may be due in part to differences in composition of the substratum. The sediment at 30m had a larger mean particle size (Fig. 1) and contained less organic detritus (visual record only) than at 35m. Preliminary experiments by Bakke (1974) have shown that the larvae of S. fiordicum require a fine sediment in which to burrow. It follows, therefore, that the larvae probably prefer the finer sediment at Station II. In addition, the greater quantity of organic matter at this Station is probably instrumental in decreasing the compactness of the sediment, thereby facilitating burrowing of the larvae. The specimens collected by Bakke for his experiments were from a finer sediment at a depth of 150m in Fanafjorden. It might be reasonable to suppose therefore, that the species occurs in even greater densities at these depths than at 35m. Even so, the number compares favourably with that recorded by Brattegard (1967) for S. ekmani Jägersten in the deep basin of Sognefjorden ($135/m^2$), although not with the dense population of S. fiordicum ($2000/m^2$) recently located by Bakke (1976) at 35m depth in Ypesund.

The preference of S. fiordicum for finer sediments may be related to its tube-forming behaviour, for George (1975) has noted that sand grains are frequently found adhering to the outer surface of tubes from the muddy-sand plain whilst those recovered from deeper water in Fanafjorden by T. Bakke (personal communication) are not encumbered by sand grains. If, as has been suggested by Little and Gupta (1969) and Southward and Southward (1970), the worm relies on the passage of amino acids through the tube wall to satisfy its nutritional requirements, then the presence of large particles adhering to the tube would seriously effect the efficiency of this operation.

Landward of Station I, the presence of a rocky substratum effectively prevented the spread of S. fiordicum into shallower water. However, should suitable sediments exist in shallow water at other localities in the fjords the seasonal variation in seawater temperature and salinity could well be greater than the species is able to withstand. Gage (1974), for example, has shown that the shallow-water zonation of the fauna associated with soft sediments in a Scottish sea loch is correlated with vertical gradients in temperature and salinity.

The small number of individuals present in each sample precluded the possibility of any statistically supportable statement regarding the microdistribution of S. fiordicum in the muddy-sand. However, careful diver observations on the protruding tubes revealed that aggregations sometimes occurred. The unevenness of distribution may have been due to the sediment-working activities of large Arenicola marina (Linnaeus) present in small numbers at the sampling site (Table 1), since George (1975) reported that pogonophore tubes were sometimes found lying on the surface of sediment near mounds recently created by the lugworm.

Distribution and abundance of other infaunal components. Analysis showed that the polychaetes were the dominant group both in number of species and number of individuals (Table 1). The molluscs were also well represented but no one species, even of the suspension-feeding bivalves, was abundant. The sole crustacean representatives were the ubiquitous ampeliscid amphipods. The small deposit-feeding holothurian, Leptosynapta decaria (Ostergren) was the only abundant echinoderm in the sand (approx. $300/m^2$).

The numerical dominance of the polychaetes was due largely to two species of the tubicolous Oweniidae, Myriochele heeri Malmgren and Owenia fusiformis delle Chiaje; the former averaging $1158/m^2$ and the latter $1421/m^2$. Both these species are suspension feeders with a capacity for indirectly feeding on the deposit by wiping their tentacles across the substratum surface. The preponderance of suspension/deposit feeders in communities inhabiting sublittoral soft substrata is well documented (Thorson 1957). Neither species is affected numerically by the difference in sediment composition between the two Stations although the larger average particle size at Station I is reflected in the larger size of the individual particles forming the tube of O. fusiformis.

Apart from the presence of S. fiordicum, the general faunal composition at the sampling site is fairly typical of the shallow water muddy-sand habitat throughout the North-east Atlantic (c.f. Southward 1957, Clark 1960,

McIntyre and Eleftheriou 1968, Gibbs 1969). The fauna contains several species characterising the boreal offshore muddy-sand association as described by Jones (1950). There is little similarity between the fauna of this community and the deep-water basins of the Norwegian fjords (c.f. Brattegard 1967, Fauchald 1972).

Predation on S. fiordicum. During dives at the Fanafjorden site no direct evidence was found to link predatory feeders with S. fiordicum, although it would be unusual for a tubiculous form to be free from predation. The errant polychaetes are the most likely infaunal group to have members preying upon the pogonophore (Table 1). Gut-contents analyses showed that the two glycerid species contained only remnants of ampeliscid amphipods, and a few errant polychaete setae. Ockelmann and Vahl (1970) who investigated the feeding of Glycera alba (Müller) in detail concluded that the worm, although a carnivore, waits in its burrow for passing prey and does not seek sedentary forms on which to feed. It seems unlikely, therefore, that glycerids attack S. fiordicum. Typosyllis armillaris (Müller) is too small an animal to be considered as a likely predator, leaving Nereis zonata and the two nephtyiid species as possible predators. Unfortunately the alimentary canals of these species were devoid of recognisable animal remains, a feature of the nephtyiids which has been reported previously (Sanders et al. 1962, Buchanan and Warwick 1974, Warwick and Price 1975). Nevertheless it would be premature to eliminate these species from a list of possible predators since rapid digestion of meals taken at infrequent intervals may account for the lack of gut contents.

Sokolova has reported (Ivanov 1963) that in the Bering Sea the epibenthic anomuran decapod, Munidopsis, feeds on the pogonophore Polybrachia annulata Ivanov. However, apart from brittlestars, epibenthic macro-invertebrates were a rarity at the Fanafjorden sampling site, although various bottom-feeding fish such as gadoids, sand eels, gobies, blennies and dragonets were seen during the dives. All these demersal fish could possibly take pogonophores as part of their regular diet, but no specimens of S. fiordicum were collected showing evidence of damage to their anterior ends. This may mean either that the pogonophore does not protrude the anterior part of its body from its tube or, as seems more likely, it has a well-developed nervous system allowing rapid contraction of the longitudinal musculature and subsequent withdrawal into the safety of its tube when danger threatens (see response to external stimuli).

In spite of the fact that pogonophore tubes are formed almost entirely from organic secretions nothing was seen to suggest that epibenthic animals were exploiting this valuable food source, for tubes long since vacated as a result of the death of S. fiordicum were still protruding from the sand. The great majority of these empty chitinous tubes were undamaged at the anterior end suggesting that they are unpalatable to raptorial feeders.

Behavioural Studies

Response to external stimuli. Tube-dwelling aquatic invertebrates have developed several different sensory mechanisms enabling them to react quickly to the approach of predators by withdrawing into their tubes. Some respond to sudden changes in light intensity others to chemical and tactile stimuli.

Simple experiments designed to test the chemical response of S. fiordicum to approaching polychaete predators were unsuccessful, the test animals giving no external sign that they were reacting to tissue extracts of Nereis zonata, Nephtys caeca or N. longosetosa.

Since S. fiordicum, unlike most pogonophores, dwells in shallow habitats where reasonable light penetration occurs, it is possible that it possesses light sensitive organs, at least at the anterior end of its body. Indeed, Nørrevang (1974) believes that he has located photoreceptors, similar to those occurring in some hirudineans, subepidermally in two lateral groups at the front end of S. fiordicum. However, in my experiments, groups of three specimens showed no reaction on five separate occasions to a sudden decrease in light intensity such as might be engendered by the approach of a fish predator: nor did they show any reaction to light after being dark adapted for 12 hrs. A similar lack of response to light stimuli has been noticed by Dr A. J. Southward and Dr E. C. Southward (personal communication).

Experiments to test the response of S. fiordicum to physical contact gave more positive results. When the anterior ends of tubes containing worms that had been left undisturbed for 12 hrs were tapped with a small implement a significant proportion (Table 2) reacted by rapidly expanding their girdles (Fig. 2A) so that their toothed setae (Fig. 2B) were thrust into the fibrous lining of the tubes, firmly anchoring their bodies at this point. This was followed by an almost simultaneous withdrawal of the anterior ends of the animals further into their tubes, although the region posterior to the girdles showed little reaction to the stimulus. The rapidity of response from the front ends of the worms would suggest that it was under nervous control and that they were reacting to predators in much the same way as many of the tube-dwelling polychaetes.

TABLE 2 Results of Experiments to test the response of S. fiordicum to physical disturbance

	Hand collected		Sieved	
	Positive response	No response	Positive response	No response
No. of specimens	14	6	8	12

A disturbing feature of the experiments was the variation in response from individuals depending on their pre-treatment. For instance, 70% of those specimens that had been carefully removed from the sand using forceps responded positively, whereas only 40% of those removed by sieving showed any response and then with a reaction time that was invariably longer than in the hand-collected specimens. These observations serve to emphasise the importance of obtaining pogonophore material in good condition if reliable results are to be obtained.

Reproductive Behaviour

Development of the eggs and larvae of S. fiordicum has been well documented by Jägersten (1957), Webb (1963, 1964), Nørrevang (1970), and Bakke (1974; 1976), but little is known of the reproductive behaviour of adults prior to fertilisation. Eggs are shed by the female from her genital apertures, situated a little in front of the girdles, and passed forward by peristaltic movements of the body to lie in the anterior end of the tube (Fig. 3A). Here the eggs develop into larvae (Fig. 3B) which are discharged from the tube mouth when they have reached a late stage of development (Fig. 3C). Carefully snipping off the anterior ends of several tubes protruding from the sediment at the Fanafjorden site revealed that the incubating eggs were frequently, although not invariably, accommodated in this part of the tube. In this position the eggs are likely to receive a better supply of oxygenated water than if they were housed in the tube deep within the substratum. Close examination in the laboratory using a suspension of colloidal graphite showed that the female appeared to maintain an effective flow of water passed the eggs. This was achieved in two ways, firstly, presumably as a result of the beating of the neural ciliated band, a small flow of water was maintained through the tube from the anterior end. Secondly, after a period of quiescence, a female would extend and retract its tentacle and the anterior part of its body several times causing relatively large water movements within the anterior part of the tube and an exchange of water at the tube mouth.

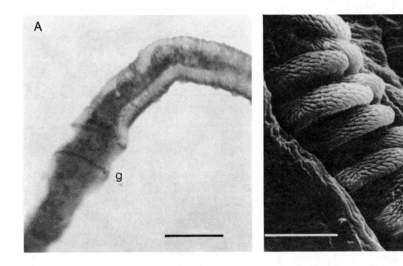

Fig. 2. A. Girdles of living S. fiordicum in a partly expanded condition; g, girdle; scale represents 250μm. B. Detailed view of the toothed heads of the girdle setae; scale represents 10μm.

Males package their sperm into large spindle-shaped spermatophores within the body (Fig. 3D) which are then shed through openings in the genital papillae situated near the anterior end of the animal (Fig. 3E). It has been assumed (Webb 1963) that spermatophores are shed whilst the males lie within the tube and are then passed forward as a result of body movements and expelled from the mouth of the tube. Spermatophores have one end drawn

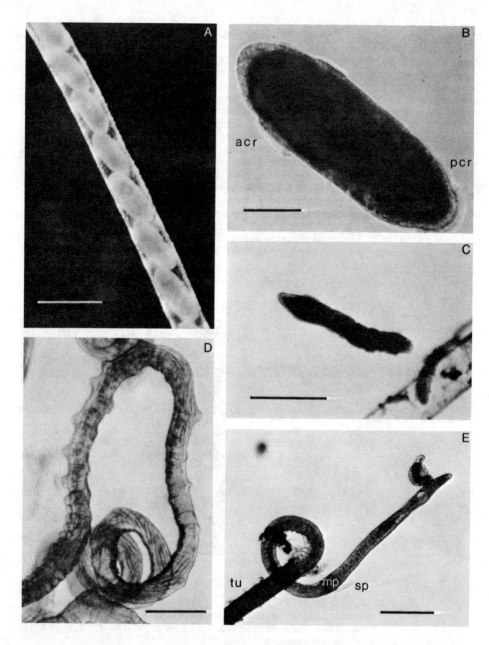

Fig. 3. S. fiordicum. A. Eggs lying in the anterior end of a tube; scale represents 500μm. B. Early larva; acr, anterior ciliary ring; pcr, posterior ciliary ring; scale represents 100μm. C. Late larva after expulsion from the maternal tube; scale represents 500μm. D. Spermatophores lying within the body of the male; scale represents 400μm. E. Male protruding from its tube and releasing spermatophores; mp, position of male genital pore; sp, spermatophore; tu, tube; scale represents 750μm.

out into a long filament which is held tightly coiled whilst they are within the body of the worm. On contact with the water this filament unravels and it is difficult to comprehend how the worm prevents the spermatophores from becoming entangled with its body. However, laboratory observations lead me to suspect that under natural conditions a male extends the anterior part of its body from the mouth of the tube sufficiently to expose the anteriorly placed genital papillae, so that spermatophores may be freely expelled into the water. These suspicions are not without foundation, for although on some occasions worms were seen to extrude spermatophores whilst still in their tubes, on others the anterior part of the body was extended from the tube in a characteristic coil and spermatophores released steadily from the pores (Fig.3E). As suspected, spermatophores extruded whilst the worms were still in their tubes became entangled with the front end of the animal and were not successfully expelled from the tube.

Under natural conditions the spermatophore presumably drifts in the water until it becomes entangled with the tentacle or exposed tube of another individual. In an earlier paper (George 1975) I produced evidence to suggest that S. fiordicum may use its long retractile tentacle as a 'cleaning brush' to prevent accummulation of plants and animals on the outside of its tube. Such cleaning movements could also prove useful to a female, allowing it to capture on its tentacle spermatophores ensnared on the outside of the tube. Withdrawal of the tentacle within the tube would then bring the spermatophores into close proximity with the incubating eggs of the female. The spermatophore wall disintegrates within a few hours (3-5 hrs under laboratory conditions) releasing sperm.

Dorsal-ventral Orientation

Since the discovery of pogonophores there has been argument amongst biologists as to the position of the phylum within the animal kingdom. Many workers believe that they are related to animals in the echinoderm-chordate branch (Deuterostomia) of the invertebrates but others argue that they should be located amongst the protostomian groups. One of the features that could play a crucial part in the argument is the dorsal-ventral orientation of the body, which is not obvious, since the animal possesses no digestive tract and is therefore without a mouth or an anus. Proponents of the deuterostome theory believe that the nerve cord is dorsal, the tentacle(s) ventral, and that the primitive heart is part of the ventral blood vessel. Advocates of protostome affinities on the otherhand reverse the orientation saying that the nerve cord is ventral, the tentacle(s) dorsal, and the heart part of the dorsal blood vessel. All discussion on the matter so far has been related to morphological features and not to behaviour of living worms. Thus it was decided to carry out some simple orientation experiments using both adults and larvae of S. fiordicum in the hope that some new light would be thrown on the problem. Vermiform errant benthic protostomes orientate themselves with their adneural surface in contact with the substratum when crawling, and this feature is retained in some tube-dwelling forms so that if removed from their tubes or their tube placed horizontally instead of vertically they will re-orientate themselves with their adneural surface downwards. The reverse is true of the worm-like enteropneust hemichordates.

Placing S. fiordicum tubes in a horizontal position, however, seemed to have no effect on the orientation of the occupant, the worm lying in whatever position it was first placed. Similarly animals removed from their tubes

showed no inclination to take up a fixed orientation to the substratum.
This situation is similar, unfortunately, to that found in a number of tube-
dwellers which, when adult, lose any tendency to orientate themselves in a
manner which gives an indication of their 'ancestral' orientation. It was
noted, however, that the natural curvature of the anterior end of living
S. fiordicum in a semi-contracted condition, was with the nerve cord on the
concave side and the tentacle on the convex side (Fig. 3E). This tendency
was even more marked in preserved animals and was due to better development
of the longitudinal musculature on the adneural side of the body. This is a
feature common to the majority of vermiform protostomians whereas in the
worm-like deuterostomians the strongest development of longitudinal
musculature is on the anti-neural side of the body. Thus, in adult
S. fiordicum, although adaptations for a tube-dwelling existence mask
responses that would be present in errant forms, the arrangement of
musculature would suggest a protostome affinity.

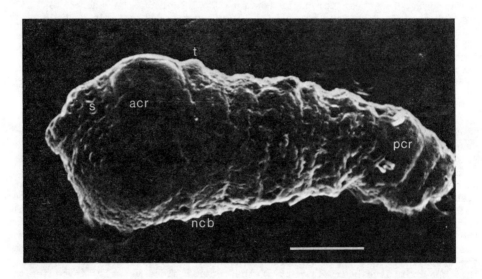

Fig. 4. Lateral view of a late larva of S. fiordicum; ncb,
position of neural ciliated band; s, sensory pit?;
t, tentacle bud; scale represents 125μm.

Larvae might be expected to display dorsal-ventral behaviour patterns more
strongly than the adults. Consequently larvae were examined in detail for
any orientation preference that they might possess. Early larvae (Fig. 3B),
which under natural conditions do not leave the maternal tube, were liberated
and allowed to swim freely in a small dish. When swimming these larvae
progressed forward rotating about their longitudinal axis as a result of
the beating of the cilia of the anterior and posterior ciliary rings (Fig.
3B) but on settling showed no preference for any particular dorsal-ventral
orientation. Late larvae (Fig. 3C, Fig. 4) shed naturally from the maternal

tube, on the otherhand, although swimming in the same way, settled on the bottom more frequently. When they did so, rotation would continue until the longitudinal ciliated band associated with the nerve cord was ventrally placed and the tentacle bud uppermost (Fig. 4). They would then glide forward seemingly testing the suitability of the bottom for commencement of burrowing (see Bakke 1974). In this connection it is interesting to note that the anterior end of the larva is well endowed with pits which may house chemical sensors (Fig. 4). Whilst the larva was moving over the bottom the main propulsive force was still from the anterior ciliary ring (Fig. 4) but the neural ciliated band seemed to be acting as a keel to stabilise the larva laterally whilst it was testing the substratum. At this stage in its life history, therefore, S. fiordicum shows a preference to orientate with its nerve cord in a ventral position, in much the same way as benthonic metatrochophore larvae of polychaetes, implying a closer relationship with protostomes than with deuterostomes.

These preliminary observations on the larvae, however, are not conclusive and it is hoped that further experiments under more closely controlled conditions, may be carried out at a later date in collaboration with Mr T. Bakke of the Espegrend laboratory.

CONCLUSIONS

The diving investigations in Fanafjorden revealed that S. fiordicum occurred at densities of approximately $60/m^2$ at Station I and over $200/m^2$ at Station II. The difference in population density at the two Stations was attributed to the composition of the substratum. Polychaetes were the dominant infaunal group, the macrofauna composition being typical of North Atlantic shallow water muddy-sand communities. No S. fiordicum predators were detected by the field studies although the pogonophores displayed a withdrawal reaction to tactile stimuli in laboratory experiments. Some new observations regarding the reproductive behaviour of S. fiordicum prior to fertilisation were made. Experiments showed that late larvae had a tendency to orientate with their nerve cords in a ventral position, implying a closer relationship with the protostomian branch of the animal kingdom than with the echinoderm/chordate group.

ACKNOWLEDGEMENTS

Dr Ulf Lie kindly gave permission for me to work at the Espegrend Biological Station. I would like to thank Mr Torgier Bakke for his help during my stay at the laboratory and for many useful discussions on the early development of pogonophores. I am most grateful to Mr Ulf Bamstedt who assisted me in my underwater investigations.

REFERENCES

Angel, H.H. and Angel, M.V. 1967 Distribution pattern analysis in a marine benthic community. Helgoländer wiss. Meeresunters. 15, 445-454.

Bakke, T. 1974 Settling of the larvae of Siboglinum fiordicum Webb (Pogonophora) in the laboratory. Sarsia 56, 57-70.

Bakke, T. 1976 The early embryos of Siboglinum fiordicum Webb (Pogonophora) reared in the laboratory. Sarsia 60, 1-12.

Brattegard, T. 1967 Pogonophora and associated fauna in the deep basin of Sognefjorden. Sarsia 29, 299-306.

Buchanan, J.B. and Warwick, R.M. 1974 An estimate of benthic macrofaunal production in the offshore mud of the Northumberland coast. J. mar. biol. Ass. U.K. 54, 197-222.

Clark, R.B. 1960 The Fauna of the Clyde Sea area. Polychaeta with keys to the British Genera. Scottish Marine Biological Association, Millport.

Fauchald, K. 1972 Some polychaetous annelids from the deep basins in Sognefjorden, Western Norway. Sarsia 49, 89-106.

Gage, J. 1974 Shallow water zonation of sea-loch benthos and its relation to hydrographic and other physical features. J. mar. biol. Ass. U.K. 54, 223-249.

George, J.D. 1975 Observations on the pogonophore, Siboglinum fiordicum Webb from Fanafjorden, Norway. Rep. Underwater Ass. 1 (NS), 17-26.

Gibbs, P.E. 1969 A quantitative study of the polychaete fauna from certain fine deposits in Plymouth Sound. J. mar. biol. Ass. U.K. 49, 311-326.

Ivanov, A.V. 1963 Pogonophora. Academic Press, London.

Jägersten, G. 1957 On the larvae of Siboglinum, with some remarks on the nutrition problems of the Pogonophora. Zool. Bidr. Upps. 32, 67-80.

Jones, N.S. 1950 Marine bottom communites. Biol. Rev. 25, 283-313.

Little, C. and Gupta, B.L. 1969 Studies on Pogonophora III. Uptake of nutrients. J. exp. Biol. 51, 759-773.

McIntyre, A.D. and Eleftheriou, A. 1968 The bottom fauna of flatfish nursery ground. J. mar. biol. Ass. U.K. 48, 113-142.

Nørrevang, A. 1970 On the embryology of Siboglinum and its implications for the systematic position of the Pogonophora. Sarsia 42, 7-16.

Nørrevang, A. 1974 Photoreceptors of the phaosome (hirudinean) type in a pogonophore. Zool. Anz. 193, 297-304.

Ockelmann, K.W. and Vahl, O. 1970 On the biology of the polychaete Glycera alba, especially its burrowing and feeding. Ophelia 8, 275-294.

Sanders, H.L., Goudsmit, E.M., Mills, E.L. and Hampson, G.E. 1962 A study of the intertidal fauna of Barnstable Harbor, Massachusetts. Limnol. Oceanogr. 7, 63-79.

Southward, E.C. 1957 The distribution of Polychaeta in offshore deposits in the Irish Sea. J. mar. biol. Ass. U.K. 36, 49-75.

Southward, A.J. and Southward, E.C. 1970 Observations on the role of dissolved organic compounds in the nutrition of benthic invertebrates. Experiments on three species of Pogonophora. Sarsia 45, 69-96.

Thorson, G. 1957 Bottom communities (sublittoral or shallow shelf). In Hedgpeth J.W. (ed.). Treatise on marine ecology and paleoecology 1 (Ecology). Ged. Soc. Amer. Mem. 67, 461-534.

Warwick, R.M. and Price, R. 1975 Macrofauna production in an estuarine mud-flat. J. mar. biol. Ass. U.K. 55, 1-18.

Webb, M. 1963 Siboglinum fiordicum sp. nov. (Pogonophora) from the Raunefjord, Western Norway. Sarsia 13, 33-44.

Webb, M. 1964 The larvae of Siboglinum fiordicum and a reconsideration of the adult body regions (Pogonophora). Sarsia 15, 57-68.

THE RE-ESTABLISHMENT OF AN *Amphiura filiformis* (O.F. Müller) POPULATION IN THE INNER PART OF THE GERMAN BIGHT

Dieter Gerdes

Institut für Meeresforschung, Bremerhaven, Germany

ABSTRACT

Since 1969 bottom samples have been taken from several fixed stations in the inner part of the German Bight. One of these stations, a "silty sand station", is situated in the south eastern part of the *Echinocardium - filiformis* Community, near the location of the former light-vessel "P 12". The samples taken from the silty-sand station in 1969, together with the findings of Stripp (1969), showed a well-established population of *Amphiura filiformis* (O.F. Müller) in this part of the German Bight. After the cold winter of 1969/70, no individuals of this species were found in samples from this station or its vicinity. The disappearance in spring of 1970 and the re-establishment in the summer of 1974 are described and discussed in relation to hydrography and temperature. The long repopulation phase is explained by transport velocities of current and hydrographic conditions in the German Bight.

INTRODUCTION

Changes in macrobenthos can be correlated with several hydrographic factors. Ziegelmeir (1953, 1964, 1970) has demonstrated and discussed the effects of very cold winters on macrobenthos in the German Bight. As he has pointed out, extremely cold water can destroy a substantial part of the near-shore-sublittoral macrobenthos, primarily species belonging to the infauna - e.g., *Tellina fabula* Gmelin and *Lunatia nitida* Donavan among the molluscs, *Echinocardium cordatum* (Pennant) and especially *Amphiura filiformis* (O.F. Müller) among the echinoderms. The destruction of a population of *Amphiura filiformis*, as a result of the cold winter of 1969/70, and its subsequent re-establishment was traced in monthly or quarterly bottom samples from several stations in the German Bight. These were taken by Rachor (IFMB) since 1969.

Study Area

The area of the investigation is roughly delimited by Helgoland in the North, the light-vessel "P 8" in the West, and the silty sand station in the East. The study area is occupied by an *Echinocardium-filiformis* community (Hagmeier 1925; Stripp 1969). The water depth at the silty sand station is 35 m; the depths of the other stations range from 35 m to 50 m. The bottom sediment is classified 1975 (see also Juario, 1975) as very fine silty sand. For a detailed picture of the types of sediment and their distribution in part of the German Bight see Reineck, 1963; Stripp 1969; Gadow & Schafer, 1973. Detailed information on the hydrography of the area was published by Goedecke (1952, 1955, 1968).

278 D. Gerdes

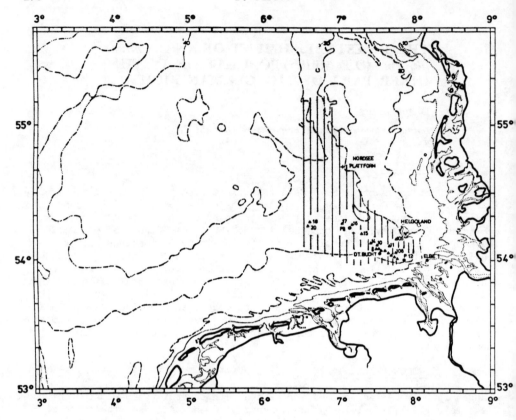

Fig. 1. Study area in the German Bight with the positions of the sampling
stations and the four light-vessels "Elbe I", "P 12", "Dt. Bucht",
and "P 8"; shaded: area of the *Echinocardium filiformis*
Community.

MATERIALS AND METHODS

The samples were taken with a Van Veen type grab covering 0.1 m^2 and were
treated as described by Rachor & Gerlach (in press).

RESULTS AND DISCUSSION

Disappearance of *Amphiura filiformis*

A. filiformis is found at depths of 15 to 70 m in sediments which can be
regarded broadly as silty sand (Buchanan 1967); it is found from Norway to
the Mediterranean (Mortensen 1927). According to Ursin (1960), *A. filiformis*
does not occur in areas with winter temperatures below 4°C, although in
Helgoland waters apparently tolerates winter temperatures as low as 3.5°C.

The echinoderm fauna of the area investigated in 1975 is tabulated in Table 1.

Table 1. Frequency and abundance of the most common echinoderm species in the Echinocardium-filiformis Community of the German Bight.

	Frequency (%)	Ind./m²
Echinocardium cordatum (Pennant)	95	43,2
Amohiura filiformis (O.F. Müller)	85	147,8
Ophiura albida Forbes	85	136.0
Ophiura texturata Lamarck	25	1,7
Echinocyamus pusillus (O.F. Müller)	10	4,6

For the period 1966 to 1968, Stripp (1969) recorded a mean density of 9 A. filiformis /m² for the area. Bottom samples taken in 1969 at the silty sand station A. filiformis occurring regularly with a max. abundance of 36/m². After February 1970 A. filiformis was absent from this station.

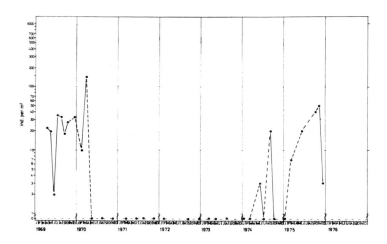

Fig. 2. Occurrence of A. filiformis at a silty sand station from 1969 to 1975 (ind./m²).

As a consequence of the cold winter of 1969/70, water temperature and salinity in the southern German Bight decreased (see Table 2).

As the water column in the study area is thoroughly mixed by various currents in winter (see below), we can assume that the temperature of the surface and bottom water is the same. The salinity in the south-eastern part of the German Bight was far below 30 o/oo. At "Elbe I", a value of 27.64 o/oo was recorded, whereas in the winter of 1962/63 the salinity was never below 31 o/oo.

At the "Elbe I" light-vessel in the east, temperature and salinity increased towards the western edge of the study area (light-vessel "P 8"), where conditions normal for the German Bight obtained. Fig. 3 shows the highest and lowest mean monthly temperatures from 1962 to 1975, measured at "Pegel Helgoland" and at "P 8" (in nov. 1972 "P 8" changes position - 54° 10' N and 6° 21' E - and is now called "TW EMS").

Table 2 Salinities (o/oo S) and temperatures (°C) (monthly mean) of the surface water, measured from three light-vessels in the German Bight from Oct. 1969 to May 1970 (according to "Meereskundliche Beobachtungen und Ergebnisse").

		"Elbe I"		"Dt Bucht"		"P 8"	
		o/oo S	°C	o/oo S	°C	o/oo S	°C
1969	Oct.	30,58	13,5	31,21	-	32,70	14,8
	Nov.	31,55	9,6	31,55	10,8	32,96	10,8
	Dec.	32,16	5,0	33,80	6,8	34,33	6,9
1970	Jan.	33,39	1,9	34,00	3,3	34,25	3,6
	Febr.	32,22	0,0	33,43	1,6	33,46	1,6
	March	29,64	1,0	32,74	2,1	33,27	1,9
	April	27,64	4,0	-	-	31,56	4,4
	May	27,68	9,1	-	-	31,44	9,0

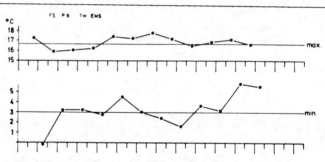

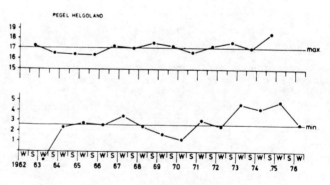

Fig. 3. Temperatures of the warmest and coldest months (monthly means) from 1962 to 1975 (according to "Meereskindliche Beobachtungen und Ergebnisse").

The comparatively low mean winter temperature for the whole period reflects the cold winters of 1962/63 and 1969/70.

In spring of 1972, Salzwedel (IFMB) tried (on two occasions) to find A. *filiformis* over a total of 13 stations. He did not find the ophiuroid at the silty sand station and only 4 specimens were recorded at some distance (about 10 km). A well established population of this species was located to the west of light-vessel "P 8", about 55 km north-west of the silty sand station (Station 17 with 70 ind./m^2, station 18 with 300 ind./m^2, and station 20 with 75 ind./m^2). A. *filiformis* was not found at the silty sand station until the summer of 1974 (see Fig. 2).

A range of abiotic factors influences the distribution and density of marine benthic organisms. These include temperature, salinity, bottom type, depth, distance from shore, erosion and deposition of bottom sediments.

The disappearance of A. *filiformis* in the inner part of the German Bight can be explained as a consequence of the cold winter of 1969/70. In the winter of 1962/63 temperatures below $3.5°C$ lasted for three and a half months while in the winter of 1969/70 such temperatures lasted for three months.

According to Ursin (1960), a period of about three months with temperatures below $3.5°C$ can be regarded as being sufficient to destroy populations of A. *filiformis*. In the cases of the two winters discussed here, however, it cannot be decided whether the destruction of the population can be attributed to the actual temperatures minima of $0°C$ (winter 1969/70) and $-1.4°C$ (winter 1962/62) or to the length of the periods with temperatures below $3.5°C$. The low salinities of the 1969/70 winter may have also contributed to the destruction of A. *filiformis*, for they would further lessen the ophiuroid's tolerance of low temperatures.

Re-establishment

The repopulation of the study area by A. *filiformis* lasted four years and, it is assumed, was carried out by pelagic larvae. Larval dispersal by currents takes place at low annual speeds (Mileikovsky, 1971).

The area of the investigation is part of a "mixing zone" where two different water bodies converge:

(1) North Sea water coming from the west and northwest into the German Bight. According to Kautsky (1973), the transport velocity of these water masses is probably about 1.9 km per day.

(2) water coming from the southeast, out of the rivers Elbe and Weser, and flowing northwards.

The mean transport velocity of the tidal currents in the "mixing zone" ranges from 3.5 km to 8.5 km (Goedecke, 1968). The area is influenced by up to five eddies and these may interfere with the influx of larvae-carrying water. No data are available on the length of the pelagic larval phase of A. *filiformis*. Fenaux (1963, 1970) found the larval phase of A. *chiajei* Forbes to be short and, in laboratory experiments (temp. $18 - 20°C$), larval development and the formation of the young bottom-dwelling form (larval appendages completely reduced) was completed in 8 days. Assuming that the larval phase of A. *filiformis* is not longer than 20 days (more than twice as long as the larval

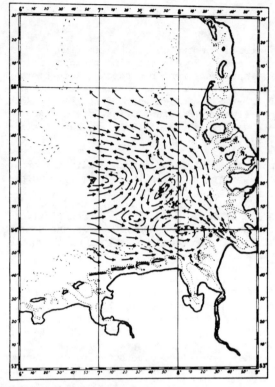

Fig. 4. Eddies in the "mixing zone" of the German Bight (according to Goedecke, 1968).

life of A. *chiajei*), the author submits the following hypothesis:

In 1970, larvae, from the population to the east of the light-vessel "P 8", were transported into the western part of the study area. The highest possible speed of distribution, calculated on the basis of the current velocity under optimal conditions, is

(a) about 38 km, due to the movement of water masses from the west
(b) and 8.5 km, due to the tidal currents. This gives a total of about 46 km for the assumed duration of the larval phase. Under these conditions, a location about 10 km west of the silty sand station is within reach of the larvae. According to Buchanan (1967) breeding of A. *filiformis* takes place in the three-year-old age group, which has a disc diameter of 6 mm. After breeding, the majority of these animals soon die. Thus, the larvae, which settled in the summer or autumn of 1970, bred in the summer of 1973. The new generation of larvae was again transported to the east and settled there.

The animals found in the summer of 1974, at the silty sand station, had a disc diameter of 1 - 2 mm, i.e. these animals were probably spawned the preceeding year.

CONCLUSION

Ziegelmeier (1970) has shown that the repopulation of a partly depleted area by larvae can be hindered by predatory filter-feeding or deposit-feeding species, e.g., *Tellina fabula* Gmelin.

In this case, the time required to re-establish the population of *A. filiformis* is mainly dependent on the velocity of the different currents in the study area. Based on the results of the present study, this hypothesis is compatible with the findings of other workers, e.g.,

(1) The assumption of Buchanan (1967) that breeding takes place in the three-year-old age group,

(2) current velocities, as assumed by Kautsky (1973) and Goedecke (1968), would permit the re-establishment of the population, in this part of the German Bight, to have taken place in the manner suggested,

(3) Rachor & Gerlach (in press) showed that species which breed after the first year of life can be successful in re-establishing a depleted area within one year.

REFERENCES

Buchanan, J.B. 1967: Dispersion and demography of some infaunal Echinoderm populations. Sym. Zool. Soc. Lond. 20: 1-11.

DHI: Meeresk. Beob. Ergebn. dt. hydrogr. Inst., No. 20-25, 30-31, 34-36, 38, 40. Beobachtungen auf den deutschen Feuerschiffen der Nord- und Ostsee. 1962-1974.

Fenaux, D. 1963. Note preliminaire sur le developement larvaire de *Amphiura chiajei* Forbes, Vie Milieu XIV (1): 91-96.
1970. Maturation of the Gonads and seasonal cycle of the planktonic larvae of the ophiuroid *Amphiura chiajei* Forbes. Biol. Bull. 138 (3): 262-271.

Gadow, S. & Schäfer, A. 1973. Die Sedimente der Deutschen Bucht: Korngrößen, Tonmineralien und Schwermetalle Senckenberg. marit. 5: 165-178.

Goedecke, E. 1952. Uber intensität und Jahresgang der thermologischen Schictung in der Deutschen Bucht. Veröff. Inst. Meeresforsch., Bremerh. 1: 236-246.
1955. Uber Intensität der Temperatur-, Salzgehalts- und Dichteschichtung in der Deutschen Bucht. Dt. hydrograph. Z. 8: 15-28.
1968. Uber die hydrographische Struktur der Deutschen Bucht im Hinblick auf die Verschmutzung in der Konvergenzzone. Helgoländer wiss. Meeresunters. 17: 108-125.

Hagmeier, A. 1925. Vorläufiger Bericht über die vorbereitenden Untersuchungen der Bodenfauna der Deutschen Bucht mit dem Petersen-Bodengreifer. Ber. dt. wiss. Kommn. Meeresforsch. N. F. 1: 247-272.

Juario, J.V. 1975. Nematode species composition and seasonal fluctuation of a sublittoral meiofauna community in the German Bight. Veröff. Inst. Meeresforsch. Bremerh. 15: 283-337.

Kautsky, H. 1973. The distribution of the radio nuclide Caesium 137 as an indicator for North Sea watermass transport. Dt. hydrogr. Z. 26 (6): 241-246.

Mileikovsky, S.A. 1971. Types of larval development in marine bottom invertebrates, their distribution and ecological significance: a re-evaluation Mar. Biol. 10: 193-213.

Mortensen, T. 1927. Handbook of the echinoderms of the British Isles. Oxford Univ. Press. 471 pp.

Rachor, E. & Gerlach, S.A. (in press): Variations in macrobenthos in the German Bight. Int. Counc. Explor. Sea. Symp. "Changes in the North Sea Fish Stocks and their Causes", 11: 2-16.

Reineck, H.E. 1963. Sedimentgefüge im Bereich der südlichen Nordsee. Abh. senckenb. naturforsch. Ges. 505: 1-138.

Stripp, K. 1969. Die Assoziationen des Benthos in der Helgoländer Bucht. Veröff. Inst. Meeresforsch. Bremerh. 12: 95-142.

Ursin, E. 1960. A quantitative investigation of the echinoderm fauna of the central North Sea. Meddr. Danm. Fisk. - og Havunders. N. S. 2(24): 1-204.

Ziegelmeier, E. 1953. Quantitative investigations of the bottom fauna (macrobenthos) in the Helgoland Bight. Annls. biol. Copenh., 9 (1952): 140-141.
        1964. Einwirkungen des kalten Winters 1962/63 auf das Makrobenthos im Ostteil der Deutschen Bucht. Helgoländer wiss. Meeresunters. 10: 276-282.
        1970. Uber Massenvorkommen verschiedener makrobenthaler Wirbelloser während der Wiederbesiedlungsphase nach Schädigungen durch "katastrophale" Unwelteinflüsse. Helgoländer wiss. Meeresunters. 21: 9-20.

AN ECOPHYSIOLOGICAL APPROACH TO THE MICRODISTRIBUTION OF MEIOBENTHIC OLIGOCHAETA. I. PHALLODRILUS MONOSPERMATHECUS (KNÖLLNER) (TUBIFICIDAE) FROM A SUBTROPICAL BEACH AT BERMUDA*

Olav Giere

Zoological Institute and Zoological Museum, University of Hamburg, Martin-Luther-King-Platz 3, 2000 Hamburg 13, Fed. Rep. Germany

ABSTRACT

Phallodrilus monospermathecus from a Bermuda beach showed a distinctly inhomogenous distribution due to dense aggregations of worms in the subsurface layers of the high-water line, whereas the lenitic eulittoral sand flats were only scarcely populated. The abiotic factor system in the various sample sites and layers was analyzed using mainly electrometrical in-situ methods. In order to find out the ecophysiological background of this distributional pattern, the tolerance of the worms to temperature, salinity, alkalinity, oxygen supply, and combinations of these parameters, was studied.
Results show that Ph. monospermathecus is extremely tolerant to each factor per se, even if it exceeds the naturally occurring range. However, rigorous parameter combinations often acted as multiple stressors and reduced survival time of the animals markedly. Judging from these experiments, the microdistribution in the field can be explained by restriction to layers with moderate temperatures and relatively stable salinity, by preference for well oxygenated sand and avoidance of unusually high alkalinity.
These studies represent the first part of an investigation comparing ecological conditions for meiobenthic oligochaetes in different climates.

INTRODUCTION

Marine oligochaetes from eulittoral beaches exhibit a markedly inhomogenous distributional pattern (Jansson 1962, 1968 ; Giere 1970, 1971, 1973; Lassèrre 1971, 1975). The complicated interaction of abiotic ecological parameters on the species distribution can be elucidated by ecophysiological experiments which achieve a more general aspect if conspecific populations with different environmental background are studied. The interstitial tubificid Ph. monospermathecus is common both in temperate European as well as in the subtropical to tropical beaches of Ber-

---

*Contribution no. 679 from the Bermuda Biological Station

muda, and, therefore, is an apt representative for these problems. Thus, it is hoped to get more information about the divergences in meiobenthic ecosystems from different climates, and also, to estimate the range by which a meiofauna species shifts its physiological capacity to adapt to different ecological situations.

MATERIAL AND METHODS

During two stays at the Bermuda Biological Station I had the occasion to study the species in a shallow tidal cove. This report is the first part of a comparative investigation to be later completed by similar studies on Baltic and North Sea shores. The Bermuda work was connected with a research program on the ecophysiology of meiofauna in a subtropical beach initiated by Wieser (Wieser et al. 1974; Wieser 1975).
The investigated beach consisted of rather uniform calcareous sand both in the slope (about 6 m wide) and in the shallow intertidal flat (Fig. 1). Average grain size was 250 μm. Exposure to atmosphere was much influenced by a sand bar which extended into a slightly elevated southwestern part. Especially on the low-lying northeastern part of the cove ("sand flat"), many puddles of highly concentrated sea water were developed during ebb-tide. More physiographical details of Tucker's Town Cove will be given as part of the results (see also Wieser et al. 1974; Wieser 1975; Farris 1975; Gnaiger 1976).
10 cm_2 deep sediment cores were taken with a perspex corer of 50 cm^2 surface area. From this the oligochaetes were quantitatively extracted by elutriation in subsamples of 50 to 100 cm^3. Temperature, alkalinity and redox-potential were recorded with thin electrodes inserted into the different layers of the sediment. Salinity was measured by a refractometer. In addition to this, representative samples were analysed for grain size composition. The tolerance experiments were conducted with 20 specimens each in small vials kept in water baths at different temperatures. Salinity gradients were established by adding to normal, filtered sea water (35 %o S) rain water or sea salt, increased alkalinity by addition of NaOH. Fully oxygenated (normoxic) sea water was attained by intensive aeration, oxygen-deficient (hypoxic) sea water (7 % saturation) by flushing with nitrogen under continuous control of pH-values (Wieser et al. 1975).
Tolerance is defined as lethal time for 50 % or 100 % of the animals tested (LT 50, 100). Differences in survival times (ΔLT) are in most cases based on the averaged results of replicate experiments. The worms were regarded as dead when no movement was detected after gentle shaking of the vials and additional exposure to light stimuli. Experiments were usually discontinued after 12 h, in some cases after 24 h, in order to rule out undesirable secondary effects due to accumulation of metabolic wastes and lack of food in the vials. Thus, the results account only for short-term reactions.

RESULTS

Field Studies on Distribution of Factors and Fauna

Based on numerous vertical cores along a grid of sampling stations, a characteristical pattern of abiotic factors and oligochaete fauna became apparent (Fig. 1 and 2):

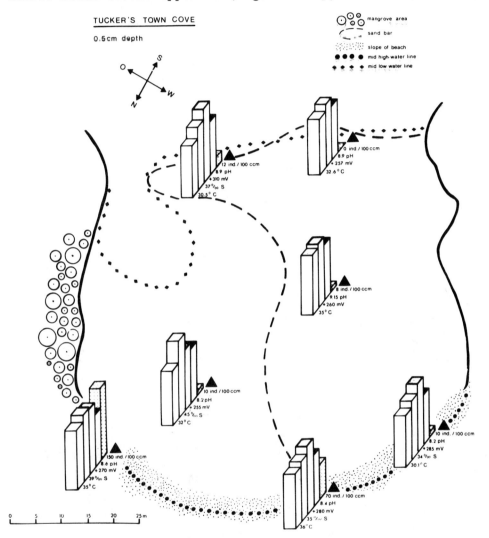

Fig. 1. Distribution of abiotic factors and <u>Phallodrilus monospermathecus</u> in surface sediments

In the superficial sediment layer (0.5 cm depth, Fig. 1) the abiotic system at ebb-tide was fairly similar both in the sand flat, the sand bar and the high-water line: a combination of markedly high pH-values (mostly exceeding 9.0, in some recordings up to 9.5; see Wieser et al. 1974; Gnaiger 1976) with relatively high salinities (35 ‰ to 40 ‰S, sometimes up to 45‰S warm temperatures (30° to 35°C) and good oxygen supply(Eh-values about +250 to +300 mV). The slope of the beach differed from the tidal reaches mainly in having lower pH-values, probably due to less assimilation activity of diatoms in the dryer surface sediments.

However, the distribution of Ph. monospermathecus displayed a distinct difference between slope and lower tidal areas: where-

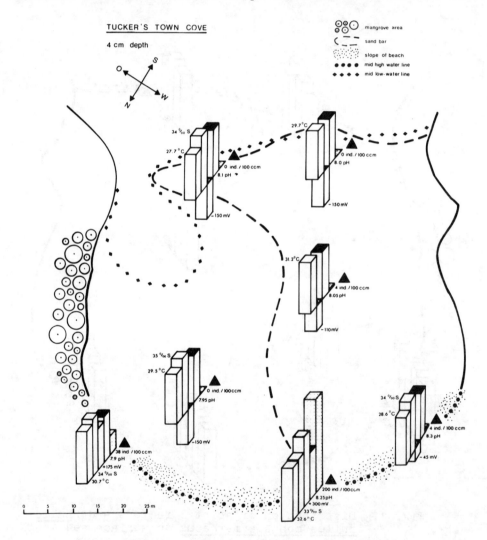

Fig. 2. Distribution of abiotic factors and Phallodrilus monospermathecus in 4 cm depth

as most parts of the high-water line were inhabited by dense populations (often >100 ind./100 cm^3), the mid- and low-tide samples contained very poor numbers of this tubificid (0-10 ind./100 cm^3). Also in the southwestern corner of the high-water area, populations were similarly small.
Examining the situation in a depth of 4 cm (Fig. 2), marked differences between the lower flats and the higher slope of the cove became visible both in the abiotic system and in the oligochaete distribution. In this depth, the almost horizontal mid-tide and low-tide parts (decline about 20 cm per 70 m), being very smoothly overflown by the tidal currents, proved to be totally anoxic (greyish colour, Eh-values mostly appr. -50 mV) and contained high concentrations of H_2S. The same was valid for the exceptional southwestern cite of the high-water line where a high input of decaying plant material combined with a diminishing slope resulted in a negative redox-potential.
In clear contrast were most of the slope stations. Here, the tidal waves percolate at high-water: good to sufficient oxygen conditions (positive Eh-values) throughout the upper 8 to 10 cm despite a rich content of organic debris. The generally lower and fairly stable overall range of the other abiotic factors evidently reflected the subsurface position of the layers examined. Salinity proved to be fairly consistent throughout the sediment column due to a neglectable influx of fresh water. Temperature regularly decreased with depth in a smooth gradient of not more than 3° to $4^{\circ}C$. Reduction of alkalinity corresponded with the diminishing assimilation of diatoms.
The response of Ph. monospermathecus to the abiotic situation in the deeper horizons (Fig. 2) is an almost complete avoidance of the lower tidal areas. This is in striking contrast to the rich aggregations of worms in the aerated subsurface layers along the high-water line (up to 200 ind./100 cm^3). Only in the southwestern sample, where sediment layering was basically identical with that of the flat, tubificids were rare.

Resistance Experiments

The experimental work should answer two main questions: 1) What impact have combined (stress-)factors compared to that of one single stressor? 2) Does a specific factor dominate other parameters in mainly determining the animals' distribution?
To start with temperature (Fig. 3), Ph. monospermathecus showed an extremely wide tolerance range. In normal oxygen-, salinity- and alkalinity-conditions, it seemed to cover a span from unnaturally low $6^{\circ}C$ up to $30^{\circ}C$ without any apparent impairment. Below $30^{\circ}C$, hypoxia had little effect on the viability of the animals (see p. 7 and Table 1). Even beyond $35^{\circ}C$, the decrease in survival rates was not very distinct before approaching the $39^{\circ}C$-mark which evidently was the upper thermal resistance limit even for short-term exposure. $40^{\circ}C$ caused severe damage and almost instantaneous death.
In order to test the combined effect of temperature with other abiotic parameters, salinity- and pH-conditions were modulated in further experiments. The extremely high tolerance of this species even in a multifactorial stress situation was emphasized by the fact that up to $30^{\circ}C$ Ph. monospermathecus kept its full viability during almost the entire experimental time, even in

water of adverse salinity- or pH-range if oxygen was present (Fig. 3). Beyond 35°C, the combined effect of relatively high temperature and low salinity, a situation well possible after heavy showers, resulted in a rapid decrease of life time and terminated at 39°C in LT-values of approximately 1 h only (solid triangles). High temperature combined with high alkalinity (pH 9.5), often to be recorded after long insolation at ebb-tide (Gnaiger 1976), caused a similarly high mortality even at normal salinities and had an even aggravated impact in the warm, highly

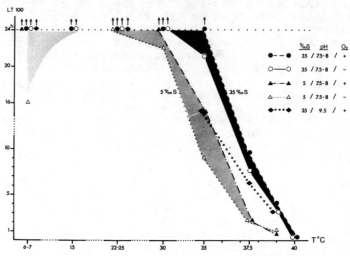

Fig. 3. Thermal resistance of Ph. monospermathecus from Bermuda at different environmental conditions

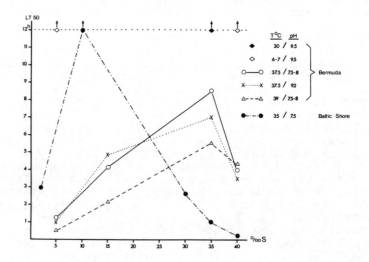

Fig. 4. Salinity tolerance of Ph. monospermathecus in normoxic water of varying temperature and alkalinity

saline water of the surface layers at ebb-tide exposed to the sun (diamonds). Finally, extremely low temperatures plus oxygen deficiency and low salinity reduced life time too (open triangles). But, whatever factor combination was tested, the upper thermal threshold remained rather constantly at 39°C as in the one-stressor experiment. Hypoxic conditions caused only a slightly more pronounced decline in survival times (greyish and black fields in Fig. 3.; see Table 1).
Plotted against salinity gradients, Fig. 4 shows Ph. monospermathecus again to be not severly affected by just one stress factor (e.g. pH 9.5, solid diamond) if temperature stayed at 30°C. Still at 6°C, the survival rate was amazing, since even extreme salinities like 5 %o and 40 %oS did not heavily deteriorate viability (open diamonds). However, beyond the 30°- to 32°C-threshold, even good oxygen supply and normal pH conditions could not rule out a loss in viability. At 37.5°C, this reduction was lowest in natural sea water of 35 %oS, increased in the range of 15 %o to 5 %oS, and heaviest approaching 40 %oS (open circles). Correspondingly, at higher alkalinities (pH 9.2, crosses) or temperatures (39°C, triangles), the deterioration of survival was even more pronounced.

Compiling the impact of oxygen within this factor pattern, a decrease of LT-times due to oxygen deficiency was particularly noticeable in a range slightly beyond "normal" environmental conditions (i.e. 30°-35°C; 35 %oS; 8.0-8.5 pH). Near the upper threshold of viability for each single factor (i.e. 39°C; about 40 %oS; 9.5 pH), life-spans uniformly approached to minutes or some few hours (Fig. 3 and Table 1). These results on the affect of hypoxia, however, are still tentative and need further examination and differentiation.

TABLE 1   Survival of Ph. monospermathecus in Different Oxygen Concentrations

	temp.	normoxic	hypoxic	LT 100
	30°	24 h	24 h	-
	35°	24 h	21 h	3 h
pH 7.5-8/35 %oS	37.5°	9 h 30'	7 h 30'	2 h
	39°	4 h 45'	3 h 15'	1 h 30'
	40°	15'-20'	10'-20'	5'-10'
pH 9.5/35 %oS	39°	2 h	1 h 20'	40'

DISCUSSION

Coincident occurrence of several abiotic stressors seems to be much more pronounced in many subtropical and tropical low-energy-beaches than in boreal shores. Higher temperatures accelerate development of oxygen deficiency and heighten alkalinity. Increased evaporation augments salinity, and leads to water losses in the sediment. Sudden heavy showers dilute the sediment water at ebb-tide drastically. In these situations, survival of Ph. monospermathecus will be restricted to a much narrower range of resistance than evident from one-stressor-experiments. This explains the distinct distributional pattern of the species in

Bermuda which can well be linked together with the ecophysiological short-term reactions revealed in the experiments (Fig. 5).

Though in all areas of the beach the surface layers harboured basically the same positive and negative factors, their influence on the settlement of Ph. monospermathecus is different. In the mid- and low-tide regions of the cove, the worms in the uppermost layers cannot avoid the adverse impact of extreme factors which prevails the positive moment of good oxygen supply.

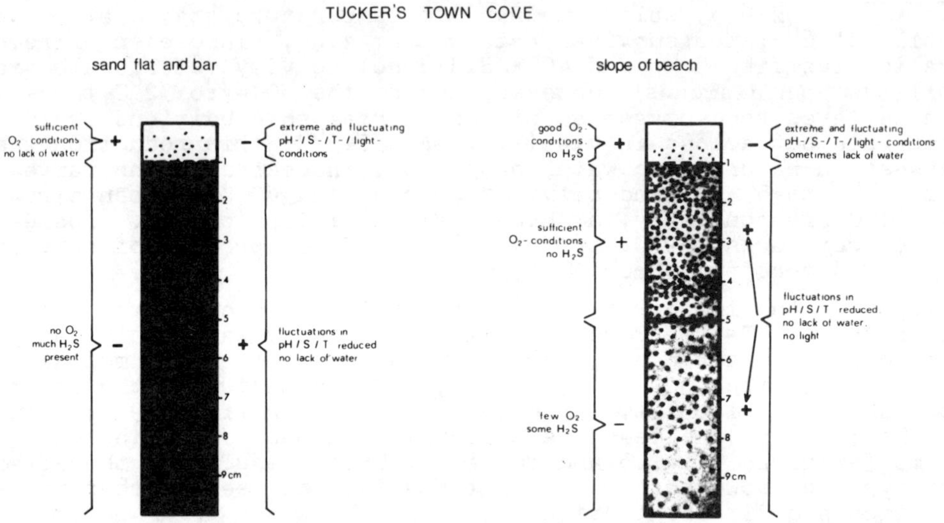

Fig. 5. Schematic interpretation of the factor system determining distribution of Ph. monospermathecus on a subtropical beach of Bermuda

In the deeper horizons, lack of oxygen and high concentrations of poisonous H_2S rule out every other possible favourable influence and strictly limit any vertical extension of the population. Thus, in the intertidal reaches and in the southwestern part of the slope, occurrence is restricted to the uppermost layers and, due to the suboptimal conditions here (multiple-stressor situation), few worms are to be found. In the better aerated slope, the subsurface layers give shelter from the rigorous fluctuations of most abiotic factors. Since, here, the worms are still sufficiently supplied with oxygen, they can readily utilize the richly accumulated organic matter.

A limiting factor, characteristic for tropical and subtropical calcareous sands and to that extent unknown in the mostly silicious beaches of boreal regions, is high alkalinity of the surface sediments which, at summer conditions, regularly exceeds the threshold of good viability for Ph. monospermathecus (Fig. 4). The alkalinity to be one of the most important abiotic factors in the Bermuda beaches was already stated by Wieser (1975) for other animals. Even in temperate areas, v. Oertzen and Schlungbaum (1972) and Theede (1973) found a close correlation

between high alkalinity and reduced resistance to oxygen deficiency.
It can be interpreted as an additional adaptation to overcome sudden changes in the abiotic environment that Ph. monospermathecus has a well developed "immediate response" reaction (Kinne 1970) to physiological shocks. This was found already by Jansson (1962) for Baltic populations and in the Bermuda-experiments it was indicated by similar survival times independent of previous acclimation (24 h).
However, if acclimation acts for longer periods than seem to apply here, it is still possible that Ph. monospermathecus exhibits also a "meaningful" reaction which would correspond with the results of Kähler (1970). The basis for this potential would be a well developped ability of osmoregulation, found also in other marine interstitial oligochaetes (Lassèrre 1969, 1971, 1975; Tynen 1969) and indicated by the effective regulation of body volume observed in my experiments.
The resistance data confirm the euryoecous nature of Ph. monospermathecus to be expected from a cosmopolitan species (Tynen 1969). But they also reveal that the Bermuda-animals had a maximal resistance to temperature and alkalinity variations at 35 %oS which is about the average salinity in their natural environment, whereas at low salinities survival in warm weather was reduced (Fig. 4). In contrast to this, results from the Baltic shore showed those populations to have optimal resistance just in the mesohaline range (about 10 %oS), their original habitat salinity and a substantially lower survival rate at 35 %oS (Fig. 4). Since maximal resistance and optimal range usually are closely correlated (Kinne 1964, 1970, 1971), the two populations probably have a divers preferential behaviour for salinity. The same seems valid for temperature, since the resistance of Baltic populations at 35°C was very reduced, even in their optimal salinity, whereas the Bermuda-animals at this temperature and in their usual salinity range showed no impairment at all. On the other hand, it was not before 6°C that temperature acted as an additional stressor. Survival at these very low temperatures (which the worms will never encounter in their subtropical environment) means that at its lower range thermal resistance is almost the same as in the boreal populations (see also Jansson 1962, 1967, 1968).
The indications of different preference ranges between populations of diverse environmental background do not necessarily contradict Jansson's data (1962, 1968) which signified a stable preference of Ph. monospermathecus for low salinities despite a one week's treatment in high salinity. It is possible (Kinne 1964, 1970), that animals possibly attain only the initial phase of stabilization during this short period of acclimation which might not be sufficient for a shift of their preference behaviour. However, the population on Bermuda certainly could establish by long-term isolation a steady state (Kinne op. cit.) and has acclimated to the subtropical conditions. This would not support the general ranking of Ph. monospermathecus as a brackish-water species, as suggested by Pfannkuche (1974).

CONCLUSIONS

The results for Ph. monospermathecus corroborate the generalized statement of Wieser (1975) for other meiofauna members from subtropical regions: the upper temperature tolerance is closely adjusted to the highest temperature to be expected in the field. As with temperature, tolerance in general was very wide if just one parameter acted as stressor. But, apparently, there is no single master-factor predominantly responsible for the viability of the worms. It is the interdependent and interacting effect of several main parameters aggravated by subtropical conditions which confines the animals' distribution. Thus, interpreting the abiotic system in terms of tolerance limits, certain principles of field distribution in subtropical climate could be derived.

It remains an interesting task to further correlate these results with the distribution and ecophysiology of corresponding populations from boreal climates. Judging from the data on thermal resistance, subtropical and boreal populations seem to have kept the same extremely wide range of genetically fixed tolerance. However, exposure to different environments for long periods of time, apparently caused new and diverse steady state reactions with probably also different stable preferences. This leads in each population to a meaningful resistance adaptation adjusting them to their respective ecological situation.

Though the nature of this adaptive process is yet unknown, it is assumed to be non-genetic (Kinne op. cit.), since the fairly uniform euryoecous background is an optimal genetical basis for the world-wide distribution of this meiobenthic oligochaete. Maintenance of a wide physiological and ecological potential and ability of non-genetic adaptation to the array of local life conditions seems to be a significant predisposition for any wide-spread distribution, so typical for this species and many other members of the interstitial fauna.

ACKNOWLEDGEMENTS

It is gratefully acknowledged that the work at the Bermuda Biological Station For Research was supported by the Deutsche Forschungsgemeinschaft (1975) and by a fellowship from Exxon (1976). I thank Mrs. L. Seitz-Hildebrand for her careful drawing of the figures.

REFERENCES

Farris, R.A. 1975. Systematics and ecology of Gnathostomulida from North Carolina and Bermuda, Diss. Univ. North Carolina, Chapel Hill.

Giere, O. 1970. Untersuchungen zur Mikrozonierung und Ökologie mariner Oligochaeten im Sylter Watt. Veröff. Inst. Meeresforsch. Bremerh. 12, 491-529.

Giere, O. 1971. Beziehungen zwischen abiotischem Faktorensystem, Zonierung und Abundanz mariner Oligochaeten in einem Küstengebiet der Nordsee. Thalassia jugosl. 7, 67-77.

Giere, O. 1973. Oxygen in the marine hygropsammal and the vertical microdistribution of oligochaetes. Mar. Biol. 21, 180-189.

Gnaiger, E. 1976. pH and temperature fluctuations in an intertidal beach in Bermuda: An experimental approach. (in press).

Jansson, B.O. 1967. The importance of tolerance and preference experiments for the interpretation of mesopsammon field distributions. Helgol. wiss. Meeresunters. 15, 41-58.

Jansson, B.O. 1968a. Quantitative and experimental studies of the interstitial fauna in four Swedish sandy beaches. Ophelia 5, 1-71.

Kähler, H.H. 1970. Uber den Einfluss der Adaptationstemperatur des Salzgehaltes auf die Hitze- und Gefrierresistenz von Enchytraeus albidus (Oligochaeta). Mar. Biol. 5, 315-324.

Kinne, O. 1964. Non-genetic adaptation to temperature and salinity. Helgol. wiss. Meeresunters. 9, 433-458.

Kinne, O. 1970. Temperature: Animals - Invertebrates. In: O.Kinne (ed.), Mar. Ecol. 1, Environmental Factors, Pt. 1. Wiley, London, 407-514.

Kinne, O. 1971. Salinity: Animals - Invertebrates. In: O. Kinne (ed.), Mar. Ecol. 1, Environmental Factors, Pt. 2. Wiley, London, 821-995.

Lassèrre, P. 1969. Régulations énergétiques entre le métabolisme respiratoire et la régulation ionique chez une Annélide Oligochète euryhaline, Marionina achaeta Hagen. C.R. Acad. Sci. Paris, 268, 1541-1544.

Lassèrre, P. 1971. Données écophysiologiques sur la répartition des Oligochètes marins meiobenthiques. Incidence des paramètres salinité, température, sur le metabolisme respiratoire de deux espèces euryhalines du genre Marionina Michaelsen 1889 (Enchytraeidae, Oligochaete). Vie Milieu Suppl. 22, 523-540.

Lassèrre, P. 1975. Métabolisme et osmorégulation chez une Annélide Oligochète de la méiofauna: Marionina achaeta Lassèrre. Cah. Biol. mar. 16, 765-798.

Oertzen, J. -A. v. and Schlungbaum, G. 1972. Experimentell-ökologische Untersuchungen über O_2 -Mangel und H_2S-Resistenz an marinen Evertebraten der westlichen Ostsee. Beitr. Meereskunde 29, 79-92.

Pfannkuche, O. 1974. Zur Systematik und Ökologie naidomorpher Brackwasseroligochaeten. Mitt. Hamburg. Zool. Mus. Inst. 71, 115-134.

Theede, H. 1973. Comparative studies on the influence of oxygen deficiency and hydrogen sulphide on marine bottom invertebrates. Neth. J. Sea Res. 7, 244-252.

Tynen, M.J. 1969. Littoral distribution of Lumbricillus reynoldsoni Backlund and other Enchytraeidae (Oligochaeta) in relation to salinity and other factors. Oikos 20, 41-53.

Wieser, W. 1975. The meiofauna as a tool in the study of habitat heterogeneity: Ecophysiological aspects. A review. Cah. Biol. mar. 16, 647-670.

Wieser, W. et al. 1974. An ecophysiological study of some meiofauna species inhabiting a sandy beach at Bermuda. Mar. Biol. 26, 235-248.

PHOSPHOGLUCOISOMERASE ALLELE FREQUENCY DATA IN *Mytilus edulis* (L) FROM IRISH COASTAL SITES: ITS ECOLOGICAL SIGNIFICANCE

Elizabeth Gosling and N.P. Wilkins

Zoology Department, University College, Galway, Ireland

ABSTRACT

Mussels produce pelagic larvae which may be dispersed over great distances from their point of origin. After metamorphosis they attach to a firm substrate and from then on lead an essentially sessile existence. The extended dispersal stage in the larvae suggests an important role for interpopulation migration. On the other hand, during sedentary adulthood, if particular environmental factors favour particular genotypes, the role of differential selection becomes increasingly important.

This paper presents frequency data on the allozymes of phosphoglucoisomerase (PGI) in *Mytilus edulis* from exposed and sheltered shores on the Irish coasts. The results indicate a significant correlation between PGI allele frequency and degree of exposure of the shore. A significant deficiency of heterozygotes was found in eight of the nine exposed sites sampled and the implication of these results is discussed with reference to the respective roles of natural selection and interpopulation migration in mussels on exposed shores. The results suggest that interpopulation migration plays a major role in the population dynamics of mussels on exposed sites.

Seed (1974) found evidence for the occurrence of the Mediterranean mussel - *Mytilus galloprovincialis* (Lmk.) - intermixed with *M. edulis* in varying proportions, along the Atlantic coasts of Ireland. The significant deficiency of heterozygotes found in exposed shore populations of mussels is discussed with reference to Seed's evidence.

INTRODUCTION

During the past 25 years electrophoretic analysis of animal proteins has revealed a large amount of genetic variation. Earlier electrophoretic studies were chiefly concerned with estimating the extent of enzymic polymorphisms (Lewontin and Hubby 1966; Selander and Kaufman 1973). However, to-day, population geneticists are questioning the biological implications of enzymic polymorphisms. While some workers believe that most of these polymorphisms are selectively neutral (Kimura and Ohta 1971a, 1971b and 1971c) there is a growing body of evidence to support the selectionist point of view i.e. that enzymic polymorphisms are of adaptive value and are maintained by some form of balancing selection (Lewontin 1974; Kojima and Yarbrough 1967; Koehn 1969; Sved and Ayala 1970; Powell 1971; Koehn and Mitton 1972; Koehn et al 1973; Koehn et al 1976; Levinton 1973; Levinton and Fundiller 1975; Boyer 1974; Murdock 1975).

How can the selective forces maintaining genetic variation within natural populations be determined? In the case of marine organisms the problem might be approached in the following ways:-

1. One can measure the change in enzymic allele frequency over the geographic range of the species and also note any deviations from Hardy-Weinberg distributions.

2. One can attempt to correlate the variation in allele frequency, if found, to some measureable component of the physical environment, such as wave action, temperature and salinity fluctuations, or to mode of breeding and dispersal of the species.

3. One can determine whether there is a regular change in allele frequency with increasing age of individuals of a population.

4. Biochemical investigations can be carried out on the functional differences that may exist between allozymes - different enzyme forms produced by different alleles at the same locus (Prakash et al 1969).

Mytilus edulis (Linne) is particularly well suited for genetic study in that it occupies a wide variety of habitats and extends from the intertidal zone to several metres below low water. In addition, the two distinct stages in the life cycle of *M. edulis* - an extended larval stage and a sedentary adult phase - afford the opportunity of studying the respective roles of migration and natural selection in mussel populations.

This study presents allele frequency data for phosphoglucoisomerase (PGI; E.C. 5.3.1.9.) in *M. edulis* from exposed and sheltered shores on Irish coasts and relates these data to population structures.

MATERIALS AND METHODS

Samples of mussels were collected from the intertidal region of nine exposed sites and twelve sheltered sites. Two samples which are asterisked, were taken from the sublittoral region (Fig.1). The posterior adductor muscle was excised from live animals and homogenized in approximately two volumes of 50mM Tris-HCl, pH 7.8. Samples were centrifuged at 4000 rev/min for 15 min at 4°C and the supernatant solution was used for electrophoresis, which was performed at 2°C in horizontal starch gel using a Tris-Maleic buffer, pH 7.4. Gels were run for approximately 16 hours at 6.5V/cm. Sliced gels were stained for PGI using the paper overlay method described by Scopes (1968).

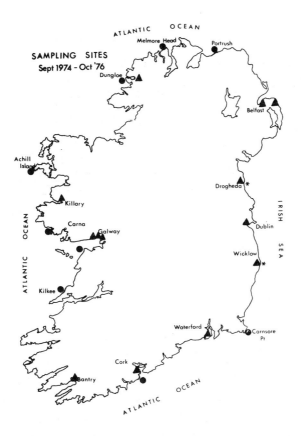

Figure 1. Location of exposed (●) and sheltered (▲) intertidal sampling sites on the Irish coast; * indicates sublittoral sites sampled.

STATISTICS

Magnitudes of heterozygote deficiency were calculated using the formula (Koehn et al 1971)

$$D = \frac{\Sigma H_o - \Sigma H_e}{\Sigma H_e},$$

where ΣH_o is the total number of heterzygotes observed and ΣH_e is the number expected using the Hardy-Weinberg binomial expansion. Negative values of D indicate a deficiency of heterozygotes while positive values indicate an excess. The data in Table 1 and PGI genotype frequency of 376 M. edulis in the length range 0.3 - 7.7cm were analysed by single factor analysis of variance (anovar) - Campbell (1967).

RESULTS

Seven alleles and a total of seventeen phenotypes have been observed at the PGI locus. The common phenotypes are illustrated in Fig. 2. These results are consistent with a dimeric structure for PGI; individuals homozygous at this locus exhibit a single-banded pattern while heterozygotes are three-banded. Table I shows the frequency of the most common allele (pA) in populations from exposed and sheltered sites on the Atlantic coasts and from sheltered sites on the Irish Sea coast.

TABLE I

Frequency of PGI allele A from exposed and sheltered sites on the Atlantic coasts, and sheltered sites on the Irish Sea coast.

ATLANTIC COASTS			IRISH SEA COAST
Exposed	Sheltered		Sheltered
0.35	0.41		0.61
0.41	0.59		0.60
0.40	0.53		0.66
0.31	0.55		0.61
0.49	0.50		0.60
0.31			0.67
0.43			
0.36			
0.44			
0.39	Mean 0.52		0.63
±0.02	S.E.M. 0.03		0.01

There are no exposed sites containing mussels along the Irish sea coast. The frequency of PGI A was lowest on exposed shores and highest on the sheltered Irish Sea sites, with intermediate frequencies on sheltered Atlantic sites.

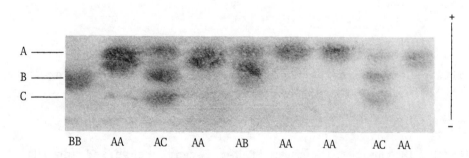

Fig. 2. PGI Phenotypes observed in *Mytilus edulis*

A significant deficiency of PGI heterozygotes ($P<0.001$) was found on eight of the nine exposed istes; the deficiency of heterozygotes found on some sheltered Atlantic sites was not statistically significant. The frequency of PGI A on exposed shore sites was found to differ significantly (P<0.001) from pA on both sets of sheltered sites. When both sets of sheltered sites were compared they also differed significantly (P<0.01) from one another.

There is a great disparity in length between a sheltered and an exposed shore mussel. Mussels in the intertidal region of sheltered shores may reach a length of 6-7cm after several years growth. On the other hand, mussels on exposed shores rarely exceed 2.0cm and may take more than 20 years to reach this length (Seed 1969b). There is the possibility that if the small mussels on exposed shores were raised under better feeding conditions and allowed to grow larger, the heterozygote deficiency might disappear with increase in length of mussel. In other words, there may be increased survival of heterozygotes in larger mussels. This hypothesis was tested by analysing a sample of 376 M. *edulis* from the intertidal region of Killary Harbour - a sheltered fjord on the west coast of Ireland (Table 2).

TABLE 2

Comparison of PGI allele frequency and D values in four 2cm length groups in *Mytilus edulis*, taken from the intertidal region of Killary Harbour.

Length (cm)	N	Allele frequency			
		A	B	C	D
0-1.9	22	0.614	0.318	0.068	-0.035
2.0-3.9	84	0.524	0.441	0.036	-0.150
4.0-5.9	169	0.515	0.435	0.050	-0.150
6.0-7.7	101	0.589	0.346	0.060	-0.143

No correlation was found between PGI genotype frequency and length over the range 0.3 - 7.7cm. With the exception of the 0-2.0cm size group, which consisted of only 22 individuals, values of D were found to be similar in the different size classes over the range 2.0 - 7.7cm. No deviations in genotype frequencies from Hardy-Weinberg expectations were found in any of the size groups analysed.

Mussels of two size groups 0.5 - 0.8cm and 1.8 - 2.2cm were taken from Blackhead- an exposed shore near Galway - and were analysed at the PGI locus. A significant deficiency of heterozygotes ($P<0.001$) was found in both groups and the deficiency was approximately the same for both groups. Since a deficiency of heterozygotes is typical of exposed shore mussels, individuals taken from different levels on the exposed shore might be expected to exhibit different degrees of heterozygote deficiency. To test this, mussels from the upper and lower intertidal zone of Blackhead were analysed. The sampled areas differed vertically by 3 meters. The deficiency of heterozygotes was found to be approximately the same at both levels on the shore.

A study of French populations of *M.edulis* and the closely related *M. galloprovincialis* (Lamarck) was undertaken during the summer of 1976. A pure population of *M.edulis* from Charente on the west coast of France and a pure population of *M. galloprovincialis* from Cannes in the South of France were analysed at the PGI locus (Table 3). The difference of PGI allele frequency between the two species was significant ($P<0.001$). *M.galloprovincialis* showed a lower pA than *M.edulis*. A sample of mussels was analysed from Granville on the Cherbourg peninsula where the two species were found intermixed on the same shore but could be easily differentiated morphologically. A significant difference in PGI allele frequency was found between the two species and the pA was once more lower in *M. galloprovincialis* than in *M.edulis*. The pA in *M.edulis* from this sample and the sample from Charente is very similar to the p̄A (0.63) found in populations of *M.edulis* along the Irish Sea coast.

DISCUSSION

How can the significant deficiency of heterozygotes found in samples of mussels from exposed shores be explained? A heterozygote deficiency can be generated through:

    a) inbreeding within the parental population,
    b) selection against heterozygotes
and c) mixing of populations (Wahlund effect).

a) <u>Inbreeding</u>: In a species, such as *M.edulis*, with an extended larval dispersal stage it seems highly unlikely that inbreeding plays a role in producing the significant heterozygote deficiency found on exposed shores.

b) <u>Selection against heterozygotes</u>: If selection is acting against heterozygotes at the post-larval stage in exposed shore populations of mussels, a greater deficiency of heterozygotes might be expected among larger (older) individuals than in smaller (younger) individuals. The data from Killary Harbour (Table 2) and from the exposed shore at Blackhead would suggest that deficiency of heterozygotes is neither correlated with absolute or relative size of individual

TABLE 3

Distribution of genotypes and allele frequency data for the more common PGI alleles in *Mytilus edulis* and *Mytilus galloprovincialis* (France).

O = Observed number of genotypes
E = Number expected from the Hardy-Weinberg formula
χ^2 = Goodness-of-fit of observed to expected frequencies
NS = Not significant at 5% level

Site/Species	N		AA	AB	Genotypes BB	AC	Others	A	Allele frequency B	C	χ^2		D
CHARENTE *M. edulis*	103	O	51	31	12	5	4	0.67	0.277	0.034	6.6	N.S.	−0.138
		E	42.6	38.2	7.9	4.7	1.4						
GRANVILLE *M. edulis*	61	O	25	24	9	1	2	0.615	0.361	0.008	2.84	N.S.	−0.049
		E	23.1	27.1	8.0	0.6	0.7						
GRANVILLE *M. galloprovincialis*	27	O	2	8	12	1	4	0.25	0.692	0.077	0.31	N.S.	−0.065
		E	1.7	9.3	12.9	1.0	3.6						
CANNES *M. galloprovincialis*	94	O	1	17	71	−	5	0.101	0.867	0.027	0.25	N.S.	−0.070
		E	1.0	16.5	70.9	0.5	5.8						

mussels nor, as the Blackhead data suggests, is it correlated with position of the mussels on the shore i.e. duration of feeding time or exposure to air.

c) <u>Mixing of populations</u> (Wahlund effect): Exposed shore populations because of their situation are more likely to be interbreeding with more distant populations while populations of mussels from sheltered harbours and inlets are less likely to do so. On exposed shores, a population of settling larvae might consist of individuals from various parental populations each with differing allele frequencies at the PGI locus. When sampled, such a mixed population will show a net deficiency of heterozygotes relative to Hardy-Weinberg expectations and this is known as the Wahlund effect. However, if there is significant gene flow occuring between exposed shore populations we would expect the deficiency of heterozygotes to be eroded over evolutionary time. A deficiency of heterozygotes will persist, however, if our exposed shore population consists of two distinct populations of mussels with limited gene flow between them i.e. if on exposed shores a mixture of two coexisting but non-interbreeding species or subspecies of *Mytilus* is being sampled.

What evidence is there to support this hypothesis? Seed (1974) carried out an extensive survey of mussel populations around the Irish coast. He found the Mediterranean mussel, *M. galloprovincialis* intermixed with *M. edulis* in varying proportions along the south, west and part of the north coast of Ireland i.e. on the Atlantic coasts (Fig. 3). He found no evidence for the occurrence of *M. galloprovincialis* along the east coast of Ireland i.e. on the Irish Sea coast. Seed separated *M. edulis* and *M. galloprovincialis* on the basis of five morphological characters but he found a considerable degree of overlap in morphological characters along the Atlantic coasts. There were large numbers of intermediate forms. It was easier to separate the two forms on sheltered shores but

> "the problem was especially acute on exposed shores which often supported dense populations of small, yet frequently quite old, badly eroded mussels and it is perhaps worth noting that it is in this type of habitat that the majority of intermediate mussels were encountered during this survey." (Seed, 1974).

The systematic position of *M. galloprovincialis* has been the subject of considerable discussion in previous literature. Some authorities (review, Lubet 1973) give it specific status while others regard it as a variety of a larger *M. edulis* superspecies. Whether one regards *M. galloprovincialis* as a species or a variety possibly depends, to some extent, on the geographical region from which samples are collected. In the Mediterranean *M. galloprovincialis* is easily separated from *M. edulis* on morphological and physiological characteristics. However,

Figure 3. Map showing in heavy outline the geographical areas in which *Mytilus galloprovincialis* will probably occur provided local conditions are favourable (taken from Seed, 1974).

towards the limits of the range of M. *galloprovincialis* - along the Atlantic and English Channel coasts - it can be difficult to distinguish the two forms due, possibly, to some degree of hybridization between them (Seed 1976).

In France pure populations of M.*edulis* and M.*galloprovincialis* were analysed (Table 3). The mussels were taken from areas where the two forms show marked morphological differences and hence can be easily distinguished. Table 3 illustrates the difference in PGI allele frequency found in M. *edulis* and M. *galloprovincial* -*is* when pure populations of each species were analysed from separate areas (Cannes, Charente) and when a mixed population of two species were analysed (Granville). Allele frequency of PGI is sufficiently different in the two species to give a deficiency of heterozygotes plus a lowering of pA when equal proportions of M.*edulis* and M. *galloprovincialis* are mixed.

Fig. 4, shows the distribution of M.*galloprovincialis* (Seed 1974) superimposed on D values obtained from exposed and sheltered

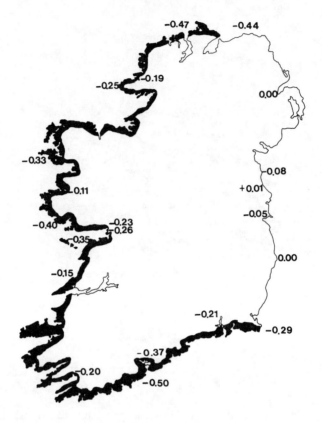

Figure 4. The distribution of *Mytilus galloprovincialis* (Seed, 1974) together with values of D from exposed and sheltered shores. Negative values indicate a deficiency of heterozygotes, positive values indicate an excess.

shores. If a deficiency of heterozygotes (negative values of D) is a measure of the proportion of *M. galloprovincialis* intermixed with *M. edulis* in our samples, the following points can be made:-

1. Along the Irish Sea coast D values in all samples are close to zero, indicating a pure population of *M. edulis*. Seed (1974) found no evidence for *M. galloprovincialis* along the Irish Sea coast.

2. The remainder of Irish coastal sites samples showed negative values of D; exposed shore sites showed more negative D values than sheltered inshore sites.

Seed (1974) found the proportions of *M. galloprovincialis* to be approximately 15% on exposed sites and 20% on sheltered sites. Our results would suggest, however, that *M. galloprovincialis* is more abundant on exposed shores than on sheltered shores. The discrepancy between Seed's results and those obtained by us may be accounted for by recalling the difficulty Seed had in separating the two forms on exposed shores.

CONCLUSIONS

Using a deficiency of heterozygotes (negative values of D) as a measure of the proportion of *M. galloprovincialis* intermixed with *M. edulis* in our sample we propose that:

a) the Irish sea coastal populations appear to be pure *M. edulis*
b) on the Atlantic coastal sites we have a mixture of *M. edulis* and *M. galloprovincialis*, the latter being more abundant on exposed than on sheltered shores.

ACKNOWLEDGEMENTS

This work was supported by a grant from the National Science Council of Ireland to Mr. N.P. Wilkins. We gratefully acknowledge Bernadette Sherlock for her technical assistance and all those who helped in collection of material. Sincere thanks are also due to Professeur Pierre Lubet, Department de Zoologie, Université de Caen, Normandie, France, for providing me with research facilities during my visit there.

REFERENCES

Boyer, J.F. 1974. Clinal and size-dependent variation at the LAP locus in *Mytilus edulis*. Biol. Bull. 147, 535-549

Campbell, R.C. 1967. Statistics for Biologists. Cambridge University Press.

Kimura, M. and T. Ohta. 1971a. Theoretical aspects of population Genetics. Princeton, Princeton,N.J.

Kimura, M. and T. Ohta. 1971b. Protein polymorphism as a phase of molecular evolution. Nature 229, 467-469.

Kimura, M. and T. Ohta. 1971c. On the rate of molecular evolution. J. Mol. Evol. 1, 1-17.

Koehn, R.K. 1969. Esterase heterogeneity: dynamics of a polymorphism. Science 163, 943-944.

Koehn, R.K., Milkman, R. and J.B. Mitton. 1976. Population Genetics of Marine Pelecypods. IV. Selection, migration and genetic differentiation in the blue mussel *Mytilus edulis* Evolution 30, 2-32.

Koehn, R.K. and J.B. Mitton. 1972. Population Genetics of Marine Pelecypods. 1. Ecological heterogeneity and evolutionary strategy at an enzyme locus. Amer. Natur. 106, No. 947, 47-56.

Koehn, R.K., Perez, J.E. and R.B. Merritt. 1971. Esterase enzyme function and genetical structure of populations of the freshwater fish, Notropsis stramineus. Amer. Natur. 105, 51-69.

Koehn, R.K., Turano, F.J. and J.B. Mitton. 1973. II. Genetic differences in microhabitats of M. demissus. Evolution 27,100-5.

Kojima, K. and K.N. Yarbrough. 1967. Frequency-dependent selection at the Esterase - 6 locus in Drosophila melanogaster. Proc. Nat. Acad. Sci. 57, 645-649.

Levinton, J.S. 1973. Genetic variation in a gradient of environ-mental variability: Marine Bivalvia (Mollusca). Science 130, 75-76.

Levinton, J.S. and D. Fundiller, 1975. An ecological and physiological approach to the study of biochemical polymorphisms. Proceedings of the ninth European marine Biology Symposium (ed. H. Barnes). 165-176, Aberdeen University Press, Aberdeen.

Lewontin, R.C. 1974. The Genetic Basis of Evolutionary Change. Columbia University Press, New York.

Lewontin, R.C. and J.L. Hubby. 1966. A molecular approach to the study of genic heterozygosity in natural populations. II. Amount of variation and degree of heterozygosity in natural populations of Drosophila pseudoobscura. Genetics 54, 595-609.

Lubet, P. 1973. Exposé synoptique des données biologiques sur la moule *Mytilus galloprovincialis* (Lamarck 1819).Synopsis FAO sur les pêches No. 88 (Sast-Moule, 3, 16 (10) 028, 08, pag. var.) FAO. Rome.

Murdock, E. A., Ferguson, A. and R. Seed. 1975. Geographical variation in LAP in *Mytilus edulis* from the Irish coasts. J. exp. mar. Biol. Ecol. 19, 33-41.

Powell, J.R. 1971. Genetic polymorphisms in varied environments. Science 174, 1035-1036.

Prakash, S., Lewontin, R.C. and J.L. Hubby. 1969. A molecular approach to the study of genic heterozygosity in natural populations. IV. Patterns of genic variation in central, marginal and isolated populations of Drosophila pseudoobscura. Genetics. 72, 169-175.

Scopes, R.K. 1968. Methods for starch-gel electrophoresis of sarcoplasmic proteins. Biochem. J. 107, 139-150.

Seed, R. 1969b. The ecology of *Mytilus edulis* L.(Lamellibranchiata) on exposed rocky shores. II. Growth and mortality. Oecologia, 3, 317-350.

Seed, R. 1974. Morphological variation in *Mytilus* from the Irish coasts in relation to the occurrence and distribution of M. *galloprovincialis* Lmk. Cahier de Biologie Marine, Roscoff. 15, 1-25.

Seed, R. 1976. Ecology, In: Marine mussels: their ecology and physiology, IBP Handbook No. 10 (Ed. B.L. Bayne). 13-65, Cambridge University Press.

Selander, R.K., D.W. Kaufmann. 1973. Genic variability and strategies of adaption in animals. Proc. natn. Acad. Sci. U.S.A. 70, 1875-1877.

Sved, J.A. and F.J. Ayala, 1970. A population cage test for heterosis in Drosphila pseudoobscura. Genetics 66, 97-113.

AN *IN SITU* STUDY OF THE PRIMARY PRODUCTION AND THE METABOLISM OF A BALTIC *Fucus vesiculosus* L. COMMUNITY

Björn Guterstam

Askö Laboratory, S-150 13 Trosa, Sweden

ABSTRACT

The metabolism of a Fucus vesiculosus community at 1 m depth was studied using 150 l capacity plastic bags. Oxygen content, pH, nutrients in the water (inside and outside the bags) and surface insolation were measured over a 30 hr. period.

The net primary production of Fucus vesiculosus was 30 and 15 mg O_2/g algae x day in August and September respectively. The influence of O_2 concentration on the respiration was shown by using light and dark plastic bags. A 24 hr period budget for the enclosed Fucus community in September gave a photosynthetic efficiency of 1.8% and a consumer respiration of 1.2% of photosynthetically active radiation at the surface. These results stress the importance of the macroalgal contribution to the primary production of the coastal areas in the Baltic proper.

INTRODUCTION

The Baltic is a shallow sea with large archipelagos along the coasts which have an extensive growth of vegetation. The bulk of the benthic algal biomass is comprised of Fucus vesiculosus which is found from about 0.5 m down to 5 m depth as a dense belt all around the Baltic proper. Investigations of the Fucus biomass over larger areas have been made by Hoffman (1952) in the southwestern Baltic and by Jansson and Kautsky (in press) in the northwestern Baltic. The high diversity and biomass of the Fucus fauna has been pointed out by Jansson (1972) and Haage (1975). The fact that Fucus occurs with a high biomass throughout the year makes it a very important primary producer in the Baltic proper. The aim of this study was to get a quantitative value of the metabolic rates of the Fucus community.

Earlier investigations of the metabolism in phytobenthic communities in the Baltic have been made by measuring oxygen in the free water (A-M. Jansson 1974), in plexiglass cylinders and plastic tubs (Elmgren and Ganning 1974) and in plastic bags (Jansson and Wulff in press, Schramm and Guterstam, in press).

MATERIAL AND METHODS

The plastic bags (Fig. 1) constituted a closed system with a transparent plexiglass lid at the top and walls of a double layered plastic material (polyamid-polyethylen) which was essentially impermeable to gases and allowed 95% light transmission. The volume

(around 150 l) could be chosen by cutting the plastic in different lenghts. At top and bottom of the bags the plastic was secured to PVC-rings with rubber bands. The enclosed bottom-area was 0.18 m^2. Three types of plastic bags were used:
1. with an artificial bottom plate (PVC) for studies of algal plants alone. 2. with an open bottom ring which could be pushed into the natural substrate for studies of the entire enclosed communitiy. 3. with dark plastic bags outside the ordinary bags for respiration studies. Two replicates were run for each type of bag (except on one occasion see Table 1.).

The metabolism was determined by measuring the changes in oxygen content of the water every 3rd hour during 30 hrs (Odum and Hoskin 1958). As a control, the O_2 content of the free water outside the plastic bags was measured. For this purpose an O_2-electrode (YSI model 54) which also measured the water temperature was used. The net primary production was calculated as the difference in O_2 concentration between the max. and min. points of the diurnal oxygen-curve (McConnel 1962). The respiration during the dark period (when R > P) was calculated in the same way. In order to get the 24 hr respiration, the dark-respiration obtained from the light bags was extrapolated over this period (Odum and Hoskin op. sit.) for comparison with the respiration of the dark bags.

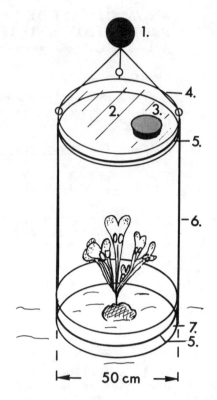

Fig. 1. Plastic bag 1: buoy 2: plexiglass lid 3: stopper 4: upper ring 5: rubber band 6: plastic 7: lower ring (open).

The insolation (300 - 2 500 nm) was recorded as the sum for every hour by an integrator coupled to a solarimeter (Kipp & Zonen) placed on the roof of the lab. about 50 m from the experimental site. The photosynthetically active radiation (400 - 700 nm) was calculated by multiplying these data by 0.45 (v. Bröckel 1975). Chemical analyses of the nutrients in the water inside and outside the plastic bags were made directly after sampling using standard colorimetric analytical methods described by Carlberg (1972) modified for automatic analysis on samples taken with a 50 ml plastic syringe connected to a plastic tube. pH was measured with a laboratory pH electrode (Radiometer).

The experiments were made in August and September 1973 at a moderately exposed Fucus community at 1 m depth, close to the Askö

TABLE 1 Respiration of Fucus vesiculosus in light- and dark plastic bags during the 2nd night of experiment.

Day	n	Dark R	O_2%	n	Light R	O_2%	L:D	ΔO_2%	T°C
17/8	2	0.71	89	2	1.08	177	1.5	88	19
21/8	2	0.17	68	1	0.45	179	2.6	111	14
12/9	2	0.15	84	2	0.37	125	2.8	41	8
					mean:		2.3	80	

Notes: n = number of plastic bags, R = respiration (mg O_2/g algae x hour).

Laboratory in the NW Baltic (58°49´N, 17°38´E). Plants of Fucus, growing on pebbles (to avoid any damage which might occur when removing them from the substrate) and weighing about 50 g dry weight were selected. These were cleaned of most epifauna and put into the plastic bags with artificial bottoms. The community bags were placed over single uncleaned Fucus plants which had been moved to a part of the Fucus belt where the bottom ring could be pushed down into the substrate. It was found essential not to enclose too high an algal biomass to avoid too large a change in oxygen content during the experiment. The bags were set out in the late afternoon and the first O_2 measurement was made at 9 p m. By doing so, the O_2 concentration within the bags was kept low at the start so that the increase during the following day could take place with a minimized risk of supersaturation and formation of O_2 bubbles.

After the experiments the algae were dried to constant weight at 60°C. No analyses of the associated fauna was made. The conversion factor for produced and respired oxygen was set to be 3.34 kcal/g O_2 (Ivlev, in Crisp 1971). The energy content of Fucus was based on analyses of plants from Askö (Guterstam, unpubl.).

RESULTS

The daily light period net primary production of F. vesiculosus without visible epifauna varied between 23 and 37 (30) mg O_2/g algae in August and 14 - 15 (15) mg O_2/g algae in September (Fig. 2). The total insolation during these two periods was 497 and 67 cal/cm^2 x day respectively. The respiration (Fig. 2) of Fucus during the dark period (when R > P) varied between 7 - 12 (9) mg O_2/g algae in August and was 5 mg O_2/g algae in September. The 24 hr respiration of the light bags was more than double that of the dark bags (31 to 12 in August and 8 to 4 in September Fig. 2).

The influence of oxygen concentration on the respiration rate was studied by experiments where light and dark plastic bags with enclosed Fucus plants as well as parts of the Fucus community were run parallel. The results from bags with single Fucus plants (Table 1) showed that the respiration during the 2nd night of the experiment was 2.3 times higher in the light bags (mean O_2 concentration 160 % of saturation) than in the dark bags (mean O_2

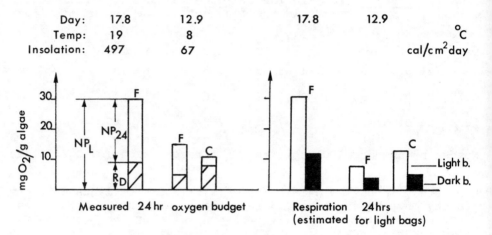

Fig. 2. The 24 hr metabolism of solitary Fucus plants (F) and of the Fucus community (C) in light and dark plastic bags, NP_L = Light period net primary production, NP_{24} = 24 hr NP, R_D = Dark period (when R > P) respiration.

concentration 80% of saturation).

The respiration of the Fucus community over a 24 hr period in September was 2.5 times higher in the light bags (13 mg O_2/g algae at a mean O_2 concentration of 106% of saturation) compared to the dark bags 5 mg O_2/g algae at a mean O_2 concentration of 82% of saturation). The diurnal oxygen concentration in the free water outside the bags was intermediate between that in the light and dark bags (Fig. 3).

Analyses of nutrients and pH were made during September in order to follow changes during the incubation period. The concentrations of nitrate + nitrate and ammonium (Fig. 4) in the plastic bags were similar to the levels in the water outside the bags. The phosphate values measured in the water outside the bags seemed to be unrealistically high, probably due to stirring up of sediment in the water during sampling. In the plastic bags the concentration was lower (about 10 µg PO_4-P/l). All nutrient concentrations during the experiment were typical for the free water in the Askö area at this time of year (Hobro and Nyqvist unpubl.).

The pH in the light bags followed the diurnal variation of the surrounding water outside but was somewhat higher at the end of the experiment (8.1 as compared to 7.8). A PQ of 1.3 was calculated (Fucus bags, for a light period from 0900 - 1500 hrs at a salinity of 7.0 °/oo) by using the changes in pH to calculate the CO_2 uptake (Buch 1945) at the same time as the O_2 output was measured (Fig. 3). In the dark bags the pH decreased to a value of about 7.4 after the first night. This underlines the considerable differences that soon develop between the light and dark bags.

The mean weight distribution (of 16 Fucus plants from August and September) on top (down to the first vesicle), middle and base (dark brown stalk) was 61 ± 5, 35 ± 5 and 4 ± 1% of total plant biomass respectively. This was an index for the plant condition.

By enclosing Fucus plants and parts of the Fucus community in separate plastic bags at the same time, an attempt was made to quantify and to separate the community metabolism into primary production and respiration of Fucus and the respiration of the consumers for a 24 hr period. Assuming that the production of the 4 enclosed plants was the same the Fucus community metabolism was calculated for September 12, 1973 (calculated on 28.4 g algae, which was the mean weight of the 2 plants in the community bags).

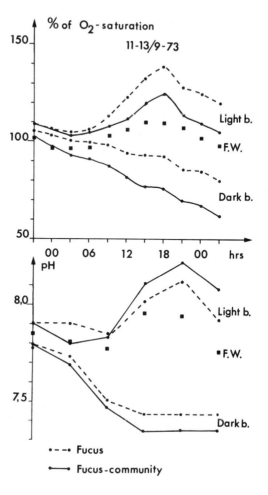

Fig. 3. The diurnal change in oxygen concentration and pH in light and dark plastic bags and in the free water (F.W.). Each curve represents the mean of two bags.

The net primary production of Fucus was 423 mg O_2 compared to that of the Fucus community, which was 306 mg O_2. The difference (117 mg O_2) then represented the O_2 consumption of the consumers (assuming no photorespiration or chemical oxygen demand of the bottom sediments) during the day (Fig. 5). During the dark period (when R > P), the respiration was 212 mg O_2 in the community and 133 mg O_2 in the Fucus bags. The total 24 hr respiration of the consumers was then (117 + 212 - 133) = 196 mg O_2 or 47% of the Fucus net production. 93 mg O_2 (or 22%) remained as a surplus after the studied period (Fig. 5).

The same results were also put together in an energy-flow model (Fig. 6) using the symbols of Odum (in Odum 1971). The flows and the storage of Fucus were calculated on a basis of the Fucus biomass density of 158 g dry weight/m^2, that occurred in the plastic bags. (This is a low value compared to about 500 g dry weight/m^2 found for this area by Jansson and Kautsky (op.cit.). The Fucus light period net primary production was 2.6% of the insolation at the surface (or 1.8% for 24 hrs), while the consumer respiration was 1.2% and the surplus stored was 0.6%.

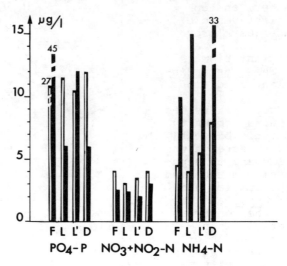

Fig. 4. The nutrient concentration at the start and after 30 hrs in the free water (F.) in light Fucus bags (L), light community bags (L') and dark community bags (D). (Mean of two bags).

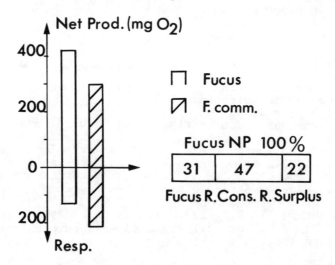

Fig. 5. A 24 hr oxygen budget for a Fucus community (158 g algae / m^2, NP = Light period net primary production, Fucus R = Fucus dark period, when R > P, respiration, Cons. R = Consumer 24 hr respiration).

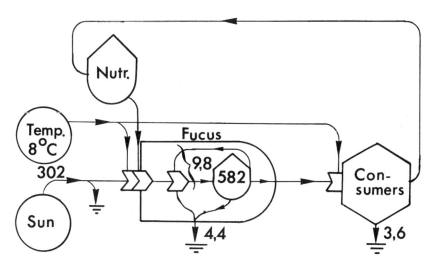

Fig. 6. A simplified energy-flow model for a baltic Fucus community in September. (Flows in kcal/m^2 x 24 hr, storage in kcal/m^2).

DISCUSSION

The main advantage with plastic bag enclosure is that it forms a closed system (no diffusion correction between water and atmosphere), and that the water within it moves in the same way as the surrounding water (Gust and Schramm pers. comm.) giving the algae optimal conditions for nutrient uptake and gaseous exchange (minimizing the "Warburg effect" pointed out by Turner and Brittain 1962). This may be one reason for the decidedly lower production values found by Elmgren and Ganning (op. cit.) in plastic tubs lacking the normal littoral turbulence. Their values are only about 50% of those reported here.

Compared to other ecosystems the photosynthetic efficiency of the Fucus community was found to be high. For example, the corresponding efficiency of a phytoplankton community in SW Baltic was 0.78% (v. Bröckel 1975). Odum (1971) gives a maximum efficiency of 8% of the photosynthetically active radiation for high productive ecosystems (net primary production).

Using the estimations of the plant biomasses within the primary research area of the Askö Laboratory (160 km^2) investigated by Jansson and Kautsky (op. cit.), the primary production of this area was calculated. The net production of F. vesiculosus in September between 1 - 2 m (with 254 tons of algae dry weight and a net production of 15 mg O_2/g algae x day or 4.3 mg C when PQ = 1.3) was 1.1 tons of C or 2.3% of the daily phytoplankton production (300 mg C/m^2 x day for the same period and 160 km^2, Hobro and Nyqvist unpubl.). The corresponding production for the total plant biomass (2 163 tons) then becomes about 10 tons or 20% of plankton production.

The yearly productive importance of the macrophytes is probably higher than for the period studied since many of these plants occur when plankton biomass is at a minimum.

In order to get more information about the production and metabolism of the phytobenthic communities over the year and their contribution to the energy flow in the Baltic (which is studied at the Askö Laboratory), this study has been continued in cooperation with the research project SFB 95 in Kiel.

CONCLUSIONS

The plastic bag method is very well suited for shorter (1 - 2 day experiments), benthic, metabolism studies, especially in the Baltic where the lack of tidal influence makes it possible to study even the most shallow parts of the littoral.

Only when the metabolic rates and the biomasses of different parts of the phytobenthic communities are known throughout the year will it be possible to evaluate fully the role of these communities in the Baltic.

ACKNOWLEDGEMENTS

This study was made possible by the cooperation between the Marine Botanical Department at Institute für Meereskunde in Kiel and the Askö Laboratory, University of Stockholm. I would especially like to thank Prof. H. Schwenke and Dr. W. Schramm for excellent technical and theoretical advice in Kiel and Prof. B-O. Jansson, Dr. A-M. Jansson, Dr. R. Elmgren, Dr. N. Kautsky and Miss I. Andersson for help and valuable discussions at the Askö Laboratory

REFERENCES

v. Bröckel, K. 1975. Der Energie-fluss im pelagischen Ökosystem vor Boknis Eck (westl. Ostsee). Reports SFB 95. 10 Univ. Kiel, 1-96.

Buch, K. 1945. Kolsyrejämvikten i Baltiska havet. Fennia 68, 5, 1-292.

Carlberg, S. 1972. New Baltic Manual - ICES Cooperative Research Report Series A 29, 1-145.

Crisp, D.J. 1971. Energy flow measurements. In Methods for the study of marine benthos. Ed. Holme, N.A. & McIntyre, A.D. IBP Handbook 16, 197-279. Oxford.

Elmgren, R. and B. Ganning 1974. Ecological studies of two shallow brackish water ecosystems. Contr. Askö Lab. 6, 1-56.

Haage, P. 1975. Quantitative seasonal fluctuations. Contr. Askö Lab. 9, 1-88.

Hoffman, C. 1952. Über das Vorkommen und die Menge industriell
   verwertbarer Algen an der Ostseeküste Schleswig-Holsteins.
   Kieler Meeresf. 9, 5-14.

Jansson, A-M. 1974. Community structure, modelling and simulation of the Cladophora ecosystem in the Baltic Sea.   Contr.
   Askö Lab. 5, 1-130.

Jansson, A-M. and N. Kautsky 1976.   Quantitative survey of hard
   bottom communities in a Baltic archipelago. In Proceedings
   of the 11th Symposium on Marine Biology ed. Keegan, B.F., P. O Ceidigh
   and P.J.S. Boaden.

Jansson, B-O, 1972. Ecosystem approach to the Baltic problem.
   Bull. Ecol. Res. Comm. NFR 16, 1-82.

Jansson, B-O. and F. Wulff 1976.   Ecosystem analysis of a shallow
   sound in the northern Baltic: a joint study by the Askö
   group.   Contr. Askö Lab. (in press).

McConnel, W.J. 1962.   Productivity relations in carboy microcosms.   Limnol. Oceanogr. 7, 335-343.

Odum, E.P. 1971.   Fundamentals of Ecology.   Saunders, Philadelphia, London, Toronto, 1-574.

Odum, H.T. and C.M. Hoskin 1958.   Comparative studies on the
   metabolism of marine waters.   Publs. Inst. Mar. Sci. Univ.
   Texas 5, 15-46.

Schramm, W. and B. Guterstam 1975.   Studies of the metabolic
   activity in Fucus communities from the Baltic. IV:
   BMB-symposium, Gdansk (Poland), Oct. 1975. (in press).

Turner, J.S. and E.G. Brittain 1962.   Oxygen as a factor in
   photosynthesis.   Biol. Rev. 37, 130-170.

REPRODUCTIVE STRATEGY IN TWO BRITISH SPECIES OF *ALCYONIUM*

Richard G. Hartnoll

Department of Marine Biology, University of Liverpool, Port Erin, Isle of Man, U.K.

ABSTRACT

Alcyonium digitatum and Alcyonium hibernicum co-exist in the area around the Isle of Man and have very different reproductive strategies. A. digitatum has equal numbers of male and female colonies, spawns in mid-winter, and discharges the gametes into the water where fertilisation and embryogenesis occur to produce pelagic planulae. A. hibernicum has only female colonies and spawns in late summer : the ova develop parthenogenically, the embryos are brooded and the planulae are benthic. It is postulated that the prime cause of these differences is the smaller size of A. hibernicum, whereby the reduced reproductive resources are insufficient to ensure replacement by the mechanisms adequate for the larger A. digitatum. To compensate it has improved reproductive efficiency, but only by sacrificing the major mechanisms of genetic variation and species dispersal.

INTRODUCTION

Alcyonium digitatum (L.) and A. hibernicum (Renouf) are the only members of the genus found around the Isle of Man. The latter was initially described as Parerythropodium hibernicum by Renouf (1931), but has subsequently been correctly placed in Alcyonium by Tixier-Durivault and Lafargue (1966, as A. pusillum) and Beldam and Robins (1971). It will be shown that the reproductive strategies of these two species are remarkably divergent, but first some background information must be provided so that these differences may be discussed in proper perspective.

A. digitatum occurs along the Atlantic coast of Europe from Portugal ($41^{\circ}N$) to Norway ($70^{\circ}N$), in Iceland and in New England (Robins, 1968). There is some doubt as to the conspecificity of the American and European populations (Feldman, pers. comm.), but that is not important in the present context. This is an extensive distribution, embracing the whole of the Eastern Atlantic Boreal Region and extending well into the Lusitanian Region. The known distribution of A. hibernicum is much more restricted, with records only from Lough Ine in Eire, Îles de Glénan in Brittany, Lundy Island and the Isle of Man (Beldam and Robins, 1971) and from Pembroke (Hiscock, pers. comm.). This distribution lies within that of A. digitatum and straddles the boundary between the Lusitanian and Boreal regions. The largest and most flourishing population is at the northernmost location, suggesting that the northern limit of this small and rare species has yet to be defined.

A. digitatum is the larger species, and fully contracted colonies can reach a height of 100 mm, a width of 200 mm and a wet weight of 1000 g. The mean wet weight of a large random sample from Port Erin was 73 g per colony. It is also very common, and in suitable conditions it is the dominant member of the community over a large part of its distribution. A. hibernicum is smaller, and its maximum contracted dimensions are a height of 40 mm, a width of 50 mm and a wet weight of 13 g. The mean wet weight of a large random sample from Port Erin was 1.6 g per colony. Even in its known locations it is restricted to a few sites, although at these the colonies may densely cover several square metres of rock. Both species favour steep or overhanging rock with at least a moderate degree of water movement, and A. digitatum extends into quite turbulent locations. The recorded depth ranges are 0 - 100 m for A. digitatum and 1 - 30 m for A. hibernicum.

RESULTS

Alcyonium digitatum

A detailed account of the reproductive cycle of this species is already available (Hartnoll, 1975), and only a summary of the relevant points, with some additional information, will be presented here.

There are approximately equal numbers of separate male and female colonies, and no evidence of sex change. Less than 1% of colonies are hermaphrodite, but these have apparently functional ova and testes (Fig. 1B) which may develop within the same polyp. Some colonies mature in their second year, the smallest of these having a wet weight of 1 g, but in others maturity is delayed until the third or subsequent years when they may weigh over 20 g.

The annual reproductive cycle commences when the gametes begin to develop in December and January, and lasts until spawning occurs some twelve months later. In both sexes the gonads develop on the edges of the mesenteries, and lie within the gastric cavity, attached to the mesentery, until spawning occurs. In males the testes are white and consist of a single layer of epithelium within which a very large number of sperms develop (Fig. 1A). In females each ovary consists of a layer of follicle cells enclosing a single ovum which becomes orange as it matures. The follicular epithelium is columnar, and against its inner face lies an amorphous layer which stains blue with Mallory. The cytoplasm of the mature ovum is densely packed with yolk granules, and there is a large peripheral nucleus with a prominent nucleolus. The gonads of both sexes reach their full development by August (Hartnoll, 1965, Fig. 2), and in this condition effectively occlude the gastric cavities of the polyps. The gonads are stored for some four months, and during this period a large proportion of the colonies enter a quiescent phase when they remain permanently contracted and do not feed, presumably as a response to the congestion of the gastric cavities. Many of the colonies become covered by an extensive surface film of epibiota.

During December and January the colonies resume activity and slough off the surface film. The ova and sperms are shed into the water by the respective colonies, fertilisation occurs externally, and the developing embryos float freely for 5 to 7 days. They give rise to actively swimming lecithotrophic planulae which may have an extended pelagic life before they eventually settle and metamorphose. The female reproductive cycle is summarised in Fig. 2.

Alcyonium hibernicum

The reproductive biology of this species was studied from February 1974 to October 1975 at a site south of Port Erin in a depth of 15 to 17 m. Numerous colonies were growing on vertical and overhanging rock on the landward side of

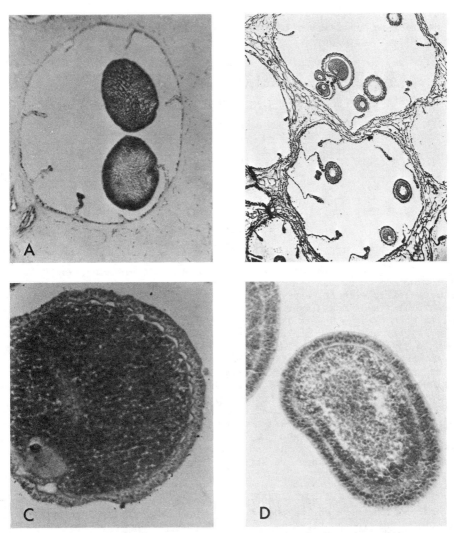

Fig. 1. <u>A. digitatum</u>. A. T.S. of polyp in November with two ripe testes (x28). B. T.S. of hermaphrodite colony in June, ova in upper polyp and testes in lower (x20). <u>A. hibernicum</u>. C. T.S. of maturing ovum in February (x150). D. T.S. of brooded planula in August (x85).

an underwater reef, which provided a degree of shelter from wave action. A small sample from Martins Haven, Pembroke, was also examined.

Several hundred colonies were sexed, and all contained abundant and obvious ova. Study of both fresh and stained material has not revealed any trace of testes, and it must be concluded that this species consists only of female colonies. In material from Lough Ine Renouf (1931) similarly observed many ova but no testes. The onset of sexual maturity was investigated from April to June, when the ova are prominent, by examining 128 colonies. All were mature including the smallest colonies which could be found which had a height of 5 mm, a width of 4 mm and a wet weight of 0.08 g. On this evidence all colonies breed in their first year.

TABLE 1 Mean and range of maximum ova diameter for samples of A. hibernicum from Port Erin

DATE	MEAN DIAMETER (mm)	RANGE (mm)
1 Sep	<0.10*	-
13 Oct	0.24 ± 0.06	0.14 - 0.44
7 Nov	0.24 ± 0.04	0.14 - 0.48
3 Feb	0.38 ± 0.06	0.22 - 0.64
17 Apr	0.42 ± 0.02	0.34 - 0.56
20 May	0.43 ± 0.02	0.34 - 0.52
30 June	0.53 ± 0.02	0.50 - 0.64
30 July	<0.10*	-

*These samples also included brooded planulae with a volume equivalent to a diameter of 0.55 to 0.65 mm

Regular samples were collected to examine ovarian development, and the results are summarised in Table 1 and Fig. 2. The ova are surrounded by a single layer of follicle cells, and from a diameter of 0.04 mm upwards there is a blue-staining amorphous layer lining it internally, as in A. digitatum. Ova smaller than 0.15 mm have a large central nucleus with a prominent nucleolus, and a lightly staining cytoplasm. Above 0.15 mm densely staining yolk granules begin to appear in the cytoplasm and the nucleus becomes peripheral, and by 0.3 mm the cytoplasm is densely packed with yolk and the appearance is typical of the mature ovum (Fig. 1C). In early September, when the planulae from the previous year have just been released, there are only a few very small ova present. During September and October there is a rapid proliferation and growth of ova, so that by the middle of October they are abundant and the mean maximum size is 0.24 mm. Growth continues slowly through the winter and spring, and by May the ova have reached 0.43 mm, and then there is a final short period of faster growth so that by the end of June they reach a mean maximum size of 0.53 mm. At this time they are still enclosed within the follicles and attached to the mesenteries.

During July the ova are released from the follicles into the gastric cavity, and are brooded there whilst they develop parthenogenically into typical ciliated planulae. The cytological details of this process are not known. The planulae have a ciliated columnar ectoderm and a layer of rather irregular endoderm, with a mass of enucleate yolk filling the centre (Fig. 1D) : this is essentially identical to the structure of the planula of A. digitatum described by Matthews (1917). The planulae are brooded for some time, and are ellipsoid with dimensions of 0.7 mm by 0.5 mm, but if released from the polyps they rapidly elongate and become mobile. Most of the planulae are released in the latter part of August, and although not seen in the field this release was observed in the laboratory. The planulae are orange-red, elongate, rounded anteriorly and tapering posteriorly, and measure up to 2 mm in length with a maximum width of 0.3 mm. They seldom swim, but crawl actively with an anti-clockwise rotation when viewed from the front. In the laboratory they settle and metamorphose within a few days. At Lough Ine "the ripe ova are shed towards the end of June, and the single polyp stage and colonies consisting of two or three polyps have been found towards the end of July" (Renouf, 1931), indicating some regional variation in the spawning season.

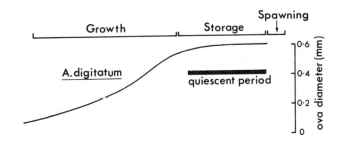
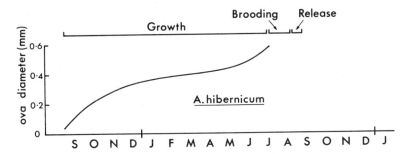

Fig. 2. The annual cycle of oogenesis and spawning in A. digitatum and A. hibernicum.

There is no quiescent period with prolonged contraction and cessation of feeding. The final growth of the ova is completed only shortly before spawning, and so there is no lengthy storage of ripe gametes like that in A. digitatum. Whenever the study populations were observed nearly all of the colonies were expanded.

DISCUSSION

A comparison of the two species reveals that they differ in almost all major aspects of reproduction, a remarkable situation for two otherwise very similar species.

A. digitatum	A. hibernicum
Separate male and female colonies	Female colonies only
Breeds in second year or later	Breeds in first year
Quiescent period Aug. to Dec.	No quiescent period
Spawns Dec. to Jan.	Spawns Aug. to Sept.
External fertilisation	Parthenogenic
External embryogenesis	Broods embryos
Pelagic larvae	Benthic larvae

In order to examine the adaptive value of these differences the reproductive biology of related species was surveyed to see if any consistent pattern exists. Unfortunately there is little such information either on the genus Alcyonium or even on other members of the Alcyonacea, and that which is available (Lacaze-Duthiers, 1865; Lo Bianco, 1909; Gohar and Roushdy, 1961) is too fragmentary to permit any useful generalisation. So attention must be confined to the two species concerned, and a search made for differences between them which might account for their divergent reproductive strategies. Ecology, geographical distribution and size are possible factors.

There are no ecological differences in such aspects as habitat preference or feeding which seem relevant to reproductive practice. In the same way geographical distribution offers no satisfactory explanation, except that the more northerly extent of A. digitatum might explain its winter spawning : this is unlikely, for it is not restricted by any means to the colder Boreal Region. On the other hand there are very obvious inter-specific size differences, with mean wet weights per colony of 73 g for A. digitatum and 1.6 g for A. hibernicum. Menge (1975) and Chia (1976) indicate how size differentials could be responsible for the evolution of reproductive differences between otherwise similar species, and Menge cites examples of several such species pairs where the larger has a pelagic larva but the smaller broods its offspring. His particular study was of two starfish, the large Pisaster ochraceus and the small Leptasterias hexactis, and his thesis is that Leptasterias has evolved small body size as the only way in which it could co-exist with the dominant Pisaster through an increased partitioning of resources. The large Pisaster is able to produce sufficient gametes to ensure replacement even after the enormous mortality of the pelagic larvae, but the limited reproductive potential of the smaller species can not accomplish this. Hence the policy of larval brooding with its higher survival rate, but with the loss of dispersive potential, has been forced upon Leptasterias as a necessary correlate of small size.

The application of a similar line of reasoning to the two species of Alcyonium permits a logical explanation to be offered for the extensive reproductive differences. A. digitatum and A. hibernicum co-exist with very similar ecological requirements, and the former is a very successful and dominant member of the community. It is a reasonable suggestion, though not one easy to prove, that the small size of A. hibernicum has evolved in order to avoid too intense a competition with A. digitatum. With that premise the following heirarchy of cause and effect is proposed.

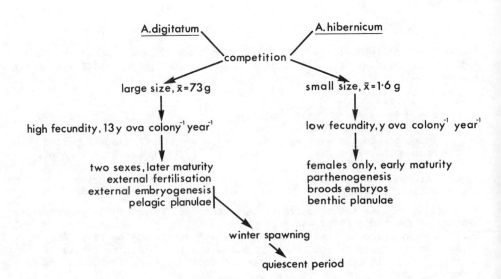

The various stages in this scheme can be considered in turn. The difference in mean colony size will obviously result in a difference in mean gamete production. The number of polyps per unit surface area of colony is similar in the two species, and the number of ova produced per polyp is only slightly greater in A. digitatum than in A. hibernicum. So the production of ova is in proportion to the surface area, and hence in proportion to the two thirds power of the weight. Hence the ratio of ova production by mean weight colonies of the two species will be :

$$(73 / 1.6)^{\frac{2}{3}} = 12.8$$

Both species breed once each year, so the annual ova production for average colonies will be roughly in the ratio of 13:1. There is no reason why the two species should experience grossly different rates of either larval or adult mortality if they adopted the same reproductive strategy, so if that were the case then a female colony of A. digitatum would have thirteen times more chance of replacing itself than would a colony of A. hibernicum. A. digitatum currently maintains a stable population with a reproductive strategy which involves separate sexes, external fertilisation and pelagic embryogenesis and planulae larvae. The small A. hibernicum could not replace its population through the use of such a strategy, because too few larvae would survive, and so various measures have evolved to increase both the production and survival rate of ova. Earlier maturity and the suppression of male colonies are steps which increase the production of ova. Parthenogenesis, the brooding of embryos and the benthic habit of the planulae combine to increase the survival rate of the offspring. Overall these factors counterbalance the smaller fecundity, but the price is the loss of the main mechanisms of genetic diversification and dispersal. A. hibernicum will have a relatively high survival rate close to the adults, which explains the observed distribution pattern of occasional dense patches. The loss of dispersal power will make the colonisation of new sites very difficult, and established populations will be vulnerable to locally adverse conditions or predator aggregation. This could account for both its rarity and its restricted geographical distribution compared with the more mobile A. digitatum.

The final differences in reproductive strategy, in breeding season and the presence of a quiescent period, may be explained as lower levels in the heirarchy. A. digitatum has pelagic embryos and larvae, and it will be advantageous to release these at the season of minimum predation by zooplankton, namely mid-winter, when there will also be least competition for settlement sites. These larvae are lecithotrophic and need no planktonic food. This winter spawning is in turn the cause of the quiescent period, since gametogenesis must occur in spring and summer when the food supply for the planktotrophic adults is plentiful. The storage of ripe gametes in the gastric cavities between gametogenesis and spawning inhibits feeding and induces the quiescent period. A. hibernicum broods its embryos and has benthic larvae, and is not subject to the pressure of zooplankton predation. Hence it spawns in late summer immediately gametogenesis is complete, and this removes the need for a non-feeding quiescent period with its attendant loss of growth potential.

In many ways the process of reproduction in A. hibernicum is now more akin to asexual than to sexual reproduction, and it is interesting that in relation to sea anemones Chia (in press) has remarked that asexual reproduction is associated with small body size and/or poor nutrition. These conditions limit reproductive resources, and asexual reproduction makes maximum short term use of them. In its reproductive strategy A. hibernicum has aquired all the advantages, together with the disadvantages, of asexual reproduction.

ACKNOWLEDGEMENTS

I am grateful to Keith Hiscock of the Oil Pollution Unit, Orielton for informing me of the presence of A. hibernicum at Martins Haven, Pembrokeshire, and for a sample of material.

REFERENCES

Beldam, G.A. and Robins, M.W. 1971 The soft-corals of Skomer. Nature Wales 12, 207-212.

Chia, F.S. 1976 Classification and adaptive significance of developmental patterns in marine invertebrates. Thalassia Jugoslavica 10, 121-130.

Chia, F.S. in press Sea anemone reproduction : patterns and adaptive radiations. Proc. 3rd International Symposium on Coelenterate Biology, Victoria, B.C., 1976. Plenum, New York.

Gohar, H.A.F. and Roushdy, H.M. 1961 On the embryology of the Xeniidae (Alcyonaria). Publs mar. biol. Stn Ghardaqa 11, 45-70.

Hartnoll, R.G. 1975 The annual cycle of Alcyonium digitatum. Estuar. coast. mar. Sci. 3, 71-78.

Lacaze-Duthiers, M. 1865 Des sexes chez les Alcyonaires. C. r. Hebd. Séanc. Acad. Sci. Paris 60, 840-843.

Lo Bianco, S. 1909 Notizie biologiche riguardanti specialmente il periodo di maturita sessuale degli animali del golfo di Napoli. Mitt. zool. Stn Neapel 19, 513-761.

Matthews, A. 1917 The development of Alcyonium digitatum with some notes on the early colony formation. Q. Jl microsc. Sci. 62, 43-94.

Menge, B.A. 1975 Brood or broadcast ? The adaptive significance of different reproductive strategies in the two intertidal sea stars Leptasterias hexactis and Pisaster ochraceus. Mar. Biol. 31, 87-100.

Renouf, L.P.W. 1931 On a new species of alcyonarian Parerythropodium hibernicum. Acta Zool. 12, 205-223.

Robins, M.W. 1968 The ecology of Alcyonium species in the Scilly Isles. Underwater Ass. Report 1968, 67-71.

Tixier-Durivault, A. and Lafargue, F. 1966 Quelques Alcyonaires des Îles de Glénan. Bull. Mus. natn. Hist. nat., Paris 38, 456-460.

OBSERVATIONS ON THE BEHAVIOUR & DISTRIBUTION OF *Virgularia mirabilis* O.F. MULLER (COELENTERATA: PENNATULACEA) IN HOLYHEAD HARBOUR, ANGLESEY

R. Hoare* and E.H. Wilson**

*Marine Science Laboratories, Menai Bridge, Anglesey, U.K.
**West Indies Laboratory, P.O. Box 4010, Christiansted, St. Croix, U.S. Virgin Is.

ABSTRACT

Virgularia mirabilis occurs throughout the shallow muddy part of Holyhead harbour. It appears to be excluded in areas where there is pollution, dredging and disturbance of the bottom by waves. The animals show a patchy distribution, probably related to larval settlement. Virgularia uses its nematocysts to capture small organisms and probably also exhibits a type of filter feeding. This is evidenced by the orientation of many colonies at right-angles to the current. Predation pressure on the Holyhead harbour population appears to be low. The animals construct mucus-lined burrows in the mud into which they can withdraw fully but the mechanisms of withdrawal and extension are not the same. The possibility of rhythmic feeding and burrowing behaviour has been investigated but results so far are not in agreement. However, the Pennatulids are apparently not sensitive to light and individual colonies are not synchronised in their behaviour.

INTRODUCTION

There appears to be very little information on the ecology of Virgularia mirabilis probably due largely to its inaccessability. The species has been obtained mostly from dredges in muddy areas at depths of about 40m. Dredged specimens were invariably damaged and because no direct observations were possible much speculation arose concerning the animals' way of life. Notwithstanding, many of the ideas of the earlier workers such as Dalyell (1848) and especially Marshall & Marshall (1882) concerning the behaviour of Virgularia seem to be substantially correct.

The new harbour at Holyhead is a sheltered area consisting largely of soft mud and rarely exceeds depths of 12m below chart datum. Virgularia is abundant over a large area of the mud especially below parts of the north breakwater. The combination of shelter and shallow depths afford a good environment for direct observations on a population of Virularia using agualung techniques.

MATERIALS AND METHODS

Standard aqualung techniques were used throughout. Length measurements were dictated into an underwater tape recorder via a bone conducting microphone.

The method used to investigate possible patchiness in distribution involved laying two 32m long transects southwards from the mid-region of the breakwater (see Fig. 1). Transect lines were nylon marked off at 1m intervals by plastic tags and held above the mud by small pegs. A diver counted the pennatulids every 1m along the line and for 1m either side of it by swimming along with a metre rule. This resulted in 32 contiguous squares on each side of the line. The data was subjected to a pattern analysis (Greig-Smith 1957). Diving was carried out at high water when most of the pennatulids seemed to be up.

Laboratory filming was done with a Bolex cine camera operated by a variable time-lapse system. Activity recording employed a photocell and level device attached by a thread to the axial rod of the animal and connected to a potentiometric chart recorder.

RESULTS

Distribution

Spot diving has shown that <u>Virgularia</u> is found in varying densities over most of the muddy part of the new harbour (Fig. 1).

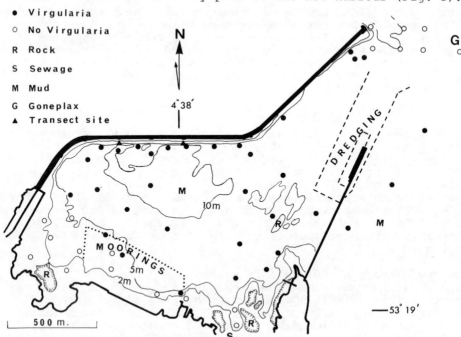

Fig. 1. Distribution of <u>Virgularia</u> within Holyhead Harbour.

It appears that the species is not found in areas where there is:
(a) much coarse material such as shells and stones within the mud. Presumably such conditions prevent successful burrowing.
(b) mechanical disturbance caused by dredging and boat moorings. The burrowing activities of a large population of Goneplax rhomboides in the north-east part of the harbour may also come into this category.
(c) sewage pollution which results in black anoxic mud in the south corner of the harbour.
(d) exposure to wave action. A large proportion of winds blow from the north east and north west, affecting that area of Holyhead Bay not protected by the breakwater.

In general it appears that Virgularia are smaller and become less common with increasing distance eastwards from the breakwater. Although the data was limited, the results of the pattern analysis indicate that the animals are patchy and the dimension of the clumps along the lines investigated is about 8 sq. m. Clumping is most obvious in the more sheltered parts of the harbour and is probably due to limited larval dispersal in the absence of strong water movements.

Length Frequency Distribution
Although the axial rod of Virgularia is partly calcareous it has no obvious growth marks. So, as with all soft-bodied animals, there is a problem in determining the age and growth rate of individuals.

Examination of 36 individuals covering a range of sizes showed

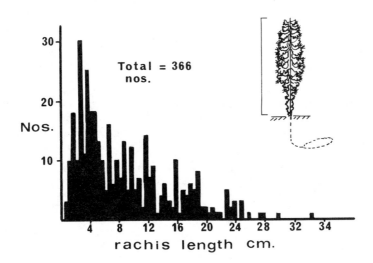

Fig. 2. Length frequency distribution

that the ratio of rachis length to total length remained fairly
constant at about 0.60. Thus, rather than destructively sample
large numbers of individuals the rachis length measured in situ
was used as an estimate of total length. Only whole animals
fully extended were measured in a series of random swims. The
resulting length frequency distribution is shown in Fig. 2.

Probably due to spatial differences in the supply of suspended
food and a long spawning period (June - October?) there appears
to be a marked variation in individual growth rates which mask
the occurrence of any clear size modes. The largest individuals
(>22cm rachis length) were found mostly near the breakwater
indicating that mortality was lowest in this area. The results
might be improved by measuring total length of individuals removed from the sediment and/or measuring the growth of marked
individuals over a period of time.

Feeding
Contrary to the observations of Marshall & Marshall (1882)
Virgularia mirabilis does possess nematocysts. These are shown
in Plate Ia to be of one simple type - atrichous isorhizas
(Hyman 1940) and are very similar to the nematocysts of
Alcyonium digitatum described by Hickson (1901). Plate I b, c
and d shows scanning electron micrographs of the external structure of a Virgularia autozoid. Cilia are extremely dense around
the mouth slit and are present on the pinnules which also bear
the nematocysts.

As suggested for Alcyonium digitatum, by Roushdy & Hansen (1961),
Virgularia is probably capable of both active predation on small
zooplanktonic organisms and filtration of suspended material.

Laboratory experiments have shown that the polyps are capable of
capturing (by trapping with the nematocysts) and ingesting small
active organisms such as Artemia nauplii. However, larger forms
such as barnacle nauplii although temporarily held by the tentacles usually break free or are rejected. A "feeding response"
characterised by dilation of the pharynx and bending of the
tentacles towards the mouth was shown to attractive food particles. The same response could be elicited by a free amino acid
extract of Artemia in the absence of particulate matter. The
amino acids alanine, aspartic acid and glutamic acid each produced the feeding response but only in artificially high concentrations. Also, apparently the same response was produced by
applying dilute inorganic acids although such low pH values are
unlikely to occur naturally. The "feeding response" did not
occur with inert particles such as carmine powder and carborun-
dum, the material being seized and quickly rejected. Thus it
appears that the polyps are sensitive to both mechanical and
chemical stimulation although the exact mechanisms have not been
elucidated.

Filtration may occur by very small particles being trapped in
mucus on the pinnae and transported to the mouth by cilia (see
Plate I). The water is probably not truly filtered, particles

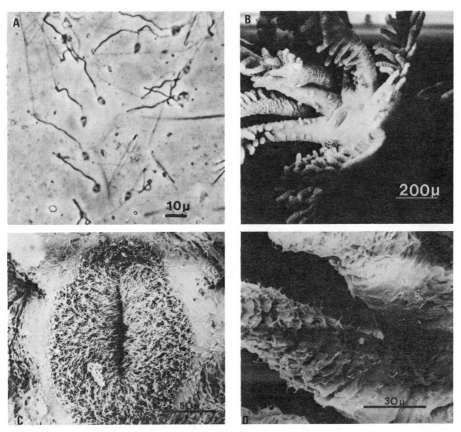

Plate I A: Virgularia nematocysts embedded in agar.
B: Polyp showing arrangement of pinnules (scanning electron micrograph).
C: Highly ciliated mouth slit (s.e.m.).
D: Pinnules showing cilia (s.e.m.).

only being retained when they contact the tentacles. When stationary there is a gap of 0.5mm between each tentacle and 0.1mm between each pinnule. The process is probably unselective, indigested material being rejected later from the pharynx. Grigg (1972) inter alia showed that Alcyonarians orientate themselves to water movements. If Virgularia does filter the water then it would be expected to orientate itself at right angles to the current for maximum efficiency.

Orientation in situ On six different occasions the orientation of Virgularia colonies was measured with respect to magnetic north using a divers compass. Each time the current speed and direction were measured using a simple current vane (see Wilson, 1975). Animals in which polyps were spiralled in different planes around the axial rod (Dalyell 1848) were not included.

The results are shown in Table 1. The angles were grouped into twelve 30° sectors such that the current direction was at the mid-point of one of the sectors. The number of animals per sector were plotted at the midpoint of that sector. All significance tests were performed according to Zar (1974) and Table 1 shows that the orientation of the colonies was not randomly distributed about a circle ($p = 0.01$). Furthermore, it was found that the observed orientations clustered around the direction of current ($p = 0.05$).

TABLE 1 Statistical analysis of orientation measured in situ on six different occasions (A-F)

	A	B	C	D	E	F
State of tide	ebb	ebb	flood	ebb	flood	ebb
Current direction	88°	148°	240°	120°	268°	98°
Water velocity	0-5cm/s	0-5cm/s	0-5cm/s	0-5cm/s	5cm/s	5cm/s
n	57	65	42	50	46	60
Orientation mean, $\bar{x}$	128°	167°	254°	131°	270°	114°
Mean angular deviation, s	57.44°	43.96°	47.12°	53.94°	9.96°	11.80°
Goodness of fit, x^2	86.40	287.99	64.28	151.47	230.89	361.20
P (% pt. x^2)	<0.001	<0.001	<0.010	<0.001	<0.001	<0.001

Mortality

Evidence of predation upon Virgularia seems limited to a report by Marshall & Marshall (1882) that the species was found in the stomach of Haddock, Gadus aeglifinus. Apart from the occasional plaice, Pleuronectes platessa large fish were not observed in the harbour and there is no evidence that plaice eat Virgularia.

Of the species commonly found amongst the Virgularia only 2 species, Pomatoschistus minutus and Macropipus depurator were considered as likely predators. In laboratory experiments where these animals were kept in tanks with Virgularia (and filmed in the case of the fish) for several weeks neither species was observed to attack the pennatulids. Consequently, combined with the fact that most of the colonies in the harbour appeared intact observations suggest that predation pressure is low.

Occasionally colonies were observed to be broken, presumably as a result of accidental contact with large animals.

Burrowing Activity

The highly muscular peduncle of Virgularia is capable of active burrowing, presumably by peristaltic contractions (see Hyman 1940). If the animals is placed upright in a container of mud in the laboratory it can construct a full-size burrow for itself in about one week.

Plate II Virgularia (a) extended from, and (b) withdrawn into its burrow.

By sawing frozen cores into sections there appear to be two types of burrow depending upon the sediment. Where stones are found beneath the layer of mud the foot bends round to form a J-shape while on soft mud straight burrows are usually constructed. Plate II a and b shows the upper part of a Virgularia burrow, where the animal has burrowed against the wall of a glass container. Plate II b shows that the polyps are withdrawn and that the burrow is lined with mucus. The animal seems to remain inactive inside the burrow, no water movements being detected when carmine particles were introduced at the top. Presumably, however, there is a slow exchange with the overlying water.

The activity record in Fig. 3 shows that the manner of withdrawal into the burrow is not the same as emergence. Withdrawal into the mud is preceded by closure of the polyps and expulsion of water from the colony. This is followed by 2 or 3 large contractions with "rest periods" between. If the animal is mechanically disturbed it is capable of more powerful contractions with a very short interval between them. Usually two such contractions are sufficient for it to disappear into the mud, thereby avoiding predators. The escape reaction takes about 30 sec while the normal activities take much longer (see Fig. 3).

There is conflicting evidence concerning the possibility of rhythmic burrowing behaviour. "In situ" time lapse photography was made difficult by turbidity maxima at certain states of the tide. However, the animals appeared to be influenced by tidal conditions, being fully extended over the period of high water when the water was clearest but withdrawing again when the suspended sediment load became too great. Wilson (1975) postulated a circadian rhythm based on time-lapse photographs taken under constant conditions in the laboratory. This has not been substantiated using activity recorders (Fig. 3) which show the strength and duration of the animals' activity to be variable.

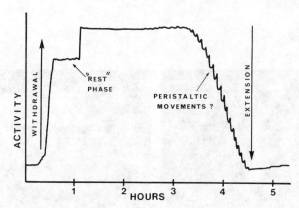

Fig. 3. Activity record of burrowing and extension.

This technique may, however, disturb the animal, photography being more reliable.

Two things seem certain. Firstly the animals are insensitive to light and secondly they are not synchronised. Some colonies are retracted into their burrows while others are extended. Clearly more work is necessary on this important aspect of behaviour.

ACKNOWLEDGEMENTS

The authors are grateful to all those people who assisted with the diving. Mr. P.G. Knight carried out the biochemical work on feeding. We also wish to thank the Torrey Marine Laboratory, Aberdeen for loan of an underwater time-lapse camera system.

REFERENCES

Dalyell, J.G. 1848. Rare and remarkable Animals of Scotland Vol. II. London.

Greig-Smith, P. 1957. Quantitative Plant Ecology. Butterworth London.

Grigg, R.W. 1972. Orientation and Growth Form of Sea Fans. Limnol. Oceanogr. 17(2), 185-192.

Hickson, S.J. 1901. Alcyonium. Liverpool Marine Biology Committee Memoirs V. Williams & Norgate, London.

Hyman, L.H. 1940. The Invertebrates Vol. I McGraw-Hill, New York.

Marshall, A.M. & W.P. Marshall, 1882. Report on the Pennatulida collected in the Oban dredging excursion of the Birmingham Natural History and Microscopical Society. Part III *Virgularia mirabilis* Lamarck. Midland Naturalist.

Roushdy, H.N. & V.K. Hansen 1961. Filtration of Phytoplankton by the Octocoral, *Alcyonium digitatum* L. Nature London 190, 649-650.

Wilson, E.H. 1975. Biology of *Virgularia mirabilis* Univ. of Wales M.Sc. Thesis. (unpubl.).

Zar, J.H. 1974. *Biostatistical Analysis* Prentice-Hall, N.J.

MEIOBENTHIC SUBCOMMUNITY STRUCTURE: SPATIAL VS. TEMPORAL VARIABILITY

William D. Hummon and Margaret R. Hummon

Department of Zoology, Ohio University, Athens, Ohio 45701 USA

ABSTRACT

Annual and seasonal variabilities in an assemblage of subcommunity properties obtained from one whole-beach transect site are compared with values from a second transect site further along the same beach for four pairs of sites, two from western Scotland and two from the eastern United States. Properties involve interspecific interaction, diversity-evenness, similarity, numbers and estimated biomass of the Gastrotricha subcommunity of meiofauna, one of the more important herbivore elements of the decomposer-based food web.

Sedimentary properties differed significantly between paired transects on three of four beaches, generally involving mean grain size, sorting and roundness but not pore space. Little consistency occurred throughout the series. No significant differences were found between subcommunity parameter values in annual vs. summer collections from the same transect, for each of the three cases where annual data were available. Yet percent composition of several species and one or more other parameters differed between spatially neighboring transects on each of the four beaches. Similarly, while no significant differences were found between permutations of annual and those of seasonal collections from the major collection transects on each beach with respect to shared species diversity, a strong difference (.001>P) occurred between spatially differentiated transect pairs.

Conclusions are that longshore gradients may play an important part in determining subcommunity structure and that such gradients warrant serious attention in future studies of beach meiofauna.

INTRODUCTION

The intertidal marine-beach habitat is well known for its dynamic physical, chemical and biological gradients. For meiofauna these have generally been studied by means of whole-beach transect planes, extending up beach from low to high tide levels and vertically from surface to low tide ground water depths in the sand. It was suggested independently by Wieser et al. (1974) and by Hummon (1975) that motile meiofauna may be treated as 'finely-tuned' integrated sensors of the environment and that their distribution and interactions can tell us about properties of the environment that cannot be measured as readily in any other way. An assemblage of such subcommunity properties, having to do with interspecific interaction, diversity-evenness, similarity, numbers and estimated biomass of the Gastrotricha subcommunity of meiofauna, have been treated by Hummon (1976) for bimonthly collections from two Scottish beaches over an annual cycle. The purpose of this paper is to compare annual and seasonal variability in these properties at one transect site with values obtained from a second transect site further along the same beach. In addition to the two beaches located on the west coast of Scotland, two beaches located on the east coast of the United States will also be treated.

Beach locations are as follows: Irvine Bay Transect A (55°37'N lat, 4°45'W long) and Transect C (3 km south along 21 km of semi-continuous beach) facing the Firth of Clyde near Irvine, Ayrshire, Scotland; Firemore Bay Transect A (57°50'N lat, 5°40'W long) and Transect D (600 m northwest along 900 m of semi-continuous beach) facing Loch Ewe near Inverasdale, Wester Ross, Scotland; Buzzards Bay Transect A (Crane's Beach of Pollock and Hummon 1971 and Hummon 1974: 41°30'N lat, 70°40'W long) and Transect B (MBL Beach of Hummon, 1974: 200 m east along 500 m of semi-continuous beach) near Woods Hole, Massachusetts, USA; and Delaware Bay Transect A (Roosevelt Inlet of Hummon et al. 1976: 38°47'N lat, 75°10'W long) and Transect B (10 m east along 55 m of continuous beach) near Lewes, Delaware, USA.

METHODS

Collection methods and procedures for faunal counts and sedimentary analyses were given by Hummon (1974, Buzzards Bay sites) and Hummon (1976, other sites). A series of 10 cm^3 sediment samples was taken from a grid set up along the vertical transect plane at each site. Number of sampling stations varied with the length of the transect profile (15-30 stations for beaches of 110-280 m length; 5-9 stations for beaches of 7-14 m length); number of vertical sampling points per station varied with the groundwater depth in the beach. Living fauna were extracted from the 53-180 sediment samples per transect collection by means of a sea-water-ice or 7% $MgCl_2$-decantation method and Gastrotricha were tallied and identified by sample and species. Sediments were wet sieved, the 0.5 to 1.0 phi ($\emptyset$) interval fractions dried, weighed, the results plotted as cumulative weight curves on probability paper and mean grain size and sorting coefficients in phi units calculated. Porosity was measured by the volume displacement method of Hummon (Hulings and Gray 1971, Sect. 2.6.2) and roundness by the method of Powers (Hulings and Gray 1971, Sect. 2.8.2).

Subcommunity properties were analyzed as in Hummon (1976). Observed data refer to those obtained directly from sand samples; adjusted data were obtained after modification to approximate continuous cores with 10 sq. cm surface area. Weighted mean number of species co-occurring in the sample per individual and its standard deviation, along with a transect covariance measure (Jumars 1975), were used as indices of interspecific interaction. H' diversity was measured after Lloyd et al. (1968), J' evenness after Pielou (1969) and S_H' similarity based on shared species diversity after Hummon (1974). The H'max used in calculating J' evenness for a given transect pair was based on the total number of species encountered in all collections on that beach (24 species for Irvine Bay transects, 41 for Firemore Bay, 16 for Buzzards Bay and 18 for Delaware Bay). Density-free dispersion of individuals among species, after Green (1966), is comparable to evenness as a measure, but is not dependent upon assumptions regarding the value of H'max. Geometric mean individuals per 10 sq. cm surface was calculated using log transformations of adjusted data from all stations of a collection and is supplemented by 95% confidence limits. Geometric mean µg biomass followed the same procedure after first converting adjusted numbers of individuals into dry weight estimates of biomass.

TABLE 1 Profile and Sedimentary Data for Eight Whole-Beach Study Transects in Scotland (Irvine Bay Transects A and C; Firemore Bay Transects A and D) and the United States (Buzzards Bay Transects A and B; Delaware Bay Transects A and B)

	Irvine Bay		Firemore Bay	
Transect	IBTA	IBTC	FBTA	FBTD
Profile distance, m	200	110	280	110
% slope (upper 1/4)	4.2	8.4	2.4	6.2
% slope (lower 3/4)	1.5	1.9	1.6	3.9
Grain Size, n	85	23	117	16
Mean $\emptyset$ (S.D. $\emptyset$)	+2.06(0.44)	+1.84(0.24)	+1.93(0.18)	+2.17(0.08)
Mean µm	240	279	262	222
Sorting $\emptyset$ (S.D. $\emptyset$)	0.49(0.23)	0.49(0.10)	0.46(0.12)	0.32(0.07)
Pore space, n	16	8	16	4
% (S.D.)	39.7(1.3)	39.7(0.6)	37.6(0.9)	38.2(0.6)
Roundness, discr.	angular	subangular	subangular	subangular
Powers scale, $\bar{X}$(S.D.)	0.22(0.05)	0.26(0.09)	0.31(0.06)	0.27(0.08)

	Buzzards Bay		Delaware Bay	
	BBTA	BBTB	DBTA	DBTB
Profile distance, m	11	12	14	7
% slope (upper 1/4)	10.9	8.0	7.0	18.5
% slope (lower 3/4)	11.3	10.1	11.0	18.6
Grain size, n	224	46	25	28
Mean $\emptyset$ (S.D. $\emptyset$)	+1.18(0.48)	+1.06(0.37)	+1.01(0.34)	+1.22(0.41)
Mean µm	443	481	495	430
Sorting $\emptyset$ (S.D. $\emptyset$)	1.04(0.26)	1.05(0.25)	0.75(0.12)	0.80(0.11)
Pore space, n	27	4	4	4
% (S.D.)	36.5(0.3)	36.8(0.6)	36.6(0.8)	37.4(0.5)
Roundness, descr.	subangular	subangular	subangular	subangular
Powers scale, $\bar{X}$(S.D.)	0.34(0.09)	0.33(0.10)	0.29(0.07)	0.29(0.08)

RESULTS

Table 1 presents profile and sedimentary data for the eight study transects from four Scottish and American beaches. Among Irvine Bay transects, IBTC is shorter and, in its upper quarter, steeper than IBTA. Means tests indicate that mean grain size (.01>P) and roundness (.001>P) were significantly different between the two transects, but not sorting or pore space (P>.05). On the other hand, the IBTC sediments were quite heterogeneous. Contrasting the upper quarter with the lower three-quarters of the transect yields the following data: mean $\emptyset$ (S.D.), +0.37 (0.40) vs. +1.98 (0.22); mean µm, 774 vs. 253; sorting $\emptyset$ (S.D.), 1.05 (0.20) vs. 0.41 (0.19); % pore space, 35.9 (0.9) vs. 39.9 (0.4); and Powers scale, subrounded 0.41 (0.18) vs. angular to subangular

0.25 (0.06). All data from the upper quarter of IBTC were significantly different from those of IBTA (.001>P), whereas contrasts between the lower three-quarters of IBTC and IBTA indicate that only sorting (.05>P) and roundness (.001>P) differed.

Among Firemore Bay transects, significant differences were observed with respect to mean grain size, sorting and roundness (all .001>P). No differences were observed between sediment data of Buzzards Bay transects, while only sorting and roundness differed (both .05>P) among Delaware Bay transects.

Subcommunity data for Gastrotricha from the two Scottish beaches are given in Table 2, those from the two American beaches in Table 3. Means tests indicate that there were no significant differences between any of the mean annual and mean summer temporal parameter values within Irvine Bay TA, Firemore Bay TA or Buzzards Bay TA (all P>.10). Comparable data were not available from Delaware Bay TA, as this transect was sampled only in the summer season. In this and the next four paragraphs, data are being reported only at the threshold levels of significance (.10>P or .05>P), even if greater levels of significance (.01>P or .001>P) are present.

Comparisons of single parameter values from Irvine Bay TC with means of annual or summer values from IBTA, on the other hand, revealed a host of significant differences (all .05>P). In IBTC vs. IBTA annual, all parameter values differed except percent composition of Xenotrichula beauchampi and those involved with individuals or biomass per 10 sq. cm surface; in IBTC vs. IBTA summer, all parameter values differed except those just listed, number of species, values involved with weighted species co-occurrence, and percent composition of Paraturbanella dohrni.

Single parameter values from Firemore Bay TD, compared with mean annual or summer values from FBTA, showed fewer significant differences. Percent composition differed among Cephalodasys turbanelloides, Turbanella hyalina and Xenotrichula beauchampi (all .05>P) when FBTD was compared with FBTA annual and summer values, and among Other Macrodasyida (.10>P) in FBTD vs. FBTA annual values. Standard deviation variability in co-occurrence differed in FBTD vs. FBTA summer (.05>P), geometric mean numbers (.10>P) and biomass (.05>P) in FBTD vs. FBTA annual values, and geometric mean numbers (.05>P) in FBTD vs. FBTA summer values.

Comparing Buzzards Bay TB with BBTA, annual or summer, resulted in the same set of significant parameter value differences, although the threshold level of significance varies from one to another. In the following, after listing the parameter, the first level of significance will refer to BBTB vs. BBTA annual, and the second to BBTB vs. BBTA summer. Differences were found among weighted mean co-occurrence (both .05>P), covariance (.05>P)(.10>P), percent composition of Pseudostomella roscovita (.05>P)(.10>P), Tetranchyroderma papii (both .05>P), Xenotrichula beauchampi (both .05>P) and Other Chaetonotida (.05>P)(.10>P), and geometric mean numbers (both .05>P) and biomass (both .10>P).

Delaware Bay TB vs. DBTA summer resulted in the fewest significantly different parameters. Included were number of species (.10>P), standard deviation variability in co-occurrence (.10>P), and percent composition of Cephalodasys sp. A, Tetranchyroderma papii and Turbanella cornuta (all .05>P).

TABLE 2 Data and Parameters Related to Biological Accomodation, Structure and Secondary Production Within the Intertidal Gastrotricha Subcommunity. Values for Single Collections at One Site are Spatially Compared With Ranges or Means (With Standard Deviation in Parentheses) of Values for Multi-temporal Collections Taken During the Same Season, or Over an Annual Cycle at a Nearby Site

Transect site - Scotland	Irvine Bay			Firemore Bay		
Season, No. collections	IBTA Annual,6	IBTA Summer,2	IBTC Summer,1	FBTA Annual,6	FBTA Summer,2	FBTD Summer,1
OBSERVED DATA						
No. stations sampled	20-30	20-30	26	24-36	24-36	15
No. samples	70-113	72-113	111	95-153	120-153	79
No. species	11.7(2.0)	13.5(0.7)	18	28.7(2.8)	30.0(2.8)	24
Wx species co-occurring in same sample/individual; S.D.	1.22(0.31) 0.76(0.12)	1.56(0.17) 0.71(0.17)	2.48 1.37	2.76(1.19) 1.88(0.31)	4.11(0.71) 2.28(0.01)	2.53 1.88
Transect covariance, sign	neg	neg	neg	pos	pos	pos
Ratio ΣV's/$V\Sigma$'s	1.004(0.007)	1.004(0.001)	1.041	0.988(0.205)	0.813(0.141)	0.947
ADJUSTED DATA						
No. individuals X 10^3	4.5-52.4	23.2-52.4	6.1	4.8-9.9	6.1-9.9	2.2
H' species diversity (in bits)	0.67(0.20)	0.49(0.01)	3.06	3.12(0.48)	3.46(0.46)	3.25
J' evenness	0.16(0.05)	0.10(0.01)	0.67	0.58(0.09)	0.65(0.09)	0.61
Dispersion: individuals among spp	+0.80(0.06)	+0.86(0.00)	+0.12	+0.18(0.08)	+0.12(0.06)	+0.12
Percent composition:						
Cephalodasys turbanelloides (Boaden)	2.32(0.67)	1.96(1.05)	25.00	0.57(0.35)	0.32(0.09)	3.35
Pleurodasys megasoma Boaden	–	–	–	31.52(13.80)	18.52(2.21)	23.37
Paraturbanella dohrni Remane	3.52(1.21)	3.42(1.97)	15.92	–	–	–
Turbanella cornuta Remane	0.07(0.17)	–	7.29	26.85(6.75)	28.38(11.05)	23.83
Turbanella hyalina Schultze	90.38(3.18)	93.17(0.28)	24.77	0.14(0.15)	0.05(0.07)	1.15
Other Macrodasyida	1.67(2.39)	0.30(0.01)	10.71	14.78(4.40)	19.58(3.49)	4.65
Xenotrichula beauchampi Levi	0.66(0.80)	0.05(0.04)	0.21	2.04(2.00)	0.17(0.25)	17.81
Other Chaetonotida	1.38(0.44)	1.10(0.62)	16.10	24.10(11.57)	32.98(16.46)	25.84
$\bar{G}x$ individuals/10 sq. cm surface (= X $10^3/m^2$)	277(178)	412(220)	184	182(29)	211(3)	106
2.5% confidence limit	123(93)	187(139)	115	136(26)	160(4)	62
97.5% confidence limit	634(348)	929(279)	296	244(31)	278(2)	184
$\bar{G}x$ μg dry wt biomass/ 10 sq. cm surface (= mg/m^2)	133(86)	204(100)	83	78(13)	85(14)	40
2.5% confidence limit	59(44)	91(61)	48	54(12)	62(11)	22
97.5% confidence limit	305(175)	460(136)	145	115(16)	118(19)	72

TABLE 3 Data and Parameters Related to Biological Accomodation, Structure and Secondary Production Within the Intertidal Gastrotricha Subcommunity, as in TABLE 2

	Buzzards Bay			Delaware Bay		
Transect site - United States	BBTA	BBTA	BBTB	DBTA	DBTA	DBTB
Season, No. collections	Annual,18	Summer,5	Summer,1	Annual,0	Summer,2	Summer,1
OBSERVED DATA						
No. stations sampled	5-7	5-7	9	-	8	8
No. samples	81-180	81-180	204	-	53	53
No. species	9.3(1.4)	9.6(0.9)	10	-	10.5(0.7)	16
$\bar{W}x$ species co-occurring in same sample/individual; S.D.	1.78(0.48)	1.96(0.29)	0.69	-	1.21(0.17)	2.31
	0.85(0.30)	0.96(0.20)	1.04	-	0.71(0.04)	1.09
Transect covariance, sign	pos	pos	neg	-	neg	pos
Ratio ΣV's/$V\Sigma$'s	0.957(0.059)	0.997(0.039)	1.093	-	1.101(0.102)	0.972
ADJUSTED DATA						
No. individuals X 10^3	0.9-4.2	1.2-4.2	4.2	-	1.4-3.1	2.1
H' species diversity (in bits)	1.35(0.40)	1.65(0.32)	2.03	-	2.43(0.13)	2.12
J' evenness	0.35(0.10)	0.42(0.08)	0.51	-	0.58(0.03)	0.51
Dispersion: individuals among spp	+0.52(0.15)	+0.41(0.14)	+0.29	-	+0.15(0.03)	+0.37
Percent composition:						
Cephalodasys sp A	8.20(5.45)	11.18(7.20)	1.62	-	-	4.11
Pseudostomella roscovita Swedmark	1.64(2.43)	2.86(4.22)	14.77	-	-	-
Tetranchyroderma papii Gerlach	72.90(10.59)	64.72(10.53)	4.51	-	-	0.51
Turbanella ambronensis Remane	11.44(4.11)	14.10(5.04)	10.67	-	29.48(9.48)	5.09
Turbanella cornuta Remane	0.12(0.28)	0.15(0.21)	-	-	12.81(2.48)	62.14
Other Macrodasyida	0.13(0.22)	0.02(0.04)	-	-	0.38(0.25)	2.25
Heteroxenotrichula variocirrata d'Hondt	-	-	-	-	16.01(3.07)	1.08
Xenotrichula beauchampi Levi	1.52(2.65)	1.69(0.70)	56.22	-	1.14(0.33)	1.67
Xenotrichula punctata Wilke	-	-	-	-	9.16(2.18)	1.64
Aspidiophorus mediterraneus Remane	-	-	-	-	27.26(15.08)	14.37
Other Chaetonotida	4.05(2.52)	5.28(2.82)	12.21	-	3.76(3.42)	7.14
$\bar{G}x$ individuals/10 sq. cm surface	116(65)	135(40)	361	-	93(6)	73
2.5% confidence limit	17(15)	19(10)	191	-	13(5)	10
97.5% confidence limit	946(439)	1099(467)	678	-	724(331)	515
$\bar{G}x$ μg dry wt biomass/10 sq. cm surf.	51(27)	60(20)	108	-	33(3)	25
2.5% confidence limit	7(7)	8(3)	69	-	6(2)	3
97.5% confidence limit	387(181)	486(193)	170	-	188(78)	155

Table 4 shows levels of shared species diversity ($S_{H'}$) among permutations of annual and summer temporal collections and among permutations of summer spatial collections on each of the four study beaches. A two way analysis of variance indicates no difference in mean shared species diversity between transect A sites on the first three beaches, no difference between annual and summer temporal collections and no significant interaction between the two sets of data (all $P > .05$). And a second two way analysis of variance indicates that no difference occurred between the four beaches with respect to summer temporal or spatial collections and no significant interaction term (both $P > .05$) but that shared species diversity of summer temporal collections differed significantly from those of summer spatial collections on the four beaches ($.001 > P$). This means that values for shared species diversity between the two study sites on each beach were consistently lower than temporal variability in shared species diversity within one (and by inference within both) of the two study sites.

DISCUSSION

Differences between the Gastrotricha meiofauna along semi-continuous beaches have been noted by Schmidt and Teuchert (1969), Hummon (1974) and Nixon (1976). But such differences have not been quantified before with respect to subcommunity parameters.

Data presented in this paper indicate significant differences in profile and sedimentary parameters between sites on three of the four study beaches, but fail to show consistency throughout the entire series. And the Delaware Bay transects, which were specifically chosen to assess the effect of a sharp hydrographic gradient on subcommunity parameters, differed no more with respect to sedimentary parameters than either pair of Scottish transects. It is clear that these profile and sedimentary data do not point to a single, simply measured parameter that can account for the complexity of subcommunity alterations along beaches.

No significant differences were found between subcommunity values in annual vs. summer collections from the same transect, for each of the three cases where annual data were available. Yet the percent composition of several species, and generally at least one parameter involving biotic interaction, diversity-evenness or numbers-biomass, differed between transects collected but once and nearby transects collected at least twice during the same season. Differences between transects collected but once and nearby transects collected at least bimonthly through an annual cycle were typically as numerous if not more numerous. Reflecting much the same phenomena, no significant differences were found between permutations of annual and those of seasonal collections from the major collection transects on each beach with respect to shared species diversity, though a highly significant difference occurred between spatially differentiated transect pairs.

Greatest differences were found in the Irvine Bay transects, the principal collection site, IBTA, being associated with large scale domestic as well as industrial pollution, and the second site, IBTC, being some 3 km away on the same continuous beach. On IBTA, the Gastrotricha showed a typical eutrophication syndrome: increased numbers and biomass but with seasonal instability (Hummon 1976), dominance of a single species, <u>Turbanella hyalina</u>, and reduced species diversity-evenness and interspecific interaction. And, although shared species diversity was high among collections from the polluted transect, IBTA shared the least species diversity with its relatively unpolluted

TABLE 4 Shared Species Diversity ($S_{H'}$) Among the Intertidal Gastrotricha Subcommunity. Values Between Single Collections from Nearby Transect Sites are Spatially Compared with Means (Standard Deviations) of Values Within Multitemporal Collections Taken From One of the Two Collections Sites of a Given Locale

Season	Temporal		Spatial
	Annual	Summer	Summer
Transect sites	IBTA	IBTA	IBTA X IBTC
No. Collections (no. perm.)	6(15)	2(1)	2X1(2)
A$\bar{x}$ (S.D.), $S_{H'}$ values	50.6(8.4)	58.7(-)	13.0(0.8)
Transect sites	FBTA	FBTA	FBTA X FBTD
No. Collections (no. perm.)	6(15)	2(1)	2X1(2)
A$\bar{x}$ (S.D.), $S_{H'}$ values	59.2(8.9)	53.8(-)	38.5(1.5)
Transect sites	BBTA	BBTA	BBTA X BBTB
No. Collections (no. perm.)	18(153)	5(10)	5X1(5)
A$\bar{x}$ (S.D.), $S_{H'}$ values	54.3(12.7)	58.3(11.3)	26.5(7.6)
Transect sites	DBTA	DBTA	DBTA X DBTB
No. Collections (no. perm.)	- (-)	2(1)	2X1(2)
A$\bar{x}$ (S.D.), $S_{H'}$ values	- (-)	67.7(-)	33.8(1.1)

partner transect, IBTC, of any pair in the study. In fact during summer IBTC shared nearly as much species diversity with Firemore Bay transects, some 260 km distant along highly discontinuous habitat ($\bar{x}$ = 10.2, SD = 2.2, n = 3), as it did with its partner transect IBTA (see Table 4). By contrast, IBTA shared a mean of only 1.2% species diversity (SD = 0.5, n = 6) with Firemore Bay transects.

Fewest differences were found in the Delaware Bay transects, where the steep hydrographic gradient appeared to have little significant effect other than in shifting species composition. As with sedimentary parameters, there was little consistency shown throughout the series, and there are few satisfactory explanations for the often sizable changes that were noted. We are led to ponder how little we really understand of the structure and dynamics of the Gastrotricha subcommunity, which of course is only a small albeit an important portion of the intertidal marine community.

From data presented, we conclude that movement along a beach of relatively uniform appearance can reveal large alterations in subcommunity structure and that there is less variation at one place (monitored over a period of time) than between that place and its spatial neighbors (sampled on a single occasion). Longshore gradients may be nearly as powerful as those oriented in an upbeach-downbeach direction and may warrant serious attention in future studies of beach meiofauna.

ACKNOWLEDGEMENTS

Aspects of this study were supported by NDEA and NASA Fellowships, through the University of Massachusetts, Amherst, by the Systematics-Ecology Program of the Marine Biological Laboratory, Woods Hole, Massachusetts, by the College of Marine Studies, University of Delaware, and by a NATO Postdoctoral Fellowship through Dr. A. D. McIntyre and the DAFS Marine Laboratory, Aberdeen, Scotland.

We express gratitude to each for making the work possible.

REFERENCES

Green, R.H. 1966. Measurement of non-randomness in spatial distributions. Res. Pop. Ecol. Kyoto Univ. (Ser. 1) 8, 1-7.

Hulings, N.C. and J.S. Gray, Eds. 1971. A manual for the study of meiofauna. Smithsonian Contrib. Zool. 78, 1-83.

Hummon, W.D. 1974. $S_{H'}$: A similarity index based on shared species diversity, used to assess temporal and spatial relations among intertidal marine Gastrotricha. Oecologia 17, 203-220.

Hummon, W.D. 1975. Habitat suitability and the ideal free distribution of Gastrotricha in a cyclic environment. In: H. Barnes (Ed.), Proc. 9th Europ. Mar. Biol. Symp., pp. 495-525. Aberdeen University Press, Aberdeen.

Hummon, W.D. 1976. Seasonal changes in secondary production, faunal similarity and biological accomodation, related to stability among the Gastrotricha of two semi-enclosed Scottish beaches. In: G. Personne (Ed.): Proc. 10th Europ. Mar. Biol. Symp. (In press).

Hummon, W.D., J.W. Fleeger and M.R. Hummon. 1976. Meiofauna-macrofauna interactions: 1. Sand beach meiofauna affected by maturing Limulus eggs. Chesapeake Sci. (In press).

Jumars, P.A. 1975. Methods of measurement of community structure in deep-sea macrobenthos. Mar. Biol. 30, 253-266.

Lloyd, M., J.H. Zar and J.R. Karr. 1968. On the calculation of information-theoretical measures of diversity. Amer. Midl. Natur. 79, 257-272.

Nixon, D.E. 1976. Dynamics of spatial patterns for a gastrotrich in the surface sand of high energy beaches. Intern. Rev. ges. Hydrobiol. (In press).

Pielou, E.C. 1969. An Introduction to Mathematical Ecology. Wiley-Interscience, New York.

Pollock, L.W. and W.D. Hummon. 1971. Cyclic changes in interstitial water content, atmospheric exposure, and temperature in a marine beach. Limol. Oceanogr. 16, 522-535.

Schmidt, P. and G. Teuchert. 1969. Quantitative Untersuchungen zur Ökologie der Gastrotrichen im Gezeiten-Sandstrand der Insel Sylt. Mar. Biol. 4, 4-23.

Wieser, W., J. Ott, F. Schiemer and E. Gnaiger. 1974. An ecophysiological study of some meiofauna species inhabiting a sandy beach at Bermuda. Mar. Biol. 26, 235-248.

HABITAT AREA, COLONIZATION, AND DEVELOPMENT OF EPIBENTHIC COMMUNITY STRUCTURE

J. B. C. Jackson

Department of Earth and Planetary Sciences, The Johns Hopkins University, Baltimore, Maryland 21218, U.S.A., and Discovery Bay Marine Laboratory, Box 35, Discovery Bay, Jamaica, West Indies

ABSTRACT

Experiments demonstrate significant variations in recruitment rates, faunal composition and diversity of artificial cryptic reef communities as a function of substratum area. Substrata smaller than about 200 square cms are potentially important refuges from spatial competition for cryptic reef organisms. Habitat selection, apparently related to differences in water movements around small and large substrata, is interpreted as a mechanism for location of refuges from spatial competition. Environmental or regional comparisons of fouling studies based upon data from different-sized substrata are subject to serious misinterpretation.

INTRODUCTION

Most of the hard substratum surface area of coral reefs lies within crevices and other cavities of the reef framework. One of the commonest of these cryptic environments on Caribbean fore reef slopes consists of the skeletal undersurfaces of foliaceous corals such as Agaricia, Montastrea, and Mycetophyllia (Goreau and Goreau 1973). These corals support a highly diverse encrusting biota of some 300 species (Jackson in press, unpubl. data). More than 50 species commonly occur under a single coral. The presence or absence of individual species under any particular coral varies enormously. However, groups of morphologically and/or taxonomically similar species commonly exhibit a simple predictable distribution pattern as a function of coral size.

Foliaceous corals project over the reef surface from limited zones of attachment to the solid reef framework. Points of attachment are usually surrounded by a veneer of sediments unsuitable for habitation by epibenthic organisms. Thus foliaceous corals form a unique environment of physically discontinuous, discrete substrata for colonization by cryptic organisms. The abundance, spatial distribution, and size-frequency distribution of these substrata often varies considerably from reef to reef.

The role of habitat scale as a determinant of community structure and diversity has been a subject of intense ecological study. In particular, island biogeography theory (MacArthur and Wilson 1967) has met with considerable success in the prediction of distribution and abundance patterns of numerous terrestrial faunas (Simberloff 1974). This theory, however, is based on the assumption of exploitative competitive interactions (sensu Miller 1967) whereas attached epibenthic organisms interact so far as is known via interference competition for space (Bryan 1973 ; Buss 1976; Chiba and Kato 1966; Connell 1961a, b; Gordon 1972; Jackson in press; Jackson and Buss 1975; Kato et al. 1963, 1967; Lang 1973; Rutzler 1970; Sara 1970; Stebbing 1972, 1973). Models based on the assumption of exploitatative competition are therefore not applicable to such systems (Gilpin 1975) and, not surprisingly, have contributed little to the understanding of epibenthic communities (e.g., Schoener 1974).

Here I present preliminary results of a field experiment designed to evaluate the importance of habitat area to the development and maintence of community structure and diversity in subtidal, marine, epibenthic communities. A more general discussion of this topic will be presented elsewhere (Jackson and Buss ms).

MATERIALS AND METHODS

The development of epibenthic communities on 6 different sizes of asbestos-cement panels has been followed for 1 year. The experiment is being conducted at a depth of 40 m in a sand channel 25 m west of Buoy Reef (Kinzie 1973) at Discovery Bay, Jamaica. Dimensions and areas of the panels are given in Table 1. There are 6 replicates of each panel size. English units were used to allow direct comparison with previous fouling studies in the same area.

TABLE 1 Dimensions, Area, and Number of Random Points Censused for Experimental Panels

Panel Dimensions (inches)	Panel Area		Number of Random Points
	in^2	cm^2	
3/4x3/4	0.6	3.6	5
1½x1½	2.3	14.4	10
3x3	9.0	57.8	20
6x6	36.0	231	40
12x12	144	924	80
24x24	576	3820	160

These panels are horizontally suspended approximately 0.6 m above the bottom from a 6x6x1½ m steel frame. Panels are arranged on the frame in a 6x6 Latin Square array. The entire frame is covered by a 10x10 m black plastic sheet which is staked to the bottom so as to allow approximately 0.3 m open space around the enclosed area. Light levels are thus uniformly low under all substrata regardless of their size. Although water movements are reduced by the plastic cover, they almost certainly vary around the panels as a function of their size, and to this extent substratum size is unavoidably not the only variable in the design (nor is it in any natural cryptic environment).

The panels were attached to the frame in late July, 1975. The first panel census was in January, 1976; the second in August, 1976. For safety reasons, the plastic sheet was removed from the frame for 3 weeks during each census. Panels are removed from the frame by divers and brought to the laboratory for analysis. After photographing and censusing, always in seawater, the panels are returned to the frame within 24 hours of collection. The great majority of encrusting organisms have withstood this treatment surprisingly well.

After each census, ascidians and sponges appeared to be pumping normally, serpulids extended their fans, etc.

For each panel size a fixed number of random points are censused under the microscope to obtain estimates of species abundances and unoccupied space (Table 1). Definite identifications of many species would require killing at least part of the organism to obtain setae or spicules or prepare histological sections. Such procedures are obviously undesirable for this experiment and I must therefore rely on more subjective descriptions. Although I cannot name some 50% of the species (mostly sponges), I am confident that I can accurately distinguish among them on the basis of tube morphology, ornamentation, surface texture, color, etc. Photographic records (7.5x7.5 cm fields) have proved valuable for taxonomic comparisons between censuses.

RESULTS

Recruitment and Percent Cover

Rates of occupation of space (larval colonization and animal growth) differ significantly ($P<0.001$) with panel size (Table 2, Fig. 1). This is most evident for the January census. Then, 69% or more of the surface area of 6 inch or smaller panels was covered whereas only a third or less of the area of the larger panels had been colonized and overgrown. These differences were much reduced after 12 months submergence ($P=0.11$) but there was still a tendency for lower cover of the larger panels. Increase in percent cover over this interval occurred by both new colonization and growth of previously established organisms. One 6 inch panel was entirely overgrown (by census estimates) after 12 months submergence. Twelve of the 18 smaller panels, but none of the larger panels were similarly covered.

There are also variations in the rates of space occupation by major faunal groups on different-sized panels (Table 2). The 2 most abundant groups are sheet-like encrusting sponges and serpulids. After 12 months submergence these sponges exhibited significant differences in percent cover ($P=0.01$) with panel size, with maximum cover on 1½ inch and 3 inch panels. In contrast, serpulids were most abundant on the smallest and larger panels ($P=0.09$).

Mortality

Comparison of census data and photographs for individual species indicates low mortality between the January and August censuses except for that due to competitive overgrowth interactions. Overgrowths of 1 species by another were of frequent occurrence on virtually every panel in August, and in January on those panels with high percent cover. Microscopic examination commonly revealed dead skeletal structures of previously recorded animals (e.g., ectoproct zooid outlines, worm tubes, etc.) beneath the live tissue of the superior forms (usually sheet-like sponges and colonial ascidians). The abundance of overgrowths clearly demonstrates that interference competition for space is of major importance in structuring these experimental communities. This conclusion is further suported by the inverse patterns of abundance of serpulids and sheet-like sponges in the August census (Table 2). Predation is apparently uncommon on the panels except for high mortality of bivalves (Echinochama, Spondylus, various dimyids) due to drilling gastropods. Occasionally, fresh gastropod or polychaete feeding trails were observed on

TABLE 2 MEANS AND STANDARD ERROR FOR PERCENT COVER AND NUMBERS OF SPECIES ON SETTLEMENT PANELS

Panel Size (inches)	Percent Cover						Number of Species			
	Total Fauna		Serpulids		Sheet Sponges		Total Fauna			
	Jan.	Aug.	Jan.	Aug.	Jan.	Aug.	Jan.	Aug.		
3/4	80.0±10.3	90.0±6.8	26.7±13.3	36.7± 8.0	36.7±17.5	16.7±10.9	1.7±0.2	2.7±0.5		
1½	80.0±10.7	96.7±3.3	21.7±12.2	15.0±11.5	56.7±18.7	81.7±12.2	2.8±0.4	2.2±0.3		
3	70.8± 4.4	93.3±3.3	40.0± 6.8	19.2± 6.1	18.3± 7.4	50.8±16.2	4.8±0.3	4.7±0.9		
6	69.2± 4.6	90.4±2.5	44.6±11.9	44.6± 8.3	22.5±11.5	36.3±10.3	6.5±1.2	10.2±0.7		
12	33.1± 4.7	88.6±5.5	12.9± 6.2	44.6±11.1	12.5± 4.2	33.4±10.5	9.3±1.3	16.0±2.4		
24	22.1± 0.9	75.7±7.1	5.9± 2.4	36.5± 6.1	7.6± 1.7	29.4± 9.6	13.0±1.6	23.7±1.7		
F Ratio*	13.4	2.00	2.43	2.10	2.29	3.68	18.2	42.3		
P	<0.001	0.11	0.06	0.09	0.07	0.01	<0.001	<0.001		

* 1 way ANOVA

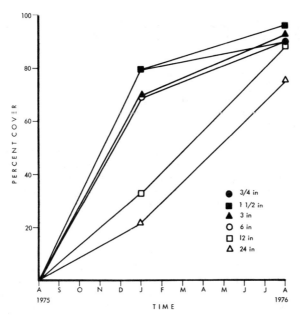

Fig. 1. Temporal variations in percent faunal cover for 6 panel sizes. Values are means of 6 replicates.

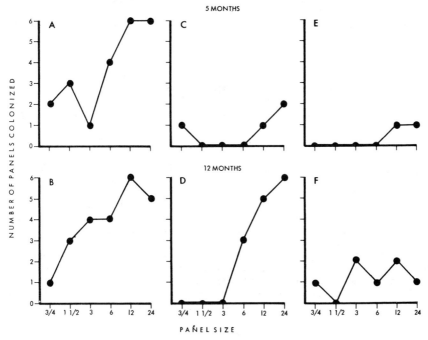

Fig. 2. Frequency of colonization of 3 species as a function of panel size at the January (5 month) and August (12 month) census periods. A,B Unidentified sheet-like demosponge; C,D bivalve Echinochama arcinella; E,F colonial ascidian Diplosoma macdonaldi.

sponges, along with the predators responsible. These are easily recognized, and rare, as they are in natural cryptic reef environments (Jackson in press; Jackson and Buss 1975).

Species Diversity and Distributions

Plots of the number of species versus number of points censused for individual panels tend to level off well before the last points are counted. Thus the panels appear to have been censused adequately for meaningful estimates of diversity (Pielou 1969).

The number of panels of 1 size colonized by individuals of any 1 species which survive long enough to be censused increases with the size of the panels in most cases (e.g., the sponge and Echinochama arcinella in Fig. 2). This was not true for the colonial ascidian Diplosoma macdonaldi, which occurred as frequently on smaller as larger panels after 12 months submergence (Fig. 2). Most species were quite rare, however, so that for any panel size, the majority of species present occurred on a single panel only.

The number of species per panel increases with substratum area in a decidedly non-linear fashion (Table 2, Fig. 3). Panels 3 inches or smaller attained a maximum number of species after 5 months submergence (Fig. 4). In contrast, the number of species on 6 inch and larger panels increased greatly between the censuses and will presumedly increase even more.

Two panels (1 each of 3/4 and 3 inches) were entirely overgrown by D. macdonaldi at the August census. One 1½ inch panel was entirely overgrown by a thin sheet-like sponge. None of the panels 6 inches or larger was close to being overgrown by a single species after 12 months submergence.

DISCUSSION

The results demonstrate that substratum area is a fundamental factor influencing the development and maintenance of community structure and diversity in cryptic reef environments. Clarification of many questions, in particular the comparative persistence stability of faunas on large versus small substrata, await further data. Nevertheless, 4 important consequences of substratum area effects are already apparent.

Importance of Small Substrata as Refuges from Spatial Competition

In order to survive, epibenthic organisms must withstand the potentially deleterious activities of their neighbors. On large substrata, any species is likely to come into contact with numerous other species, many of which may be superior spatial competitors. Overgrowth or other forms of interference are highly probable, and the chances of survival relatively small. In order to survive in evolutionary time, species inhabiting large substrata must either be good space competitors or 'weed' species capable of settling and reaching reproductive maturity on recently cleared substrata before encroachment and overgrowth by competitively superior forms. This is the situation outlined by Levin and Paine (1974) in their analysis of the importance of disturbance-generated patches to rocky intertidal community structure.

The experiments reveal a different situation for small substrata. Very few species settle, and the probability of deleterious competitive interactions is decreased. For species with indeterminate growth, the chances of

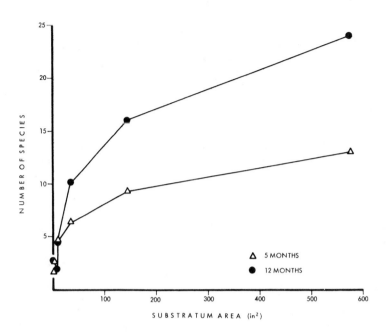

Fig. 3. Number of species per panel as a function of panel size at the January (5 month) and August (12 month) census periods. Values are means of 6 replicates.

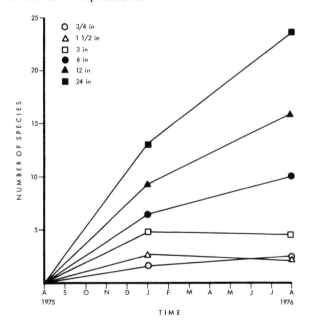

Fig. 4. Temporal variations in numbers of species per panel for 6 panel sizes. Values are means of 6 replicates.

occupying the entire area of small, discrete substrata before other species can become established are therefore relatively high. This has already occurred on 3 of the smaller substrata after 12 months submergence. Once the entire substratum is overgrown, the possibility of larval recruitment by other organisms is further reduced (Jackson in press). So long as a species produces large numbers of larvae, the presence of small substrata in the environment should allow the species to survive regardless of its interference competitive ability. If this is true, variations in the size-frequency distributions of substrata in an environment, and particularly the abundance of small substrata (corals, shells, boulders), are fundamental to the maintenance of epibenthic diversity patterns. This refuge effect should be especially important in low disturbance environments such as the cryptic reef system where the probability of patch formation on large substrata is relatively small.

Unpredictability of Species Distributions

More than half of the foliaceous corals on the Jamaican reefs are smaller than the 12 inch panels. The unpredictability of species distributions under corals, therefore, is clearly related to decreasing probabilities of colonization by any species with decreasing substratum size (Fig. 2). In contrast, the development of significant differences in percent cover of serpulids and sheet-like sponges on different-sized panels occurred after the January census (Table 2). This suggests that the predictability of faunal groups under different-sized corals are due more to post-settlement interactions than to colonization processes.

Flow Effects

The experimental design precludes compositional differences in substrata or light stimuli as factors responsible for variations in recruitment rates onto different-sized panels. Presumably larvae can detect differences in water movements around small versus large substrata. The adaptive significance of such habitat selection is readily apparent as a mechanism for location of refuges from spatial competition (Buss in press). The preferential, often species-specific, recruitment of numerous species of ectoprocts, sponges, and hydroids onto mollusc shells (e.g., Gordon 1972) may be of similar significance.

Comparative Fouling Studies

Experimental studies of fouling communities throughout the world employ a vast size range of substrata. The present results clearly demonstrate the problems inherent in comparison of data from different localities unless the size, shape, and composition of experimental substrata are similar. Two-fold differences in recruitment rates, for example, may be as much a consequence of substratum size as variations in larval densities or physical conditions between localities.

ACKNOWLEDGEMENTS

The impetus for these experiments came from a seminar discussion of MacArthur's book with C. Schnell and L. W. Buss. L. W. Buss has contributed fundamentally to the further development of the ideas presented in this paper. R. Karlson, B. A. and J. L. Menge, J. R. Sutherland, and S. A. Woodin provided much useful criticism. B. D. Keller and N. Knowlton assisted in all aspects of the diving work. Numerous others at the Discovery Bay Laboratory have provided further field assistance. This work was supported by NSF Grant DES-72-01559-401. To all I am grateful.

REFERENCES

Bryan, P. G. 1973 Growth rate, toxicity, and distribution of the encrusting sponge Terpios sp: (Hadromeda: Suberitidae) in Guam, Mariana Islands. Micronesica 9, 237-242.

Buss, L. W. 1976 Better living through chemistry: the relationship between allelopathy and competitive networks. In F. W. Harrison and R. R. Cowden, eds. Aspects of Sponge Biology Academic Press, New York.

Buss, L. W. in press On the optimal exploitation of spatial refuges. In B. Rosen and G. P. Larwood, eds. Biology and Systematics of Colonial Animals Academic Press, London.

Chiba, Y. and M. Kato 1966 Interspecific relation in the colony formation among Bougainvillia sp. and Cladonema radiatum (Hydrozoa: coelenterata). Sci. Rep. Tohoku U., Ser. 4 (Biol.) 32, 201-206.

Connell, J. H. 1961a The influence of interspecific competition and other factors on the distribution of the barnacle Chthamalus stellatus. Ecol. 42, 710-723.

_____ 1961b The effects of competition, predation by Thais lapillus, and other factors on natural populations of the barnacle Balanus balanoides. Ecol. Monogr. 31, 61-104.

Gilpin, M. E. 1975 Limit cycles in competition communities. Amer. Natur. 109, 51-60.

Goreau, T. F. and N. I. Goreau 1973 The ecology of Jamaican coral reefs. II. Geomorphology, zonation and sedimentary phases. Bull. Mar. Sci. 23, 399-464.

Gordon, D. P. 1972 Biological relationships of an intertidal bryozoan population. J. Nat. Hist. 6, 503-514.

Jackson, J. B. C. in press Competition on marine hard substrata: the adaptive significance of solitary and colonial strategies. Amer. Natur.

Jackson, J. B. C. and L. W. Buss 1975 Allelopathy and spatial competition among coral reef invertebrates. Proc. Nat. Acad. Sci. U.S.A. 72, 5160-5163.

Kato, M., E. Hirai, and Y. Kakinuma 1963 Further experiments on the interspecific relation in the colony formation among some hydrozoan species. Sci. Rep. Tohoku U., Ser. 4 (Biol.) 29, 317-325.

_____ 1967 Experiments on the coaction among hydrozoan species in colony formation. Sci. Rep. Tohoku U., Ser. 4 (Biol.) 33, 359-373.

Kinzie, R. A. 1973 The zonation of West Indian gorgonians. Bull Mar. Sci. 23, 93-155.

Lang, J. C. 1973 Interspecific aggression by scleractinian corals. 2. Why the race is not only to the swift. Bull. Mar. Sci. 23, 260-279.

Levin, S. A. and R. T. Paine 1974 Disturbance, patch formation, and community structure. Proc. Nat. Acad. Sci. U.S.A. 71, 2744-2747.

MacArthur, R. H. and E. O. Wilson 1967 The Theory of Island Biogeography Princeton Univ. Press, Princeton.

Miller, R. S. 1967 Pattern and process in competition. Adv. Ecol. Res. 4, 1-74.

Pielou, E. C. 1969 An Introduction to Mathematical Ecology Wiley-Interscience, New York.

Rutzler, K. 1970 Spatial competition among Porifera: solution by epizooism. Oecol. 5, 85-95.

Sara, M. 1970 Competition and cooperation in sponge populations. Symp. Zool. Soc. Lond. 25, 273-284.

Schoener, A. 1974 Colonization curves for palnar islands. Ecol. 55, 818-827.

Simberloff, D. S. 1974 Equilibrium theory of island biogeography and ecology. An. Rev. Ecol. Syst. 5, 161-182.

Stebbing, A. R. D. 1972 Some observations on colony overgrowth and spatial competition. In G. P. Larwood, ed. Living and Fossil Bryozoa Academic Press, London.

_____ 1973 Competition for space between epiphytes of Fucus serratus L. J. Mar. Biol. Ass. U.K. 53, 247-261.

QUANTITATIVE SURVEY OF HARD BOTTOM COMMUNITIES IN A BALTIC ARCHIPELAGO

Ann-Mari Jansson and Nils Kautsky

Department of Zoology and the Askö Laboratory, University of Stockholm, Sweden

ABSTRACT

With the aim of modelling the hard bottom communities of the Askö area in the northern Baltic proper, a stratified random sampling program was designed to quantify the macroscopic fauna and flora of the system over an area of 160 km^2.

Sampling was made along 32 randomly selected transects. Composition, density and vertical extension of the vegetation belts were mapped and characterized by SCUBA divers. Sampling frames were placed at random along the marked line in a stratified manner, ensuring that at least 1-2 quantitative samples were obtained from each defined belt. By relating the collected data to physical features such as exposure, depth and bottom type, it was possible to calculate the abundance and dry weight biomass distribution of the main macroflora and macrofauna components for the inner and outer parts of the investigation area.

Maximum plant biomass, 610 g · m^{-2}, was found at a depth of 1-2 m in the inner area. In the outer area a maximum of 383 g · m^{-2} was reached at 3-4.5 m depth. _Fucus vesiculosus_ constituted 33% of the plant biomass. The zone of annual filamentous algae (_Pilayella littoralis_, _Ceramium tenuicorne_, _Cladophora glomerata_ etc.) situated above the _Fucus_ belt had an average biomass of 78 g · m^{-2}. Below the _Fucus_ vegetation the plant biomass was dominated by the red algae _Furcellaria fastigiata_, _Phyllophora truncata_ and _Rhodomela confervoides_.

Maximum animal biomass, ca 490 g · m^{-2}, was found in the 3-4.5 m depth zone. _Mytilus edulis_ was the most important species constituting 90% of the animal biomass (incl. shells).

The total figures for standing crops and standing stocks of the investigated area amount to 2163 tons of plants and 8618 tons of animals.

Using available data from the area and from literature for turnover times and production of different species, estimates of the total production of the hard bottom communities are presented.

INTRODUCTION

This survey was carried out with the purpose of quantifying in terms of abundance, biomass and production, the relative importance of the hard bottom benthos within the primary research area of the Askö project, "Dynamics and energy flow in a Baltic ecosystem" (Jansson 1972). The pelagic (Hobro & Nyqvist 1973) and the soft bottom (Ankar & Elmgren, 1976) systems have already been quantified. This study now completes the knowledge of the major subsystems of the Askö project by giving a first rough estimate of biomass and production magnitudes. A more precise presentation will follow after the huge amount of collected data has been further analysed.

Wallentinus (1976) has given a detailed description of species composition of the macroflora in the Askö area. Zonation and structure of phytobenthos in the Baltic has earlier been studied by Waern (1952), Schwenke (1966, 1969) and Ravanko (1972).

MATERIAL AND METHODS

The area studied (Fig. 1) comprises 160 km^2 in a comparatively unpolluted archipelago of the Swedish Baltic coast. It has been divided into an inner, sheltered and an outer more exposed part which are treated separately in the analyses.

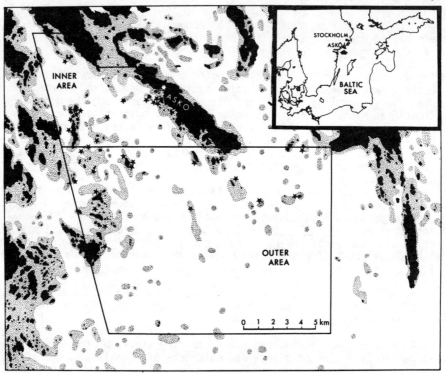

Fig. 1. Primary research area of the Askö Laboratory. Bottoms down to 6 m spotted. Stars indicate position of sampling sites.

The salinity is quite stable around 6-6.5‰. The morphology of the sea bottom is irregular with a mean depth of 25-30 m. Data on the areal distribution of different depth intervals and bottom types shown in Fig. 2 have been compiled from detailed sea charts made by hand sounding. It is most probable that the hard substrates at greater depths consist mainly of hard glacial clay. A more realistic hardbottom distribution below 30 m depth is indicated in the diagram. This estimated value has been used for the later calculations. Water temperature varies from below 0°C in winter, with an average of 1-3 months of ice cover, to a maximum of about 20°C in summer. This investigation was concentrated over the summer period. During this season, changes in plant biomass and composition are comparatively small and changes between sampling occasions within and between the sampling years were thus minimized.

As a basis for the selection of sampling localities, detailed sea charts
(1:6.750) were used. The 3 m depth curve on the charts was divided into 1 mm
intervals which gave about 15.000 consecutively numbered points with an equi-
distance of about 7 m in nature. From these points 32 sites were randomly chosen
and visited during the middle of June till the end of August in 1974 and 1975.
At each site sampling was made along a line transect at right angles to the
depth curves from the shore line down to the vegetation free soft bottom.

Fig. 2 shows the number of transects covering each depth interval. By choosing
the 3 m depth curve instead of the actual shore line in the randomization over-
representation of the shallow semi-enclosed soft bottom bays was avoided. These
bays could for many reasons be regarded as separate systems and are now subjected to special investigations.

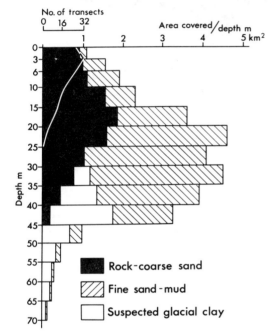

Fig. 2. Areal distribution of bottom
types within different depth intervals
in the Askö area. White line indicates
number of transects reaching each depth.

Of 32 transects commencing at 3 m,
16 reached to a depth of more than
12 m. For the large areas of hard
bottoms present below 25 m depth,
a complementary sampling program
has to be designed to get a better
estimation of the total Mytilus
biomass.

Sampling was made by SCUBA diving
along a nylon line marked with
coloured tape at every half meter.
At least two divers were swimming
along the transect registering
depth, distance from shore, species
composition and abundance of macro
constituents by estimating the
percentage of bottom surface
covered. The divers also defined
the limits of the different homo-
genous zones and belts. All observations were taken down underwater
with lead pencil on white roughened
plastic sheets. After surfacing
the divers compared and discussed
their observations and determined
the sampling positions by randomization, ensuring that at least
1-2 samples were obtained from
each defined belt. Sampling frames
15 x 15 cm, 20 x 20 cm or 50 x 50 cm
were used separately or in combinations depending on structure and density of
the community.

The samples were scraped into net bags and brought to the laboratory for analyses. Most samples were subsampled according to Haage (1975) before further
treatment. Plants and macrofauna (>1 mm) were sorted to species, counted and
dried at 60°C to constant weight. Dry weight biomasses have been roughly recalculated into ashfree dry weight using conversion factors of 0.8 for plants,
0.21 for Mytilus, 0.7 for Amphipoda (Lappalainen & Kangas. 1975) and 0.3 for
"other" animals since this group mainly consists of molluscs. The essential
parts of the used sampling techniques have recently been included in the Baltic

Marine Biologists' recommendations for phytobenthos studies. (Dybern et al. 1976).

RESULTS AND DISCUSSION

Vertical variation of plant and animal biomasses for the inner and outer areas is shown in Figs. 3 and 4. The values express a mean of all observations (totally 289 samples) within the respective depth interval. The distribution of samples within the intervals has not been accounted for.

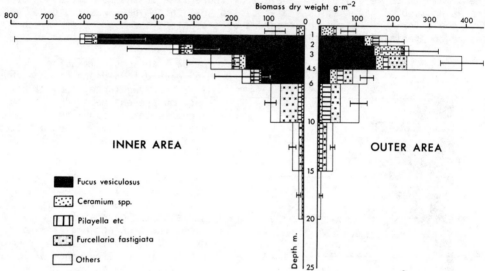

Fig. 3. Vertical distribution of plant biomass (g dryweight $\cdot$ m^{-2}) for outer and inner area. Standard error of mean indicated by horizontal lines.

Maximum plant biomass, 610 g $\cdot$ m^{-2}, was found between 1-2 m depth in the inner area. In the outer area a lower maximum of 383 g $\cdot$ m^{-2} was found at 3-4.5 m depth.

At 0-1 m depth, both subareas had an average biomass of 78 g $\cdot$ m^{-2}. This zone is normally completely cleaned by ice during winter and thus mainly consists of annual filamentous algae, which show a marked seasonal succession e.g. Cladophora glomerata (L.) Kütz., constituting the main part of "Others" within this depth interval, Ceramium tenuicorne (Kütz).) Waern (perennial in deeper parts) and Pilayella littoralis (L.) Kjellm. The annual peak biomass of the filamentous algae occurs in June and was probably caught in the early part of this investigation.

Below this belt a vegetation of perennial Fucus vesiculosus L. with filamentous epiphytes dominates, having a pronounced maximum between 1-2 m depth in sheltered areas. In exposed areas the maximum lies at 2-4 m depth. Generally, though, as Fucus is sensitive to strong wave action, it is replaced by Ceramium tenuicorne between 1-3 m and Furcellaria fastigiata (L.) Lamour. between 3-6 m depth at exposed localities.

Below the Fucus vegetation the plant biomass is dominated by the red algae Furcellaria fastigiata, Phyllophora truncata (Pallas) Newroth & Taylor and

Rhodomela confervoides (Huds.) Silva with the percentage of Phyllophora and Rhodomela increasing with depth.

Rhodomela and Phyllophora make up the main part of "Others" below 6 m. Pilayella etc. (= Pilayella and other filamentous brown algae) and Ceramium spp. occurring below 10 m depth to a large extent consist of loose lying algae. The fraction "Others" between 3 and 4.5 m is mainly made up by Chorda filum (L.) Stackh. and vascular plants especially Potamogeton pectinatus L., Ruppia spiralis L. and Zostera marina L. due to a higher percentage of sandy bottom samples obtained in this depth interval.

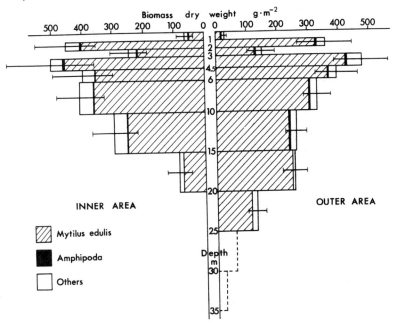

Fig. 4. Vertical distribution of animal biomass (g dry weight · m^{-2}) for outer and inner areas. Values below 25 m are estimated. Standard error of mean indicated by horizontal lines.

The vertical distribution of animal biomass (Fig. 4) shows a similar pattern in both areas with a maximum at 3-4.5 m depth, reaching a value including shells of 505 g · m^{-2} in the inner and 477 g · m^{-2} in the outer area. From this peak the biomass decreases with depth. Below 15 m the outer area shows higher biomass values than the inner area, probably because stronger currents keep the hardbottoms free from sediment and thus allow better conditions for Mytilus edulis L. which completely dominates the animal biomass at all depths.

No samples were obtained deeper than 25 m in this investigation but observations of Mytilus beds at greater depths made by divers show substantial biomasses also further down as indicated in Fig. 2.

The pronounced minimum at 2-3 m depth occurring in both areas, cannot yet be fully explained. The fact that Mytilus has essentially the same demands of bottom type, exposure etc. as Fucus and uses this alga as a substrate and shelter, however, indicates a positive correlation between these two species down to 3 m. Below this depth the distribution of Mytilus is not dependent on Fucus.

In the uppermost zone, 0-4.5 m depth, Amphipoda (mainly Gammaridae) had an average biomass of 3.9 g·m^{-2}. The fraction "Others" is dominated by Theodoxus fluviatilis L. down to 2 m, and Hydrobia spp. from 2-10 m together with Macoma baltica L. from 6 m on downwards.

Total biomass and production

Since every step in the sampling was randomized, the total biomass can be calculated by multiplying the biomass·m^{-2} (Figs. 3 and 4) by the hard bottom area within each depth interval (Fig. 2). Table 1 presents values of standing crops and standing stocks within each depth interval along with total biomass values for the area. The total standing crop amounted to 2163 tons dryweight (1730 tons ashfree dry wt.), with Fucus being the most important species constituting 33.2% of the biomass.

TABLE 1. Total calculated biomass (tons dryweight incl. shells) within different depth intervals and the total investigation area for Pilayella and other filamentous brown algae, Fucus vesiculosus, Ceramium spp., Furcellaria fastigiata, Mytilus edulis and Amphipoda. Values marked ★ are estimated. Total annual production is given for plants and animals.

Depth interval m	Pilayella etc.	Fucus ves.	Ceramium	Furcellaria	Others	Total Plants	Mytilus edulis	Amphipoda	Others	Total Animals	N
0-1	5.7	4.1	34.2	0	49.9	93.9	33.6	5.2	8.8	47.6	37
1-2	18.3	253.8	21.1	0.3	20.0	313.5	347.8	5.2	34.0	387.0	36
2-3	6.1	170.8	47.3	4.2	10.2	238.6	133.6	2.9	17.7	154.4	22
3-4.5	16.6	197.4	23.9	47.3	143.3	428.5	547.2	5.7	63.5	616.4	44
4.5-6	20.2	77.7	17.2	30.4	36.5	182.0	457.6	2.2	39.5	499.3	39
6-10	96.4	14.7	34.9	160.5	215.3	521.8	1602.1	9.3	145.9	1757.3	58
10-15	54.1	0	31.5	76.3	134.7	296.6	1906.0	16.1	207.5	2129.6	31
15-20	8.4	0	13.7	4.1	50.6	76.8	1835.6	1.1	66.0	1902.7	15
20-25	4.1	0	1.4	2.2	3.1	10.8	662.3	0.7	110.5	773.5	9
25-30	0	0	0	0	0	0	221.1★	0	45.3★	266.4★	0
30-35	0	0	0	0	0	0	49.5★	0	34.3★	83.8★	0
Total	229.9	718.5	225.2	325.3	663.6	2162.5	7796.6	48.4	773.0	8618.0	289
%	10.6	33.2	10.4	15.1	30.7	100	90.5	0.6	8.9	100	
Ash-free drw						1730	1637	34	232	1903	
Production tons ash-free dry wt.·year^{-1}						1500				1900	

The ratio of annual net production to total measured biomass was assumed to be 3 for filamentous brown and green algae, 1 for Ceramium and Chorda, 0.7 for vascular plants, 0.5 for Fucus and Rhodomela and 0.3 for Furcellaria and Phyllophora. This gives a total net production of 1500 tons ashfree dry wt.·year^{-1} in the area. Pilayella and filamentous brown algae being most important with 500 tons. These values are to be regarded as minimum estimates as fragmentation and exports of spores and dissolved organic matter are not accounted for.

Annual phytoplankton production is about 36 800 tons ashfree dry wt. (recalculated from Hobro and Nyqvist 1973). The annual macroscopic plant production of hard bottoms thus constitutes ca 4% of the total primary production. Guterstam (1976) estimated the total Fucus production in the 1-2 m depth interval to amount to 2.3% of the daily phytoplankton production in September.

Total standing stock amounted to 8618 tons dry wt. incl. shells (1903 tons ashfree dry wt.). Mytilus edulis was the most important species constituting 90.5% of the animal biomass or 86% if calculated as ashfree dry weight. Animals were assumed to have an average P/B of 1 which gives an annual production of about 1900 tons ashfree dry wt.·year^{-1} for hardbottoms. The corresponding production of softbottom macrofauna is at least 1220 tons (recalculated from Ankar and Elmgren 1976).

CONCLUSIONS

This investigation showed that:

1. The methods used for sampling and randomization were very well suited for obtaining good quantitative information of hardbottom communities in a complex archipelago area.

2. The survey was representative of hardbottom communities down to 25 m.

3. The total plant biomass was dominated by Fucus vesiculosus while filamentous brown algae were probably the most important contributors to annual production.

4. Among animals Mytilus edulis completely dominated both biomass and production.

ACKNOWLEDGEMENTS

It would not have been possible to complete this study without the enthusiastic help of many students, scientists and technicians at the Askö Laboratory. We are especially grateful to Hans Kautsky for leading the diving team and coordinating much of the sorting work. We also wish to thank Inger Wallentinus who took active part in the investigation as our botanical expert, Bibbi Mayrhofer for drawing the illustrations and Bibbi Berggren for typing the manuscript.

REFERENCES

Ankar, S. and Elmgren, R. 1976. The Benthic Macro- and Meiofauna of the Askö-Landsort Area (Northern Baltic Proper). A Stratified Random Sampling Survey. Contr. Askö Lab. 11, 1-115.

Dybern, B.I. Ackefors, H. and Elmgren, R. Editors 1976. Recommendations on Methods for Marine Biological Studies in the Baltic Sea. The Baltic Marine Biologists Publ. No. 1, 77-98.

Guterstam, B. 1976. An In Situ Study of the Primary Production and the Community Metabolism of a Baltic Fucus vesiculosus L. Community. 11th European symposium on marine biology, Galway (Ireland) 1976. (in press).

Haage, P. 1975. Quantitative Investigations of the Baltic Fucus Belt Macrofauna 2. Quantitative Seasonal Fluctuations. Contr. Askö Lab. 9, 1-88.

Hobro, R. and Nyqvist, B. 1973. Studies of Primary Production and Relevant Environmental Factors in the Northern Baltic Proper. (Abstract). Thalassia Jugosl. 7(1), 115-116.

Jansson, B.-O. 1972. Ecosystem Approach to the Baltic Problem. Bull. Ecol. Res. Comm. NFR 16, 1-82.

Ravanko, O. 1972. The Physiognomy and Structure of the Benthic Macrophyte Communities on Rocky Shores in the Southwestern Archipelago of Finland (Seili Islands). Nowa Hedwigia 23, 363-403.

Schwenke, H. 1966. Untersuchungen zur Marinen Vegetationskunde. I. Über den Aufbau der Marinen Benthosvegetation im Westteil der Kieler Bucht (Westliche Ostsee) Kieler Meeresforsch. 22:2, 163-170.

Schwenke, H. 1969. Meeresbotanische Untersuchungen in der Westlichen Ostsee als Beitrag zu einer marinen Vegetationskunde. Int. Revue ges. Hydrobiol. 54:1, 35-94.

Waern, M. 1952. Rocky Shore Algae in the Öregrund Archipelago. Acta Phytogeogr. Suec. 30, 1-298.

Wallentinus, I. 1976. Environmental Influences on Benthic Macrovegetation in the Trosa-Askö Area I. Hydrographical and Chemical Parameters and the Macrophytic Communities. Contr. Askö Lab. (in press).

ASPECTS OF THE ECOLOGY OF *SARGASSUM MUTICUM* (YENDO) FENSHOLT, IN THE SOLENT REGION OF THE BRITISH ISLES. I. THE GROWTH CYCLE AND EPIPHYTES

Nicholas A. Jephson and Peter W. G. Gray

Portsmouth Polytechnic, Department of Biological Sciences, Marine Laboratory, Ferry Road, Hayling Island, Hants., England

ABSTRACT

The arrival of *S. muticum* in the Solent in 1973 caused considerable concern amongst marine biologists. A programme was established to monitor its spread and effects on the indigenous species of the area.

The seasonal growth cycle of *S. muticum* in the Solent would appear to differ from that of the same species in Japan and parts of North America. In the British Isles there is no dormant period, new growth starting before the last season's lateral branches have degenerated. Extension growth of the primary laterals, though rapid (up to 4 cm/day) is maintained for a comparatively short period (May-June). Reproductive maturity occurs from July to September with a concomitant reduction in the rate of elongation. In the Solent this species behaves mainly as a coloniser, although its ability to perennate can enable it to produce more permanent stands.

Comparisons have been made between the epiphytes of *S. muticum* and other dominant macrophytes growing at similar sites. The diversity and abundance of species epiphytic on *S. muticum* are comparatively large. During the period of rapid extension growth the plants remain relatively free of epiphytes becoming increasingly more heavily colonised with the onset of fertility and then senescence. There is little carry over of epiphytic species from one year to the next as most of the tissue that is colonised is lost during the winter. A few species maintain their presence over the winter attached to the basal system.

INTRODUCTION

In 1973 *Sargassum muticum* (Yendo) Fensholt, a species of brown alga indigenous to Japan, was found to be growing attached on the Bembridge Ledges, a shore on the Isle of Wight, Southern England (Farnham, *et al.*, 1973). The same species had previously become established in North America, arriving in British Columbia in the 1940s (Scagel, 1956) and spreading southwards some 3,000 km along the west coast as far as Baja California by 1974. In the British Isles attached *S. muticum* is still largely confined to the Solent region. Despite extensive attempts to eradicate the species (Farnham and Jones, 1974) the populations have extended locally and at the original sites of infection its establishment has become consolidated.

At Bembridge, and elsewhere in the Solent, *S. muticum* is largely confined to lagoons, the lower eulittoral and the sublittoral fringe. Where that portion of the shore is extensive, large areas of dense *S. muticum* have become established. A summer population of plants 'wandering' (Nicholson, et al., in press) from the eulittoral into the sublittoral also develops and some of these plants are able to overwinter in the sublittoral.

In Japan *S. muticum* accomplishes its growth and reproduction by the middle of the summer, followed by a dormant stage during which the plants die back to their perennial basal systems. The growth cycle recommences again in the late autumn (Deysher, pers. comm.). In North America, lateral branches are also shed after fertility leaving the perennial holdfast. Some degree of dormancy is also intimated though it is unclear whether this is a local or general phenomenon as it is also recorded that plants may be found fertile at all times of the year (Nicholson, pers. comm.). The maximum size of adult plants would also appear to vary with geographical location. Yendo (1907) gives the size of plants in Japan as 75-120 cm. At Santa Catalina, in Southern California, plants of up to 5 metres in length are not uncommon (Nicholson, et al., in press), whereas in the British Isles plants of 3-4 metres in length are frequent and a few plants of up to 5 metres have been observed.

A description of the individual and population growth-cycles of *S. muticum* in the Solent may help to clarify some of these differences in its biology and may aid in our ability to predict how the species will perform at new sites that it colonises.

MATERIALS AND METHODS

Data on the development of individual *S. muticum* plants have been collected by tagging a number of plants at Portsmouth Harbour, Bembridge and St. Helens. The loss rates of plants at Bembridge and St. Helens were very high and at these sites it was only possible to follow individuals for periods of about 2-3 months. The survival rate at Portsmouth was considerably better and some plants there have been followed for up to three growth cycles. This reflects the more stable nature of the substrate and more sheltered conditions at Portsmouth, where the plants were growing on pontoons. At Bembridge and St. Helens they were attached to friable bedrock, small stones and old oyster shells. The tags were made from small strips of "Cobex" plastic sheeting coiled to form 'bird rings', numbers and symbols being engraved into the plastic. These rings were clipped around the bases of the lateral branches where they apparently caused little damage and could be easily seen.

Measurements were made of the length of the dominant primary laterals, a convenient measure of growth in *S. muticum**. The reproductive state of the plants was recorded for each primary lateral on an arbitrary 5-point scale, I = vegetative tissue only; II = development of receptacles beginning; III = well developed receptacles present, conceptacles becoming obvious; IV = receptacles with mature conceptacles, zygotes sitting on the receptacle surface and V = receptacles, though still active, showing obvious signs of senescence e.g. epiphytes growing from conceptacles.

Population data was gathered by clearing all vegetation from sets of five contiguous 50 x 50 cm quadrats run through the original site of dense *S. muticum* colonisation. The overall length of the primary laterals, the fresh

* Terminology after Jensen, J.B. (1974).

weight and fertility status of individual plants being recorded. Data was also collected on the associated flora and fauna. A comparison of the epiphyte species commonly found on *S. muticum* and *Cystoseira nodicaulis* (With.) Roberts was undertaken. *C. nodicaulis* is the most closely associated large macrophyte with *S. muticum* at sites on the Isle of Wight. Less extensive collections of *Halidrys siliquosa* (L.) Lyngb. were also made for comparative purposes. Collections of individual plants were made by carefully inserting them into mesh bags before cutting them from the substrate. This was done so as to retain the associated fauna.

"Tufnol" fouling test panels (20 cm x 30 cm and 15 cm x 13 cm) were placed within a stand of *S. muticum* at Portsmouth Harbour and of *S. muticum* and *C. nodicaulis* at Bembridge. Panels were placed in the field at approximately three weekly intervals and brought back to the laboratory for inspection after about six weeks. A low power binocular microscope was used to examine the panels for settlement and the presence and abundance of settling species were recorded. The larger panels were subsampled using a grid of 5 x 5 cm squares in which the numbers of organisms present were counted.

Average monthly surface water temperatures have been taken from data recorded at the Hayling Island Laboratory nearby. Daylength was calculated from Science Research Council data for the times of sunrise and sunset at Greenwich, London.

RESULTS

The Growth Cycle. Fig. 1 shows the first season's lateral branch development of a typical tagged plant from Portsmouth Harbour. The relationship between the dominant primary lateral (PL1) and the subservient lateral (PL2) is interesting. The development of PL2 would appear, at least partially, to be dependant on the physiological state of PL1. The development of PL2 was initiated later than PL1, the growth rate was then similar with PL2 remaining shorter than PL1 until the latter became damaged at the end of June. The onset of fertility also occurred earlier in PL1. PL1 had reached reproductive stage IV by the end of May, the same stage was not arrived at by PL2 until about one month later, after PL1 had become damaged. This 'delayed' maturity in PL2 would not appear to be due to a requirement for a minimum branch size. PL1 became fertile by a length of 150 cm whereas PL2 was not at the same reproductive stage until it had reached about 210 cm in length. Once the apical region of PL1 was damaged there was little further extension growth. PL2 continued to increased in length but at a reduced rate of growth becoming fertile fairly quickly.

The start of development in the second growth cycle is also shown in Fig. 1. It can be seen that in the second year the initiation of the new primary lateral branches occurs before the old ones have decayed. Another noteworthy feature is that in the second season the number of lateral branches produced by the plant has increased from two to five, the same plant in its third season produced nine laterals.

Fig. 2 shows the growth of the *S. muticum* population at Bembridge during 1975. The average length of the longest branch is given for the population in each month. The average monthly temperature and the hours of daylight are also given. In 1975 the average length of the population was at a maximum during the early part of July. A considerable amount of growth had been accomplished whilst sea temperatures were quite low. For instance, during April and May the average sea temperatures were 7.8°C and 11.0°C with average plant lengths of 26 cm and 70 cm respectively. In fact the average length of the popula-

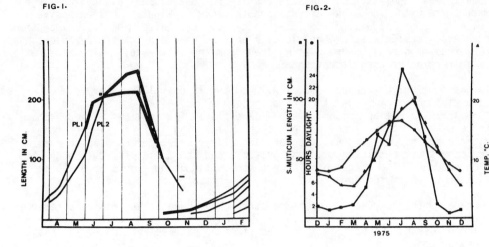

Fig. 1. Lateral Branch Development in a tagged Plant from Portsmouth Harbour. PL1 = Dominant Primary Lateral; PL2 = Subservient Primary Lateral. Fertile stages IV and V are represented by the double black lines. ■ = The time at which the apex of PL1 was damaged.

Fig. 2. The Development of the *S. muticum* Population at Bembridge during 1975. ■ = The average length of the population; ● = Hours of daylight; ▲ = The average monthly surface water temperatures at Hayling Island.

tion had begun to increase from January onwards, at a time when the temperature was still dropping. The average length also began to decrease before the temperature had reached its maximum. Population growth more nearly followed the curve of daylight hours. Both began to increase in January reaching peaks in July and then decreasing.

TABLE 1   The Proportion of the *S. muticum* Population in Reproductive Stages IV and V.

1975	Mar.	Apr.	May	June	July	Aug.	Sept.	Oct.	Nov.
%	0.0	0.40 (IV)	2.67 (IV)	6.65 (IV)	39.78 (IV)	68.10 (IV/V)	64.78 (IV/V)	35.71 (V)	0.0
n	748	553	337	376	186	316	316	28	9

Data on the proportion of the population reaching reproductive stages IV and V at Bembridge are presented in Table 1. Though there were occasional plants bearing active receptacles in April and May, the bulk of the population was actively fertile during July, August and September. There was thus very little release of zygotes before the average temperatures reached 15°C with the greatest activity occurring during the period of maximum water temperatures. The population's average length was also decreasing by the time maximum reproductive activity was reached.

Epiphytes

The distribution of epiphytic species on *S. muticum* is related closely to the growth cycle. Fig. 3 shows the settlement of the colonial ascidian *Trididemnum tenerum* Varrill on first season *S. muticum* plants and "Tufnol" panels at Portsmouth Harbour during 1975. The *S. muticum* population supported a small amount of *T. tenerum* during the winter, at a time when settlement did not occur on the panels. The low degree of colonisation on the *S. muticum* during the early part of the year was reflected on the panels (see Fig. 4) with a build-up in the amount of colonisation during the later part of the summer. Settlement of *T. tenerum* is extended on the panels, reaching its peak about one month after that on the plants. This timing corresponds with the growth cycle of *S. muticum* (Figs. 1 and 2). The early part of the cycle remains relatively free of *T. tenerum* with a build up of the epiphytes during the reproductive phase of *Sargassum*. The loss of tissue and increase in abundance of other epiphytic species accompanying senescence in *S. muticum* would appear to account for the later peak in the settlement of *T. tenerum* on the panels.

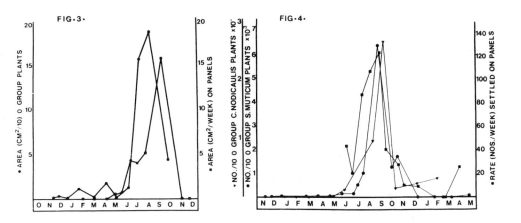

Fig. 3. The settlement of *Trididemnum tenerum* on first season *S. muticum* plants. (●) and panels (■) at Portsmouth Harbour during 1975.

Fig. 4. The occurrence of *Janua (Janua) pagenstecheri* on *S. muticum* (●), *C. nodicaulis* (▼) and panels (■) at Bembridge during 1975.

At Bembridge, *S. muticum* is closely associated with *Cystoseira nodicaulis*. The species occurring epiphytically on the two macrophytes are generally similar, as are those on panels immersed in the same area. Fig. 4 shows the occurrences of the spirorbid worm *Janua (Janua) pagenstecheri* (Quatrefages) on *S. muticum*, *C. nodicaulis* and panels at Bembridge during 1975. As with *T. tenerum* a small population of *J. pagenstecheri* persisted on the *S. muticum* while none was recorded on the panels during the winter months. There was a rapid build-up of the epiphyte during June and August reaching a peak in September. The picture was essentially the same on *C. nodicaulis* and the panels, though on both of them the peak was later, namely in October.

Table 2 presents data on the total number of epiphytes recorded on both *S. muticum* and *C. nodicaulis*, those which were common for at least part of the year, and those restricted to either *S. muticum* or *C. nodicaulis*.

TABLE 2   Epiphytes on *S. muticum* and *C. nodicaulis* from Bembridge.

	Plant and animal species recorded on:			
	S. muticum	*C. nodicaulis*	*S. muticum* and not on *C. nodicaulis*	*C. nodicaulis* and not on *S. muticum*
Total numbers	animals 82	animals 72	animals 5	animals 4
	algae 61	algae 60	algae 8	algae 7
Species abundant for at least part of the year	animals 35	animals 32	animals 1	animals 0
	algae 27	algae 34	algae 1	algae 1

C. nodicaulis is a comparatively long-lived species (Roberts, pers. comm.) and though it too loses much of its tissue over the winter it may nevertheless provide more of an inoculum of epiphyte species from season to season than does *S. muticum*. Few of the common epiphytes are restricted to either one of the host species. Species such as *Sphacelaria cirrosa* (Roth) C.Ag., *Elachista flaccida* (Dillw.) Aresch. though present on both are more abundant on *C. nodicaulis* and others such as *Polysiphonia urceolata* (Lightf. ex Dillw.) Grev. are more abundant on *S. muticum*.

As well as the sessile epiphytes there are a number of free-living species which appear to gain shelter from *S. muticum*. For instance juvenile prawns, *Palaemon serratus* (Pennant); juvenile wrasse, *Crenilabrus melops* (L.); and adult gobies *Chaparrudo flavescens* (Fabricius) are numerous.

H. siliquosa, which is also commonly associated with *S. muticum* at Bembridge has, in the intertidal, relatively few epiphytes for much of the season. A small number of algal species colonise the receptacles as they degenerate in the late summer and there is also some animal cover especially in the basal regions at this time. The species involved include spirorbids, colonial tunicates such as *T. tenerum* and *Diplosoma listerianum* Milne-Edwards and some hydroids.

Generally, species that settled on the panels also occured on *S. muticum* and at Bembridge usually on *C. nodicaulis* too. Exceptions to this were the barnacle *Elminius modestus* Darwin and *Spirorbis tridentatus* Levinsen, both found to be quite common on the panels but not on the plants. The spirorbids *Janua (Dexiospira) braziliensis* (Grube) and *Pileolaria rosepigmentata* (Uchida), newly recorded for the British Isles (Knight-Jones, *et al.*, 1975) have so far been found only on *S. muticum* at Portsmouth Harbour. If these species become established elsewhere in the Solent it will be interesting to see if they are able to colonise *C. nodicaulis* and other species associated with *S. muticum*.

DISCUSSION

The seasonal pattern of development of *S. muticum* in the British Isles would appear to be prolonged when compared to the same species in Japan and possibly parts of North America. The greater overall length attained by plants and the continuous nature of the growth cycle would appear to represent the behaviour of the species in a less hostile environment. Whether this is due to more favourable physical conditions or to less competition from closely related species is unclear. Certainly in the Solent S. *muticum* is growing in habitats which were generally sparsely colonised by other large macrophytes. Where S. *muticum* germlings settle in unfavourable dry areas a shortened growth cycle, similar to that in Japan ensues. Growth which is possible during the winter is soon curtailed with the warmer weather of the spring. The lateral branches, which remain short, die back to the basal systems. These, if they are not desiccated, may survive the summer in an apparently inactive state. In Japan, competing species or some other factors may push S. *muticum* into habitats in which it is unable to fulfil its full growth potential, thereby accounting for its smaller size.

The effect of damage to the apex of a dominant primary lateral is quite complex and depends to some extent on the stage of the growth cycle. Where damage occurs late in the growth cycle (Fig. 1) both laterals tend to become fertile, with little extra growth of the damaged lateral. The onset of fertility in the new 'dominant' lateral is also followed by a reduction in the rate of growth. The relationship between the onset of fertility and a decreasing elongation rate has been shown for cultured material (Norton unpublished). In segments of S. *muticum* in culture, growth rates and the degree of fertility were found to be inversely proportional. Where damage occurs in the earlier part of the rapid growth phase one of the subservient laterals generally becomes dominant, carrying on growth. There is again a tendency for fertility to be reached earlier. Where branches are very small (< 50 cm) one of the most apical secondary laterals may take over the role of apex and carry on growth. If the subservient laterals are little developed the 'repaired' dominant primary may resume its role as overall dominant.

In North America S. *muticum* has not spread to any extent in a northerly direction from the sites of initial colonisation. It has been suggested by Norton (in press) that restricted growth due to low temperatures may be the causative factor. He found that in cultures the rate of linear growth increased with rising temperatures up to $25^{\circ}C$. However there is sufficient growth at lower temperatures to produce plants in the field capable of becoming fertile and of considerable length. We consider that temperature is more important in, at least partially, controlling the onset of fertility though daylength effect cannot be ruled out. At Bembridge, during 1975, there was little reproductive activity in the population before June, during which month the average seawater temperature was $15.9^{\circ}C$ (Table 1, Fig. 2). It has often been noted with segments of S. *muticum* kept in culture for experimental purposes that those kept at temperatures of $15^{\circ}C$ and above became fertile whilst those at $10^{\circ}C$ and below did not. It may also be pertinent to note that there does not appear to be a particular plant length required for the initiation of fertility. In the early part of the cycle lateral branches of 150 cm and more (Fig. 1) are often not fertile whilst later in the year (August) plants as small as 30 cm may have active receptacles. It would thus appear that the inability of S. *muticum* to penetrate further north in America may not be due directly to inhibited growth but to the inability of the plant to reproduce. This accords well with the distribution of S. *muticum* compared to the ranges in seawater temperature for the west coast of North America (Norton, in press) where S. *muticum* is found only at sites with maximum temperatures above $12^{\circ}C$. Maximum settlement of zygotes occurs during July and

August when in the vicinity of adult plants a blanket cover of germlings is produced. These germlings begin growth immediately and continue to develop throughout the winter. New primary laterals are initiated in adult plants before last season's primaries have decayed (Fig. 1). Relatively few second season plants survive the winter but where they do they begin the new rapid growth phase in a more advanced state than the germlings. The process of shedding of branches reported for the species in North America (Norton, in press; Nicholson, et al., in press) is not general in the Solent. There is considerable fragmentation of the plants in the autumn but this occurs as the branches become inactive and covered by epiphytes and would appear to be due to a general process of senescence and not to systematic abscission of the lateral branches.

The suggestion that the variety of epiphytes colonising S. muticum may be limited compared to indigenous species (Druehl, pers. comm.) does not apply to Britain. Not only are the species numerous as shown by Withers et al., (1975), but they are also similar to those on C. nodicaulis. Generally the species are very similar though their abundance may be different. C. nodicauli by its truly perennial nature and more bushy habit would appear to be a more stable host for epiphytes to overwinter on. The possibility that the tannin-like substances produced during the rapid growth phase of S. muticum may have an anti-algal effect cannot be discounted. For instance the ephemeral Eudesme virescens (Carm. ex Harv. in Hoek.) J.Ag. which is abundant in the lagoons at Bembridge during the rapid growth phase of S. muticum is commonly found as an epiphyte on C. nodicaulis but has not been found on Sargassum muticum. Antibiotic effects have been shown for the pelagic Sargassum natans (L) J. Meyen and S. fluitans Børgesen (Conover and Sieburth, 1964) and this may be the case with S. muticum. Colonisation is limited during the period of rapid growth though the plants are quite sizeable for a considerable time. Generally settlement by epiphytes takes place in the basal regions and continues along the primary laterals, presumably any antibiotic effect being reduced away from the active apices. There are exceptions to this, for example, the colonial diatom Amphipleura rutilans (Trent) Cleve and the nest building amphipod Pleonexes gammaroides Bate which are abundant in the early part of the cycle and become attached throughout the length of the plant. Data on the pattern of settlement on the inert panels and C. nodicaulis (Figs. 3 and 4) suggest tha the relatively low colonisation of S. muticum, in the early part of the growth cycle, may be due primarily to the lack of larvae, etc., available at this time

The extended growth period of S. muticum away from the constraints of its original environment in Japan and the degree of flexibility within the growth cycle may explain its success in colonising the west coast of North America and South of England. The ability of comparatively large numbers of indigenous species to colonise Sargassum would suggest that should it become more widespread the diversity of associated species need not necessarily be affected.

ACKNOWLEDGEMENTS

We would like to thank Dr. E.B.G. Jones and Mr. W.F. Farnham for encouraging this work and reading the manuscript. We are grateful to the Natural Environment Research Council (N.A.J.) and the Department of the Environment (P.W.G.) for funding this work (the latter under Contract DGR 483/2) and for permission to publish.

REFERENCES

Conover, J.T., and Sieburth, J. McN. 1964. Effect of *Sargassum* distribution on its epibiota and antibacterial activity. Botanica Mar., 6, 147 - 157.

Farnham, W.F., Fletcher, R.L. and Irvine, L.M. 1973. Attached *Sargassum* found in Britain. Nature, Lond., 243, 231 - 232.

Farnham, W.F. and Jones, E.B.G. 1974. The eradication of the seaweed *Sargassum muticum* from Britain. Biological Conservation, 6, 57 - 58.

Jensen, J.B. 1974. Morphological Studies in Cystoseiraceae and Sargassaceae (Phaeophyceae). Univ. Calif. Publ. in Bot., 68.

Knight-Jones, P., Knight-Jones, E.W., Thorp, C.H. and Gray, P.W., 1975. Immigrant Spirorbids (Polychaeta Sedentaria) on the Japanese *Sargassum* at Portsmouth, England. Zoologica Scripta, 4, 145 - 149.

Nicholson, N., Hosmer, H., Bird, K., Hart, L., Sandlin, W., Shoemaker, C., and Sloan, C.,(In press). The Biology of *Sargassum muticum* (Yendo) Fensholt at Santa Catalina Island, California. Proc. Eighth Int. Seaweed Symp.

Norton, T.A. (In press). *Sargassum muticum* on the Pacific coast of North America. Proc. Eight Int. Seaweed Symp.

Scagel, R.F. 1956. Introduction of a Japanese alga, *Sargassum muticum* into the north-east Pacific. Fish. Res. pap. Wash. Dep. Fish. 1, 1 - 10.

Withers, R.G., Farnham, W.F., Lewey, S., Jephson, N.A., Haythorn, J.H., and Gray, P.W., 1975. The Epibionts of *Sargassum muticum* in British Waters. Marine Biology, 31, 79 - 86.

Yendo, K. 1907. Fucaceae of Japan. J. College Sci., Tokyo Imp. Univ., 21, 1 - 174.

THE EFFECT OF DEPTH ON POPULATIONS OF *LAMINARIA HYPERBOREA*

Joanna M. Kain (Mrs. N.S. Jones)

Department of Marine Biology, University of Liverpool, Port Erin, Isle of Man

ABSTRACT

Populations of Laminaria Hyperborea (Gunn.) Fosl. were sampled in the Isle of Man (54° 3-5'N, 4° 44-7'W) western Scotland (Connel Sound, 56° 27'N, 5° 24'W; Cuan Sound, 56° 16'N, 5°38'W) and the Outer Hebrides (Muldoanich, 56° 55'N, 7° 26'W; McKenzie Rock, 57° 8'N, 7° 14'W; Griean Head, 57° 1'N, 7° 32'W). The total biomass of the species was removed from 1 m^2 quadrats, 1-5 quadrats comprising a sample. Normally 4 samples were taken at each depth and a 95% confidence limit found for the main parameters.

The mean standing crop varied from 0.37 to 23 kg fresh weight per m^2. A rapid drop with depth in the Isle of Man was attributed to high Echinus density and at Connel Sound to high light attenuation. At Muldoanich there was little change above 11 m below lowest astronomical tide (LAT).

The annual production of dry matter was calculated from the frond fresh and dry weights and stipe and holdfast annual increments. It is postulated that production is related partly to irradiance and partly to nutrient supply, the latter directly affected by water movement. Both irradiance and wave action decrease logarithmically with depth; it is not expected that production would be directly proportional to either but that threshold values would be essential at the limiting depth. A curve calculated on this basis fits the observed production between 11 and 19 m below LAT off Muldoanich. Between 4 and 11 m, however, production (about 1 kg dry matter/m^2 year) did not differ significantly. It is assumed that some unidentified factor was limiting at these depths because a saturation effect is not expected in a layered forest where excess light or nutrients can be used by undergrowth plants. The production on a submerged rock 2 km south of Muldoanich, at 20 and 24 m below LAT, was higher than expected from the inshore results. In view of the low Echinus density and similar substratum, this is tentatively attributed to water movement. It seems likely that at any one site one factor could limit production at one time of year (e.g. light in the winter) and another at a different season (e.g. nutrient supply in summer). Thus although irradiance must partly limit production in deep water, improved nutrient supply can increase it. The fact that, compared with Muldoanich, production was significantly lower on McKenzie Rock, which is relatively sheltered from waves and tidal stream, and significantly higher off Griean Head, which is exposed to full Atlantic swell, supports the beneficial effect of water movement.

The frond area index, or ratio of frond to substratum area, was usually 3 to 5 in shallow water. At Muldoanich it dropped below unity only at 17 m. The effect of the canopy on stipe lengths, previously observed, was confirmed. This causes a clustering of stipe length frequency at each end of the range by inhibition of the young plants until there is a break in the canopy allowing fast growth. On flat substrata the critical frond area index associated with this effect was found to be about 2.

In most populations the maximum age was 12 years (up to 15 being recorded) but in shallow water with severe wave action there were few plants over 5 years. Longevity was also reduced in deep water on solid rock and at only 9 m off NE Muldoanich on an unstable bottom.

ACKNOWLEDGEMENTS

Mr. M.J. Bates very ably assisted in all diving work and much of the material handling, as did my husband, Dr. N.S. Jones, on the first Hebrides visit. On the second, Messrs R. Dalley, D.J. Morris and J.D. Paul worked extremely hard both underwater and at the tedious chopping and measuring operations.

SUBLITTORAL TRANSECTS IN THE MENAI STRAITS AND MILFORD HAVEN

E. W. Knight-Jones and A. Nelson-Smith

Zoology Department, University College of Swansea, U.K.

ABSTRACT

The Menai Straits just east of the suspension bridge yielded 231 species, including 27 Algae. The most varied and colourful fauna was found in the algal zone (depths less than 7m). Tidal currents are strong in the main channel, on one side of which there is gravel abrasion with a Flustra assemblage and, on the other, stable rocks with a Halichondria assemblage. On open coasts, exposure to swell increases the downward extension of some shallow-water species such as Mytilus edulis, but the light gradient is everywhere important in separating a substantial minority of species between algal and subalgal zones. Numerous groups can be recognised, each with a primary species and several dependent ones, the interaction between groups mainly involving competition between the primary species.

INTRODUCTION AND METHODS

The Menai Bridge transect (OS grid ref. 558714), which was studied twenty years ago (Knight-Jones, Jones and Lucas, 1957) but never fully described, followed a submarine telephone cable about 80m east of the suspension bridge. Sampling stations were marked out along this cable every 20m across the channel and every 10m as the water shallowed towards each side, the end stations being amongst the Laminaria (Fig. 1). The depth was recorded every 10m (Fig. 2) and, at each sampling station, rocks, shells and as many different species as

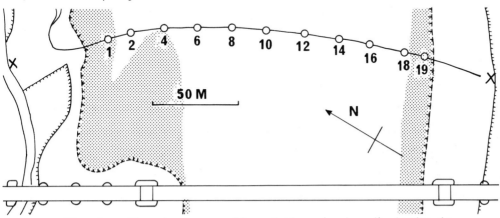

Fig. 1. Plan showing sampling stations (numbered), Laminaria zones (stippled), low-water line (broken), high water line (continuous), cable-marking posts (crosses) access road (on left) and bridge (bottom).

could be seen or felt were put into sacks and hauled to a dinghy overhead. Laboratory sorting followed within 36 hr and species were recorded (see Appendices) as present (p), common (c, indicating a judgment that some could be obtained on a subsequent visit to the same site), abundant (A, indicating that plenty could be thus obtained) or dominant (D, for a few strikingly abundant species). A subsequent survey recorded the percentage cover of some of these abundant forms, at and around each station (Table 1). Stations are here compared using the similarity index of Sørensen (1948) including first, only records at "abundant" or "dominant" levels; secondly, those which were at least common and thirdly, all records (Fig. 3A, B & C respectively). Enterocola bilamellata Sars (1890-1921) was identified from the original description. Otherwise the authors of species and works of identification used are listed by the Marine Biological Association (1957); Bruce, Colman and Jones (1963) and/or Crothers (1966).

In Milford Haven our sublittoral studies so far have been less detailed. The transect of Fig. 4 was at OS grid ref. 980044, just east of the new bridge which has replaced Pembroke Ferry. It was sampled by a line of bags, as described previously (Bailey, Nelson-Smith and Knight-Jones, 1967).

RESULTS

Considering conspicuous species (Fig. 2) the sampling stations at Menai Bridge fell into three broad groups. Those of the algal zone (1, 2, 4, 18, 19) had diverse faunas (Table 1 and Appendices 1 & 2), particularly on the north side, when stations 2 & 4 lay on a shelf protected from strong currents. Fish were mostly confined to this shelf. Beyond the shelf (6, 8) was an assemblage dominated by Flustra and Tealia (Appendix 3), which must occasionally have suffered abrasion, for there were banks of gravel around station 8. Most of the deep channel and the south bank of the straits (12, 14, 16) were dominated by Halichondria and Tubularia, with crowds of Caprella linearis and other planktotrophic amphipods (Appendix 4). If comparisons between stations are based only upon abundant occurrences the resulting dendrogram (Fig. 3A) illustrates clearly enough this triple grouping. There, circles mark the algal stations and squares those dominated by Halichondria. The main Flustra stations are left unmarked, but station 10 is underlined, for this was intermediate between the Flustra and Halichondria assemblages. A broader based comparison, however, detaches 8 from the rest (Fig. 3B). Its greatest affinities remain with its neighbours, but they have closer links elsewhere. If all records are considered the range of similarities is reduced and the groupings are indistinct (Fig. 3C), for many species were widely or sporadically distributed, more or less independently of the dominant forms.

The greatest diversity was around the lower fringes of the algal zone, the least was deeper in the channel and amongst the gravel banks of station 8 (Table 1). Ranking classes and phyla etc. by numbers of species represented (Table 1) gives prominence to predaceous groups of invertebrates, but these were represented sporadically, apart from the planktotrophic amphipods etc. which were so abundant amongst the Halichondria and Tubularia. Only about 15% of polychaetes and molluscs were anywhere rated as common, but over 40% of algae and sponges and about 60% of coelenterates, bryozoans and tunicates came into that category.

The Milford Haven transect (Fig. 4) also shows maximum diversity in the algal

Table 1. Numbers of species per station at Menai Bridge and (below) percentage cover, with densities of *Tealia*.

	1	2	4	6	8	10	12	14	16	18	19	Total	% common
Crustacea	5	21	22	9	1	8	14	17	22	15	6	47	40
Polychaeta	13	12	12	10	4	5	10	10	10	6	7	30	17
Mollusca	7	12	7	6	-	3	4	2	12	7	7	29	14
Algae	16	5	4	-	-	-	-	-	-	9	13	27	44
Porifera	4	11	18	10	?	4	5	9	14	13	11	22	41
Bryozoa	12	8	9	8	9	5	3	9	11	9	8	22	59
Coelenterata	5	6	7	9	6	4	5	6	11	8	9	21	57
Pisces	2	4	3	1	-	-	-	-	1	-	-	8	25
Platyhelminthes	-	3	2	-	1	-	-	-	-	1	-	5	-
Echinodermata	3	4	4	5	2	3	2	2	3	3	3	5	80
Tunicata	2	3	3	2	1	-	-	-	1	2	2	5	60
Pycnogonida	2	2	1	4	-	1	1	1	3	1	2	4	-
Nemertini	2	2	2	1	1	2	2	3	2	2	2	3	66
Protozoa	-	-	-	-	-	-	-	1	-	18	-	2	-
Kamptozoa	-	-	-	-	1	-	-	-	-	-	1	1	-
Totals	68	93	96	65	28	35	46	60	90	76	70		
Percentage cover													
Algae	10	12	5	-	-	-	-	-	-	10	50		
Halichondria	-	1	25	-	2	13	30	20	40	6	2		
Flustra	-	-	5	12	8	-	-	-	-	-	-		
Tealia	-	0·4	1·4	8	8	4·4	1	-	-	-	-		
Nr Tealia/m²	-	2	4	28	40	14	2						

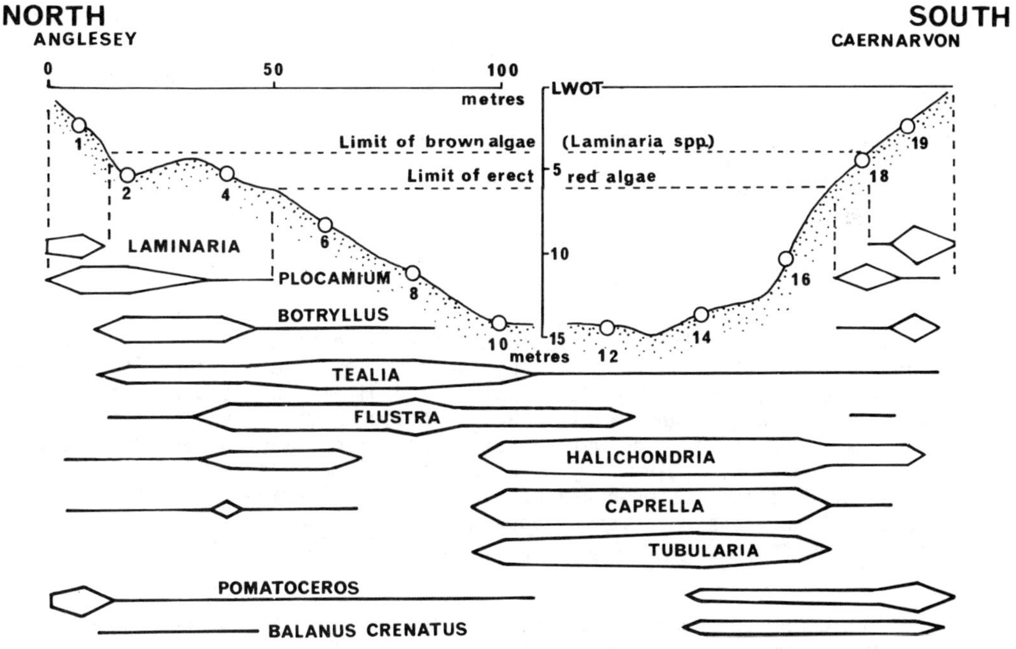

Fig. 2 The Menai Straits transect in section, showing the distribution of common plants and animals.

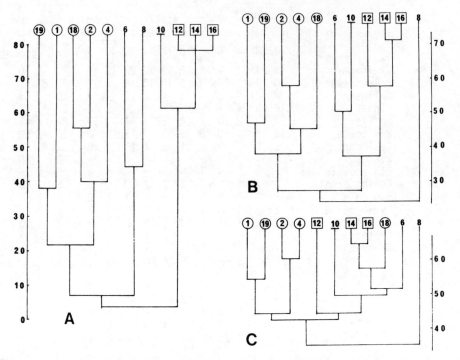

Fig. 3 Dendrograms of similarity between stations, judged by A, records of species which were abundant or dominant; B, those which were at least common and C, all species.

zone, particularly at its lowest fringe, and illustrates the contrasting distribution of Aglaophenia pluma and Aglophenia tubulifera, one of several pairs of closely related species which have quite different vertical ranges (Knight-Jones and Jones, 1955).

DISCUSSION

The diversity of the algal zone fauna compared with that at greater depth may be explained at least partially by the range of illuminations in different microhabitats and by the additional microhabitats and food resources provided by the algae themselves. Sciatophilic animals may find dim light under rocks, even in shallow water, but deeper places must necessarily lack photophilic species. Diversity of microhabitats may also explain the blurring of divisions between assemblages when all species are taken into account, with no consideration of abundance. Even in current-swept areas there are sheltered crevices where polychaetes for instance may take refuge. In such respects all stations are somewhat similar.

On the inside of the bend there must be a lesser range of current velocities and there is probably some shifting of gravel banks by spring-tide currents. Here lesser diversity, which is often regarded as being ipso facto ecologically unstable, may in fact result from unstable physical conditions. It may be noted that Flustra, Abietinaria and Hydrallmania are well adapted to withstand shifting gravel by their toughness and erect form, but there would appear to be no interaction between these forms. They are together because they are favoured by and adapted to the same physical conditions.

Halichondria and Tubularia indivisa live together remarkably well, with some mutual support, but they seem to be distributed independently. On the north side there was plenty of Halichondria without Tubularia, whereas the lower slopes of Chicken Rock, south of the Isle of Man, seemed covered by Tubularia without Halichondria, to judge from a dive there in 1973. There always seem to be great numbers of Caprella, Jassa etc. amongst the Tubularia and the amphipods are probably protected by the nematocysts around them, for no fish were seen in the channel under Menai Bridge. It would be interesting to dive in such places at dawn and dusk, when fish might be feeding more actively. Around station 4, where Tubularia was scarce and fish were common, Caprella was mostly replaced by Tritaeta, which lives protected in holes which it digs out of the surface of sponges and ascidians.

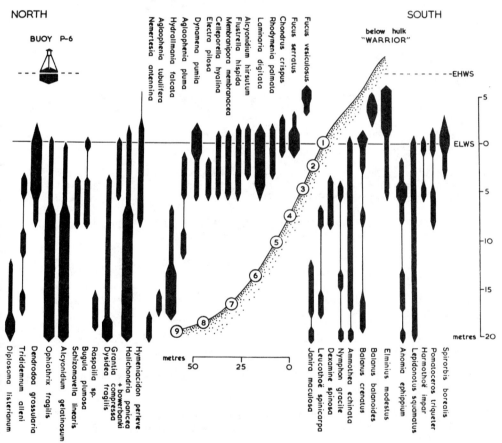

Fig. 4   Transect of a steep shore between Pembroke Ferry and the hulk "Warrior", Milford Haven, showing collecting stations and the vertical distribution of some organisms present. Note that the vertical scale is exaggerated.

The behaviour of Tritaeta recalls the concept of interdependence within a community. Such behaviour is seen in numerous small associations of species. Botryllus for instance may be regarded as the hub species in an association which included Cycloporus, Ancula, Goniodoris, Enterocola and Musculus marmoratus. Such associations usually involve feeding or protection, but sometimes merely provision of substrata. An associated group may be large or small and the links may be obligatory, facultative or casual, but usually they are fairly strong, involving special patterns of behaviour.

In contrast the interaction between hub species seems generally to be limited to competition altering the physical environment. This may nevertheless be important where populations are dense, as was well illustrated by the sharpness of a horizontal boundary between Mytilus and Antedon on a wreck half a mile off Dale Point, opposite the mouth of Milford Haven. The Mytilus extended down to 6m depth on the sheltered north side of the wreck, but down to 8m on the south side, which was exposed to swell entering the Haven. On the south-west face of the Mewstone off Skomer, where the swell was much stronger, there was a similar sharp line between Mytilus and Ophiothrix, at a depth of 15m.

ACKNOWLEDGMENTS

We are grateful to Professor D.J. Crisp, F.R.S. for facilities at Menai Bridge and to Miss Doreen Lucas, Mr Alan Osborn and Drs W.C. Jones, W.E. Jones and T.E. Thompson for assistance with diving and/or identifications of species.

REFERENCES

Bailey, Julie H., Nelson-Smith, A. and Knight-Jones, E.W. 1967. Some methods for transects across steep rocks and channels. Underwater Association Report 1966-67, 107-111.

Bruce, J.R., Colman, J.S. and Jones, N.S. 1963. Marine Fauna of the Isle of Man, 2nd ed. 307pp. Liverpool University Press.

Crothers, J.H. 1966. Dale Fort Marine Fauna, 2nd ed. Field Studies 2 (suppl.), 1-169.

Knight-Jones, E.W. and Jones, W.C. 1955. The fauna of rocks at various depths off Bardsey. Bardsey Observatory Report 3, 1-8.

Knight-Jones, E.W., Jones, W.C. and Lucas, Doreen. 1957. A survey of a submarine rocky channel. Rep. Challenger Soc. 3(9), 20-22.

Marine Biological Association, 1957. Plymouth Marine Fauna 3rd ed. 453 pp. Plymouth.

Sörensen, T. 1948. A method of establishing groups of equal amplitude in plant sociology based on similarity of species content. Biol. Skr. 5(4), 1-34.

APPENDIX I. Species restricted to algal zone of Menai transect

Inside of bend only (north side)

ALGAE Stations	1	2	4
Chaetomorpha melarginium	P	-	-
Corallina officinalis	P	-	-
Dilsea edulis	P	-	-
Gracilaria verrucosa	P	-	-
Halidrys siliquosa	P	-	-
Heterosiphonia plumosa	-	P	-
Laminaria saccharina	A	-	-
Polyneura gmelinii	-	-	P
Polysiphonia elongata	A	-	-
Ulva lactuca	P	-	-

COELENTERATA

	1	2	4
Aglaophenia pluma	A	P	-

PLATYHELMINTHES

	1	2	4
Cycloporus papillosus	-	P	-

POLYCHAETA

	1	2	4
Nicolea zostericola	P	P	-
Spirorbis corallinae	P	-	-

CRUSTACEA

	1	2	4
Enterocola bilamellata	-	A	A
Corophium bonelli	-	P	P

MOLLUSCA

	1	2	4
Ancula cristata	-	P	-
Lacuna vincta	-	P	-
Acmaea virginea	P	-	-
Tricolia pullus	-	P	-

BRYOZOA

	1	2	4
Alcyonidium mytili	P	-	-
Celleporaria pumicosa	A	-	-
Hippothoa hyalina	A	-	-

TUNICATA

	1	2	4
Botrylloides leachii	A	C	C

PISCES

	1	2	4
Acanthocottus scorpius	P	-	P
Pollachius pollachius	-	C	P
Gobius flavescens	-	C	-
Diplecogaster bimaculatus	P	-	-

Both sides or (below) outside only

ALGAE	1	2	4	18	19
Ceramium rubrum	A	-	-	-	P
Cladophora rupestris	P	-	-	-	P
Delesseria sanguinea	A	-	-	-	P
Cryptopleura ramosa	-	A	A	A	P
Ectocarpus confervoides	A	-	-	-	P
"*Lithothamnion*"	P	P	P	-	A
Phycodrys rubens	-	P	-	P	-
Phyllophora membranifolia	P	-	P	P	-
Plocamium coccineum	A	A	P	A	P

COELENTERATA

	1	2	4	18	19
Diphasia attenuata	P	-	-	-	P

PLATYHELMINTHES

	1	2	4	18	19
Fecampia erythrocephala	-	P	-	P	-

POLYCHAETA

	1	2	4	18	19
Odontosyllis gibba	P	-	P	-	P

CRUSTACEA

	1	2	4	18	19
Aora typica	-	P	P	P	-
Caprella acanthifera	-	P	-	-	C
Carcinus maenas	A	A	C	C	-

MOLLUSCA

	1	2	4	18	19
Doto coronata	P	P	-	-	P

ALGAE

	18	19
Desmarestia ligulata	C	-
Gigartina stellata	-	P
Griffithsia flosculosa	C	P
Hypoglossum woodwardii	P	-
Laminaria hyperborea	P	D
Rhodomela subfusca	P	P
Rhodymenia palmata	P	-
Sphacelaria cirrosa	-	P

COELENTERATA

	18	19
Obelia geniculata	-	C

MOLLUSCA

	18	19
Hermaea bifida	-	P

BRYOZOA

	18	19
Membranipora membranacea	-	A

APPENDIX 2. Species mostly in algal zone of Menai transect.

		1	2	4	6	8	10	12	14	16	18	19
Mostly inside bend												
PORIFERA	*Leucosolenia complicata*	C	P	P	P	-	-	-	-	-	-	C
CRUSTACEA	*Phtisica marina*	C	A	P	P	-	-	-	-	-	P	-
	Hippolyte varians	-	C	-	-	-	-	-	-	P	-	-
	Macropipus puber	-	C	C	C	-	-	-	-	P	-	-
MOLLUSCA	*Goniodoris nodosa*	-	P	P	P	-	-	-	-	-	-	-
BRYOZOA	*Alcyonidium gelatinosum*	P	P	P	P	-	-	P	-	-	-	P
	Alcyonidium hirsutum	P	A	A	-	-	P	-	-	P	P	P
ECHINODERMATA	*Amphipholis squamata*	C	A	A	A	P	C	-	-	P	A	C
	Henricia sanguinolenta	-	P	C	P	-	-	-	-	-	-	-
TUNICATA	*Botryllus schlosseri*	A	A	A	P	P	-	-	-	-	P	A
On both sides												
PORIFERA	*Grantia compressa*	A	A	C	P	C	-	-	P	C	P	A
	Mycale macilenta	-	C	C	-	-	-	-	-	P	P	P
	Sycon coronatum	A	P	C	P	P	-	-	-	P	P	A
COELENTERATA	*Garveia nutans*	C	-	P	-	-	-	-	-	P	P	P
POLYCHAETA	*Nereis pelagica*	P	P	P	-	-	-	-	P	P	-	P
	Pomatoceros triqueter	A	P	P	P	P	P	-	C	C	C	A
CRUSTACEA	*Cancer pagurus*	C	C	C	C	-	-	-	C	C	-	C
	Inachus dorynchus	-	C	-	-	-	-	-	-	P	P	-
PYCNOGONIDA	*Achelia echinata*	P	P	-	P	-	-	-	-	P	P	P
MOLLUSCA	*Gibbula cineraria*	-	P	P	P	-	-	-	-	P	-	P
	Nucella lapillus	-	P	-	-	-	-	-	-	P	-	-
	Brachystomia lukisii	-	C	-	-	-	-	-	-	P	C	-
	Onchidoris muricata	P	-	-	-	-	-	-	-	P	P	-
	Polycera quadrilineata	P	C	-	-	-	-	-	-	P	P	-
	Rissoa parva	A	A	C	P	-	-	P	-	P	A	C
BRYOZOA	*Bugula flabellata*	-	P	P	P	-	-	-	P	-	P	-
	Crisia eburnea	A	A	A	P	C	-	C	P	C	P	A
	Cryptosula pallasiana	A	-	C	-	-	-	-	P	C	C	P
	Escharoides coccineus	A	-	P	P	C	P	-	C	C	A	-
ECHINODERMATA	*Ophiothrix fragilis*	-	C	C	C	-	P	P	P	P	C	P
TUNICATA	*Polyclinum aurantium*	-	A	A	-	-	-	-	-	P	A	-

APPENDIX 3. *Flustra* assemblage, withstanding gravel abrasion.

		1	2	4	6	8	10	12	14	16	18	19
COELENTERATA	*Abietinaria abietina*	-	-	P	A	C	-	-	P	C	-	-
	Alcyonium digitatum	-	P	P	C	-	-	-	-	-	P	-
	Hydrallmania falcata	C	P	C	C	C	C	-	-	P	P	P
	Tealia felina	-	C	C	A	A	C	P	P	P	P	P
POLYCHAETA	*Circeis spirillum*	-	-	-	P	-	-	-	-	P	-	-
PYCNOGONIDA	*Pycnogonum littorale*	-	-	-	P	-	P	-	-	-	-	-
BRYOZOA	*Crisia denticulata*	-	P	-	P	A	-	-	-	-	-	-
	Flustra foliacea	-	P	A	A	D	C	C	-	-	P	-
	Scrupocellaria scruposa	-	-	-	-	C	-	-	-	-	-	-
ECHINODERMATA	*Asterias rubens*	-	C	C	C	A	C	P	P	P	C	C

APPENDIX 4. *Halichondria* assemblage, in strong currents free from gravel.

		1	2	4	6	8	10	12	14	16	18	19
PORIFERA	*Amphilectus fucorum*	-	-	-	C	-	C	-	C	C	C	-
	Cliona celata	-	-	P	-	-	P	-	P	P	P	-
	Halichondria panicea	P	P	C	C	-	A	D	D	D	C	C
COELENTERATA	*Tubularia indivisa*	-	-	-	-	-	A	A	D	A	-	-
	Tubularia larynx	-	-	-	-	-	-	C	-	-	-	-
NEMERTINI	*Tetrastemma vermiculus*	P	P	-	-	-	C	C	C	P	-	P
POLYCHAETA	*Autolytus pictus*	-	P	P	P	-	-	P	C	C	P	-
	Syllis gracilis	-	-	-	P	-	-	-	C	P	P	P
CRUSTACEA	*Caprella linearis*	P	P	C	P	-	D	D	D	D	P	-
	Colomastix pusilla	-	-	C	-	-	-	-	C	P	-	-
	Jassa falcata	-	P	A	-	-	A	A	A	A	P	-
	Neopleustes monocuspis	-	-	-	-	-	-	A	-	-	-	-
	Stenothöe monoculoides	-	-	C	-	-	-	A	A	A	-	-
	Tritaeta gibbosa	-	P	A	-	-	-	A	A	A	P	-
	Idotea pelagica	-	-	P	-	-	-	P	P	P	-	-
	Janira maculosa	-	-	C	-	-	-	A	A	A	P	-
	Janiropsis breviremis	-	-	-	-	-	-	P	-	P	-	-
	Pagurus bernhardus	-	-	-	-	-	-	P	P	P	-	-
	Parapleustes bicuspis	-	-	-	-	-	-	-	P	P	-	-
MOLLUSCA	*Archidoris pseudoargus*	-	-	-	-	-	-	-	-	P	-	-
	Dendronotus frondosus	P	-	P	-	-	A	A	A	C	-	-

APPENDIX 5. Other subalgal species commoner in deeper water.

		1	2	4	6	8	10	12	14	16	18	19
COELENTERATA	*Plumularia setacea*	-	-	P	-	P	C	-	A	P	P	-
	Sertularia argentea	-	-	C	P	A	-	A	P	A	-	P
POLYCHAETA	*Eusyllis blomstrandi*	-	P	C	-	P	-	C	P	P	-	-
MOLLUSCA	*Mytilus edulis*	-	-	-	-	-	C	A	A	-	-	-
	juvenile mytilids	-	P	A	P	-	P	P	A	P	P	-

APPENDIX 6. Widely distributed species.

		1	2	4	6	8	10	12	14	16	18	19
PORIFERA	*Hymeniacidon sanguinea*	-	P	P	-	-	-	P	-	P	P	P
	Microciona atrasanguinea	-	P	P	P	-	-	-	P	P	P	P
	Mycale littoralis	-	P	P	-	-	-	P	P	P	P	P
	Myxilla rosacea	-	C	C	C	-	-	C	C	C	C	-
	Oscarella lobularis	-	-	P	P	-	-	-	C	P	P	P
COELENTERATA	*Calycella syringa*	-	P	A	P	P	-	A	-	P	P	P
	Sertularella polyzonias	P	-	-	P	-	-	-	P	-	P	P
NEMERTINI	*Oerstedia dorsalis*	P	C	P	P	C	P	C	P	C	C	C
POLYCHAETA	*Eulalia viridis*	C	P	P	P	P	P	P	P	P	C	P
	Harmothoe imbricata	-	P	-	P	-	-	-	-	P	-	P
	Lagisca extenuata	-	P	P	P	-	P	-	P	-	-	-
	Lepidonotus squamatus	P	P	-	P	P	P	P	P	P	P	P
	Pholoë minuta	P	-	P	-	-	-	P	-	P	-	-
CRUSTACEA	*Balanus crenatus*	P	P	P	-	-	-	-	C	C	P	C
	Verruca stroemia	-	-	P	-	P	P	-	P	C	P	-
	Lysianassa ceratina	-	P	-	P	-	-	P	-	P	-	-
	Parajassa pelagica	-	P	P	P	-	P	C	P	P	P	-
	Porcellana longicornis	P	C	C	P	-	C	C	P	C	-	C
PYCNOGONIDA	*Nymphon brevirostre*	P	P	P	P	-	-	P	P	P	-	P
BRYOZOA	*Diastopora patina*	A	-	-	-	-	C	-	P	P	-	-
	Electra pilosa	A	P	C	P	P	P	A	C	C	P	A
	Lichenopora hispida	-	-	P	C	P	-	-	C	P	P	-

APPENDIX 7. Scarcer species, with station numbers

PROTOZOA
  Ephelota gemmipara   18
  Gromia oviformis   14

PORIFERA
  Adocia cinerea   2, 4
  Adocia simplex   19
  Dysidea fragilis   2, 4
  Haliclona oculata   4, 16
  Hymeniacidon sanguinea   12
  Leucandra nivea   6
  Microciona armata   4
  Ophlitaspongia seriata   18
  Prosuberites epiphytum   4, 16
  Tethyspira spinosa   4

COELENTERATA
  Eudendrium capillare   19
  Halecium halecinum   16
  Metridium senile   2
  Nemertesia antennina   6
  Obelia dichotoma   19
  Sagartie elegans   16, 18
  Sertularia operculata   16

PLATYHELMINTHES
  Leptoplana tremellaris   4, 8
  Stylochoplana maculata   2
  Stylostomum ellipse   4

NEMERTINI
  Lineus longissimus   4, 14, 18

POLYCHAETA
  Autolytus prolifer   4, 12
  Dasychone bombyx   1
  Eulalia sanguinea   1
  Exogone gemmifera   1*, 12
  Flabelligera affinis   12
  Harmothoë impar   4
  Kefersteinia cirrata   4
  Leptonereis glauca   1, 12
  Nereis fucata   6
  Nereis zonata   12
  Odontosyllis ctenostoma   2, 4
  Ophryotrocha puerilis   18
  Sabella pavonina   2, 6
  Scalisetosus pellucidus   2
  Sthenelais boa   14

CRUSTACEA
  Apherusa jurinei   2
  Balanus balanus   4, 14, 19
  Cytherura cellulosa   4
  Dulichia porrecta   12
  Elminius modestus   18*
  Gammarellus angulosus   16
  Homarus gammarus   6
  Hyas coarctatus   16
  Jassa dentex   4
  Jassa pusilla   2
  Macropipus depurator   6
  Macropodia longirostris   10
  Munna fabricii   4
  Munna kroyeri   12
  Panoploea minuta   2
  Pilumnus hirtellus   2, 10, 14
  Porcellana platycheles (juv.)   14

PYCNOGONIDA
  Nymphon gracile   6, 16

MOLLUSCA
  Aeolidia papillosa   16
  Buccinum undatum   2
  Calliostoma zizyphinum   4
  Coryphella pedata   19
  Coryphella verrucosa   18
  Cuthona concinna   19
  Eubranchus pallidus   12
  Facelina auriculata   6, 10, 18
  Hiatella arctica   16
  Lepidopleurus asellus   4
  Musculus discors   16
  Musculus marmoratus   19
  Venerupis rhomboides   6

BRYOZOA
  Bugula avicularia   1
  Callopora lineata   1
  Crisidia cornuta   8, 16, 18
  Microporella ciliata   16
  Tubulipora liliacea   1, 14, 16

KAMPTOZOA
  Pedicellina cernua   8, 19

TUNICATA
  Dendrodoa grossularia   6
  Sidnyum turbinatum   8, 19

ECHINODERMATA
  Solaster papposus   6

PISCES
  Agonus cataphractus   6
  Centronotus gunnellus   2, 16
  Crenilabrus melops   4
  Taurulus bubalis   2

* Common at this station

EPIBENTHIC ASSEMBLAGES AS INDICATORS OF ENVIRONMENTAL CONDITIONS

G. Könnecker

Zoology Department, University College, Galway, Ireland

ABSTRACT

Natural assemblages of attached macro-organisms are introduced as dependable indicators of environmental conditions. Three main regimes, reflecting conditions of salinity and temperature, are recognised for the west coast of Ireland. These regimes, and their constituent associations, are as follows:

    I.   Stenohaline, stenothermal, offshore.

        (a) *Tethyopsilla - Tetilla* Association (below 40 m.)
        (b) *Axinella dissimilis* Association (25 - 40 m.)

    II. Stenohaline, eurythermal, offshore-inshore.

        (a) Upper *Laminaria hyperborea* Association (0 - 15 m.)
        (b) Lower *Laminaria hyperborea* Association (15 - lower limit of *Laminaria*)

    III. Euryhaline, eurythermal, inshore.

        (a) *Lithothamnium* Association (0 - 20 m.)
        (b) *Laminaria saccharina* Association (0 - 10 m.)
        (c) *Raspailia - Stelligera* Association (below 10 m.)
        (d) *Musculus discors* Association (No depth limits, strong currents)

The associations are discussed in their environmental context, particularly in relation to turbidity and water-movements.

INTRODUCTION

Many recent benthic ecological studies focus exclusively on macro-elements of the infauna as indicators of environmental conditions. This is indeed regrettable when one considers the indicator potential of the 'attached' epibenthos - due primarily to its fixed state and the longevity of many of its components.

Taxonomical familiarity with a broad range of epifaunal organisms led the author to consider their responses - as constituents of natural associations - to such environmental features as salinity, temperature, water-movement and turbidity. This work was centered mainly on the Galway Bay area of the west coast of Ireland (Fig. 1).

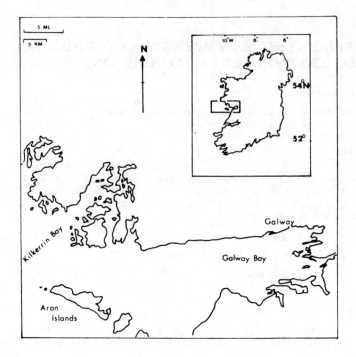

Fig. 1.   Location Map of Study Area.

Remote and *in situ* survey methods were employed and, for the following reasons, the emphasis in sampling was on qualitative description:

(a) truly quantitative work is not possible outside of diving limits, and, even within these limits, it is so time consuming as to inhibit significant spatial surveying;

(b) even where intercomparison of sites is logistically possible, the variability of microhabitats, the small number of species and low biomass in some associations, and factors such as larval dispersal makes quantitative assessment very difficult.

RESULTS

To-date, concentrating on Porifera, Coelenterates (particularly Hydroids), Bryozoans and Tunicates, eight major associations have been identified. Conveniently, these can be ascribed to three regimes.

The first regime, *i.e.* comprising the *Thetyopsilla-Tetilla* Association and the *Axinella dissimilis* (Bowerbank) Association, is indicative of stenohaline and stenothermal conditions.   Named after the two sponges *Thetyopsilla zetlandica* (Carter) and *Tetilla cranium* (O.F. Müller), the *Thetyopsilla - Tetilla* Association is found below 40m and, during the present study, was traced to a depth exceeding 200 m.   Sponges dominate and some 60% of those recorded do not occur in shallow water or inshore.   Generally, they have a

wide geographical and bathymetrical range with their lower known depth limits
often exceeding 1000 m (Arndt, 1934). The more common species include
Phakellia ventilabrum (Johnston), *Axinella pyramidata* Stephens, *Jophon piceus*
(Vosmaer), *Myxilla fimbriata* (Bowerbank), and members of the genera *Hymedesmia*
and *Eurypon*. Also common are the brachiopods *Crania anomala* (O.F. Müller)
and *Terebratulina caput-serpentis* (L.). The temperature range apparently
lies between 8 and 11°C with full oceanic salinity. No information on
water-movement is presently available. *T. zetlandica* and *T. cranium* were
selected as characterising spp. because of their regular occurrence in dredge
samples and also because their spiculae were commonly found adhering to other
material. This is taken to suggest a high density of the contributing
species and, as a matter of interest, the author found it to apply exclusively
to calcareous spicules in shallow water.

The *Axinella dissimilis* Association can be found above the *Thetyopsilla Tetilla*
Association and extends upwards to 25m. First described from the English
Channel (Cabioch, 1968), it shows, by comparison with the latter association,
an increase in the importance of groups other than the Porifera. Especially
noticeable is the reduction in the number of small encrusting sponges. It
occupies rocky patches which are typically surrounded by shell gravel or
coarse sand and which are undoubtedly subjected to abrasive scouring by these
sediments during gales. The characterising species are as follows: the
sponges *Axinella dissimilis*, *A. infundibuliformis* (L.), *Phakellia ventilabrum*
and *Desmacidon fruticosum* (Montagu); the tunicate *Diazona violacea* (Savigny);
the octocorals *Eunicella verrucosa* (Pallas) and *Alcyonium glomeratum* (Hassal),
and the bryozoans *Pentapora foliacea* (Ellis and Solander) and *Porella
compressa* (Sowerby). Depending on water-clarity, algal growth can sometimes
extend into this association. Temperatures range between 8 and 14°C.

The second regime also comprises two associations, each of which features
the alga *Laminaria hyperborea* (Gunn.) Fosl. Both require the following
conditions: (i) good light penetration, and (ii) rather strong water-
movement, generally through wave-surge but, under inshore conditions, strong
currents. Temperatures between 6 and 18°C. The Lower *Laminaria hyperborea*
Association (Künnecker, 1973) occurs between 15 m and the lower limit of the
alga, *i.e.* 25-30 m in the study area. In the main, the epifauna is poor.
Tunicates may be absent altogether. Hydroids are scarce except on
Laminaria stipes and sponges are mostly represented by massive specimens of
Cliona celata (Grant) and *Pachymatisma johnstonia* (Bowerbank). The
following echinoderms are typically well represented: *Holothuria forskali*
Delle Chiaje, *Cucumaria normani* Pace and *Echinus esculentus* L. It is
presumed that the feeding activity of *Holothuria* and *Echinus* contributes
significantly to the overall barrenness of the association.

The Upper *Laminaria hyperborea* Association is markedly richer in species.
Here the most conspicuous species include the sponges *Myxilla incrustans*
(Johnston) *Amphilectus fucorum* (Esper), *Ophlitaspongia seriata* (Grant) and
Scypha compressa (Fabricius); the hydroids *Obelia geniculata* (L.) and
Aglaophenia pluma (L.), and the bryozoans *Electra pilosa* (L.) and *Membranipora
membranacea* (L.) which are found on the *Laminaria* stipes and fronds respectively. Amongst the numerous tunicates, *Distomus variolosus* Gaertner is easily
the most prominent and is followed by *Tridemnium* and *Didemnium* spp., and by
members of the genus *Polycarpa*. Competition with algae can be a limiting
factor especially on upwards facing surfaces.

The third regime consists of four associations, all found inshore, often under conditions of reduced salinity and experiencing considerable fluctuations in temperature within the range 4-20°C. In their distribution, the associations exhibit distinct preferences in relation to current strength and water turbidity. The *Laminaria saccharina* (L.) Association and the *Raspailia-Stelligera* Association (Könnecker 1973) are found in moderate to slack current flow and apparently thrive in conditions of high turbidity.

Within the *L. saccharina* Association, which extends downwards to a depth of 10-15 m, heavy algal growth limits the epifauna to vertical or overhanging rockfaces and to the algal surfaces themselves. The dominant faunal elements are the tunicates, particularly *Dendrodoa grossularia* (van Beneden) and *Ascidia* and *Ascidiella* spp. These seem to be immune to excessive sediment precipitation and probable stifling effects of downhanging curtains of *Laminaria*.

Below the *L. saccharina* is found the *Raspailia - Stelligera* Association. It is named after the sponges *Raspailia ramosa* f. *pumila*, (Montagu) and *Stelligera stuposa* (Montagu), both upright branching forms. Within the study area, it was found to extend to a depth of 40 m but the author believes that it has a greater depth range. Sponges and tunicates feature prominently in a rich and varied fauna which contains, as characterising elements, the following: the sponges *Raspailia ramosa* f. *pumila*, *Stelligera stuposa*, *Polymastia boletiformis* (Lamarck), *Tethya aurantium* (Pallas), *Poecilastra compressa* (Bowerbank) and *Stryphnus ponderosus* (Bowerbank); the tunicates *Ascidia virginea* (O.F. Müller), *Corella paralellogramma* (O.F. Müller), *Ciona intestinalis* (L.) and, notably, *Pyura tesselata* (Forbes). The hydroids *Nemertesia antennina* (L.), *N. ramosa* (Lamouroux), *Hydrallmania falcata* (L.) and *Tubularia indivisa* (L.) are quite common, as is also the bryozoan *Alcyonidium gelatinosum* (L.). Most of the sponges, because of their upright stance, can live on horizontal surfaces, and this despite heavy sedimentation. More sensitive species are confined to vertical or overhanging rocks and the undersides of stones.
The *Lithothamnium* Association colonises branched or massive calcareous algae of the genera *Lithothamnium* and *Phymatolithon*. In the main it comprises sereral sponges the more diagnostic being *Stylostichon plumosum* (Montagu), *Mycale rotalis* (Bowerbank), *Amphilectus fucorum* and *Haliclona* spp. Within the area of the association, rocky outcrops and other hard surfaces can be heavily settled by the feather-star *Antedon bifida* (Pennant), by the octocoral *Alcyonium digitatum*, the bryozoan *Flustra foliacea* (L.) and by a variety of hydroids. The *Lithothamnium* Association is found in strong currents - up to 1.25m/sec in the study area - and extends to depths of some 20 m. Rhodolitic *Phymatolithon* is apparently associated with whirlpool action and is typically found near the outlets of channels with strong tidal flow.

The *Musculus discors* (L.) Association, originally described by Cabioch (1968), colonises hard substrates swept by strong currents (3m/sec in the study area). Characteristically the bivalve *Musculus discors* exhibits extreme aggregation. Covering most available surfaces, the epifauna is exceptionally rich, with the sponge *Pachymatisma johnstonia* (Bowerbank) being especially prominent. In one localised area, the association extends to a depth of 25 m.

DISCUSSION

In the main, the separation of the three regimes is based on temperature. It is the author's opinion that in their vertical depth ranging, the regimes reflect thermal stratification of the water body and not the actual pressure gradient. Whilst present knowledge of the physical and biological factors which influence the deeper offshore associations is understandably sketchy, the inshore associations are much better understood. Obviously water turbidity and sediment precipitation exert a major control on epifaunal distribution patterns. This must be especially so in the case of the more sensitive organisms like the Porifera which are particularly prone to clogging of their incurrent canals, (viz. Boury-Esmault 1971, Reiswig 1971 and Kűnnecker 1973 regarding anti-clogging defense machanisms in sponges). Tunicates are immune to sedimentation whilst hydroids and bryozoans seem to be able to cope.

It seems that light penetration has little direct influence on the bulk of the epifauna. Obviously where it allows algal growth there will be a varying degree of spatial competition between flora and fauna. Over its distributional range, the *Raspailia - Stelligera* Association extends from bright sunlit waters to the dark depths without noticable change in its faunistic composition. This also applies to the *Musculus discors* Association. As defined, the *Laminaria saccharina* and the *Lithothamnium* Associations require light for the characterising species to flourish.

ACKNOWLEDGEMENTS

This contribution reports on part of an on-going programme of benthic studies which is directed by Brendan F. Keegan and financed by the National Science Council of Ireland.

REFERENCES

Arndt, W. 1934. In Tierwelt der Nord - und Ostsee, Teil 3, a 1, pp 140 (lief. 27).

Cabioch, L. 1968. Contribution à la Connaissance des Peuplements Benthiques de la Manche Occidentale. Cah. Biol. mar. 9 : 493-720.

Gislen, I. 1930. Epibioses of the Gullmar Fjord. 11. Marine sociology, skr. K. svenska Vetensk. Kristinebergs. Zool. Stat. 1877-1927 4: 1-380.

Kűnnecker, G. 1973. Littoral and Benthic Investigations on the west coast of Ireland - 1 (Section A: Faunistic and Ecological studies). The sponge fauna of Kilkieran Bay and adjacent areas. Proc. R. Ir. Acad. 73(B): 451-472.

Reiswig, H.M. 1971. In situ pumping activities of tropical Demospongiae. Mar. Biology 9: 38-50.

Boury-Esmault, N. Une structure inhalante remarquable des spongiaires: le crible. Etude morphologique et cytologique. Arch. Zool. Exp. gen. 113 (1) : 7-23.

RECHERCHES SUR LE RÉGIME ALIMENTAIRE ET LE COMPORTEMENT PRÉDATEUR DÉS DECAPODES BENTHIQUES DE LA PENTE CONTINENTALE DE L'ATLANTIQUE NORD ORIENTAL (GOLFE DE GASCOGNE ET MAROC)

Jean Paul Lagardère

Station Marine d'Endoume - Antenne de La Rochelle, C.R.E.O., Allée des Tamaris, 17000 La Rochelle, France

ABSTRACT

This paper presents the main results of a study of biocenotic structure and predator behaviour analysed through the nutrition of bathyal decapods.

Existence and maintenance of bathyal decapod populations, along the continental slope, are supported by migrating crustaceans (copepods, euphausiids : Meganystiphanes norvegica, and shrimps: Sergestes arcticus and Pasiphaea sivado) able to feed by night upon near-surface organisms.

These investigations suggest also that on the continental margin, the vertical distribution and migration of pelagic prey are the main factors determining the vertical range of their benthic predators. On one hand, faunal change at 400 meters reveals the partial inadaptability of shelf decapods, because of the disturbance of their feeding rhythms by the darkness, to collect pelagic prey which come close to the bottom during the day. On the other hand, the faunal change between 1000m and 1200m seems highly correlated with an important decrease in direct energy transfer from the plankton to the bottom. The specific composition of the food of the bathyal decapods shows the existence of three categories of predator behaviour: species which search for their food by odour (Plesionika heterocarpus, Geryon longipes ...), species which hunt by sight (Bathypelagic shrimps) and species which locate their prey by the sounds or vibrations caused by their movements (Plesiopenaeus edwardsianus, Aristaeomorpha foliacea, Aristeus antennatus, Plesionika martia, Solenocera membranacea ...).

INTRODUCTION

Longtemps considérés comme des limivores, les Crustacés du talus continental attiraient peu l'attention sur leurs problèmes alimentaires. Pourtant les travaux de Brian (1931), Massuti (1953) et de Maurin et Carries (1968) dessinaient un profil assez carnassier et euryphage des grands Pénéides bathyaux.

Concernant les espèces bathypélagiques, les observations de Renfro et Pearcy (1966) et d'Omori (1969) soulignent un régime alimentaire carnivore basé sur divers organismes planctoniques (Copépodes, Ostracodes, Chaetognathes ...).

La pluspart des espèces de Décapodes bathypélagiques et benthiques, recueillis sur les fonds meubles du talus continental de l'Atlantique Nord oriental, forment des populations d'individus souvent très denses, et cette prolifération est incompatible avec la faiblesse des ressources en invertébrés benthiques disponibles sur le fond. Quelles sont alors les ressources alimentaires où ces animaux puisent leur dynamisme? La réponse à cette interrogation exigeait une analyse détaillée et précise des contenus stomacaux de chacune des espèces rencontrées. Mais, très vite, cette recherche a posé de nouveaux problèmes. La cohésion des groupements carcinologiques de la pente continentale est-elle, au moins partiellement, de nature trophique? Quel est l'impact de l'alimentation sur les mécanismes de distribution verticale? Comment enfin, ces Crustacés profonds détectent et capturent-ils leurs proies dans l'obscurité des profondeurs?

Ce sont là les problèmes abordés par cette étude de l'alimentation des Décapodes bathyaux de l'Atlantique Nord oriental, étude poursuivie sur les fonds meubles du talus continental du Golfe de Gascogne et sur la vasière bathyale de la côte atlantique marocaine ($35°$ $50'N$ à $33°$ $40'N$). La présente note a pour but d'en souligner les principaux résultats et acquis.

METHODES

L'échantillonnage des Crustacés bathyaux a été réalisé à l'aide d'un chalut Marinovitch de 13m d'ouverture et d'un maillage de 10mm en cul; Sur la poche principale est maillé un petit filet à plancton (maille 0.5mm) destiné à recueillir la petite faune carcinologique mise en suspension par le bourrelet (Lagardère 1969).

Les fonds prospectés s'étendent entre 100 et 1300 m de profondeur dans le Golfe de Gascogne, et 100 - 700m sur la côte marocaine.

Ces prélèvements qualitatifs furent complétés par des pêches expérimentales destinées à analyser le comportement prédateur des animaux étudiés. Les casiers employés étaient de deux types. L'un, en forme de nasse, est la réplique de celui préconisé par Massuti (1967). L'autre s'apparente à une trappe (Fig. 1) dont la fermeture est assurée par rupture d'un barreau de magnésium, calibré suivant le temps d'ouverture que l'on désire programmer. Les parois du piège sont faites en filet de 10mm de maille. Ce casier présente l'avantage de permettre un libre accés vers le leurre. Il supprime

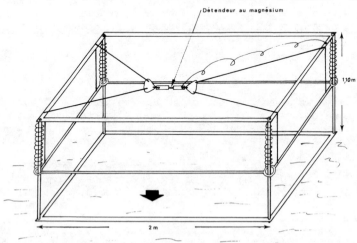

Fig. 1. Casier trappe schématisé en position ouverte.

la contrainte possible que respresente le passage dans l'entonnoir grillagé du modèle classique.

RESULTATS

Distribution verticale

Les captures réalisées le long de la pente continentale du Golfe de Gascogne permettent d'isoler trois grands groupements carcinologiques (utilisation du coeffecient de distinctiveness de Menzies et al. 1973) qui se superposent assez bien aux limites d'étagement proposées par Pérès et Picard (1964), Reyes (1970) et Carpine (1970). C'est ainsi que l'on peut distinguer:

une zone de transition (200 - 400m) où dominent pour les Décapodes : Sergestes arcticus Kröyer, Pasiphaea sivado (Risso), Dichelopandalus bonnieri Caullery, Nephrops norvegicus (Linné), Munida sarsi Brinkmann; pour les Euphausiacés : Nyctiphanes couchii (Bell) et Meganyctiphanes norvegica (M. Sars) ; pour les Mysidacés : Gastrosaccus lobatus Nouvel et Anchialina agilis (G.O. Sars) entre 200 et 300m, Mysideis cf parva et Hypererythrops serriventer Holt et Tattersall entre 300 et 400m; pour les Isopodes : Cirolana borealis Lilljeborg; pour les Cumacés : Campylaspis glabra G.O. Sars, Hemilamprops cristata G.O. Sars et Diastyloides serrata (G.O. Sars), pour les Amphipodes: Tryphosites longipes (Bate et Westwood), Hippomedon denticulatus (Bahe), Acidostoma obesum (Bate) et Perioculodes longimanus (Bate et Westwood).

un horizon supérieur de l'étage bathyal (400 - 1000m) où dominent chez les Décapodes : Sergestes robustus Smith, Pasiphaea tarda Kröyer, P. multidentata Esmark, Pontophilus norvegicus (M. Sars), Polycheles typhlops Heller, Pagurus variabilis (Milne-Edwards et Bouvier), Munida perarmata Milne-Edwards et Bouvier, Geryon longipes Milne-Edwards; pour les Euphausiacés : Nematoscelis megalops G.O. Sars; pour les Mysidacés : Boreomysis arctica (Kröyer), Parapseudomma calloplura Holt et Tattersall, Paramblyops rostrata (Holt et Tattersall) et Mysidetes farrani Holt et Tattersall; pour les Tanaidacés et Isopodes : Apseudes grossimanus Norman et Stebbing, A. spinosus (M. Sars) et Bathycopea typhlops Tattersall; pour les Cumacés : Procampylaspis armata Bonnier, Cyclaspis longicaudata G.O. Sars et Leptostylis macrura (G.O. Sars); pour les Amphipodes : Tryphosites alleni Sexton, des Eusiridae, Nicippe tumida Bruzelius et Syrrhoë affinis Chevreux.

un horizon inférieur de l'étage bathyal (1200 - 2500m) où dominent chez les Décapodes : Gennadas valens (Smith), Nematocarcinus sp., Glyphocrangon sp., Stereomastis sculpta (Smith), Munnidopsis sp., Neolithodes grimaldii Milne Edwards et Bouvier et Parapagurus pilosimanus Smith, chez les Mysidacés : les adultes de Gnathophausia zoea W.-Suhm, Eucopia hanseni Nouvel, Paramblyops bidigitata Tattersall et Michthyops parva (Vanhöffen).

Sur les côtes marocaines, la zone de transition est marquée par l'abondance de Plesionika heterocarpus (Costa) et de Parapenaeus longirostris (Lucas). Quant à l'horizon supérieur de l'étage bathyal, il abrite d'importantes populations de Plesionika martia (Milne Edwards), d'Aristeus antennatus (Risso), d'Aristaeomorpha foliacea (Risso) et de Plesiopenaeus edwardsianus (Johnson).

Si la cohésion des groupements carcinologiques ainsi mis en évidence est de nature trophique, celà doit se retrouver dans l'analyse des contenus stomacaux des principales espèces de Décapodes. Cette analyse réalisée sur 25 espèces et plus de 2500 individus, n'a pas abouti au résultat escompté (Lagardère

1972, 1973, 1976a et b). L'énoncé qui va suivre et qui rapporte les principales proies ingérées par ces grands Crustacés, va nous permettre d'en juger.

Régime alimentaire des crevettes bathypélagiques

Sergestes arcticus : proies pélagiques à dominance de Copépodes; Alimentation principale nocturne près de la surface, alimentation secondaire diurne près du fond;

Sergestes robustus : proies pélagiques à dominance de Chaetognathes, d'Ostracodes et de Copépodes. Chasse nocturne très probable mais remontée moins proche de la surface que l'espèce précédente.

Pasiphaea sivado : proies pélagiques à dominance d'Euphausiacés. Chasse nocturne près de la surface.

Pasiphaea tarda : proies pélagiques à dominance de Mysidacés. Chasse nocturne probablement en pleine eau.

Pasiphaea multidentata : proies pélagiques à dominance d'Euphausiacés. Chasse nocturne probable.

Régime alimentaire des Décapodes benthiques de la zone de transition
 - Golfe de Gascogne

Solenocera membranacea (Risso) : proies benthiques à dominance de Polychètes (Glycera sp., Nephthys sp., Lumbriconereis sp., Spionidés) et d'Amphipodes (Acidostoma obesum, Hippomedon denticulatus), quelques proies pélagiques également (Sergestes arcticus). Chasse nocturne.

Plesionika heterocarpus : proies benthiques variées + nécrophagie. Chasse diurne.

Dichelopandalus bonnieri : alimentation à base de Crustacés pélagiques et benthiques (Sergestes arcticus, Meganyctiphanes norvegica, Cirolana borealis), de Polychètes (Aphroditidés), d'Ophiures plus un complément nécrophagique. Chasse diurne très étalée.

Processa canaliculata Leach : proies benthiques à dominance de Polychètes et Proies pélagiques, exclusivement des Copépodes. Chasse diurne.

Pontocaris lacazei (Gourret) : proies benthiques à dominance de Polychètes.

Pontophilus spinosus (Leach) : proies benthiques à dominance de Polychètes Sabellidés (Allen 1964).

Philocheras echinulatus (M; Sars) : proies benthiques variées (Mysidacés, Amphipodes, Polychètes errantes, Pélécypodes). Deux phases nutritionnelles, l'une à l'aube, l'autre au crépuscule.

Nephrops norvegicus : proies benthiques et pélagiques diverses (Sergestes arcticus, Pasiphaea sivado, Meganyctiphanes norvegica, Cirolana borealis, Aphroditidés, Opheliidés, Glycera sp., Mollusques, Ophiures), nécrophagie importante. Activité de chasse soumise à un rythme endogène unimodal. Elle se manifeste aussi bien le jour que la nuit;

Pagurus variabilis : proies benthiques + nécrophagie.

Munida sarsi : proies pélagiques à dominance de Crustacés (Meganyctiphanes norvegica et Copépodes) et nécrophagie.
 - Maroc

Parapenaeus longirostris : proies benthiques variées (Philocheras bispinosus (Hailstone), Lophogaster typicus M. Sars, Gastrosaccus lobatus, Anchialina agilis, Pélécypodes). Chasse nocturne.

Plesionika heterocarpus : proies benthiques et pélagiques (Meganyctiphanes norvegica, Nyctiphanes couchii, Euphausia krohnii (Brandt), Mysidacés, Pélécypodes) plus nécrophagie. Activité diurne.

Chlorotocus crassicornis (Costa) : proies pélagiques (Meganyctiphanes norvegica).

Régime alimentaire des Décapodes benthiques de l'horizon supérieur de l'étage bathyal.
 - Golfe de Gascogne.

Plesionika martia : proies à dominance de Polychètes (Aphroditidés) + nécrophagie.

Pontophilus norvegicus : proies benthiques où dominent les Polychètes (Glycéridés et Spionidés) et les Mollusques Opisthobranches, Philine quadrata(Wood). Chasse diurne, peut-être aussi nocturne.

Geryon longipes : nécrophagie importante complétée par la capture de proies benthiques diverses.
 - Maroc

Plesionika martia : proies pélagiques (Pasiphaea sivado, Meganyctiphanes norvegica) représentant plus des trois quarts du volume de nourriture ingérée, complément nécrophage. Chasse diurne.

Aristeus antennatus : proies benthiques et pélagiques (Crustacés divers et Polychètes errantes). Activité nocturne probable.

Aristaeomorpha foliacea : proies benthiques et pélagiques (Plesionika martia, Pasiphaea sivado, Meganyctiphanes norvegica). Chasse diurne.

Plesiopenaeus edwardsianus : proies benthiques et pélagiques (Polycheles typhlops, Plesionika martia, Pasiphaea sivado, Sergestes sp., Meganyctiphanes norvegica). Activité diurne, plus accusée à l'aube et au crépuscule.

DISCUSSION

Liens d'interdépendance
Ce bref énoncé du régime alimentaire des Décapodes évoluant le long de la pente continentale met en évidence l'électisme dont ils font preuve dans le choix de leurs proies. On ne constate jamais, en dehors de Pontophilus spinosus vis à vis des Sabellidés, de dépendance précise et exclusive de tel prédateur à l'égard de telle ou telle espèce proie benthique. La cohésion des groupements carcinologiques benthiques de la pente continentale ne peut donc s'expliquer par des liens d'interdépendance de nature trophique.

Cepedant, une dépendance manifeste s'affirme entre ces mêmes Décapodes et les Crustacés pélagiques qui viennent chercher un refuge diurne au voisinage du fond. Elle est d'ailleurs d'autant plus importante que le prédateur vit plus profond. Ainsi les espèces de la zone de transition n'utilisent ces ressources d'origine pélagique que comme un appoint nutitionel, la base de leur alimentation demeurant des proies d'origine benthique. Par contre, dans l'horizon supérieur de l'étage bathyal, les grands Décapodes carnassiers (Plesionika martia, Aristaeomorpha foliacea, Plesiopenaeus edwardsianus, et Polycheles typhlops) s'alimentent de manière assez exclusive de Crustacés pélagiques, sans toutefois pousser cette exclusivité jusqu'à un niveau spécifique. Pour ces prédateurs, il ne fait aucun doute que les apports énergétiques provenant du plancton de surface et transitant à travers les différents maillons migratoires que sont les Crustacés infra et bathypélagiques (Copépodes, Mysidacés Euphausiacés et Natantia), sont le fondement de leur maintien et du développement de leurs populations dans ces biotopes profonds où la production benthique s'affaiblit.

Ce cycle direct d'enrichissement des fonds du haut de la pente continentale rejoint le schéma vinogradovien (Vinogradov 1961) du transfert de l'énergie vers les zone profondes. Il marque fortement la physionomie de l'étage bathyal et souligne la dimension verticale des communautés benthiques de la pente continentale.

Influence du facteur trophique sur la distribution verticale des Décapodes bathyaux.   Influence du facteur éclairement.

Les limites bathymétriques de la zone de transition (200 - 400m) peuvent s'interpréter comme représentant, d'une part, la limite de pénétration en profondeur de la faune littorale (400m) et d'autre part, le niveau supérieur de remontée de la faune profonde (200m). L'affrontement de la faune du domaine phytal avec celle du domaine aphytal devrait, logiquement, se régler sous le contrôle du facteur éclairement. Nous allons voir comment ce facteur peut influencer le comportement des divers constituants du peuplement carcinologique de la zone de transition.

Dans cette zone de pénombre, l'apport énergétique lié à la consommation de Crustacés infra et bathypélagiques assure à la plupart des Décapodes, colonisant ces fonds, de bonnes conditions de vie. Cet apport exogène, bien que régulier, n'est pas constant tout au long de la journée. Seule, la période diurne, permet aux divers prédateurs benthiques d'y puiser. Il y a là une première influence de l'éclairement qui, réglant le va et vient des organismes pélagiques, impose un rythme bien précis aux Décapodes benthiques qui en dépendent.

Mais l'éclairement agit aussi sur le rythme d'activité des prédateurs. On sait qu'un éclairement important entraine une réaction d'inhibition chez les espèces bathyales. A l'inverse, une obscurité totale peut provoquer l'arrêt de la photocinèse chez les espèces circalittorales. Cependant des travaux récents, dont ceux de Macquart-Moulin (1975), montrent que dans ce cas des rythmes endogènes se manifestent et maintiennent l'activité du Crustacé. L'animal peut donc être actif malgré l'obscurité permanente, mais à la longue son rythme se dérégle. Il y a alors soit étalement dans le temps des heures d'activité (cas des Isopodes étudiés par Macquart-Moulin), soit décalage progressif des heures d'activité lié à l'augmentation de la période (cas des Amphipodes). Ces deux cas s'observent chez les Décapodes où Dichelopandalus bonnieri serait du type "Isopode" et Nephrops norvegicus du type "Amphipode".

Le premier manifeste une très longue période de chasse (Lagardère 1973), alors qu'on remarque surtout un fort décalage dans le rythme d'activité du second. Bref, à travers ces deux exemples, on peut mesurer les problèmes auxquels vont se heurter les Décapodes d'origine littorale, au cours de leur tentative de colonisation de la pente continentale. En effet, pour eux, la survie dépend de la bonne synchronisation entre leur phase d'activité et le temps de disponibilité, près du fond, des proies pélagiques. De ce fait la transition entre l'étage circalittoral et l'étage bathyal se place, chez les Décapodes, sous le double contrôle de l'éclairement et des relations trophiques.

Le renouvellement faunistique qui intervient au delà de 1000m et qui scinde l'étage bathyal en deux horizons, procède d'un mécanisme purement trophique. Nous le pensons lié à la diminution massive des transferts énergétiques les plus directs du domaine pélagique vers le domaine benthique. En deçà de 1000m la relation directe plancton-benthos permet encore le maintien et l'épanouissement des grands carnassiers que sont : Plesionika martia, Aristaeomorpha foliacea, Plesiopenaeus edwardsianus ... Au delà de 1000m, cette liaison directe unissant production planctonique et production benthique devient impossible car la profondeur excède alors les possibilités migratoires quotidiennes de la plupart des organismes pélagiques. A ce moment, on note l'établissement d'un relai en pleine eau qui absorbe une large part de l'énergie acheminée vers le fond. Les carnassiers deviennent de plus en plus pélagiques alors que l'alimentation des Décapodes benthiques tend à s'uniformiser dans l'utilisation des détritus et dans l'ingestion de petites proies mortes ou vivantes comme le constatent Dayton et Hessler (1972).

Comportement prédateur des Décapodes bathyaux

L'analyse précise des proies capturées par ces grands Crustacés apporte de précieuses indications sur leur comportement prédateur. Le tableau de la page suivante regroupe les pourcentages numériques de capture des diverses proies appartenant aux principaux groupes d'invertébrés. Il mentionne également le pourcentage de capture des proies mobiles au sein de l'alimentation normale de chaque espèce.

L'odorat. La rubrique Poisson du tableau suivant permet d'emblée d'isoler les principales espèces nécrophages évoluant sur les substrats meubles de la pente continentale. Ce sont: Dichelopandalus bonnieri, Plesionika heterocarpus, Munida sarsi, Pagurus variabilis et Geryon longipes. Les essais de pêche au casier réalisés dans l'Atlantique Nord oriental, les observations de Bomace, celles de Massuti (1967) confirment l'exactitude de ce comportement nécrophage pour les espèces suivantes: Dichelopandalus bonnieri, Plesionika martia, Aristeus antennatus, Nephrops norvegicus, Pagurus arrosor, Macropipus tuberculatus (Roux), Bathynectes superbus (Costa). Toutefois il convient d'indiquer que le faible rendement des captures de ces espèces parait souligner une certaine passivité de leur nécophagie.

Cette diminution des formes nécrophages, surtout les plus actives et les plus spécialisées, au fur et à mesure de l'accroissement de la profondeur est également sensible devant les résultats obtenus sur les fonds abyssaux du Golfe de Gascogne (Rannou et Nouguier 1974). Il semble donc que ce mode alimentaire, lié à l'odorat, soit en régression dans le domaine aphytal malgré "l'importante éconimie énergétique" qu'il représente par rapport à la prédation (Arnaud 1974).

Deux facteurs importants desservent la nécrophagie en profondeur. D'abord l'

1 - GOLFE DE GASCOGNE
ESPECES DU HAUT DE LA PENTE CONTINENTALE

Pourcentage des captures	Solenocera membranacea	Processa canaliculata	Dichelopandalus bonnieri	Plesionika heterocarpus	Pontophilus spinosus	Philocheras echinulatus	Nephrops norvegicus	Munida sarsi
Poissons	2 %	5 %	28 %	19,5 %	~	4 %	9,5 %	25 %
Crustacés	27,5 %	35 %	38,5 %	36,5 %	7,5 %	42 %	24 %	57 %
Polychètes errantes sédentaires	38 % 90,5 '' 9,5 ''	40 % 67 '' 33 ''	12 % 100 '' ~	18 % 48 '' 52 ''	79,5 % 20 '' 80 ''	21 % 89 '' 11 ''	40,5 % 96 '' 4 ''	3 % 100 '' ~
Mollusques mobiles sédentaires	25,5 % 52 '' 48 ''	~	7,5 % 29,5 '' 70,5 ''	19 % ~ 100 ''	2,5 % ~ 100 ''	28 % 33 '' 67 ''	17,5 % 56,5 '' 43,5 ''	~
Echinodermes Ophiurides autres	3,5 % 60 '' 40 ''	1,5 %	13 % 75 '' 25 ''	2,5 %	2,5 %	2 % 50 '' 50 ''	7 % 80 '' 20 ''	4 % 50 '' 50 ''
Divers	3 %	18 %	0,5 %	4 %	7,5 %	2 %	1 %	10 %
Cumul des pourcentages de capture des proies mobiles	80 %	65 %	64 %	47 %	26 %	69 %	78 %	62 %

ESPECES DE LA PENTE CONTINENTALE

Pourcentage des captures	Plesionika martia	Pontophilus norvegicus	Polycheles typhlops	Stereomastis sculpta	Pagurus variabilis	Munida perarmata	Geryon longipes
Poissons	*	~	10 %	*	20 %	~	20,5 %
Crustacés	*	14,5 %	69,5 %	*	49 %	*	30,5 %
Polychètes errantes sédentaires	*	70,5 % 71,5 '' 28,5 ''	20 % 100 '' ~	~	11 % 100 '' ~	*	12,5 % 60 '' 40 ''
Mollusques mobiles sédentaires	~	10,5 % 91 '' 9 ''	~	~	11 % ~ 100 ''	~	28,5 % 36 '' 64 ''
Echinodermes Ophiurides autres	~	1 % ~ 100 ''	~	~	9 % 50 '' 50 ''	*	2,5 % ~ 100 ''
Divers	~	3 %	~	~		*	5 %
Cumul des pourcentages de capture des proies mobiles		75 %	90 %		64 %		48 %

2 - MAROC

Pourcentage des captures	Parapenaeus longirostris	Aristeus antennatus	Aristeomorpha foliacea	Plesiopenaeus edwardsianus	Plesionika heterocarpus	Plesionika martia
Poissons	3,5 %	6 %	13 %	7 %	10 %	9 %
Crustacés	62,5 %	39,5 %	67 %	87 %	37,5 %	84 %
Polychètes errantes sédentaires	6,5 % 100 '' ~	35 % 98 '' 2 ''	2 % 100 '' ~	3 % 100 '' ~	3 % 70 '' 30 ''	~
Mollusques mobiles sédentaires	25,5 % 30 '' 70 ''	13 % 20 '' 80 ''	13 % 84,5 '' 15,5 ''	1,5 % 67 '' 33 ''	41 % ~ 100 ''	5,5 % 60 '' 40 ''
Echinodermes Ophiurides autres	1 %	5 % 75 '' 25 ''	5 % 80 '' 20 ''	1,5 % 100 '' ~	8 % 5 '' 95 ''	~
Divers	0,5 %	1 %	~	~	0,5 %	1 %
Cumul des pourcentages de capture des proies mobiles	76 %	91 %	83 %	92,5 %	40 %	87 %

* Pourcentage non chiffré : échantillonnage insuffisant

absence de populations denses de vertébrés au delà de 800m, ce qui a pour corollaire la raréfaction des cadavres, des déchets de leur métabolisme, et l'inconstance de cet apport.   Ensuite, le fait que les courants profonds ont des oscillations moins importantes que les courants littoraux (Le Floch 1969) ce qui aboutit à des écoulement unidirectionels.   On conçoit alors que la dispersion des émanations issues d'un cadavre soit beaucoup plus restreinte en profondeur.

<u>La vue</u>;   Cet organe sensoriel semble assez peu mis en oeuvre par les Décapodes bathyaux dans la recherche de leur nourriture.   L'obscurité du milieu lui est certes peu favorable.   Mais il faut reconnaitre que cet organe est aussi rarement sollicité, dans la détection des proies, par les Décapodes littoraux qui jouissent pourtant d'un éclairement convenable.

Pourtant, on ne peut, chez les crevettes bathypélagiques, écarter totalement l'utilisation de la vue dans la recherche des proies.   Ces crevettes chassent la nuit, en pleine eau, en se déplaçant continuellement ce qui ne s'accorde ni avec une détection olfactive, ni avec une analyse acoustique.   De plus, on relève chez <u>Pasiphaea sivado</u> et chez <u>Sergestes articus</u>, un <u>très fort pourcentage de capture de proies bioluminescentes</u> :   Euphausiacés pour la première, Copépodes chez la seconde.   Ce pourcentage parait supérieur à l'importance relative des espèces bioluminescentes au sein de leur communauté pélagique.   On ne peut non plus invoquer une prédation par simple contact mécannique car cette dernière aboutirait à une sélection moins accusée.

Un processus expérimental est actuellement en cours de réalisation (leurres mécaniques et Euphausiacés vivants stimulés chimiquement), pour démontrer l'existence d'une perception des signaux bioluminescents chez les Pasiphaeidés et les Sergestidés.

<u>L'ouie</u>.   Nous venons de voir que l'utilisation de l'odorat, ou de la vue, dans la   détection de la nourriture reste l'apanage d'une nombre assez restreint d'espèces profondes.   Quel est donc le support sensoriel mis en oeuvre par la majorité des Décapodes bathyaux?

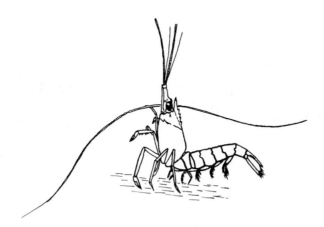

Fig. 2.   Posture d'affût de <u>Solenocera membranacea</u> (d'après Heegaard, 1967).

Si l'on suppose, chez certaines espèces, que la détection des proies se fait à l'aide des récepteurs de vibrations, il faut s'attendre à trouver dans leurs contenus stomacaux un fort pourcentage de proies mobiles dont les ondes de vibration, liées à leurs déplacements, ne manqueront pas d'éveiller l'attention du prédateur.

Avant de nous reporter au tableau qui donne, pour chaque espèce, le pourcentage de capture des proies vagiles, il convient d'envisager le cas de Solenocera membranacea. Le comportement de cette espèce a fait l'objet, de la part de Heegaard (1967), d'une analyse très fine. L'auteur nous décrit en particulier la posture d'affût (Fig. 2). A ce moment, la crevette est parfaitement immobile, tendue à l'extrême, seules les antennes sont animées d'un mouvement lent et rotatif. Cette attitude n'est pas sans rappeler la posture d'écoute des mammifères, d'autant que Heegaard précise plus loin "When the shrimp was not swimming in the free water-mass, but was walking or sitting on the mud in its search for food, it was very easily frightened by either a sound or unusual movements". Cette sensibilité aiguë à toutes vibrations ou ébranlements intervenant au moment où la posture d'affût est adoptée, s'inscrit dans la logique d'un animal aux écoutes.

On pourrait penser que le balayage antennaire correspond à une recherche tactile analogue à celle décrite chez Crangon crangon (Linné) par Dahm (1975). Il en diffère nettement en ce qu'il apparait totalement indépendant d'une perception chimique initiale déclenchant les processus de localisation de la proie émettrice. Par ailleurs, Heegaard souligne que le Solénocère enfoui dans le sédiment, à la différence de Crangon crangon, ne prête aucune attention aux proies environnantes. On remarque aussi que l'immobilité attentive de Solenocera membranacea sur le fond et la rapidité de ses réactions d'attaque n'ont aucune similitude avec la quête olfactive lente et tâtonnante du hommard américain étudié par Mc Lesse (1973).

Enfin, il convient de souligner que, pour nourrir ses crevettes et décrire leur attitude de recherche de la nourriture, Heegaard a utilisé des Nephthys, Polychète certainement parmi les plus "remuantes". Or, dans les conditions de vie normale de Solenocera membranacea, les formes vagiles représentent 80% du nombre des proies capturées (cf tableau précédent). Ce pourcentage élevé tend à démontrer l'existence d'une relation particulière entre l'efficacité de la prédation et la mobilité des proies.

Qu'en est-il des autres espèces de Décapodes ? Le tableau précédent nous permet de constater que les espèces suivantes : Plesiopenaeus edwardsianus (91%), Aristeus antennatus (91%), Polycheles typhlops (90%), Plesionika martia (87%), Aristaeomorpha foliacea (83%), Solenocera membranacea (80%), Nephrops norvegicus (78%), Parapenaeus longirostris (76%), Pontophilus norvegicus (75%), Philocheras echinulatus (69%), réalisent un pourcentage de capture de proies mobiles élevé, voisin de celui de Solenocera membranacea, ce qui incline à leur prêter un comportement analogue, donc une utilisation très vraisemblable des récepteurs de vibrations dans la détection et la localisation de leurs proies.

On pouvait s'attendre à l'emportance numérique, largement dominante, de cette catégorie de prédateur. En effet, elle marque vraisemblablement l'adaptation compensatoire des espèces bathyales chez lesquelles les récepteurs de vibrations prennent le relai de l'odorat, moins utile en profondeur du fait de la rareté des cadavres et de la moins bonne dispersion des émanations. Il est à remarquer que cette voie est aussi celle empruntée par certains Crustacés cavernicoles, tel l'Amphipode Niphargus virei Chevreux (Turquin 1973).

REFERENCES

Allen J.A. 1964. On the biology of Pontophilus spinosus (Leach). Cah. Biol. mar. 5, 17-26;

Arnaud, P.M. 1974. Contribution à la bionomie marine benthique des régions antarctiques et subantarctiques. Téthys 6 (3), 465-656;

Brian, A. 1931. La biologia del fondo a "scampi" nel Mare Ligure. V. Aristeomorpha, Aristeus ed atri macruri natante. Boll. Mus. Zool. Anat. comp. A. Univ. Genova 2(45), 1-6.

Carpine, C. 1970. Ecologie de l'étage bethyal dans la Méditerranée occidentale. Mém. Inst. océanogr. Monaco 2, 1-146.

Dahm, E. 1975. Untersuchungen zum Nahrungserwerb von Crangon crangon Linné. Ber. dt. wiss. Kommn. Meeresforsch. 24, 105-133.

Dayton, P.K. et Hessler, R.R. Role of biological disturbance in maintaining diversity in the deep-sea. Deep-Sea Res. 19, 199_208.

Heegaard, P. 1967. On behaviour, sex-ratio and growth of Solenocera membranacea (Risso) (Decapoda - Penaeidea). Crustaceana 13 (2), 227-237.

Lagardère, J.P. 1969. Les crevettes du Golfe de Gascogne (région Sud). Téthys 1 (4), 1023-1048.

--- , 1972. Recherches sur l'alimentation des crevettes de la pente continentale marocaine. Ibid. 3(3); 655-675.

--- , 1973. Données sur la biologie et sur l'alimentation de Dichelopandalus bonnieri (Crustacé - Natantia) dans le Golfe de Gascogne. Ibid. 5 (1), 155-166.

--- , 1976a. Recherches sur l'alimentation des crevettes bathypélagiques du talus continental du Golfe de Gascogne. Rev. Trav. Inst. Pêches marit. 39 (2), 213 - 229.

--- ,1976b. Recherches sur la distribution verticale et sur l'alimentation des Crustacés Décapodes de la pente continentale de l'Atlantique Nord-oriental. Thèse Univ. Aix-Marseille II, Arch. orig. C.N.R.S., 12.237.

Le Floch, J. 1969. Surla circulation de l'eau d'origine méditerranéenne dans le Golfe de Gascogne et ses variations à courte période. Cah. océanogr. 21 (7), 653-661.

Macquart-Moulin, C. 1975. Les Péracarides benthiques dans le plancton nocturne Amphipodes, Cumacés, Isopodes, Mysidacés. Analyse des comportements migratoires dans le Golfe de Marseille. Recherches expérimentales sur l'origine des migrations et le contrôle de la distribution des espéces. Thèse Univ. Aix-Marseille II, Arch. orig. C.N.R.S., 10.864.

Massuti, M. 1953. Bionomia de los fondos de 300 a 600 metros en el Sur y Suroeste de Mallorca. Bol. Inst. esp. oceanogr. 63, 1-20.

--- , 1967. Resultados de las pruebas experimentales effectuades en agua de Mallorca para la pesca con nasas de las gambas de profundidad. Publ. Tec. Junta Est. Pesca 6, 19-85.

Maurin, C; et Carries, C. 1968. Note préliminaire sur l'alimentation des crevettes profondes. Rapp. P.V. Comm. int. Mer Médit., 19 (2), 155-156.

Mc Leese, D.W. 1973. Orientation of lobsters (Homarus americanus) to odor. J. Fish. Res. Bd. Canada 30, 838-840.

Menzies, R.J., George, R.Y. et Rowe, G.T. 1973. Abyssal environment and ecology of the world oceans. Wiley, New York.

Omori, M. 1969. The biology of a Sergestid shrimp Sergestes lucens Hansen. Bull. Ocean. Res. Inst. Univ. Tokyo 4, 1-83.

Pérès, J.M. et Picard, J. Nouveau manuel de bionomie benthique de la Mer Méditerranée. Rec. Trav. Sta. mar. Endoume 47 (31), 5-137.

Rannou, M. et Nouguier, J. 1974. Pêches abyssales aux casiers. Ann. Inst. océanogr. 50 (2), 139-143.

Renfro, W.C. et Pearcy, W.G. 1966. Food and feeding apparatus of two pelagic shrimps. J. Fish. Res. Bd. Canada 23, 1971-1975.

Reyes, D. 1970. Bionomie benthique de deux canyons sous-marins de la mer catalanne : le rech du Cap et le rech Lacaze-Duthiers. Thèse Fac. Sci. Paris.

Turquin, M.J. 1973. Compensations sensorielles chez les Gammaridés hypogés. Ann. Spéléol. 28 (2), 187-191.

Vinogradov, M.E., 1961. The feeding of deep-sea zooplankton. Rapp. P.V. Cons. perm. int. Explor. Mer 153, 114-120.

FEEDBACK AND STRUCTURE IN DEPOSIT-FEEDING MARINE BENTHIC COMMUNITIES

J. S. Levinton,* G. R. Lopez,** H. Heidemann Lassen** and U. Rahn**

*Ecology and Evolution Program, State University of New York, Stony Brook, New York, U.S.A. 11794
**Institute of Ecology and Genetics, University of Aarhus, Arhus C, Denmark

ABSTRACT

We show that feeding on detritus by the amphipod Orchestia grillus enhances microbial growth and detrital mineralization. Pellet egestion and breakdown rates are measured for the deposit-feeding gastropod Hydrobia minuta. Feeding ceases when the sediment is mostly pelletized. A model assuming fecal pellet breakdown as a controlling factor for deposit-feeding populations is developed and predicts Hydrobia densities compatible with field observations.

INTRODUCTION

Deposit-feeding marine benthic communities are dominated by macroinvertebrates that ingest organic debris and assimilate the attendant microorganisms(Fenchel 1972, and references therein). Because bacteria tend to be associated with particles of high surface- area to volume ratio, food for deposit-feeders is concentrated in finer-grained sediments(Zobell 1938, Newell 1965). Thus deposit-feeding invertebrates are usually more abundant in muddy sediments(Sanders 1958, Newell 1965, Levinton 1972). The presence of dense populations exploiting organic debris at a rapid rate suggests that deposit-feeding populations should often be food-limited and undergo intra- and interspecific competition for food (Sanders 1958, Levinton 1972). Circumstantial evidence supports this conclusion (Levinton and Bambach 1975, Fenchel 1975) and Fenchel and Kofoed(1976) experimentally show that competition between species of Hydrobia leads to niche partitioning on the basis of particle size.

Deposit-feeders ingest particles and bind up the egested material into fecal pellets which are the site of microbial recolonization on sediment particles. Two possible rate-limiting steps for food

Contribution number 188 to the Program in Ecology and Evolution, State University of New York at Stony Brook availability are the breakdown rate of fecal pellets and the recolonization rate by the microbial community. Although several workers have emphasized the importance of coprophagy among deposit-feeders (Johannes and Satomi 1966, Frankenberg and Smith 1967) we have not observed ingestion of whole pellets by Orchestia grillus (Amphipoda) or Hydrobia minuta (Gastropoda) unless they have been broken down.

Deposit-feeders may reduce the size of the microbial population associated with deposits, at the same time stimulating microbial activity and removal of nutrients from detritus (Hargrave 1970, Barsdate et al. 1974, Lopez 1976). We have investigated the interaction between deposit-feeding, detrital decomposition and regeneration of food through fecal pellet breakdown. We present a model predicting maximum densities of deposit-feeders under food limitation.

Materials and Methods

The deposit-feeding gastropod Hydrobia minuta and the detritus eating amphipod Orchestia grillus were both collected from Flax Pond, a Spartina salt marsh in Old Field, New York, U.S.A. H. minuta lives in muddy sediments in ponds in the high intertidal whereas Orchestia grillus was found in wrack accumulating mainly in the upper intertidal. For Hydrobia feeding experiments, sediment was collected and passed through a 63 μ sieve and silt and clay were collected in a bowl. For Orchestia experiments, Spartina alterniflora detritus was collected at Flax Pond, ground in a mill and particles passing through a 63 μ sieve were employed.

Orchestia grillus was fed Spartina detritus at the following rations: 4, 8, and 16 mg per animal. The ingestion rate at 15C is 4 mg day^{-1} animal^{-1} (Lopez 1976). Microbial growth on the Spartina litter particles was followed using qualitative microscopic analysis and through ATP extraction, employing the boiling Tris method (Lopez 1976). ATP mg^{-1} detritus was followed for a period of 21 days at the three grazing levels.

Hydrobia was fed sediment at different snail densities corresponding to 30,000 , 15,000 , and 7,500 meter^{-2}. Sediment per snail was 1.73×10^{-3} g, 3.45×10^{-3} g and 6.93×10^{-3} for the respective densities of 8, 4 and 2 snails per chamber. Fecal pellet egestion rate was measured at 20C and assumed to be ingestion rate, as feeding was steady. Pellets were collected on a 100 μ sieve, diluted to 10 ml of seawater and counts were made of 1 ml of this mixture under a dissecting scope. The per cent pelletization was determined by separating sediment and pellets and weighing on glass fibre filters. Pellet breakdown was followed at 20C by collecting pellets and counting the number of intact pellets remaining after 2-14 days.

Results

Figure 1A shows the change in ATP on detritus grazed at different rates by Orchestia, with a control sample of detritus having no animals. In both the cases of newly egested Orchestia and Hydrobia fecal material, bacteria initiate colonization with bacterial grazers such as ciliates and flagellates appearing later. The microbial community, as estimated by ATP, increased more rapidly at high grazing levels than at other treatments. Furthermore, the per cent nitrogen present in the non-living fraction of the detrital sample was significantly less under high grazing than at low grazing (Fig. 1B). This indicates that grazing by Orchestia influences microbial growth and mineralization rates of detritus.

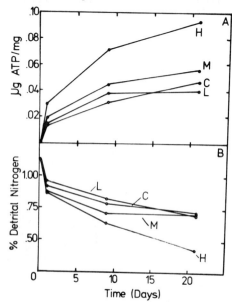

Fig. 1. Changes in ATP mg^{-1} and per cent non-living nitrogen in detritus with microbial community, over time: under low(L), medium(M), high(H) and no(C) grazing by Orchestia grillus. Each point is a mean of 5 replicates. In both A and B the results for treatment H differs significantly from other treatments (p <.05).

Pellet egestion rates are shown in Fig. 2 for the 3 experimental levels of Hydrobia feeding. The highest grazing rate shows a logistic response, reaching a plateau of about 9000 pellets after 50 hours. At this time, almost all of the snails ceased movement, with only occasional feeding observed. This same observation was made in other experiments whenever most of the sediment was pelletized. At maximal pelletization, 68.3 % (mean of 4 replicates) of the available sediment was in the form of fecal pellets. This is probably an underestimate as pellets may have broken during the process of sieving.

The effect of grazing and snail movement on the breakdown of fecal pellets was estimated by measuring fecal pellet accumulation rates at the three densities. Linear regressions were calculated for the first 24 hours of pelletization, where accumulation appeared linear in all cases. All 3 regressions are significant (p < .05). If there is an effect due to snail grazing, then the ratios of the regression coefficients of the high and

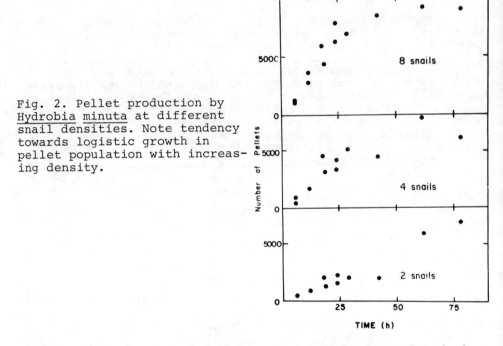

Fig. 2. Pellet production by *Hydrobia minuta* at different snail densities. Note tendency towards logistic growth in pellet population with increasing density.

medium, and medium and low treatments should be less than 2, because grazing intensities differed by this factor and higher grazing should promote pellet breakdown and proportionately fewer pellets per snail per chamber. The ratio between the regression coefficients for high and medium (8 and 4 snails) treatments was 2.1, and the ratio between medium and low (4 and 2 snails) was 1.9. Therefore there is no measurable effect of grazing intensity on pellet breakdown in the first 24 hours.

Figure 3 shows pellet breakdown rates. For the first 4 days pellet breakdown is slow and mainly consists of mechanical fragmentation. However, pellets then begin to become indistinct and breakdown rate increases. Although a semi-logarithmic plot shows an increase of breakdown rate with time, a constant percentage (about 9%) broke down per day giving a linear relationship (F_s = 652.7, n = 36, p <.01).

Discussion

The data relating grazing by the amphipod *Orchestia grillus* to the development of the microbial community shows that grazing enhances microbial growth and accelerates the removal of nitrogen from detritus. Our results parallel those of other workers who show an enhancement of bacterial activity with grazing (Hargrave 1970, Barsdate et al. 1974, Rahn 1975). Feedback between deposit-feeders and their food is thus indicated.

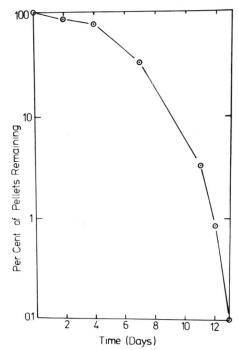

Fig. 3. Survivorship curve of intact fecal pellets of Hydrobia minuta.

The rapid rate of pelletization with subsequent cessation of feeding observed in this study is important because neither Hydrobia nor Orchestia have ever been observed by us to ingest their own fecal pellets. It is not clear whether published accounts of fecal pellet ingestion (e.g. Newell 1965, Johannes and Satomi 1966) refer to ingestion of partially disaggregated fecal paterial or the intact pellets themselves. We suggest that coprophagy, in the strict sense of the word, may not be as common as previously supposed (Frankenburg and Smith 1967). Without pellet ingestion, pellet breakdown is a necessary step before food can be reobtained. There is adaptive sense to this because freshly egested pellets are likely to be low in nutritive value. Microbial colonization is likely to occur as pellets disintegrate. We have no evidence of the enhancement of pellet breakdown by Hydrobia grazing, but this is because our data comes from pellets less than 24 hours old, when they are still fresh. Older pellets are less mechanically sound and grazing is likely to speed up their disintegration. Orchestia pellets are encased in a chitinous membrane. Without stirring or mechanical movement by Orchestia, they will remain intact for months. Grazing activity must therefore play a role in pellet breakdown as we did not observe pellets to accumulate in the Orchestia grazing experiments.

If pellet breakdown occurs during grazing, and if pellet disintegration is a rate-limiting step in renewing exploitable sediment for deposit-feeders (see Hylleberg 1975), then a model predicting the balance of pellet production and disintegration is appropriate. If the recolonization of the microbial community approximates the pellet breakdown rate or is faster, then pellet disintegration is the rate-limiting step in food production for deposit-feeding. If the microbial community colonizes much more slowly than pellet breakdown, then pellet disintegration rates are of no interest, because the newly disintegrated pellets will be devoid of exploitable food.

If p is the number of fecal pellets produced in a day and a is the fraction of pellets remaining after one day, then after 3 days the total number of fecal pellets, P, will be:

$$P = p + ap + a \cdot ap. \qquad (1)$$

The first term of the summation is the third day's accumulation, the second term the second day's accumulation, minus the fraction that has broken down in a day while the third term is the first day's accumulation, minus two days' disintegration. Generally, this is the series:

$$P = \sum_n a^n p, \qquad (2)$$

where n is the number of days. This series converges when n is infinitely large and

$$P = p/1-a. \qquad (3)$$

Thus p/1-a is the equilibrium number of pellets produced as a balance between pellet production and pellet disintegration. In simulation trials this series usually converges to within 2% of the final number of pellets in 37 days. If one knows the total number of pellets that can be made from a given amount of sediment, K, then the population density will be at carrying capacity at a pelletization rate converging on K. At K, the rate of sediment exploitation will balance the rate of renewal of food through pellet breakdown. Simulations of different pellet breakdown rates and population densities and available sediment, to be published elsewhere, allow predictions of population densities of deposit-feeders.

If *Hydrobia* exploits the sediment to a depth of 2 mm, there are 2000 cc/m^2 of sediment. If the water content of the sediment is 60%, the specific gravity of the sedimentary grains is 2.7, and a typical sediment consists of about 10% of particles small enough to be ingested (see Fenchel and Kofoed 1976), then there are 216 g/m^2 available. From our experiments we assume that feeding activity ceases at about 68% pelletization, presumably because foraging brings little return to a given snail. A *Hydrobia* fecal pellet weighs about 1 microgram, and the number of pellets at full exploitation, K, will therefore be about 140 x 10^6 pellets. From the regresstio analysis, a snail makes 41 pellets per hour. The population density that converges on this K is 14,329. This density is well within the limits observed for large adult *Hydrobia* populations (Newell 1965, Sanders et al.

1962, J. Hylleberg, personal communication) and encourages us
to believe the efficacy of the model. A similar model could be
derived employing the colonization of the microbial community
onto fecal material. If the time scale for this colonization is
similar then it may instead be the limiting factor. Our model
does not therefore prove that pellet breakdown is the limiting
step. However the large number of subtidal sediments observed
to be almost completely pelletized suggest that this factor may
be extremely important. When populations of Capitella capitata
completely pelletize the sediment, feeding ceases and signs of
starvation ensue (D. Schneider, verbal communication). In
either of the cases of pellet breakdown or microbial food limitation, however, it is now possible to predict population densities in the field with a deductive model.

We gratefully acknowledge the technical assistance and encouragement of S. S. Kirkpatrick and A. Heidemann Lassen. Discussions with J. Hylleberg were helpful.

References

Barsdate, R. J., Prentki, R. T. and Fenchel, T. 1974. Phosphorus cycle of model ecosystems: significance for decomposer
 food chains and effect of bacterial grazers. Oikos 25, 239-
 251.
Fenchel, T. 1972. Aspects of decomposer food-chains in marine
 benthos. Verh. Deutsch. Zool. Ges. 65 Jahresversamml. 14, 14-
 22.
Fenchel, T. and Kofoed, L. H. 1976. Evidence for exploitative
 interspecific competition in mud snails (Hydrobiidae). Oikos
 27, in press.
Frankenburg, D. and Smith, K. L. 1967. Coprophagy in marine
 animals. Limnol. Oceanogr. 12, 443-450.
Hargrave, B. T. 1970. The effect of a deposit-feeding amphipod
 on the metabolism of benthic microflora. Limnol. Oceanogr. 15,
 21-30.
Hylleberg, J. 1975. The effect of salinity and temperature
 on egestion in mud snails (Gastropoda, Hydrobiidae). A study
 on niche overlap. Oecologia (Berl.) 21, 279-289.
Johannes, R. E. and Satomi, M. 1966. Composition and nutritive
 value of fecal pellets of a marine crustacean. Limnol.
 Oceanogr. 11, 191-197.
Levinton, J. S. 1972. Stability and trophic structure in
 deposit-feeding and suspension-feeding communities. Amer.
 Nat. 106, 472-486.
Lopez, G. R. 1976. The role of Orchestia grillus (Talitridae,
 Amphipoda) in the decomposition of Spartina alterniflora
 litter. Ph.D. Dissertation. State Univ. of New York at Stony
 Brook. 142 pp.
Newell, R. 1965. The role of detritus in the nutrition of two
 marine deposit feeders, the prosobranch Hydrobia ulvae and the
 bivalve Macoma balthica. Proc. Zool. Soc. Lond. 144, 25-45.
Rahn, U. 1975. Phosphorus turnover in relation to decomposition
 of detritus in aquatic microcosms. Dissertation, University
 of Aarhus, Denmark. (in Danish).

Sanders, H. 1958. Benthic studies in Buzzards Bay I. Animal-sediment relationships. *Limnol. Oceanogr*. 3, 245-258.

Sanders, H., Goudsmit, E. M., Mills, E. L., and Hampson, G. R. 1962. A study of the intertidal fauna of Barnstable Harbor, Massachusetts. *Limnol. Oceanogr*. 7, 63-79.

Zobell, C. E. 1938. Studies on the bacterial flora of marine bottome sediments. *J. Sedim. Petrol*. 8, 10-18.

THE ROLE OF PHYSICAL AND BIOLOGICAL FACTORS IN THE DISTRIBUTION AND STABILITY OF ROCKY SHORE COMMUNITIES

J. R. Lewis

*Wellcome Marine Laboratory, University of Leeds, Robin Hood's Bay,
N. Yorkshire, England YO22 4SL*

ABSTRACT

The transition from physically correlated surveys to studies of biological interactions and regulation has entailed a risk that underlying physical factors are overlooked. The broad distribution pattern of European shores is described in terms of the main physical gradients and the extents to which community stability results from the physically determined composition of communities, the ensuing biological reactions and the functional performance of key species at different tide levels. Stability is least among low-level mussel communities, increases upshore and is greatest where large algae dominate. Contrasts are drawn with Pacific coasts and especially the differing emphasis attaching to competitive and predatory relationships in the two areas. The transportation of reproductive stages appears a fruitful field for future study, but even more basic to community regulation is the annual variation in reproduction/recruitment. The effects of their presumed sensitivity to slight physical changes could be enhanced by the key-species structure of shore communities.

INTRODUCTION

Initially the study of rocky shores had a strong physical bias. Surveys described distribution in terms of the obvious physical gradients, and because these may be remarkably steep and appear to produce sharp biological patterns in response, the over-riding importance of physical factors was quickly established. Biological interactions were not ignored. Limpets grazed, thaids predated and so forth ; but then, relatively suddenly and mainly in the last decade or so studies of interactions and regulatory mechanisms expanded and transformed an apparently static subject into one that is dynamic and intellectually more satisfying.

This change was inevitable in a habitat with visible and sedentary populations, and where the physical gradients appear too obvious and unsubtle for further study but paradoxically are so difficult to measure or analyse that we rely on biological indicators. Considering the time needed for such studies awareness of biological interactions has grown remarkably. Indeed it has now reached the point at which, except in the case of upper intertidal limits, it is increasingly and regrettably easy to overlook or dismiss physical influences which remain as important today as ever or still offer a research challenge.

The ultimate primacy of physical factors is irrefutable, but species are often prevented from realising the full potential of their physical tolerances by biological intervention or inadequacy. Thus upper-shore species for example may have lower limits set by predation or competition, and algal scarcity may be caused by grazers rather than excessive wave

action. Distribution that is limited by such means was initially prone to
misinterpretation. "Lives in this habitat" could falsely progress to "unable
to live elsewhere" ; and it needed clearance experiments, transplants or
chance presence in the 'wrong habitat' to show that unexpectedly wide physical
tolerances exist yet may do little to set limits or abundance. But having
realised this truth it could be erroneous to lean the other way and assume
that <u>all</u> distribution is biologically determined. There still remain, for
example, high-level species for which no convincing biological control of
their lower limits has yet been shown, and which may indeed have evolved into
truly 'intertidal' organisms. It would be equally erroneous also to
automatically ascribe, for example, all absence of algae to grazers, of
dogwhelks to bird or crab predation or of mussels to starfish, without first
considering that there might be, as indeed there are in these cases,
uncontestable local or intermittent physical causes.

Species are faced with a potentially wide range of physical and biological
factors (summarised by Connell, 1973) some of which are local and some more
general, some direct and some indirect, and with an incidence which probably
varies from place to place and time to time. We should perhaps go no further
than merely adding cautiously to a long list of individual cases and local
examples but I venture to make some new generalisations about distribution
and biological stability and in so doing to re-emphasise the points at which
physical influences assert themselves, or may do so in the future.

DISTRIBUTION AND STABILITY PATTERNS IN N.W. EUROPE

Rocky shores the world over are characterised by two physical gradients of
overwhelming importance : horizontally between wave exposure and shelter,
and vertically between tide levels. Their populations are characterised by
1) a relatively few ground-cover species such as algae, barnacles, mussels,
etc., 2) the grazers and predators of the above, and 3) a wide range of
subordinate or opportunistic species whose presence is often dependent upon
the type, abundance and age of the ground-cover species. The locally and
geographically different composition of the first two categories largely
determines whether and where abundance and stability along the two physical
gradients will be determined by direct physical or indirect biological means.

Shores around the North Atlantic and in North West Europe especially are
uniquely characterised by large intertidal fucoid algae most of which become
progressively more luxuriant with decreasing wave action. The largest species
(<u>Ascophyllum nodosum</u>) lives for probably several decades at least and forms a
climax-type vegetation which perenially determines the type and abundance of
most of the accompanying species. In particular it excludes species typically
associated with more wave-swept coasts. I know no case of the long-fronded
<u>Ascophyllum</u> in extreme shelter declining appreciably for entirely natural
reasons. Thus this and some companion species, present in abundance in
direct response to the physical regime, impose long-term community stability.

With increasing wave action the fucoid cover declines, partly because of more
grazers and partly as a direct response to the greater turbulence. The
replacing attached species are mainly barnacles, mussels and short, tufted or
cushion-forming algae with high wave tolerance. The possibility of there
being here a biological stability equal to that imposed by <u>Ascophyllum</u> depends
upon one of the smaller, replacing species being able to exert a similar
dominance, or upon the existence of a dynamic balance among the potential
occupiers of the same space. This leads to matters of size and competitive

power, potential life spans and susceptibility to predators or grazers. It also reveals that at the exposed end of the horizontal gradient the submersion/emersion gradient is much more influential than on the fucoid-dominated shores.

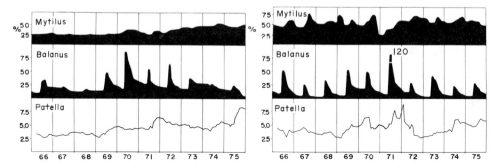

Figs. 1A and B. Abundance per m^2 of Mytilus (% cover), Balanus (1000's) and Patella (100's) at high- and mid-tide levels respectively.

Figures 1 and 2 show the temporal abundance of Balanus balanoides and Mytilus edulis, and the main littoral grazer, Patella vulgata, at three tidal levels on wave-swept sites at Robin Hood's Bay. There are no other barnacles or mussels, and cushion-forming algae appeared only intermittently at the lower levels so it is a sparse macro-community with wide tolerance of intertidal physical conditions. The highest site (1A) is marked by considerable stability when one discounts the recruitment peaks of Balanus and their rapid decline. Growth rates are low, competitive interactions slight and adult survival is high, some mussels and limpets living 20 and 15 years respectively (Seed, 1969 ; Lewis and Bowman 1975). At the middle site (1B) growth is faster, recruitment rates are higher and/or more regular and there is intermittent predation by Thais lapillus which preferentially attacks young Balanus. Increases in mussel cover result from the summer growth, and from settlement which is not very predictable in time or amount. Declines indicate Thais activity, usually in the autumn, and storm dislodgement. Over the long term this site too has remained relatively stable, although the fluctuations in mussel cover and the more annual role of Balanus give greater short-term variability than upshore.

At the low-level sites (Fig. 2) the situation is highly variable, ranging through periods of mussel or barnacle dominance, bare rock or ephemeral algae. Sudden increases clearly reflect recruitment, but the falls have various

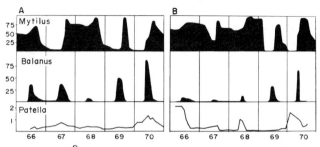

Fig. 2. Abundance per m^2 over 5 years at two low-tide sites. Scales as Fig.1.

causes : 1) Predation of barnacles and mussels by Thais and of mussels by Asterias. 2) Predation of limpets by birds. 3) Blanketing settlement of Mytilus on barnacles, limpets, Thais and algae. 4) Loss of Mytilus following the death of its barnacle or other biological substratum. 5) Self-elimination of Balanus and Mytilus following heavy settlement and rapid growth. 6) Mytilus loss during heavy storms. Apart from the last, which is obviously physical, the proximal causes of change are biological, but there are physical causes underlying the incidence and intensity of most of the biological process.

First, there is the fundamental composition of the community itself with the main predators, Thais and Asterias, being restricted by desiccation to levels 2 and/or 3. Secondly, destruction as severe as that caused by predation has resulted from the intrinsic biological properties of Balanus and Mytilus at the lower tide levels. Moving downshore an increase in recruitment and growth rates (especially in Mytilus) greatly increases the potential for intra- and interspecific interactions when space is the primary resource. Except when Mytilus settles upon and kills Balanus (and both are lost) it is the better competitor and, once established, it can crowd out and prevent settlement by all other species until removed itself by predation, self-elimination or storms. In practice, at this low level, few Mytilus live more than 2-3 years but recurrent settlements prevent the establishment of long-term alternatives. Heavy competitive recruitment allied to great competitive powers throughout life suggest that Mytilus has both 'r' and 'K' selected attributes - if the theory can encompass this apparent contradiction !

Biological chain reactions are common here and their form often depends upon the chance timing of destructive storms relative to reproductive cycles (Lewis 1970). The most tortuous chain with a specific physical origin traced a summer Balanus population to temperature conditions the previous autumn. The link was not a direct effect on Balanus reproduction but involved enhanced autumnal growth of young Thais, the inability of migratory winter birds to each such large juveniles, and the consequent predation of mussels by unusually large numbers of large juvenile Thais.

Thus from high to low level on these wave-swept shores but within the vertical range of the same dominant species biological stability is highest at the harshest physical level where diversity is low, and is least where conditions are more equable and macro-diversity greater. (There are implications for ecological theory here). This situation relates to macrospecies with annual reproduction and potentially long life. Subordinate, shorter-lived species have not been studied, but since many are dependent upon the macro-species some of their temporal patterns seem likely to harmonise with those of their 'hosts'.

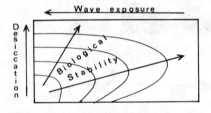

Fig.3. Stability relative to the physical gradients.

Summarising the situation over the two physical gradients we find biological stability at high level in exposure and at most levels in shelter (Fig. 3), but it has markedly different causes.

COMPARISON WITH THE PACIFIC COAST OF NORTH AMERICA

The greatest volume of data on the dynamic aspects of rocky shore ecology now relates to the Pacific coast of North America (see especially Connell 1970 ; Dayton 1971, 1975 ; Paine 1974). A resumé is out of the question here and a few only of what seem the more significant points of comparison with N.W. Europe will be made.

The same physical gradients exist, obviously, and while the same general types of organisms occur the larger number of species of all types results in greater local variation. Additionally there are two major differences in the populations that afford marked ecological contrast. Firstly, on the horizontal gradient there is no sheltered habitat alga of the size nor, presumably, with the dominating role of Ascophyllum. While smaller fucoids or cushion-forming algae may substitute to some degree, this lack suggests that interactions between algae/grazers and barnacles/thaids will create a more fluctuating community than in the European equivalent.

On the open coasts we again find high-level barnacles and mussels, slow-growing and long-lived, occupying a "refuge zone" above reach of thaids and asteroids. But in the middle and lower levels the faunistic differences assert themselves strongly. Mytilus californianus appears to have a greater ability than M. edulis to dominate the entire mid/lower shore perenially. That it does not do so everywhere is due to another major point of difference with Europe, i.e. the main mussel predator, Pisaster, occupies the lower intertidal whereas Asterias is an intermittent intruder from the sublittoral. But since Pisaster is intolerant of extreme wave crash M. californianus is able to exert, at the exposed end of the gradient, a biological dominance comparable to that exerted by Ascophyllum at the opposite end of the gradient in Europe.

In the Pisaster-predated areas the situation is highly variable, locally and temporally, with re-occupation and elimination cycles involving many species. Their detail is immaterial. The point of interest is that this instability, or imposed stability in some cases, results mainly from interaction between a potential ground-cover dominant and its predator. This contrasts strongly with Europe. Severe predation of M. edulis does indeed occur, but equal destruction results from competition with Balanus (in which both may suffer) or from self-elimination following heavy recruitment and fast growth, both of which are strongly influenced by tidal level. Even where predation is slight it is not unusual to find that the longest-lived mussel patches are in the mid/upper shore where the species' own potentially self-defeating processes proceed at slow rates.

THE FUTURE

Studies on dynamic aspects will surely expand and we shall move closer towards being able to read into more of the vagaries of today's abundance the biological and physical events that must have occurred yesterday. But some important components of our understanding will still be lacking. Before physical conditions on the shore can exert their influence or biological interactions can take place organisms must be present, however fleetingly. Many littoral species have motile reproductive stages and I wonder if we too easily assume these stages to be universally distributed in coastal waters each season. Leaving aside local patchiness due to settlement behaviour or to rapid adhesion of some algal spores, planktonic larvae and their settlement were long ago (Barnes 1956) shown to be influenced by changing winds. More locally, the age structure of sedentary populations can show recruitment to be annual in one position but intermittent nearby where the orientation to winds, waves or tides is different. Personal transplant experiments with barnacles have shown that absence may result, not from inadequacies of a site, but from failure of larvae to reach it, and some patterns of mussel settlement seem explicable only by extreme patchiness of larvae just before attachment.

I suggest that we know very little about the movement of reproductive stages along, towards or from the shore line, and that there are questions of horizontal and vertical distribution about which it is premature to reach conclusions until we know the frequency and extent to which spores or larvae are carried to and make contact with a suitable surface.

Transportation is however but one phase, and a late one too, in the fundamental reproduction/recruitment process. It is possible that the amount and frequency of recruitment, provided they are above certain minima, have little influence upon adult abundance, the population perhaps being regulated later by other means. Certainly competition has proved important in Balanus/Chthamalus relationships. (Southward and Crisp 1956 ; Southward 1967). But even if this is so, even if there is little effect upon the species itself, recruitment variation may impinge dramatically upon other species, and especially those in the unstable community of exposed coasts. For the British shores described above one can now predict the probable repurcussions on the entire community of a heavy mussel settlement, but can anyone predict when good or poor settlements will take place ? These remain 'chance events' for this and probably most species ; accepted enigmas the ecological significance of which is largely ignored. It is this basic aspect of abundance and stability that now appears most in need of attention.

The complete reproduction/recruitment process may extend over a long period and be influenced by both physical and biological factors, any one of which might be limiting. The scope for breeding and culture experiments is very wide and with the prevalence of planktonic phases we might ultimately be involved in nutrient cycles, phytoplankton productivity and their annual variation. This latter is a daunting prospect, but sooner or later we must integrate the shoreline into the wider marine system. Meanwhile the potential on the shoreline itself is appreciable and not without some encouragement.
In the first place reproductive processes and phases tend to be the more sensitive to physical conditions, and the seasonality of breeding in temperate latitudes at least narrows the field and points towards physical factors that may be reproducible in the laboratory and which will vary annually and regionally in the field. My own attempts to make progress here have so far met with success that seems to be inversely proportional to the duration of the planktonic phase - and that in itself is informative. Thus Mytilus with prolonged primary and secondary settlement phases remains an enigma. But in Patella vulgata, with a 7-10 day phase, the recruitment pattern so far points encouraging to the main source of variability being the incidence of autumnal frosts on the shoreline itself and the susceptibility thereto of newly settled spat. Should frosts not occur at the critical period the juveniles become tolerant of the later and more severe mid-winter conditions (Bowman and Lewis, in press).

The second source of encouragement to pursue such work lies in the fact that difficult though it may be to find the controlling factor(s) in some cases, the 'key-species' structure of rocky shore communities means that an understanding gained about these few species will permit prediction for the wider community. For example, in the British Isles we know that a decline in limpets leads to more algae and then to fewer barnacles and associated species. If the frost-control hypothesis is correct a slight change of autumnal climate, whereby the first intermittent frosts regularly occur 2-4 weeks earlier, would alone suffice to have disproportionately profound effects upon the character of much of the British littoral.

This limpet work has perhaps been remarkably lucky in that the control appears to lie on the shoreline, not at sea. More substantiation is required, but it suggests that an understanding of recruitment variability is perhaps possible as well as desirable. Shore ecology has passed through a first 'physical descriptive' phase (leaving some unanswered questions) and is now deep in a second phase of biological interactions. But interactions and the resultant fluctuations or imposed stability are often a background noise which obscures the potentially more fundamental source of temporal variation, i.e. recruitment variability. I suggest that the latter is the logical phase 3 target, and wonder how prominent the direct physical influences will prove to be.

REFERENCES

Barnes, H. 1956. *Balanus balanoides* in the Firth of Clyde; the development and annual variation of the larval population and the causative factors. J. Anim. Ecol., 25, 72-85.

Connell, J.H. 1970. A predator-prey system in the marine intertidal region. I. *Balanus glandula* and several predatory species of *Thais*. Ecol. Monogr., 40, 49-78.

Connell, J.H. 1972. Community interactions on marine rocky shores. Annu. Rev. Ecol. Syst., 3, 169-192.

Dayton, P.K. 1971. Competition, disturbance and community organisation. The provision and subsequent utilization of space in a rocky intertidal community. Ecol. Monogr., 41, 351-389.

Dayton, P.K. 1975. Experimental evaluation of ecological dominance in a rocky intertidal community. Ecol. Monogr., 45, 137-159.

Lewis, J.R. 1970. Problems and approaches to base-line studies in coastal communities. F.A.O. tech. Conf. mar. Pollut. Rome, pap. E-22.

Lewis, J.R. and Bowman, R.S. 1975. Local habitat-induced variations in the population dynamics of *Patella vulgata*. J. exp. mar. Biol. Ecol., 17, 165-203.

Paine, R.T. 1974. Intertidal community structure: Experimental studies on the relationship between a dominant competitor and its principal predator. Oecologia, 15, 93-120.

Seed, R. 1969. The ecology of *Mytilus edulis* on exposed rocky shores. 2. Growth and mortality. Oecologia, 3, 317-350.

Southward, A.J. 1967. Recent changes in abundance of intertidal barnacles in south-west England: a possible effect of climate deterioration. J. mar. biol. Ass. U.K. 47, 81-96.

Southward, A.J. and Crisp, D.J. 1956. Fluctuations in the distribution and abundance of intertidal barnacles. J. mar. biol. Ass. U.K., 35, 211-229.

MEIOFAUNAL COMMUNITY STRUCTURE AND VERTICAL DISTRIBUTION: A COMPARISON OF SOME CO. DOWN BEACHES

Cathy Maguire

Queen's University Marine Biology Station, Portaferry, N. Ireland

ABSTRACT

The abundance and composition of beach sand meiofauna at nine localities is compared. The species comprising the meiofauna show four main types of vertical distribution each associated with particular redox conditions. Types I, II and III are mainly restricted to surface, grey and black sand respectively; type IV species are adaptable occurring mainly at the surface but extending down into black sand. The thiobiotic community is of more constant composition than the surface dwelling community. The greatest number of species were found in the surface layers of moderately exposed beaches. In finer sediments the gastrotrichs Dolichodasys sp. and Turbanella varians replace Thiodasys sterreri and the archiannelid Protodriloides symbioticus from the deeper and surface layers respectively. The abundance of Nematoda increases with the proportion of fine particles; however the number of Turbellaria decreases with the kalyptorhynchids being more or less replaced by seriates and acoels. In general the faunal changes associated with increased shelter of the beaches parallel the changes associated with increased depth in the sediment.

INTRODUCTION

Studies at Ballymaconell Beach (Maguire 1977a, b) have shown that the meiofauna could be grouped into four main categories on the basis of their vertical distribution in relation to sediment redox characteristics. This paper examines vertical distribution in beaches of somewhat different characteristics and seeks to determine whether these categories may be applied as a more general "rule of distribution".

MATERIALS AND METHODS

The beaches studied are all in the Strangford Lough area of Co. Down, Northern Ireland. South Bay and Port Kelly are the most exposed; Millin Bay, Benderg and Ballymaconell are moderately sheltered; Kilclief and Hanna's Mill are sheltered from waves but subject to strong tidal currents; Barr Hall and Doctor's Bay are very sheltered. Sampling was during low tide from about halfway between mid-tide and low-water neap level. Three or more replicate 30 cm long, 3.2 cm diameter cores were taken from a small beach area except at South Bay and Doctor's Bay (cores 20 and 9 cm long respectively due to underlying stones) and Millin, Port Kelly and Barr Hall (non quantitative samples of surface and underlying sand). Physical measurements were according to Maguire (1977b). Turbellaria and Gastrotricha were identified to species; authorities are given as a footnote to Table 1.

TABLE 1 Relative abundance in surface+ and sub-surface sediments

	S.B.	B	Ba	HM.	K.	D.B.
GROUP I.						
Turbanella varians △	27/1		*/	90/	417/	32/
Dactylopodola baltica			102/5			
Cicerina remanei	13/	*/	231/57		*/	4/
Cheliplanilla caudata	5/*			*/	*/	
Schizochilus marcusi			57/6			
Neoschizorhynchus brevipharynggus			21/5			
Proschizorhynchus triductibus			30/9			
Macrostomid L.	10/*	*/	63/7			
Harpacticoid copepods	*/*	*/	*/		*/	40/1
Protodriloides symbioticus	*/	*/	10^4./*			
GROUP II.						
Neodasys chaetonotoides	*/	1/2	117/368		24/	*/
Neoschizorhynchus longipharynggus		*/	6/12	/*		
Proschizorhynchus bivaginatus	/*		9/16			/*
Pseudoschizorhynchoides ruber		/*	24/211		/*	
Red acoel	/*	/*	36/166			/*
GROUP III.						
Thiodasys sterreri		/294	/12 294/2784	/66		/*
Dolichodasys sp.		/4	/21	/17		/88
Paraturbanella cuanensis		2/19	2/136	/*	/3	
Macrodasys affinis			/43	/*		
Neoschizorhynchus parvorostro	4/6	5/4	/75	/2	/*	1/3
Acoel A	/*	/20	12/118	/26	/9	1/7
Acoel Bb		1/60		/192	/*	/*
Catenulid Fb		/*	/*	/*	/13	/*
Black turbellarian			/*	/*	/*	/*
Gnathostomulids			1/15		/*	/24
GROUP IV.						
Cephalodasys turbanelloides		2/1	284/317			/*
Schizochilus choriurus	1/1	/*	30/27			
Schizorhynchoides spirostylus			/*			4/4
Diascorhynchus rubrus		/*	5/21 102/5			
Coelogynopora sp.	/7	/6	75/82	/2	/1	/6

*Occasional individuals; +Surface sediment: i.e. above 3 cm. for Doctor's Bay

above 5 cm. for all other beaches; Δ Species authorities as follows:-
T. varians Maguire, D. baltica Remane, C. remanei Karling, C. caudata
Meixner, S. marcusi Boaden, N. brevipharynggus Schilke, P. triductibus
Schilke, P. symbioticus (Giard), N. chaetonotoides Remane, N. longipharyggus
Schilke, P. bivaginatus Schilke, P. ruber, Schilke, C. turbanelloides
(Boaden), S. choriurus Boaden, S. spirostylus Boaden, D. rubrus Boaden,
T. sterreri Boaden, P. cuanensis Maguire, M. affinis Remane, N. parvorostro
Schilke. P. eireanna, Maguire.

RESULTS

Some characteristics of the main beaches studied are given in Table 2. The
three additional beaches had much decaying weed along the drift line
producing a sulphide rich run-off. The black layer at Barr Hall was 2 cm
below the surface, at Millin the sand was grey at 6 cm but there was no
distinct black layer either here or at Port Kelly although the sediment
smelt strongly of H_2S.

Fauna

South Bay, Benderg, Ballymaconnell, Hanna's Mill, Kilclief and Doctor's Bay.
The faunal densities from the six main beaches are shown in Fig. 1 together
with the relative proportions of the major taxa. Densities were lowest at
South Bay, the most exposed shore and highest at Doctor's Bay, the most
sheltered, but the increases were due almost entirely to the great rise in
the numbers of Nematoda, a feature also noted by McIntyre (1971). Other
authors e.g., Eleftheriou & Nicholson (1975) have reported a positive
correlation between meiofaunal densities and degree of shelter, however the
results for Hanna's Mill and Kilclief do not appear to fit this pattern
(Fig. 1). In view of the winter maxima reported from Ballymaconnell
(see Maguire 1977a), it seems likely that the value for these beaches are a
high estimate of the population, whilst those from South Bay and Benderg
probably represent a summer minimum.

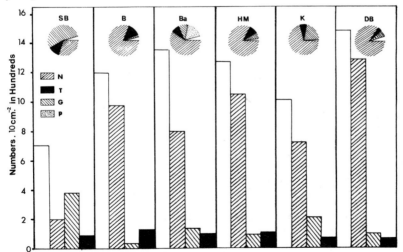

Fig. 1 Numbers and proportions of various taxa at each beach. N-Nematoda;
T-Turbellaria; G-Gastrotricha; P-Protodriloides.

As the sediment became finer the proportion of nematodes increased, (Fig. 1) whilst the turbellarian component became reduced and there were changes in the taxa represented. Nematode densities (except at Kilclief) were inversely correlated with median grain size. Species diversity as a whole was greatest in the moderately exposed beaches, Ballymaconnell and Benderg, where the reduced zones were lowest in the sediment column (Fig. 2). There was a numerous and diverse surface dwelling community at Ballymaconnell, where the archiannelid Protodriloides symbioticus was extremely abundant and the kalyptorhynchid Turbellaria were particularly diverse (see Maguire 1977a). The fauna at Benderg, however was more evenly distributed throughout the sand column and in the upper layers at the sampling site there was no single dominant species, although Protodriloides is abundant further up the beach and Tetranchyroderma sp. dominates the fauna towards low tide. Protodriloides symbioticus tends to disappear in the finer sediments; it was not recorded at all in the sample areas of the three most sheltered beaches, although very occasionally individuals have been observed further up the beach. On the other hand, this species is very numerous towards the top of the shore at Benderg and South Bay (Boaden & Erwin 1971), although it was infrequent at the sampling points. Protodriloides symbioticus is a relatively large species and is presumably excluded from the finer sediment by its size (see Gray 1966).

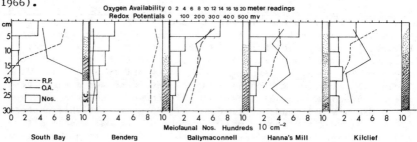

Fig. 2 Vertical distributions of faunas and physical factors. R.P. - Redox potential; O.A. - Oxygen availability; S.C. - Sediment colour.

There were distinct changes in the turbellarian fauna; the larger species, particularly the Kalyptorhynchia, tended to disappear in the finer sediments and were replaced by smaller, fast moving acoels and seriates. The variation in gastrotrich densities was not related to sediment particle characteristics, but some changes in species composition were. Turbanella varians was more numerous in the finer sediments, where it took over the position of dominant surface form from the archiannelid Protodriloides symbioticus. The numbers of T. varians were particularly high at Kilclief and Doctor's Bay, where sewage effluent runs onto the beach; but larger variants of this species were also present in the coaser sediments of South Bay and Millin Bay, (see Maguire 1976), where large quantities of decaying weed were present. It seems that this species is able to tolerate high organic content but not an anaerobic environment. Thiodasys sterreri was replaced by the more fragile and elongated Dolichodasys in finer sediments.

The vertical distributions and vertical changes in species composition of the faunas were very similar in all the above six beaches. Densities were highest at the surface and decreased with depth (Fig. 2). In general, the proportions of nematodes increased whilst the proportions of Turbellaria, particularly Kalyptorhynchia decreased. There were distinct vertical zonation patterns amongst the Gastrotricha and some of the Turbellaria. All six beaches contained very similar groups of species at equivalent depths in the sand column, under similar redox conditions. These groups were impoverished or slightly changed versions of those recorded previously at Ballymaconnell (Maguire 1977b) and could be associated with particular redox conditions, as manifested by sediment colour (Table II). The thiobiotic fauna (groups II and III) is extremely similar in each of the beaches whereas the surface fauna (group I) is much more diverse and variable in compsition. This must partly reflect the more uniform conditions and similarity of the underlying zones, but resource poverty and slower reproductive rates may also be partly responsible for the reduced species diversity (see Boaden 1976; Maguire 1977a, b).

Port Kelly, Millin Bay and Barr Hall Bay. The faunas of these three beaches were very reduced both in numbers and diversity. It is probably the high levels of hydrogen sulphide or other toxic material in the ground water which are responsible for this reduction, since other sediment characteristics were within the range found on the other beaches examined. The majority of the fauna were concentrated in the top few centimetres of sediment, most typical thiobiotic forms especially the Gastrotricha, were absent so the vertical population groups reported previously were not apparent. Nevertheless where similar species were recorded, these were found at roughly equivalent depths; for example, the fauna in the top 2 cm. of Barr Hall Bay was similar to the surface fauna at Doctor's Bay, whereas below 2 cm., only nematodes, ciliates and a gnathostomulid species were recorded. At Millin Bay, there was no distinct thiobiotic fauna, but amongst the few kalyptorhynchids recorded, vertical zonation was much the same as that recorded at the other beaches; e.g. Cicerina remanei, Thylacorhynchus sp. and Neoschizorhynchus brevipharynggus were found near the surface, whilst Proschizorhynchus bivaginatus and Schizorhynchoides spirostylus (a species known to be associated with polluted conditions) occurred below 5 cm. The results from these beaches suggest that although the thiobiotic meiofauna can tolerate anaerobic conditions and moderate levels of hydrogen sulphide etc., they are sensitive to high or fluctuating levels of sulphides and other products of decaying organic waste.

DISCUSSION

The results show that many of the changes in species composition correlated with the degree of shelter are similar to those that occur with increasing depth in the sediment. In both cases there are increases in the proportions of nematodes and decreases in the proportions of Turbellaria. The turbellarian taxa change, with the Kalyptorhynchia being replaced by other Turbellaria, many of which are not such active predators. To some extent changes are due to the contraction of the vertical zones in the more sheltered beaches which reduces the habitat space available to the aerobic species and tends to favour the deeper dwelling thiobiotic forms. There are, however, important differences in the character of the changes associated with shelter and depth in the sediment. The general increase in densities (mainly Nematoda) with increasing shelter is probably a result of increased food availability; the increased surface area of the finer sediment particles would support much higher numbers of attached bacteria and algae, on which many nematodes feed (see McIntyre & Murison 1973). The reduction of meiofaunal densities with depth in the sediment, together with the absence of predatory forms, particularly Kalyptorhynchia, suggests that reduced food availability may be partly responsible (see Maguire 1977b) although many species must also be excluded by the anaerobic nature of the environment.

TABLE 2 Comparison of main beaches

	South Bay	Benderg	Ballymaconnell	Hanna's Mill	Kilclief	Doctor's Bay
Month	August	June	15 months	November	December	April
Volume	480 ml.	720 ml.	720 ml.	720 ml.	720 ml.	220 ml.
Organic input	Moderate-high decaying weed	Moderate-high decaying weed	Moderate-sewage from Belfast Lough	Low, some fine plant debris	Fairly high sewage and plant debris	Very high raw sewage
Grain size	140 - 290	145	160 - 170	122	115	130 - 140
Slope	Fairly steep	Moderate	Moderate	Fairly steep	Fairly steep	Very flat
Dominant species	T. varians T. sterreri Harpacticoids C.remanei Nematoplana sp. N.parvorostro C.caudata Protodriloides	Acoel Bb D. rubrus P.cuanensis Acoel A Catenulid Fb T.sterreri N.parvorostro Protodriloides	Protodriloides T. sterreri C.turbanelloides N.chaetonotoides P.ruber C.remanei Red acoel Acoel A D.rubrus P.cuanensis D.baltica N.parvorostro	Acoel Bb T.varians T.sterreri Acoel A P.eireanna Dolichodasys sp.	T. varians Dolichodasys sp. Caten. Fb. Acoel A N.chaetonotoides	C. cambriensis Dolichodasys sp. T.varians Harpacticoids Macrostomid L. Pterognathia sp. S.spirostylus Acoel A
No. of taxa	31	52	+70	28	22	36

The reduced numbers of Kalyptorhynchia in finer sediments seems more likely to be due to their size; they would be less successful at moving rapidly through the sediment than the small, fast moving acoels which were more common in the sheltered beaches. Food specificity amongst the Kalyptorhynchia may also exclude them from the finer sediments. The very diverse kalyptorhynchid fauna at South Bay indicates that each species has a narrow niche specificity, Boaden & Platt (1971) suggested that deposited planktonic material could provide an important food source for these Kalyptorhynchia; this would be limited in areas of reduced water movement.

The main difference between the population changes associated with these two parameters is, however, one of degree. Those related to shelter and sediment characteristics are gradual and progressive, with certain species gradually replacing others. Those related to depth tend to be more discrete, with recognizable groups of species existing in different zones. There is some overlap of species between zones but the similarities of vertical distributions in each of the beaches, except those with very high levels of organic waste, support the view that these groups can be regarded as distinct assemblages.

ACKNOWLEDGEMENTS

My sincere thanks to Dr. Pat Boaden, Ms. Bernadette Byers and Messrs. Paddy Trainor, Michael Curran, Desmond Rogers, Philip Johnston, Ian Jenkinson and Bob Bleakley for all their help with various aspects of this work. The research was carried out at the Queen's University Marine Biology Station, Portaferry, during the tenure of a N.E.R.C. Research Assistantship.

REFERENCES

Boaden, P.J.S., 1976. Thiobiotic facts and fancies. (Aspects of the Distribution and Evolution of Anaerobic Meiofauna) Mikrofauna 1 in press.

Boaden, P.J.S. & Erwin, D.G., 1971. Turbanella hyalina versus Protodriloides symbioticus. A study in interstitial Ecology, Vie et Milieu suppl. 22, pp. 479 - 492.

Boaden, P.J.S. & Platt, H.M., 1971. Daily migration patterns in an intertidal meiobenthic community. Thalassia jug. 7 pp. 1 - 12.

Eleftheriou, A. & Nicholson, M.D., 1975. The effects of exposure on beach meiofauna. Cah. Biol. Mar. 16, pp. 695 - 710.

Gray, J.S., 1966. Selection of sands by Protodrilus symbioticus Giard. Veroff. Inst. Meeresforsch. Bremerh. 2, pp. 105 - 116.

Maguire, C., 1976. Turbanella varians n.sp., A marine gastrotrich with local morphological variants. Cah. Biol. Mar. in press.

Maguire, C., 1977a. Community structure of the meiofauna in Ballymaconnell beach I. Composition abundance and seasonal fluctuations of the Meiofauna. Est. and Coast. Mar. Sci. in press.

Maguire, C., 1977b. Community structure of the meiofauna in Ballymaconnell beach II. Vertical distributions. Est. and Coast. Mar. Sci. in press.

McIntyre, A.D., 1971. Control factors on meiofauna populations. Thalass jug. 7, pp. 209 - 215.

McIntyre, A.D. & Murison, D.J. The meiofauna of a flat fish nursery ground. J. mar. biol. Ass. U.K. 53, pp. 93 - 118.

ETUDE COMPARATIVE DE L'EFFICACITE DE DEUX BENNES ET D'UNE SUCEUSE EN FONCTION DE LA NATURE DU FOND

Henri Massé, Raphaël Plante et Jean-Pierre Reys

Station marine d'Endoume, Rue Batterie des Lions 13007 Marseille, France

ABSTRACT

The macrofauna at three stations (fine sand, muddy sand and mud) was sampled with two types of spring-loaded grab ("Smith-McIntyre" and "Briba-Reys") and an air-lift suction sampler.
The efficiencies of the samplers were compared using non-parametric statistical analyses of the number and distribution of species and of their abundance and biomass. Both grabs are significantly less efficient than the suction sampler on fine compact sand inhabited by deep-living organisms. However, on muddy substrates, the Smith-McIntyre grab is as efficient as the suction sampler.

INTRODUCTION

L'échantillonnage quantitatif du macrobenthos marin pose toujours des problèmes de choix embarassants, en raison de la diversité des engins de prélèvement proposés aux biologistes, et du manque de standardisation des méthodes. Dans le cadre du programme biologique international (I.B.P.), plusieurs revues du matériel et des méthodes ont été proposées, parmi lesquelles certaines sont très détaillées (Holme & McIntyre, 1971, Reys & Salvat, 1971), sans que l'on aboutisse pour autant à la désignation d'un engin de prélèvement standard.

En Mer Méditerranée, ce problème est également posé régulièrement, soit dans le cadre des activités du Comité Benthos de la C.I.E.S.M.M., soit dans celui du programme d'étude en commun de la Mer Méditerranée (E.C.M.).

Il nous a semblé intéressant d'apporter une nouvelle contribution pour progresser dans cette voie. Notre but principal a été d'établir un lien entre les données recueillies sur le plateau continental de Méditerranée nord-occidentale avec celles des autres mers. Mises à part les données de Guille (1971) recueillies grâce à une benne Van Veen dont l'efficacité a déjà fait l'objet de comparaison (Smith & McIntyre, 1954, Beukema, 1974), l'essentiel des études faites sur les côtes de Provence ont été réalisées à l'aide d'une benne originale appelée benne d'Endoume construite par Briba & Reys (1966) en s'inspirant du principe de la benne d'Aberdeen (Smith & McIntyre, 1954) à partir d'une benne "orange-peel". La benne d'Endoume a été utilisée régulièrement depuis 1966 et a donné lieu aux travaux suivants : Bourcier, 1970, 1976, Tahvildari, 1976, qui représentent les données quantitatives de base, d'un point de vue à la fois extensif et dynamique, sur le macrobenthos de cette zone géographique.

La benne d'Aberdeen ayant fait l'objet de nombreuses comparaisons : Smith & McIntyre, 1954, Wigley, 1967, Lie, 1968, Higgins, 1972, Smith & Howard, 1972 Dickinson & Carey, 1975, Bourcier *et al.*, 1975, est considérée comme l'une des meilleures. Toutefois, sa pénétration est considérée comme insuffisante dans les sédiments compacts (Smith & Howard,1972).Ce phénomène, général pour l'ensemble des bennes, est bien connu et il a été illustré par les auteurs utilisant des appareils à succion (Massé,1967, Amouroux et Guille, 1973, Christie & Allen,1972). Dans certains fonds compacts, la profondeur d'enfouissement des espèces peut être telle que les bennes deviennent même totalement inefficaces (Barnett & Hardy,1967, Keegan & Könnecker, 1973).

Cependant, si l'on considère le plateau continental dans son ensemble et dans la plupart des zones marines, une grande partie des fonds est constituée de sédiments plus ou moins vaseux; aussi convient-il de se pencher en détail sur ce type de substrats meubles, étant entendu qu'il faudra rechercher des solutions particulières pour tous les fonds dont les caractéristiques sont telles que leur échantillonnage par une benne est mauvais. Il faudra alors avoir une idée des rapports d'efficacité entre les engins, en fonction du type de fond. Il apparaît en effet important, et Christie (1975) l'a bien souligné, de prendre en considération la texture du sédiment lors d'une étude sur l'efficacité d'un engin de prélèvement.

MATERIEL ET METHODES

Les Bennes

Les bennes utilisées sont conformes aux descriptions données par les auteurs: Smith & McIntyre (1954) et Briba & Reys (1966). Chacune de ces bennes est composée d'un bâti assurant une bonne assise de l'engin sur le fond et pourvu d'un système de déclenchement à ressorts qui propulse l'échantillonneur dans le sédiment. L'échantillonneur est dérivé de la benne Van Veen dans le cas de la benne Smith-Mc Intyre et de l'"orange-peel bucket" dans celui de la benne Briba-Reys [1]. La B.A. prélève un échantillon d'une surface approximative de $1/10 \, m^2$ (cette surface est celle que l'on mesure quand les mâchoires de l'échantillonneur frappent le sédiment, légèrement écartées sous la poussée des ressorts). L'empreinte de la B.E. a une surface d'environ $1/12 \, m^2$ (Reys, 1968). Pour rendre les résultats aussi comparables que possible, nous avons lesté les deux bennes de manière à obtenir des poids totaux voisins de 130 kg (les poids à vide sont de 72 kg pour la B.A. et 108 kg pour la B.E.).

Les séries de coups de benne ont été faites à partir du bateau ancré, par beau temps (mer 1 à 3). Pour essayer d'éliminer dans les résultats un biais éventuel dû à une microdistribution des espèces (Bourcier et al.,1975), les prélèvements ont été faits en alternant l'une et l'autre benne. De plus, à intervalles réguliers, un léger déplacement du bateau à l'intérieur de son cercle d'évitage supprimait toute possibilité de prélever deux fois dans la même empreinte.

Les contenus des bennes ont été tamisés sur un tamis métallique à maille carrée de 1 mm de côté. Les refus de tamis, conservés au formol à 10% neutralisé, ont été triés au laboratoire après coloration au rose Bengale. Les animaux ont été comptés après identification, puis regroupés par embranchements pour être pesés après décalcification et dessication à 85° C.

---

[1] Pour la commodité de l'exposé, nous emploierons les abréviations suivantes: B.A. pour la benne d'Aberdeen (Smith & Mc Intyre), B.E. pour la benne d'Endoume (Briba &Reys) et S pour la suceuse.

La Suceuse

Les prélèvements de référence ont été effectués à l'aide d'une suceuse à benthos analogue à l'appareil décrit par Barnett & Hardy (1967) à quelques modifications près :
- le cylindre d'échantillonnage, d'une surface de section de $0,1m^2$ est enfoncé dans le sédiment à la main (Massé, 1970)
- l'air comprimé est fourni par une ou des bouteilles de plongée qui sont connectées au tuyau d'aspiration grâce à un détendeur HP qui fournit de l'air à une pression de 7 kg/cm^2
- le tuyau d'aspiration lui-même mesure 2m de long et 9cm de diamètre
- les filets de collecte sont des sacs tronconiques en filet de nylon (Massé, 1970). La maille de ce filet est de même dimension que celle du tamis utilisé pour le traitement des échantillons ramenés à la benne (1mm), dans le souci d'obtenir un tamisage uniforme quelque soit l'engin utilisé.

Equipée de cette façon la suceuse est relativement indépendante du bateau de surface, ce qui permettait de l'utiliser indépendamment de la manipulation des bennes, en se plaçant en dehors du cercle d'évitage du bateau. L'emplacement de chaque prélèvement était déterminé de façon aussi aléatoire que possible.

Le contenu de chaque filet de suceuse était, bien sûr, traité de la même façon que celui des bennes.

STATIONS ETUDIEES

Les trois points étudiés ont été choisis au voisinage d'une bouée située au large du cordon sableux qui ferme le golfe de Fos. Cette zone se trouve sous l'influence des apports sédimentaires du fleuve le Rhône, de sorte qu'on pouvait s'attendre à une suite très caractéristique de sédiments, allant des sables fins terrigènes bien triés de la côte aux vases terrigènes vers le large.

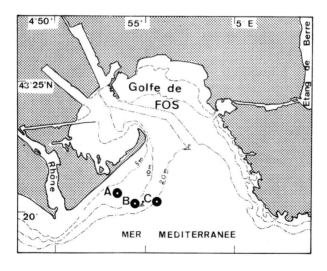

CARTE 1 : STATIONS ETUDIEES

En fait la réalité est légèrement différente :

1) la première station (A), à 6m de profondeur, est formée d'un sable fin bien classé mais légèrement envasé (7,2% constitué par une fraction inférieure à 40mm). Cet envasement était un phénomène relativement récent bien qu'antérieur à l'époque des prélèvements, ainsi qu'en attestait la présence, sous quelques cm de sable, d'abondants débris végétaux d'origine terrestre.
2) la station moyenne (B), à 10m de profondeur, se trouvait sur des sables fins encore légèrement envasés (4,5% de vase), de composition relativement uniforme sur toute l'épaisseur du prélèvement.
3) la station la plus profonde (C), à 22m de profondeur, se trouvait en bordure de la zone d'extension des vases terrigènes côtières (74,3% de vase).

	Suceuse			benne Aberdeen			benne Endoume		
	Sp	N	B	Sp	N	B	Sp	N	B
Station A −6m	77	3093	16528	80	3034	9815	74	1745	10217
Station B −10m	101	3943	3315	81	2379	2228	77	1924	2039
Station C −22m	89	2122	9555	90	2183	8196	58	613	4763

Tableau 1 Données globales : Nombres d'espèces (Sp), densités (N) et biomasses (B) par m2 (poids secs en mg)

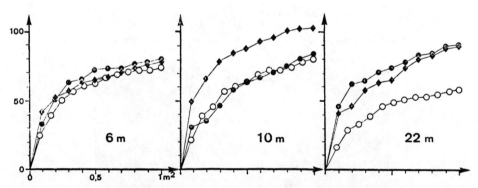

Fig. 1 Courbes cumulatives du recrutement des espèces dans chaque fond par chaque engin :
   losanges : suceuse
   cercles blancs : benne d'Endoume
   cercles noirs : benne d'Aberdeen

performances très voisines. Seules les valeurs de la biomasse donnent un avantage à la suceuse.

Les prélèvements ont été effectués les 10 et 11 juillet 1974.

RESULTATS

L'analyse des nombres croissants d'espèces recrutées par les prélèvements successifs pour chaque appareil (Fig. 1), de même que les résultats du tableau 1, montrent clairement le rendement supérieur de l'appareil à succion, par rapport aux bennes, dans les fonds sableux les plus purs (station B, -10m). Par contre, dans les fonds légèrement vaseux (station A), ou très vaseux (station C), la benne d'Aberdeen et la suceuse ont des Dans le cas de la station C, la benne d'Endoume accuse une chute importante de son efficacité par rapport aux deux autres engins. A la station B, dans le sable le moins vaseux, les deux bennes ont une efficacité comparable, ce qui confirme les résultats du test préliminaire réalisé dans des sables fins purs (Bourcier et al., 1975). Ces résultats sont confirmés par l'analyse statistique des effectifs des 10 prélèvements ajustés à 1/10m2. En raison de la non normalité des données (risque de microdistribution) la comparaison des 3 appareils se fera à l'aide d'un test non paramétrique, le test de Kruskal-Wallis (équivalent d'une analyse de variance). Pour la benne d'Endoume, on a pris les 10 premiers prélèvements ajustés à 1/10m2. Ce test sera suivi (Tableau 2) d'une comparaison multiple afin de mettre en évidence les différences entre chaque couple d'engins ; nous emploierons le "simultaneous test Procedure" (Sokal & Rohlf, 1969).

Comparaison globale	Station A	Station B	Station C
test Kruskal-Wallis	$X^2 = 73,94$ $P<0,001$	$X^2 = 75,79$ $P<0,001$	$X^2 = 71,56$ $P<0,001$
Comparaison des engins deux à deux (Simultaneous test Procedure)			
S. Versus B.A.	U = 56 NS	U = 87 $0,01<P<0,05$	U = 47,5 NS
S. Versus B.E.	U = 96 $0,01<P<0,05$	U = 98 $0,01<P<0,05$	U = 100 $P<0,01$
B.A. Versus B.E.	U = 88 $0,01<P<0,05$	U = 71 NS	U = 100 $P<0,01$

Tableau 2 Comparaison des 3 engins de prélèvement
S. : suceuse, B.A. : benne Smith-McIntyre, B.E. : benne Briba-Reys, NS : test non significatif

Afin d'entrer dans le détail des différences d'efficacité entre les trois engins, nous avons testé (X^2) les résultats fournis par les espèces numériquement dominantes. Nous avons considéré les 12 à 14 espèces qui, dans chaque fond, représentent environ 80% de l'effectif total récolté (Tableau 3). Le test du X^2 est très hautement significatif aux 3 stations. Les engins ne prélèvent donc pas dans les mêmes proportions les espèces du fond considéré. L'examen de détail est difficile à interpréter et demande une analyse par espèce. Par exemple, si aux stations A et B *Spisula subtruncata* apporte une forte contribution au X^2 pour la B.A., à la station A c'est en raison d'une "sous-capture", alors qu'à la station B, c'est en raison d'une "sur-capture". On retrouve le même phénomène pour *Chaetozone setosa*. On peut encore citer la "sous-capture" des *Owenia fusiformis* avec la suceuse, à la station A. Malgré les précautions prises lors de la récolte, il est probable que les phénomènes de microdistribution sont la cause principale de tels résultats. Par contre, pour les espèces comme *Acrocnida brachiata*, et peut-être *Mediomastus capensis*, le sur-prélèvement à la suceuse dénote une meilleure aptitude à la capture des animaux enfouis profondément, plutôt qu'un phénomène de microdistribution.

D'une manière plus générale, cela signifie que, connaissant le rapport global entre les rendements des différents appareils, on ne peut pas utiliser ce rapport pour calculer une densité supposée réelle d'une espèce, à partir de la densité observée avec l'engin le moins efficace. Il existe une interférence entre l'hétérogénéité de distribution des espèces et le rendement propre d'un engin de prélèvement, celle-là pouvant compenser la déficience de celui-ci.

	STATION A (-6m)						STATION B (-10 m)						STATION C (-22 m)					
	S.		B.A.		B.E.		S.		B.A.		B.E.		S.		B.A.		B.E.	
	N	χ^2	N	χ^2	N	χ^2	N	χ^2	N	χ^2	N	χ^2	N	χ^2	N	χ^2	N	χ^2
Cardium sp.	147	0	124	3,11	106	5,6	143	5,55	142	14,40	75	0,67						
Mactra coral.	117	1,42	101	0,05	50	1,63												
Spisula sub.	150	31,81	49	21,28	44	1,96	570	0,13	410	12,72	217	11,93						
Venus gallina	94	0,20	85	1,51	73	4,93												
Corbula gibba													69	1,39	45	0,78	2	10,83
Thyasira flex.													583	0	683	12,51	81	45,96
Turritella com.													76	0,02	53	7,28	44	22,85
Nucula sulc.													41	1,82	34	0	2	6,08
Abra nitida													38	0,19	39	0,19	5	2,71
Chaetozone set.	279	0	271	0,1	167	0,28	393	23,49	374	19,44	270	4,43						
Heterocirrus al.	227	39,20	458	42,62	192	0,08	296	1,24	152	6,74	205	20,53						
Heterocirrus caput.							83	0,11	43	1,29	52	3,09						
Lumbrineris imp.	581	5,84	462	6,04	301	0	206	2,56	81	7,39	94	0,51						
Lumbrineris frag.													50	0,01	43	1,11	21	3,12
Lumb. lat. grac.													460	0,64	401	6,37	166	10,61
Nephtys h mb.													53	1,46	82	4,92	10	3,70
Glycera rouxi													43	0,02	35	2,28	24	9,77
Tharyx multibr.													21	5,02	44	2,32	14	1,67
Tharyx marioni													21	2,58	37	1,36	11	0,63
Notomastus aber.													162	5,46	201	0	88	17,36
Ampharete grub.													52	4,46	29	2,95	9	0,46
Magelona pap.	145	3,59	156	0,75	133	13,34	338	0,49	204	0,09	184	1,88						
Mediomastus cap.	250	17,24	185	0,11	57	25,52	510	110	80	64,29	74	40,10						
Owenia fus.	204	34,10	362	12,02	217	10,05												
Paradoneis armata	192	3,03	180	1,03	61	13,23	93	13,87	17	11,31	21	2,70						
Spiophanes bomb.							58	0,43	14	9,79	38	6,51						
Spio decoratus							117	2,19	44	4,55	50	0,05						
Prionospio caspersi							203	6,37	166	3,39	132	2,58						
Chone sp.							80	0,05	48	0,01	42	0,24						
Perioculodes long.							77	10,28	106	24,59	46	0,79						
Acrocnida brac.	98	41,69	15	25,20	19	3,73												
Aspidosiphon mül.													35	0,09	23	3,61	19	9,0
Total	2484		2448		1420		3167		1881		1500		1704		1749		496	
χ^2	372,38		$\ll$ P = 0,001				453,84		$\ll$ P = 0,001				213,59		$\ll$ P = 0,001			

TABLEAU 3 Contributions des espèces numériquement dominantes à la variabilité des prélèvements.

DISCUSSION

On se rappellera que l'objet de cette étude est avant tout de situer la benne d'Endoume, par rapport à la benne d'Aberdeen, sur le plan de son efficacité, dans différents types de fond. Les prélèvements faits à la suceuse servent de base de comparaison.

Thorson (1957) critiquait sévèrement l'emploi de la benne "orange-peel" pour l'obtention d'échantillons quantitatifs comparables de la macrofaune, opposant essentiellement trois types d'arguments à l'utilisation de cet appareil.
1- La construction même de la benne implique un manque d'étanchéité des mâchoires.
2- La surface du prélèvement est difficile à définir.
3- Dans les fonds vaseux, l'appareil s'enfonce sous la surface du sédiment, les animaux vivant en surface s'échappent entre les mâchoires et se retrouvent englués dans le mécanisme de fermeture, s'ils n'ont pas été lavés pendant la remontée.

Le premier défaut a reçu une solution acceptable dans le modèle Briba-Reys, sous la forme de plaques de protection soudées à la partie supérieure des mâchoires, de manière à former un opercule complet quand la benne est fermée. Le problème d'étanchéité, lors de la fermeture, lié à la présence de gros débris entre les mâchoires est à notre avis un faux problème. Nous pensons en effet qu'il est impératif de contrôler le volume de sédiment ramené à chaque prélèvement et d'éliminer les coups de benne ramenant une quantité de sédiment s'écartant du volume moyen de sédiment susceptible d'être obtenu dans le fond considéré.

En ce qui concerne la détermination de la surface de sédiment prélevée, elle peut se faire avec précision par la méthode utilisée par Gallardo (1965). C'est par cette méthode que Reys (1968) a évalué la surface de prélèvement de cette benne à $1/12$ m^2.

La dernière critique de Thorson semble la plus difficile à résoudre et s'avère tout à fait recevable dans le cas de la B.E. Les faibles performances de cette benne dans le fond vaseux (Station C) semblent en rapport avec la forme et la contenance de l'échantillonneur (orange-peel bucket). En effet, dans les conditions de notre test de comparaison, les deux bennes sont lourdes, de plus leur système de déclenchement à ressorts les enfonce profondément dans les vases. Dans le cas de la B.A., le grand volume de l'échantillonneur permet d'emprisonner une grande quantité de sédiment (15 dm^3). De plus, la couche superficielle peut être tamisée par les volets grillagés situés à la partie supérieure des mâchoires de l'échantillonneur. Dans le cas de la B.E., les mâchoires écartées laisseront en partie s'échapper la couche superficielle du sédiment en raison de la faible capacité (5 dm^3) et de la forme de l'échantillonneur, avant sa fermeture. Celle-ci se fera au-dessous de la surface du sédiment. Thorson avait bien vu cet inconvénient. Or, c'est un fait connu, dans les fonds vaseux la plupart des espèces vivent près de la surface du sédiment (Christie, 1975), la sous-estimation du nombre d'espèces, de la densité, et de la biomasse est donc facilement explicable. Nous pouvons dire, d'une manière générale, qu'il faudra utiliser ces bennes sans lest dans les petits fonds vaseux et particulièrement la B.E. Toutefois, pour cette dernière, son poids sans lest de 108 Kg sera un handicap insurmontable.

Il semble que l'une des obsessions des constructeurs de benne soit de résoudre le problème de la pénétration et de la régularité de celle-ci dans les fonds compacts. La B.E. est satisfaisante sur ce point, cependant il ne faut pas oublier que la majorité des fonds du plateau continental sont des fonds plus ou moins vaseux où le problème majeur n'est pas un problème de pénétration, mais un problème de capacité et d'aptitude à bien envelopper l'échantillon à prélever, correspondant à la surface de la morsure de l'engin, sans perdre la couche de surface.

En résumé, il semble bien que cette critique de l'échantillonneur "orange-peel" soit l'inconvénient majeur de la B.E. Il permet, par exemple, d'expliquer une certaine déficience de la benne dans l'échantilonnage de petits inddividus vivant à la surface du sédiment: *Thyasira* et *Corbula*, Crustacés Amphipodes et Cumacés. L'effet dû à l'onde de choc (Wigley, 1967), souvent évoqué pour expliquer la perte de cette catégorie d'animaux, ne semble pas jouer un rôle capital dans le cas de la B.E. En effet, nous avons pu constater des performances tout à fait comparables entre la B.E. et la B.A. dans la capture de petits Crustacés, dans un sable fin compact (Bourcier et *al.*,1975).

En ce qui concerne les données fournies par la suceuse, nous ne nous attarderons pas sur les performances observées à la Station B, dans le sable le moins vaseux, ou à la Station A, avec l'échantillonnage des *Acrocnida brachiata* (Tableau 3), Ophiures enfouies dans le sédiment et dont la densité est sous-estimée par les 2 bennes. Par contre, du point de vue des données globales, nous pouvons souligner le peu de différence existant entre les données recueillies à la benne d'Aberdeen et à la suceuse, dans le fond vaseux (Station C).

Nous pensons que l'aspiration de la vase se fait moins bien que celle des sédiments sableux. Ceci a pour conséquence une augmentation de la durée du prélèvement et du même coup de la durée du tamisage sous pression. On peut se demander si le fait d'avoir un long tamisage sous pression de l'échantillon, avec un tamis (sac de prélèvement de la suceuse) de 1 mm de maille, ne facilite pas la perte de petits animaux. En effet, par comparaison, le tamisage des échantillons prélevés à la benne se fait sur le pont du bateau avec un tamis de même maille, mais sous une pression réduite.

Néanmoins, il est intéressant de noter que l'ordre de grandeur des données obtenues par la B.A. est comparable à celui des données de la suceuse, dans les fonds vaseux, ce qui autorise les comparaisons entre les résultats obtenus par ces deux types d'engins.

CONCLUSIONS

1 - La suceuse est l'engin le plus efficace dans les fonds sableux compacts où les bennes sous-estiment les données (nombre d'espèces, densité, biomasse).

2 - La benne d'Aberdeen est supérieure à la benne d'Endoume dans les sédiments vaseux et sablo-vaseux, dans ces fonds son efficacité est comparable à celle de la suceuse.

3 - L'efficacité de la benne d'Endoume diminue lorsque la teneur des sédiments en pélites augmente.

4 - L'hétérogénéité de la distribution des espèces sur le fond interfère avec l'efficacité propre de l'engin de prélèvement.

5 - Il convient d'ajuster le poids de la benne à la nature du sédiment.

6 - Il est indispensable de contrôler la régularité des volumes prélevés à l'intérieur d'une série pour un type de sédiment donné.

REFERENCES

Amouroux, J.M. et Guille, A. 1973 Premières estimations des biomasses dans l'infralittoral à Banyuls à l'aide d'une suceuse à pompe immergeable. Rapp. Comm. int. Mer Médit. 21 (9), 605-607.

Barnett, P.R.O. and Hardy, B.L.S. 1967 A diver-operated quantitative bottom sampler for sand macrofauna. Helgoländer wiss. Meereswaters. 15, 390-398.

Beukema 1974 The efficiency of the Van Veen grab compact with the Reineck box sampler. J. Cons. int. Explor.Mer 35 (3), 319-327.

Bourcier, M. 1970 Etude quantitative du macrobenthos de la Baie de Cassis. Tethys 2 (3), 633-638.
---------- 1976 Economie benthique d'une baie méditerranéenne largement ouverte et des régions voisines en fonction des influences naturelles et humaines. Thèse Univ. Aix Marseille, 161 p., Arch. C.N.R.S. A.O. 12.150.
Bourcier, M. Massé, H. Plante, R. Reys, J.P. et Tahvildari,B. 1975 Note préliminaire sur l'étude comparative des bennes Smith-McIntyre et Briba-Reys. Rapp. Comm. int. Mer Médit. 23 (2), 155-156.
Briba, C. et Reys, J.P. 1966 Modification d'une benne "orange-peel" pour des prélèvements quantitatifs du benthos de substrats meubles. Rec. Trav. Sta. mar. Endoume 21 (57), 117-121.
Christie, N.D. 1975 Relationship between sediment textures, species richness, and volume of sediment sampled by a grab. Mar. Biol. 30 (1), 89-96.
Christie, N.D. and Allen, J.C. 1972 A self contained diver-operated quantitative sampler for investigating the macrofauna of soft substrates. Trans. roy. Soc. S. Afr. 40, 299-307.
Dickinson, J.L. and Carey, A.G. 1975 A comparison of two benthic samplers. Limnol. and Oceanogr. 20 (5), 900-902.
Gallardo, V.A. 1965 Observations on the biting profiles of three 0,1m2 bottom samplers. Ophelia 2, 319-322.
Guille, A. 1970 Bionomie benthique du plateau continental de la côte catalane française. II Les communautés de la macrofaune. Vie et Milieu 21 (1 B), 149-280.
Higgins, R.C. 1972 Comparative efficiences of the Smith-McIntyre and Baird grabs in collecting *Echinocardium cordatum* (Permant) (Echinoidea Spatangoidea) from a muddy substrate. N.Z.O.I. Records 1 (8), 135-140.
Holme, N.A. and McIntyre, A.D. 1971 Methods for the study of marine Benthos. BP Handbook n° 16 Blackwell, Oxford and Edinburgh.
Keegan, B.F. and Könnecker, G.K. 1973 In situ quantitative sampling of benthic organisms. Helgoländer wiss. Meeresunters 24, 256-263.
Lie, U. 1968 A quantitative study of benthic infauna in Puget Sound, Washington USA in 1963-1964. Fisk. Dir. Skr. (Ser. Havunders) 14, 229-556.
Massé, H. 1967 Emploi d'une suceuse hydraulique transformée par les prélèvements quantitatifs dans les substrats meubles infralittoraux. Helgoländer wiss. Meeresunters. 15, 500-505.
-------- 1970 La suceuse hydraulique, bilan de quatre années d'emploi, sa manipulation, ses avantages et inconvénients. Peuplements benthiques. Tethys 2 (2), 547-556.
Reys, J.P. 1968 Quelques données quantitatives sur les biocoenoses benthiques du golfe de Marseille. Rapp. Comm. int. Mer Médit. 19 (2), 121-123.
Reys, J.P. et Salvat, B. 1971 L'échantillonnage de la macrofaune des sédiments marins, in: "Echantillonnage en milieu aquatique". Lamotte édit. Masson & Co. Paris.
Smith, K.L. and Howard, J.D. 1972 Comparison of a grab sampler and a large volume corer. Limnol. and Oceanogr. 17 (1), 142-144.
Smith, W. and McIntyre, A.D. 1954 A spring loaded bottom sampler. J. mar. biol. Ass. U.K. 33 (1), 257-264.
Sokal, R.R. and Rohlf, F.J. 1969 Biometry. W.H. Freeman and Co. San Francisco.
Tahvildari, B. 1976 Contribution à l'étude dynamique des variations qualitatives et quantitatives du peuplement benthique d'un fond soumis à des alternances sédimentaires. Univ. Aix Marseille II, Thèse 3ème cycle, 69 p.
Thorson, G. 1957 Sampling the benthos. Mem. geol. Soc. Amer. 67 (1), 61-73.
Wigley, R.L. 1967 Comparative efficiencies of Van Veen and Smith-McIntyre grab samplers revealed by motion pictures. Ecology 48 (1), 168-169.

ORGANIZATION IN SIMPLE COMMUNITIES: OBSERVATIONS ON THE NATURAL HISTORY OF *HYALE NILSSONI* (AMPHIPODA) IN HIGH LITTORAL SEAWEEDS

P. G. Moore

University Marine Biological Station Millport, Isle of Cumbrae, Scotland KA28 0EG

ABSTRACT

The density of a Hyale nilssoni population in Pelvetia has remained within narrow limits for three years. Fluctuations occurred, but were predictable and numbers returned to the equilibrium density. An hypothesis is advanced as to the mechanism of this regulation. Simple marine ecosystems achieve stability by different means. Biological accommodation is not an exclusive attribute of complex communities.

INTRODUCTION

Although its vertical range is extensive, Hyale nilssoni (Rathke) is particularly abundant amongst the high shore fucoids Pelvetia canaliculata (L.) Dcne. et Thur. and Fucus spiralis L., in which it nestles in virtual isolation during periods of tidal emersion. As probably the simplest phytal ecosystem in the harshest zone of the shore, its stability is of special interest.

MATERIAL AND METHODS

Duplicate samples of Pelvetia were taken at low tide from Site A1, Isle of Cumbrae (Ordnance Survey ref.NS14955484), usually between 08.00-10.00h G.M.T. Hyale were formalin extracted. A length index of individuals was measured as the lateral width of the third epimeral plate (1 unit = 0.102mm).

Intensive sampling was restricted to a small area (c.3 x 1m). Since sampling is destructive and modifies the integrity of the remaining stand, and since conservation of the population was required for other studies, sample replication was kept to a minimum.

A startling variety of descriptive parameters have been used to quantify algal faunas. Here, Hyale density has been related to wet (damp dried) weight of weed and corrected to numbers/100g. This is both simple and presumably of functional significance to the amphipods. However it is possible that the use of a particular parameter may necessarily constrain any interpretation of stability. For instance, animal density per plant could remain constant in a plot from which plants were being removed although animal density per plot would decline. No resolution of this paradox is presented.

RESULTS

Fig 1A shows the changes in mean density /100g weed (transformed log (n+1)) over a three year period. Whilst fluctuations occurred, density remained remarkably constant at c.100 individuals/100g weed. Since only two replicates were taken, confidence limits are not presented. Replication was usually within a factor of 2. On occasions density fell to zero. Emigration, not mortality was responsible. During tidal emersion, particularly in summer, Pelvetia may dry out, becoming, especially at neap tide periods, black and brittle. With the advent of drying conditions, Hyale at first congregate amongst the damper subsurface fronds but are eventually driven down to the rock surface. Continued desiccation of Pelvetia finally creates untenable conditions and Hyale move downshore under the weed canopy along a moisture gradient.

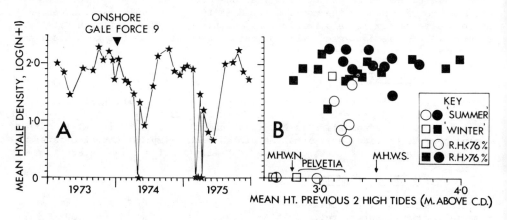

Fig. 1. Density of Hyale, A. with time, B. with habitat conditions.

To illustrate the effect of tides and climate, these density data have been rearranged and plotted (Fig. 1B) against the average height of the two high tides preceding sampling (to accommodate marked diurnal inequality). Actual tidal data were obtained from the gauge at Keppel Pier, Millport. Interpretation is assisted by distinguishing i) between 'summer' and 'winter' periods (the six warmest and coldest months based on monthly average air temperatures for Millport), and ii) between samples preceded by a period of low relative humidity.

Tidal immersion is less important in winter when increased rainfall, increased wave splash and high relative humidity combine to prevent desiccation of Pelvetia. A tenable habitat is thus maintained during neap tide periods. With progressively more desiccating atmospheric conditions, tidal inundation becomes necessary to ensure moist conditions at low tide. At spring tide periods in summer, the population of Hyale is comparable with winter levels.

It is instructive to consider the impact on Hyale of components of littoral 'stress'. Wave action is of no consequence. Density was unaffected even by a gale force 9 (46kt, gusting 72kt) blowing directly onshore. The clinging ability of Hyale in Pelvetia's branched thallus is prodigious. If displaced it will roll into a ball for protection and so facilitate sinking (cf.Jansson and Matthieson 1971). Fresh water is of no consequence, Hyale penetrates estuaries certainly up to the 17o/oo isohaline. In the laboratory, it will tolerate 1o/oo or 72o/oo for 24h. It is markedly resistant to oxygen depletion (Wieser and Kanwisher 1959). It survives freezing temperatures (24h,-5°C) and is relatively tolerant of desiccation.

When driven from Pelvetia by desiccating conditions, Hyale congregate in alternative habitats. Fucus spiralis is wetted by most neap tides, and where Pelvetia and F.spiralis abut, Hyale will move into Fucus (Fig 1B). Other 'overspill' habitats are also utilized. An association between H.nilssoni and littoral gastropods has been reported by Brattegard (1963). However, the dynamic nature of this relationship has been overlooked. Table 1 shows that Hyale nestle in the damp apertures of gastropods particularly during neap tides, vacating them at spring tide when Hyale recolonizes higher levels. Branch (1975) classified H.grandicornis as a facultative inquiline of Patella spp. in S. Africa. It vacates its host at high tide. No doubt H.nilssoni does too, only to return at low tide whilst in search of damp conditions during summertime neap periods. No difference was apparent between the size structure of Hyale under dog whelk and winkle shells and Pelvetia (Fig 2A), in contrast to Norway where the juveniles are associated with Thais (Brattegard 1963 see also Branch 1975).

TABLE 1   Incidence of H.nilssoni under Thais (n=10), site BP2, + 2.2 C.D.

Date	1.10.74	8.10.74	14.10.74	21.10.74
Tide	SPRING	NEAP	SPRING	NEAP
Average No.Hyale/shell	0.1	3.8	0.1	1.8
% Shells with Hyale	10	100	10	60

Prolonged observations on H.nilssoni have however revealed a spatial separation of post-marsupial juveniles from the adult population. At no time did post-marsupial juveniles account for a significant proportion of the population in Pelvetia. These tiny individuals are abundant (Fig 2B) in finely-branched, filamentous weeds (Cladophora sp., Polysiphonia spp.) lower down the shore. The mechanism by which this distribution is effected has not been established, but seems likely to involve migration of pregnant females since females with active juveniles in their brood pouches are surprisingly scarce in the Pelvetia samples (cf. Jansson and Matthieson 1971). Juvenile Hyale are less common in Cladophora in rockpools than in Cladophora present as an open-rock understorey. They graze the microepiflora without damaging the Cladophora. Filamentous weeds then act as 'nursery grounds' for Hyale (also for Idotea (Jansson and Matthieson 1971), Mytilus (Bayne 1964), Rissoa (Wigham 1975), Gammarus (Steele and Steele 1975)) and provide an abundant 'reserve' whose maintenance offers no threat to the adult stock, but from whose ranks, once grown to a stage when thigmotactic requirements cease to be met by filamentous weeds, replacements can be recruited.

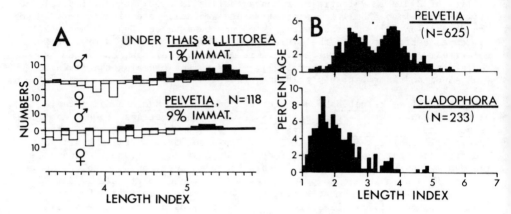

Fig. 2. Size distribution of Hyale in Pelvetia and, A. under gastropods (April 1974), B. in Cladophora (July 1972).

It seems remarkable that in animals breeding from February-October, where 50% of the population are females producing successive broods of up to 40 eggs every two weeks (during summer) that numbers remain within the narrow limits shown in Fig 1A and B. It seems most unlikely that the activity of known predators in the Pelvetia zone could so stabilize the system. Mortality is presumably highest amongst the juveniles downshore. Undoubtedly, the exploitation of local habitat heterogeneity by immatures and adults described above will have a stabilizing effect on density, but what determines the level of this equilibrium in Pelvetia? Pelvetia is hardly 'full-up' at a density of 100 Hyale/100g weed. Perhaps underlying the gross mass of weed is some true resource which may be in short supply, limiting the 'carrying capacity' of Pelvetia? Food and space are obvious possibilities.

In the laboratory, adult H.nilssoni graze different weeds at different rates (assessed as faecal pellet output). The following species are arranged in descending order of palatability; Pelvetia canaliculata, 'Pseudendoclonium' > Enteromorpha prolifera (O.F.Müll.)J.Ag., Rhizoclonium riparium (Roth) Harv., E.intestinalis (L.) Link, Blidingia minima (Näg.ex Kutz.)Kylin > 'Rosenvingiella', Cladophora rupestris (L.)Kütz, Ralfsia verrucosa (Aresch.) J.Ag. > Hildenbrandia prototypus Nardo, Callithamnion roseum (Roth) Lyngb., Bangia fuscopurpurea (Dillw.) Lyngb., Prasiola stipatata Suhr in Jessen, Pilayella littoralis (L.) Kjellm., Acrosiphonia arcta (Dillw.)J.Ag., Porphyra umbilicalis (L.) J.Ag., Rhodochorton purpureum (Lightf.) Rosenv., Porphyra linearis Grev. and Ralfsia clavata (Harv. in Hook) Crouan frat. The latter group were not grazed at all.

Consumption of Pelvetia varies according to the tissue presented (Table 2). Using a t-test approximation, pellet production in all experiments is significantly in excess of controls and consumption of basal tissue is significantly in excess of vegetative tips at 5% level. Given a choice,

TABLE 2 Faecal Pellet Production by individual, freshly-collected, male H.nilssoni (24h, dark, 15-17°C, Aug.1976, n=10)

	Controls (no weed)	Vegetative Tips	Fruiting Tips	Basal Tissue
Mean x	2.8	7.1	13.1	42.1
Standard deviation s_x	2.7	4.4	11.9	44.0

TABLE 3 Faecal Pellet Production by individual, starved, male H.nilssoni given a choice between equivalent tissue segments. Pellets counted each side of the vessel midline (24h, dark, 15-17°C, Aug. 1976)

Basal tissue	44	22	69	6	17	d = 2.2, P < 0.1
Vegetative tip	7	8	2	8	9	

Hyale prefer to graze on the basal region of the thallus (Table 3). Faecal pellets, from field collections and laboratory experiments, contain some Pelvetia meristoderm cells together with 'Pseudendoclonium', Urospora bangioides sensu Knight and Parke, blue green algae and various fungi which are basally epiphytic on Pelvetia and some diatoms. Basal tissue is rougher, easier to gouge, more channeled, of higher organic content, lower permeability, lower polyphenolic content (?) and has more epiphytes than tissue from vegetative tips. But it is not clear to which attribute or combination of attributes Hyale is responsive. The actively growing regions of Phaeophyta have been shown by Sieburth (1968) to secrete antibiotic polyphenolic substances which may deter epifloral colonization. Perhaps this is related to the apparent increase in permeability of the thallus tips? Ryland (1974) thought that the absence of the spirorbid Janua formosa (Bush) and the bryozoan Membranipora tuberculata Bosc. from the young thalli of Sargassum natans (L.) Meyen was caused by tannins inhibiting the settlement of the prerequisite microflora. In addition, however, the rougher texture of the Pelvetia basal tissue would seem to offer a better purchase for biting. Indeed the rough nature of the basal region may result from localized regenerative response to patchy grazer damage. Also, the complex basal anastomosis often encountered in holdfasts and the more channeled nature of the thallus would enhance the thigmotactic suitability of the basal region for a nestling animal.

It seems fair to assume therefore that the whole plant does not represent an equal resource to Hyale in terms either of food or shelter. It is suggested that population regulation of Hyale may be achieved by the saturation of some true resource at the equilibrium density. Verification of this hypothesis requires critical seasonal studies on algal production, amphipod food consumption and feeding preferenda and information on intraspecific aggressive behaviour.

DISCUSSION

Reviewing stabilizing mechanisms in populations, Murdoch and Oaten (1975) distinguished three main conditions producing persistence, 1) refuges for the prey, 2) an invulnerable class of prey and 3) spatial heterogeneity.

All these conditions are operative in the Hyale/Pelvetia system. Regarding refuges, the existence of Pelvetia high enough up the beach ensures that cropping is not continuous. Feeding rate in H.nilssoni is tidally rhythmic (apparently exogenously controlled, cf. Joseph 1972), the activity of Hyale being curtailed by the evaporative removal of moisture at low tide and possibly by the depletion of oxygen from the water film around the alga during darkness by algal respiration when no photosynthesis is taking place (Wieser and Kanwisher 1959; Joseph 1972). Plants at the topmost fringe of the Pelvetia zone will be subject to proportionately lower grazing pressure. As to an invulnerable class of prey, Mr. M. Schonbeck (pers. comm.) has experimented with H.nilssoni grazing sporelings of Fucus spiralis (5 days and 35 days old) in laboratory culture. Whilst Hyale did damage embryos by feeding and dislodgement, their impact after 5 days was hardly different from ungrazed control slides. Amphipods may dislodge spores, but equally their entanglement on body setae may expedite dispersal and lodgment in crevices. Thus the early life history stages of fucoids, as well as the growing tips and fruiting bodies (above) may be relatively invulnerable to grazing. The exploitation of Pelvetia by Hyale certainly appears conservative compared with the improvidence encountered in certain sub-littoral gastropods, eg. Lacuna vincta (Montagu) (Fralick et al. 1974). Habitat heterogeneity, both microscopic (within Pelvetia) and macroscopic (between habitats on a shore) is of fundamental significance as a buffer to excessive population fluctuation in this, and other highly mobile species (Huffaker 1958; Jenkins et al. 1967). Only an understanding of the total environment locally allows of a rational natural history (Lewis and Bowman 1975).

It is becoming clear that the classical intuitive relationship linking diversity and stability first proposed by Elton (1958) may not be of general applicability. Firstly, theoretical developments have led to a rejection of a simplistic view of stability. Secondly, it has been suggested that diversity in the sea is multifarious, eg. within-and between-habitat, transient and long-term, primary and secondary, taxonomic and phenologic. An attempt to establish any universal link between Diversity and Stability would seem certain to founder on theoretical grounds alone. Nevertheless, little empirical information is available to test the hypothesis.

The data here presented have a number of inherent limitations as outlined. The unavoidable conclusion remains: this effective monoculture, made up of relatively short-lived, highly fecund animals has persisted over an extensive period of time (relative to an individual's lifespan) at a predictable level. Fluctuations, in the form of emigrations, occur but the population returns to a 'normal' density when habitable conditions are re-established. Whether we consider stability to mean the maintenance of an equilibrium (here dynamic not static), or the ability to 'bounce back' (Preston 1969), the Hyale population studied is clearly remarkably stable. A population is not unstable because its members move!

The sedentary macrofauna inhabiting the high littoral of rocky shores is of low diversity, yet populations of molluscs and cirripedes are extremely stable with long-lived individuals (Lewis 1964; Lewis and Bowman 1975; Paine 1974). Other field observers have found stability in simple sub-littoral marine ecosystems (Fager 1968). Instability is not the invariable consequence of simplicity, nor an inevitable property of littoral as opposed to sublittoral systems, nor an attribute of short-lived as opposed to long lived animals, nor a property of temperate rather than tropical systems (Green 1968).

Useful generalizations about community organizations must be sought elsewhere. Margalef (1968) coined the term 'maturity' to describe the degree of organization of an ecosystem. Thus (p.43), 'the benthos of a soft, shifting bottom is less stable, has less possibilities for constructing organization, and must be considered less mature than benthos fixed to rocks'. Comparison of the Hyale data with McGrorty's (1972) observations on erratic fluctuations of high-shore populations of Bathyporeia on sandy beaches would lend support to this hypothesis, which is certainly one of some antiquity (Matthew VIII, 24-27), but the metabolic corollaries proposed by Margalef await future testing. Sanders (1968), in a classical paper, sought to explain global differences in diversity within the soft-ooze habitat by his Stability-Time hypothesis. Whilst the mechanism of diversification proposed by Sanders is disputed, the characteristics predicted by his model may bear on the present case. Sanders regarded a transect from the deep sea to the supralittoral zone as a gradient of stress, with biologically accommodated abyssal ecosystems progressively giving way to physically controlled littoral communities and ultimately to abiotic conditions. In the light of the Hyale work, any proposition that biological accommodation dwindles to insignificance in less diverse littoral ecosystems on rocky shores would clearly be untenable. Paine (1974) has argued the same case, but Paine's presumption that Sanders' hypothesis is applicable to ecosystems on rocky substrata may be at fault. Sanders' predictions have always been set within the constraints of the soft-ooze habitat and, whilst an all-embracing hypothesis of community organization in the sea would be especially welcomed, there do seem to be real differences between the scope of ecosystems on hard and soft bottoms and so Paine's criticisms of Sanders may be presently unjust.

It is perhaps tautology to say that each extant species must be biologically accommodated. We must not fall into the pitfall of viewing the 'stresses' of life in the littoral zone 'through the eyes' of an abyssal species. Organisms can adapt to the most rigorous environments (Green 1969). Perhaps the factor which de-stabilizes simple artificial ecosystems may not be their simplicity per se, but rather their lack of evolutionary pedigree (May 1975).

CONCLUSION

Hyale nilssoni dynamically achieved an equilibrium density in high shore Pelvetia which was stable over a three year period. Habitat heterogeneity proved central to this capability. Simple communities are not necessarily unstable, neither need their constituents lack biological accommodation.

ACKNOWLEDGEMENTS

I am grateful to J. Clokie, Anja Preston and Polly Thomas for assistance.

REFERENCES

Bayne, B.L. 1964. Primary and secondary settlement in *Mytilus edulis* (Mollusca). *J. Anim. Ecol.* 33, 513-523.

Branch, G.M. 1975. The ecology of *Patella* from the Cape Peninsula, South Africa, 5. Commensalism. *Zoologica Africana* 10, 133-162.

Brattegard, T. 1963. Crustaceans sheltering under shells of *Nucella*. *Sarsia* 11, 1-3.

Elton, C.S. 1958. *The Ecology of Invasions by Animals and Plants* Methuen, London.

Fager, E.W. 1968. A sand-bottom epifaunal community of invertebrates in shallow water. *Limnol. Oceanogr.* 13, 448-464.

Fralick, R.A., Turgeon, K.W. and Mathieson, A.C. 1974. Destruction of kelp populations by *Lacuna vincta* (Montagu). *Nautilus* 88, 112-114.

Green, R.H. 1968. Mortality and stability in a low diversity subtropical intertidal community. *Ecology* 49, 848-854.

Green, R.H. 1969. Population dynamics and environmental variability. *Am. Zool.* 9, 393-398.

Huffaker, C.B. 1958. Experimental studies on predation: dispersion factors and predator-prey oscillations. *Hilgardia* 27, 343-383.

Jansson, A-M. and Matthieson, A-S. 1971. The ecology of young *Idotea* in the Baltic. *Proc. IVth European Marine Biology Symposium* Cambridge University Press, 71-88.

Jenkins, D., Watson, A. and Miller, G.R. 1967. Population fluctuations in the red grouse *Lagopus lagopus* scoticus. *J. Anim. Ecol.* 36, 97-122.

Joseph, M.M. 1972. Tidal rhythm in the feeding of the intertidal amphipod *Hyale hawaiensis* (Dana) *Proc. Indian Acad. Sci.* 38B, 456-461.

Lewis, J.R. 1964. *The Ecology of Rocky Shores* English Universities Press, London.

Lewis, J.R. and Bowman, R.S. 1975. Local habitat-induced variations in the populations dynamics of *Patella vulgata* L. *J. exp. mar. Biol. Ecol.* 17, 165-203.

Margalef, R. 1968. *Perspectives in Ecological Theory* University of Chicago Press, Chicago.

May, R.M. 1975. Stability in ecosystems: some comments. In, *Unifying Concepts in Ecology* Dr. W. Junk, B.V. Publishers, 161-168.

Murdoch, W.W. and Oaten, A. 1975. Predation and population stability. *Adv. Ecol. Res.* 9, 1-131.

McGrorty, S. 1972. Factors affecting the distribution of Bathyporeia species. Ph.D., thesis, Leeds University.

Paine, R.T. 1974. Intertidal community structure. Experimental studies on the relationship between a dominant competitor and its principal predator. Oecologia (Berl.) 15, 93-120.

Preston, F.W. 1969. Diversity and stability in the biological world. Brookhaven Symp. in Biol. 22, 1-12.

Ryland, J.S. 1974. Observations on some epibionts of Gulf-weed Sargassum natans (L.) Meyen. J. exp. mar. Biol. Ecol. 14, 17-25.

Sanders, H.L. 1968. Marine benthic diversity: a comparative study. Am. Nat. 102, 243-282.

Sieburth, J. McN. 1968. The influence of algal antibiosis on the ecology of marine microorganisms. Adv. Microbiol. Sea 1, 63-94.

Steele, D.H. and Steele, V.J. 1975. The biology of Gammarus (Crustacea, Amphipoda) in the northwestern Atlantic. XI. comparison and discussion. Can. J. Zool. 53, 1116-1126.

Wieser, W. and Kanwisher, J. 1959. Respiration and anaerobic survival in some sea weed-inhabiting invertebrates. Biol. Bull. mar. biol. lab., Wood's Hole 117, 594-600.

Wigham, G.D. 1975. The biology and ecology of Rissoa Parva (da Costa) (Gastropoda: Prosobranchia). J. mar. biol. Ass. U.K. 55, 45-67.

DYNAMICS AND PRODUCTION OF *Pectinaria koreni* (Malmgren) IN KIEL BAY, WEST GERMANY*

Frederic H. Nichols

U. S. Geological Survey, Menlo Park, California 94025 U.S.A.

ABSTRACT

Samples collected periodically for 2 years (November 1973 to November 1975) along a 2.5-km transect crossing the channel system of southwestern Kiel Bay are used to describe the dynamics and production of a numerically dominant macroinvertebrate, the polychaete *Pectinaria koreni*. This species has, in recent years, increased in abundance and become an important food for demersal fish. Recruitment of young normally occurred in early summer, with an additional recruitment in late summer of some years. Population size varied widely from year to year, from less than 100 adult specimens/m^2 in autumn of 1973 and 1975 to greater than 1000 specimens/m^2 in 1974. Biomass and production reached highest levels during late summer of 1974 (from 0.7 to 2.5 g C/m^2 and from 0.5 to 1.0 g C/m^2/month respectively), with highest values recorded at deeper stations. The previous and following years showed very low levels of both, due to poor recruitment or high mortality.

The periodic development of an oxygen deficiency in bottom waters appears to be the major factor determining year to year differences in the abundance and biomass of this and other dominant species in the community. In late summer of those years having calm and warm summer weather conditions (in this case, 1973 and 1975), bottom waters below 20 m became anoxic causing nearly total destruction of the benthic community. In the autumn of 1973 the result was a greatly reduced standing stock of *Pectinaria* throughout the following winter and spring. During summer of 1974 there was much rain and wind, and similar hydrographic stability leading to oxygen deficiency in the bottom water did not develop. The recruitment in 1974 of all species, but especially that of *Pectinaria*, was heavy, and biomass and numbers of specimens were large. Although oxygen deficiency again developed during the hot summer of 1975, an autumn recruitment of *Pectinaria* allowed for reestablishment of its dominance during the same year.

INTRODUCTION

Long-term trends in the biology of the Baltic Sea (enrichment of the flora and fauna by the penetration of warm water marine species) reflect an increase in the salinity and temperature of Baltic Sea water (Nikolayev 1974). Although the productivity of the demersal fishery in the western Baltic has increased in the past several decades, there are not sufficient data to relate this increase to an appropriate increase in the productivity of the benthos (Arntz and Brunswig 1976). Assessments of medium and long-range stability of benthic biomass and productivity are hindered by the great variability in the structure of the benthic community caused, perhaps, by short-term (seasonal or annual) variations in climatic conditions which, superimposed on the long-term trends described above, affect water mass characteristics and movements.

---

*Publication No. 105 of the Joint Research Project, "SFB 95 - Interaction Sea - Sea Bottom," sponsored by the German Research Association.

As part of the Special Research Project, "Sonderforschungsbereich 95," at the University of Kiel, West Germany, interactions between the sea and sea bottom at a study site in the western Baltic near Kiel are being studied in some detail (Hempel 1975). A part of this project is the study of the relations between the distribution and abundance of benthic invertebrates and the feeding habits of demersal fish. Bottom grab samples have been collected for nearly 3 years by D. Brunswig in an effort to estimate productivity of the dominant invertebrate species along a transect adjacent to the study site and across the major channel in the southwest portion of Kiel Bay. Specimens of the dominant polychaete *Pectinaria koreni* (Malmgren), collected between November 1973 and November 1975, were examined for this paper. This species has shown a marked increase in abundance, in recent years, in the benthic community as well as in the stomach contents of demersal fish (Arntz and Brunswig 1976, in press). The data presented here are used to describe the dynamics and production of this species and to show how abundance throughout the year is closely related to summer hydrographic conditions.

Lucas and Holthuis (1975), noting past confusion in the identity of some of the northern European species of the genus *Pectinaria*, have proposed a major revision in nomenclature of the entire group. Their proposed new name for *P. koreni* is *Cistena cylindraria*. But, because other workers are appealing to the International Commission of Nomenclature to have these changes dropped (C. Nielsen, personal communication), I use the familiar name here.

MATERIALS AND METHODS

The samples were collected at near-monthly intervals at up to 11 stations, varying in depth from 15 to 29 m on both sides of the channel, along a 2.5 km transect in western Kiel Bay (Fig. 1). Station location was determined using routine navigational techniques, but the ship was not anchored. Between three and five 0.1-m^2 van Veen bottom grab samples were collected at each station. The amount of sediment recovered in each grab sample was estimated by eye. All but one of the replicate samples were washed on a 1-mm sieve to separate organisms from sediment. The remaining sample was washed on a 0.5-mm sieve. There were no noticeable differences among replicates in the mean size of specimens collected on the two sieve sizes in part because the sorting procedures used for both the 0.5 and 1.0-mm sieve samples did not differ. Therefore, all samples have been treated as if they had been washed on the 1.0-mm sieve. The animals remaining on each sieve were preserved in formalin, sorted into major taxonomic groups, and then stored in alcohol. The cephalic plate widths (CPW) of all *Pectinaria* specimens were measured, converted to ash free dry weight, then to carbon equivalents using the conversion factors determined for the similar species, *Pectinaria californiensis*, in Puget Sound, Washington (Nichols 1975).

Because of the close spacing of some of the stations, and the fact that the ship was not anchored at each station, there is a high probability that the ship drifted between stations during sampling. Ship drift is also suggested by the data concerning amount of sediment collected in each grab sample and, in some cases, by the number and mean size of the specimens collected. Mean sediment size and the distribution of the benthic fauna vary directly with depth in this area (Arntz and Brunswig in press). Because the van Veen grab sampler collects less sediment from harder substrates, sediment volume and age structure show differences that seem more reliable as indications of sampling location than the ship's position determined at the outset of sampling. The

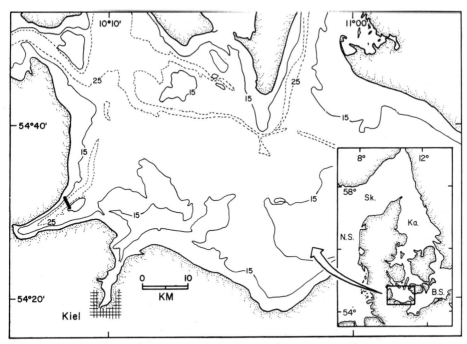

Fig. 1 Kiel Bay, West Germany, showing the major channel system (depth contours in m) and the sampling transect (indicated by a solid bar approximately 20 km to the northwest of the city of Kiel). (N.S. = North Sea, Sk. = Skagerrak, Ka. = Kattegat, and B.S. = Baltic Sea)

samples collected within the depth ranges of 15-18, 20-22, and below 24 m are generally indistinguishable both in terms of specimen number and size and sediment volume (the sampler containing generally less than 20%, between 20 and 60%, and greater than 60% of its capacity of sediment at the three depth zones respectively). Therefore, the following groups of stations, from NW to SE along the transect, are considered single sampling locations: 15/18, 20/21/22, 24, 29, 24, 21, and 15/18 m.

Mean specimen number and size for each size class were determined for each sampling date. A computer program for separating component modes of polymodal distributions (Yong and Skillman 1975) was used in those cases in which modes of the size frequency distributions were not easily distinguished. The size structures of the populations at any depth (e.g., 15 and 18 m on both sides of the channel) were similar. Therefore, the data for comparable depths were lumped in the final interpretation of growth, mortality, and production. Because of infrequent sampling and small specimen numbers at the 29-m location, these data are not considered further.

Estimates of mortality for each size class in each depth zone were obtained from equations describing the relation between mean numbers of animals and the number of days elapsed since 1 January 1973. Similar equations were derived to determine the mean body-weight increase per month. Growth slows during winter, but the variability in the mean weight data within seasons prevented making detailed interpretations of seasonal growth. Therefore, a linear growth function was used. Such a growth curve will not fit immature specimens

(Nichols 1975), so I have not extrapolated to the small animals (CPW of less than 1.5 mm). These animals are also excluded from the mortality computations. Biomass was computed for each sampling date as the product of mean size (mg C/animal) and numbers of animals/m^2, using data obtained from the growth and mortality equations. Production for each size group at each depth was determined through numerical integration of the curve defining the relation between number of specimens and mean body weight ("Allen Curve"; see Peer 1970). The mortality curve was extrapolated back for, at most, 2 months to estimate more realistic count data for the months when mean CPW was less than 1.5 mm, assuming the same exponential rate of decrease holds during the entire period.

RESULTS

When sampling began in November 1973, there were two easily distinguished size classes of *Pectinaria* at 15/18 m (Fig. 2). The low number of specimens representing the older (1972) year class (one to five specimens per sample from November 1973 to May 1974) and the variability of mean size made interpretation of growth, mortality, and production for this size class impossible. Specimens of this older year class were seen only in February 1974 at 20/21/22 m and not at all below this depth zone. The 1973 year class, resulting from an early summer recruitment period, was well represented throughout 1974 at shallow depths, but was poorly represented at 20/21/22 m after July 1974 and was absent after this time at the deeper depths. A strong size class appeared at all depths at the end of July 1974 (Figs. 2 and 3). The specimens of this year class at 24 m were larger and more numerous than those from shallow depths. For example, in November when all specimens of this size class were sufficiently large to be retained on the 1.0-mm sieve, 250 specimens with an average weight of 1.5 mg C were collected per sample at 24 m, compared with 90 specimens/sample and 0.6 mg C/specimen at 15/18 m and 170 specimens/sample and 0.7 mg C/specimen at 20/21/22m.

The 1974 year class had nearly disappeared at all depths by autumn of 1975 (Figs. 2 and 3). At the same time there was no spring recruitment of a 1975 year class, with the exception of specimens (an average of about 30/sample) of an appropriate size that appeared at 15/18 m in November 1975. Very young specimens began appearing at 15/18 m in September 1975, and at 20/21/22 m in November. This suggests a late summer or autumn recruitment which, in this case, allowed for reestablishment of the *Pectinaria* population. Without this late recruitment, there would have been few or no specimens for the remainder of the annual cycle.

Settlement of a very few small specimens (CPW < 1.0 mm) takes place sporadically throughout much of every year. But, these size classes could not be followed during successive samples: they commonly disappear by the next sampling period, perhaps due in part to the patchiness of their distribution and to the error in station finding.

The individual data points describing the increase in mean body size with time (Fig. 2) suggest that growth is most rapid in summer and slowest in winter. However, the data show wide scatter, so an average linear growth for the year has been used for computations of production. The mortality rate for even strong year classes is high, with few specimens reaching an age of 1 year (Fig. 3, the 1973 and 1974 year classes at 20/21/22 and 24 m).

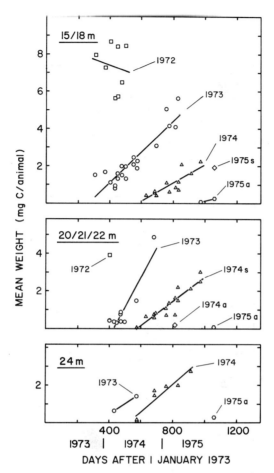

Fig. 2 Mean body weight of the various size classes of *Pectinaria* at three depth zones on each sampling date. Two size classes, spring (s) and autumn (a), are differentiated in 1974 and 1975.

These patterns in the dynamics of *Pectinaria* are reflected in the temporal distribution of biomass and in estimates of production for each month during the 2-year period (Figs. 4 and 5). The strength of the 1974 year class is apparent from the data relating mean weight and numbers/m^2 (Fig. 4), especially at 24 m where mean specimen size and specimen numbers were greatest. This year class contributed most to total biomass during late 1974 and early 1975 (Fig. 5). Production, computed for each depth zone (Fig. 5) as the sum of the values determined for each year class during each month, reached a maximum during the summer of 1974 (approaching 1.0 g C/m^2/month at 24 m and 0.5 g C/m^2/month at the shallower stations). The rather sudden increase in production shown in mid-summer (Fig. 5) is artificial, because the data do not include the early summer production associated with the small specimens that pass through the 1.0-mm sieve. Growth of newly-settled animals is not linear (Nichols 1975). Therefore, the equations for growth determined above are not

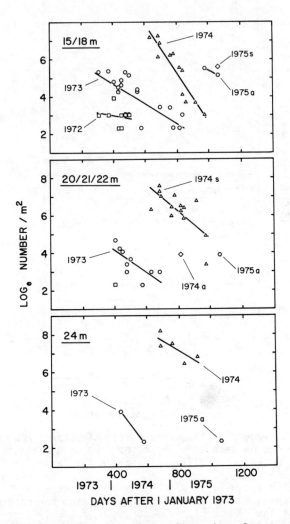

Fig. 3 Number of specimens of the various size classes of *Pectinaria* at three depth zones on each sampling date. Spring (s) and autumn (a) size classes are differentiated where necessary.

appropriate for young animals. But, maximum productivity for the related species, *P. californiensis*, occurred when the specimens were at about the minimum size retained on the sieve (Nichols 1975). Assuming the same for *P. koreni* in Kiel Bay, the maximum estimates of production (Fig. 5) may be realistic. Production each month was much lower both before and after this late summer period of 1974 due simply to the low number of specimens present during the remainder of the study period.

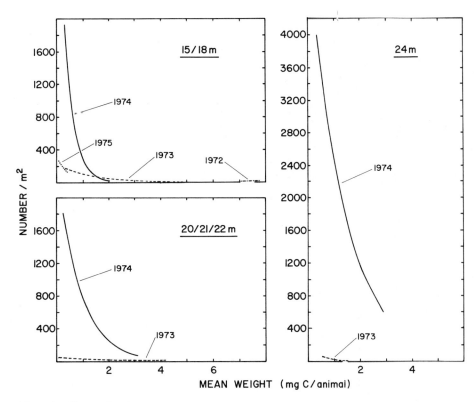

Fig. 4 The relation between mean weight and specimen number for the various size classes of *Pectinaria* at three depth zones.

DISCUSSION

Large year to year variations in the biomass and productivity of the benthos as a result of, for example, fluctuations in water temperature, success of recruitment or disturbance of sediments through biotic or abiotic processes (e.g., Eagle 1975), are to be expected. But, below 18 to 20 m in western Kiel Bay the range of variation encountered from one year to the next is unusually extreme, resulting in either a rich fauna (Fig. 5, 1974/75) or an impoverished one (Fig. 5, 1973/74). These wide fluctuations may be normal, but it is more likely that extraordinary circumstances lead to the impoverishment of the fauna in some years. The size data (Fig. 2) reveal that the *Pectinaria* populations (with the exception of the 1973 year class at the shallow depths) contain few animals older than 1 year, despite the fact that this family of polychaetes can have a life span of several years. High mortality could result from the disturbance of the bottom during commercial fishing operations as well as from size selective predation by demersal fish (in late autumn and winter, the recruits from the previous summer probably have reached a size preferred by demersal fish). But, these factors contributing to overall mortality should be roughly the same from year to year. Moreover, there were sufficient animals remaining in the population in spring of both years to allow for a normal recruitment of young. Nonetheless, there was no settlement of a major size class in the summer of 1975 (Fig. 2).

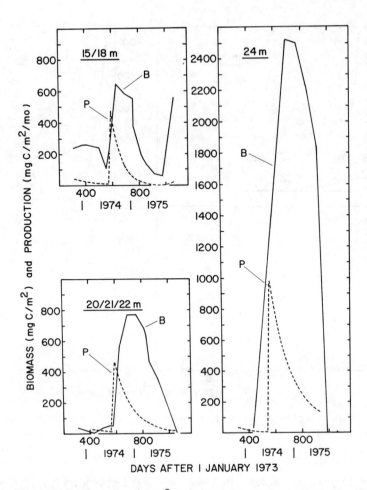

Fig. 5 Total biomass (B, mg C/m^2) on each sampling date and production (P, mg C/m^2/month) for the important size classes of *Pectinaria* at three depth zones.

When the samples were collected during mid-1975, the surface sediment was anoxic and gave off a distinct odor of hydrogen sulfide (D. Brunswig, personal communication). Oxygen data collected to 20 m two to four times a month between January 1972 and June 1975 during plankton studies in the same area (von Bodungen 1975; von Bodungen et al. 1975) show that oxygen concentrations in August and September of each year fell well below saturation values. For example, values of 0.11 to 0.37 ml/l O_2 were recorded at 20 m for the entire month of September 1973. These data, together with observations of anoxic surface sediment, indicate that the bottom water in the deeper regions (>20 m) becomes oxygen deficient during the latter half of some summers.

The source of the low-oxygen bottom water is not precisely known. Smetacek et al. (1976) have shown that periodic pulses (several per month during summer) of high salinity, low oxygen water enter the study area. This implies that bottom water entering Kiel Bay from the North Sea (see Fig. 1) is already somewhat oxygen-depleted. The water column is normally highly stratified in summer, thus preventing turnover and oxygen renewal (Arntz and Brunswig 1976). These factors, coupled with the occasion of calm, warm summers, such as in 1973 and 1975, cause stratification that permits bottom water as shallow as 18 to 20 m to become anoxic. Anoxia has led to the rapid decline of apparently strong *Pectinaria* year classes (Fig. 3) and to the decline of the entire benthic fauna below these depths (see also Brunswig, et al. in press). The timing of depletion of oxygen in bottom waters is critical to the maintenance of a normal benthic community: if anoxia coincides with, or immediately follows recruitment of young to the community, the fauna will be meager or non-existent until the following year. If, alternatively, a late recruitment occurs, as during the autumn of 1975, the population reestablishes itself. The standing crop and, concomitantly, the production of the benthos seems directly related to water mass properties in the western Baltic, and indirectly with prevailing weather conditions in summer.

The 24-m data reveal that production in deeper water can be greater than at shallow depths. This is probably due to the presence of finer sediments (a greater preponderance of silt over sand) and the associated detritus and bacteria, both potential food for *Pectinaria*. Von Bodungen et al. (1975) reported that organic matter produced in the water column during numerous summer blooms settles rapidly to the bottom where it is remineralized. If this material, as potential food for the benthos, is concentrated by watermass movement in the deeper channels, high rates of benthic production could be expected (e.g., Fig. 5, summer 1974) when oxygen concentrations are sufficiently high.

Pectinaria is an important food item for demersal fish (Arntz, personal communication), and is potentially available to these fish at all depths in Kiel Bay. Following warm, dry summers however, this and most of the other common macroinvertebrates are restricted to rather shallow depths because of oxygen deficient bottom conditions. When they are present at deeper depths (during years without anoxic bottom conditions) the specimens are necessarily young and therefore small, so they may not contribute significantly to the nutritional requirements of the fish (see Kerr 1971). The question remains as to how important this reduction in available food is to the fish populations.

CONCLUSIONS

1. Mean specimen size and abundance of *Pectinaria koreni* (Malmgren), a dominant species in the benthic community of Kiel Bay, reveal marked year to year fluctuations, a situation reflected in the benthic community as a whole.

2. During 1974 highest values of biomass and production were achieved at deeper depths, probably because the finer sediment texture provided a greater food supply, and organic matter sinking from the water column concentrated at the sediment surface.

3. An unusually sudden disappearance of *Pectinaria* in mid-1975 coincided with the onset of a very warm, calm summer, and the subsequent stagnation of bottom water.

4. Water chemistry data from past years reveal a coincidence between the occurrence of pronounced oxygen deficiency in bottom water during warm, calm summers and the decline of the benthic fauna. Climatic conditions, therefore, contribute indirectly to the richness and the productivity of the fauna.

5. Autumn recruitment of *Pectinaria* seems to be a mechanism whereby the population can maintain itself despite a summer period of anoxia. Otherwise, the species nearly disappears from the community until the following summer.

ACKNOWLEDGEMENTS

Special thanks are due D. Brunswig for permitting me access to his samples, Professor G. Hempel for inviting me to participate in the "Special Research Project" at Kiel, the U. S. Geological Survey for allowing me the opportunity to accept the invitation, and W. E. Arntz, D. Brunswig, H. Rumohr, and other colleagues for valuable discussions during my year in Kiel.

REFERENCES

Arntz, W. E., and D. Brunswig 1976 Studies on structure and dynamics of macrobenthos in the western Baltic carried out by the Special Research Programme "Interaction Sea - Sea Bottom," (SFB 95/Kiel). Proc. 10th Europ. Symp. Mar. Biol., Ostend, Belgium, 2, 17-42.

Arntz, W. E., and D. Brunswig. In press. Zonation of macrobenthos in the Kiel Bay channel system and its implications for demersal fish. Proc. IVth Symp. Baltic Sea Mar. Biol., Gdansk, Poland, October 1975.

Bodungen, B. von 1975 Der Jahresgang der Nährsalze und der Primärproduktion des Planktons in der Kieler Bucht unter Berücksichtigung der Hydrographie. Ph.D. Thesis. Univ. Kiel. 116 p.

Bodungen, B. von, K. von Bröckel, V. Smetacek, and B. Zeitzschel 1975 Ecological studies on the plankton in the Kiel Bight. I. Phytoplankton. Merentutkimuslait. Julk./HavsforskInst. Skr., Helsingf. 239, 179-186.

Brunswig, D., W. E. Arntz, and H. Rumohr. In press. A tentative field experiment on population dynamics of macrobenthos in the western Baltic. Kieler Meeresforsch.

Eagle, R. A. 1975 Natural fluctuations in a soft bottom benthic community. J. Mar. Biol. Ass., U.K. 55, 865-878.

Hempel, G. 1975 An interdisciplinary marine project at the University of Kiel, "Sonderforschungsbereich 95." Merentutkimuslait. Julk./HavsforskInst. Skr., Helsingf. 239, 162-166.

Kerr, S. R. 1971 Prediction of fish growth efficiency in nature. J. Fish. Res. Bd. Can. 28, 809-814.

Lucas, J.A.W., and L. B. Holthuis 1975 On the identity and nomenclature of "*Pectinaria belgica* (Pallas, 1976)" (Polychaeta, Amphictenidae). Zoöl. Meded., Leiden 49, 85-90.

Nichols, F. H. 1975 Dynamics and energetics of three deposit-feeding benthic invertebrate populations in Puget Sound, Washington. Ecol. Monogr. 45, 57-82.

Nikolayev, I. N. 1974 Main trends in the biology of the present-day Baltic Sea. Oceanology 14, 873-881.

Peer, D. L. 1970 Relation between biomass, productivity and loss to predators in a population of a marine benthic polychaete, *Pectinaria hyperborea*. J. Fish. Res. Bd. Can. 27, 2143-2153.

Smetacek, V., B. von Bodungen, K. von Bröckel, and B. Zeitzschel 1976 The plankton tower. II. Release of nutrients from sediments due to changes in the density of bottom water. Mar. Biol. 34, 373-378.

Yong, M.Y.Y., and R. A. Skillman 1975 A computer program for analysis of polymodal frequency distributions (ENORMSEP), Fortran IV. Fish. Bull. 73, 681.

THE EFFECTS OF STORMS ON THE DYNAMICS OF SHALLOW WATER BENTHIC ASSOCIATIONS

E.I.S. Rees, A. Nicholaidou and P. Laskaridou

Marine Science Laboratories, Menai Bridge, Gwynedd, U.K.

ABSTRACT

Muddy sands with dense benthic faunas occur patchily along the North Wales coast between tide swept grounds offshore and the wave washed sandy beaches. Although sheltered from prevailing winds the patches are intermittently disturbed by onshore gales. Monthly sampling in four localities with differing exposure over the 1975-76 winter allowed faunal changes to be compared with the relative wave energy of the storms. Wave scour washed out much of the fauna, sometimes causing mass stranding. Wave action also redistributed the fauna, thus playing a part in the formation of benthic associations as well as their destruction.

INTRODUCTION

In Liverpool Bay (S.E. Irish Sea) (Fig. 1.1) the degree of mobility of the sea bed is the most important factor governing the abundance and diversity of the benthic fauna (Rees et al. 1972). Most of the bay is floored by mobile sands and lag gravels, but around the fringe the local topography provides sufficient shelter from the tidal currents and prevailing winds for patches of muddy sand to form. These patches often carry dense benthic macrofaunas referable to various muddy sand facies of the infra-littoral etage (Glemarec 1973).

Eagle (1973, 1976) sampled the muddy sand patches in the south eastern corner of Liverpool Bay at half yearly intervals for four years and found very large temporal population changes. He concluded that intermittent disturbance by waves was the prime cause of change. During the 1975-76 winter samples were taken at approximately monthly intervals on four other patches of in-shore muddy ground off the North Wales coast during studies on Pectinaria koreni Malmgren and Abra alba (W. Wood). One or other of these species is usually numerically dominant in the muddy sands and their relative importance varies from time to time in any one patch. Severe onshore gales causing mass stranding of benthic organisms on local beaches occurred during some of the sampling intervals. Thus, the opportunity arose to confirm and

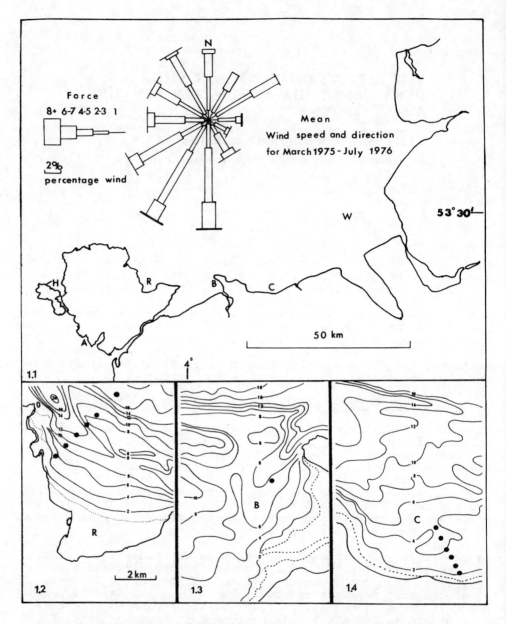

Fig. 1.1. Location of sampling areas mentioned in text. A. Aberffraw Bay, H. Holyhead Harbour, R. Red Wharf Bay, B. Beaumaris Bay, C. Colwyn Bay, W. Wirral. Wind rose shows percentage frequency of Beaufort scale speeds.
Figs. 1.2, 1.3 and 1.4. Local Bathymetry.

extend Eagle's observations by comparing four inshore muddy sand patches with differing shelter from the storms.

MATERIALS AND METHODS

Van Veen 0.1 m² grab samples were taken at the locations shown in Figs. 1.2, 1.3 and 1.4 in Red Wharf Bay, Beaumaris Bay and Colwyn Bay. The stations in Red Wharf Bay and Colwyn Bay were on transects running across the muddy sand patches. The outermost stations on each transect were on the more mobile sand offshore. The Beaumaris Bay patch south-west of the Great Orme was too small to accommodate a series of stations. Positions were normally fixed by a combination of shore mark alignment and horizontal sextant angles. Holyhead Outer Harbour (Fig. 1, Hoare and Wilson this symposium) was sampled from small boats using a 0.02 m² grab hauled by hand at sixteen stations located near buoys or other marks. At each station at least two samples were taken, on each occasion, and sieved using 1 mm nylon mesh.

For comparative purposes we have also drawn on original data from seven muddy stations sampled by Eagle between the Mersey and Dee estuaries (marked W in Fig. 1.1). Mean over winter survival rates were calculated from 1970-71, 1972-73 and 1973-74 spring and autumn samples.

After all major onshore gales in the 1975-76 winter beaches adjacent to the benthic sampling sites were searched for stranded organisms. At Colwyn Bay lower shore sand samples were sieved to extract Pectinaria that had temporarily become established intertidally.

Wave energy likely to have impinged on the north facing coasts generally, but not on the exact stations, was estimated to give comparative daily values. Using methods described by Vollbrecht (1966) and graphs of Bretschneider (1952), arbitrary energy units were calculated from the wind speed, direction and fetch. Hourly wind records from Valley (Anglesey) meteorological station were used. Since only moderate to strong winds could cause bottom disturbance, force 5 on the Beaufort scale was taken as the threshold. Total energy units per day, above a further threshold of 30 units per hour (about force 6 with 50 nautical miles fetch), were added together. These daily energy values are shown in Figs. 2 and 3.

RESULTS

The major onshore storms during the 1975-76 winter were on 16-17 November, 2-3 January and 23-26 January. In Red Wharf Bay the most numerous species found stranded after these storms were Abra alba, Spisula subtruncata (da Costa), Mactra corallina (L.), Asterias rubens L., Ophiura texturata Lamarck and Alcyonium digitatum (L). There were moderate numbers of Nucula turgida Leckenby and Marshall, Venus striatula da Costa, Dosinia lupinus (L.), Cardium echinatum L., Cultellus pellucidus (Pennant), Echinocardium cordatum (Pennant), Astropecten irregularis

TABLE 1. Mean abundance in autumn and apparent over-winter survival rate of Pectinaria koreni and Abra alba.

	Pectinaria koreni		Abra alba	
	Mean abundance per m^2	Apparent % survival	Mean abundance per m^2	Apparent % survival
C.B.	590	2	52	12
R.W.B.	975	1	6870	7
W.	672	14	615	7
H.	-	-	115	54

TABLE 2. Mean autumn abundance and apparent over-winter survival during 1975-76 at Colwyn Bay of a range of species in survival ranking order.

Species	Autumn No/m^2	Apparent % survival
Mysella bidentata (Montagu)	22	50
Magelona papillicornis F. Muller	25	50
Nephthys hombergi Lamarck	140	29
Owenia fusiformis Delle Chiaje	111	22
Eteone longa (Fabricius)	38	22
Tellina fabula Gmelin	96	14
Notomastus latericeus M. Sars	13	12
Abra alba (W. Wood)	52	12
Gyptis capensis Day	22	11
Spiophanes bombyx (Claparede)	132	11
Heteromastus filiformis (Claparede)	38	5
Pectinaria koreni (Malmgren)	590	2
Lanice conchilega (Pallas)	236	1
Eulalia (Eumida) sanguinea Oersted	56	1
Phyllodoce maculata (L.)	125	0
Ophiura albida Forbes	70	0

TABLE 3. Variations in the silt/clay fraction of the sediment at Colwyn Bay and Red Wharf Bay during 1975-76.

	APR.	OCT.	DEC.	JAN.	FEB.	MAR.	APR.	JUN.	JUL.
C.B.	-	-	20.3	2.6	2.6	4.1	11.2	7.2	6.8
R.W.B.	17.0	18.7	-	24.3	25.2	25.7	17.4	26.0	-

(Pennant), Ophiura albida Forbes, Buccinum undatum L., Natica alderi (Forbes), Eupagurus bernhardus (L.), Aphrodite aculeata L., and Pectinaria koreni. Small numbers of Pecten maximus (L.), Chlamys opercularis (L.), Mya truncata L., Gari fervensis (Gmelin), Tellina fabula Gmelin, Ensis ensis (L.), E. siliqua (L.) and Cyprina islandica (L.) were also found. In Colwyn Bay Pectinaria koreni made up the majority of the stranded specimens with fewer Asterias rubens and Ophiura albida.

Table 1 shows the mean abundance of Pectinaria koreni and Abra alba at stations in the four localities sampled, in 1975-76, during the months prior to the storms and the percentage apparently remaining in the months immediately after the storms. Figures are also included for the Wirral patches based on the means of seven stations sampled in autumn and spring of three winters. Owing to differences in sampling dates and sampling intensity, the figures are only comparable in general terms. The Holyhead figures compare the mean from sixteen stations sampled between mid August and late October with the same stations sampled between late January and mid March. Only one station in the six in Red Wharf Bay had more than a few Pectinaria so the figures compare the mean of three pre-storm samples (Aug. - Oct.) with the mean of three post storm samples (Jan. - March). Red Wharf Bay Abra figures compare the same months but are the means from the three inner stations. The Beaumaris Bay figures compare the mean from the single station in August - November with January - April. The Colwyn Bay figures are for the four middle stations on the transect sampled four times before and four times after the storms.

Table 2 shows the same pre and post storm comparison for the other less abundant species at the middle four Colwyn Bay stations. The muddy sand sediment was apparently scoured right away here so that the grab brought up gravel. Variations in the mud content of the sediments are shown in Table 3. Beaumaris Bay and Red Wharf Bay sediments remained muddy but markedly more shell debris was in the grab samples after storms.

DISCUSSION

Mass stranding on the beaches of benthic species from sub-tidal areas provides ample evidence for the disturbing effects of waves. The range of species found on the shore does not include smaller soft bodied animals. As thousands of gulls (Larus sp.) congregate on the shores only species with hard shells, that are difficult to open, will survive for long.

The north-facing bays are sheltered from the prevailing winds and they often have sand banks offshore giving added protection. Figures 1.2, 1.3 and 1.4 showing the bottom contours (interpreted from Admiralty charts numbers 1977 and 1978) give an impression of the local topography. In severe weather the waves break on the outer banks and refraction will add further complexity to the pattern of potential disturbance.

In July 1976 G. O'Sullivan (personal communication) compared the

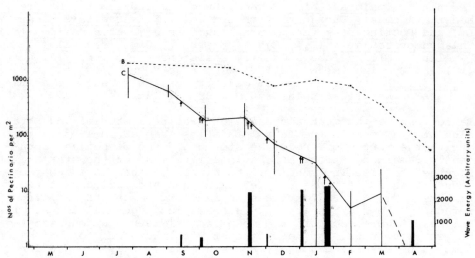

Fig. 2. Population change during 1975-76 winter of *Pectinaria koreni* at Beaumaris Bay site and middle four stations in Colwyn Bay showing means and range. The incidence and daily totals of wave energy units over a threshold are shown for Colwyn Bay.

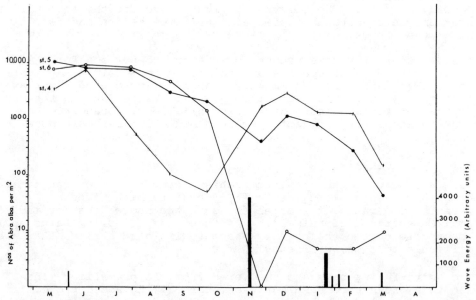

Fig. 3. Population change during 1975-76 winter of *Abra alba* at three stations in Red Wharf Bay. Station 6 is inshore. The incidence and daily totals of wave energy units over a threshold are shown.

benthos on transects running out from low water mark in Red
Wharf Bay and Aberffraw Bay on the south west coast of Anglesey.
Whereas the muddy sand fauna comes in to 3m on the sheltered
north facing shore it is found below about 8m at Aberffraw. It
is suggested that, at Aberffraw, wave activity is too frequent for
settling mud to be incorporated in the sediment, while long calm
periods allow this in the north facing bays. At times, ephemeral
patches of fluid mud even settle on the north facing beaches.
These coastal patches of mud are particularly vulnerable to
disturbance when the wind is onshore.

As the 2-3 January 1976 storm caused extensive damage to coast
protection works and washed away most of the commercial mussel
beds at Conway, it must be considered exceptional; but the other
storms were not. Orton (1929) and Lingwood (1976) record similar mass mortalities on the Lancashire coast and personal
experience in other years suggests that these events are not
uncommon. A northerly gale on 11-12 September 1976 threw piles
of Echinocardium cordatum on to the strand line in Red Wharf
Bay.

Differences in the apparent over winter survival rate at the
five localities in Table 1 suggest that exposure can reduce survival without the whole population being carried away. Eagle
(1976) described over winter contraction of the muddy sand patches . In the present sample data the inner and outer stations
showed early losses. Figures 3 and 4 show that the smallest
changes to the Abra and Pectinaria populations occurred in months
when there were no storms. The most dramatic change on the Red
Wharf Bay transect coincided with the November storm. The shoreward station (6) lost virtually all the Abra and station (4) had
a major gain. Both changes were maintained in later months confirming that they were not freak samples. Nor do length
frequencies show that spat recruitment accounted for the gain in
numbers. This indicates that movement and re-establishment of
Abra took place on a large scale. Re-establishment of displaced
animals can sometimes be seen on the shore. After the November
storm, a Pectinaria population became established in a beach
runnel at Colwyn Bay. It lasted until the January storms. After
storms, Spisula subtruncata can be seen digging themselves in on
the beach but they do not seem to survive for long. Since most
soft bottom benthic animals are capable of digging in after disturbance, re-establishment on a large scale is to be expected
unless the animals are disabled by the movement, cast ashore or
taken by predators. Predation on the Red Wharf Bay Abra population, by Asterias rubens and flatfish, is substantial. The
sudden loss of Abra in March is however more likely to be caused
by spawning mortality rather than the resumption of feeding by
plaice (Pleuronectes platessa L.) after their spawning period
fast.

Thorson (1950, 1966) stressed the importance of various selective factors, operating on the larvae and settling spat, in
determining the composition of benthic communities. In contrast
Dauer and Simon (1976) show that adult dispersal played a significant part in the repopulation of a beach after a red tide.

Redistribution, initiated by storms, of adult animals in subtidal areas is shown here. Baggerman (1953) showed that such water induced redistribution was selective, with spat sized Cardium edule L. being moved from exposed sand flats to more sheltered muddy sand localities. On the North Wales coast bottom drifter and moored current meter studies all show shoreward bottom residual movement (Harvey 1968, Halliwell 1972, Ramster 1972). Halliwell (1975) also postulates shoreward bottom mass transport in storms to explain the rapid and high return rate of drifters. Rees et al. (in press 1976) suggest that the selective deposition of detritus carried in from much wider areas accounts in part for the high density of the benthos carried by the coastal muddy sand patches. Selective movement initiated by waves of both the spat and larger sizes of benthic organisms into these sediment traps now also seems likely.

CONCLUSIONS

The strength and temporal distribution of wave activity are important, but often overlooked, factors in the establishment and maintenance of soft bottom benthic associations in coastal waters. Species with annual or shorter life-cycles normally dominate the more frequently disturbed areas and mobile predators are favoured. The animals stranded on the beach after a storm probably represent only the fringe of population changes seen in such areas.

ACKNOWLEDGEMENTS

Our thanks are due to the Captains and crews of the research vessels Lewis Morris and Prince Madog and to several colleagues including R. Hoare and A.J.M. Walker who assisted on sampling trips. R.A. Eagle and G. O'Sullivan have generously given us access to their original data.

REFERENCES

Baggerman, B. 1953. Spatfall and transport of Cardium edule L. Archs. neerl. Zool. 10, 315-342.

Bretschneider, C.L. 1952. The generation and decay of wind waves in deep water. Trans. Am. Geophys. Union 33(3), 381-389.

Dauer, D.M. and J.L. Simon, 1976. Repopulation of the Polychaete Fauna of an Intertidal Habitat Following Natural Defaunation: Species Equilibrium. Oecologia 22, 99-117.

Eagle, R.A. 1973. Benthic studies in the south east of Liverpool Bay. Estuarine and Coastal mar. Sci. 1, 285-299.

Eagle, R.A. 1976. Natural fluctuations in a soft bottom benthic community. J. mar. biol. Assoc. U.K. 55(4), 865-878.

Glemarec, M. 1973. The benthic communities of the European north Atlantic continental shelf. Oceanogr. mar. biol. Ann. Rev. 11, 263-289.

Halliwell, A.R. 1972. Sea-bed drifter study. In Out of Sight Out of Mind Vol. 2 appendix 7, 81-130. Report of a working party on sludge disposal in Liverpool Bay. Department of the Environment, HMSO London.

Harvey, J.G. 1968. The movement of sea-bed and surface drifters in the Irish Sea. Sarsia 34, 227-242.

Hoare, R. and E. Wilson 1977. Observations on the distribution and behaviour of Virularia mirabilis (Coelenterata: Pennatulacea) in Holyhead Harbour, Anglesey. In Proceedings of the 11th European Symposium on Marine Biology ed., Keegan, B.F., P. O Ceidigh and P.J.S. Borden. Pergamon Press.

Lingwood, P.F. 1976. Moribund macrofauna of the south west Lancashire coast - 1971-1975. Lancashire and Cheshire fauna Soc. 69, 6-7.

Orton, S.H. 1929. Severe environmental mortality among Abra (=Syndosmya) alba, Donax vittatus and other organisms off the Lancashire Coast. Nature 124, 911.

Ramster, J.W. 1972. Current measurements. In Out of Sight Out of Mind Vol. 2 appendix 6, 57-79. Report of a working party on sludge disposal in Liverpool Bay. Department of the Environment, HMSO London.

Rees, E.I.S., A.J.M. Walker and A.R. Ward, 1972. Benthic fauna in relation to sludge disposal. In Out of Sight Out of Mind Vol. 2 appendix 14, 297-343. Report of a working party on sludge disposal in Liverpool Bay. Department of the Environment, HMSO London.

Rees, E.I.S., R.A. Eagle and A.J.M. Walker, 1976. Trophic and other influences on macrobenthos population fluctuations in Liverpool Bay. In Proceedings of the 10th European Symposium on Marine Biology, Ostend. ed. Persoone, G. and E. Jaspers, 2, 589-599.

Thorson, G. 1950. Reproductive and larval ecology of marine
    bottom invertebrates. Biol. Rev. 25, 1-45.

Thorson, G. 1966. Some factors influencing the recruitment
    and establishment of marine benthic communities.
    Neth. J. Sea Res. 3(2), 267-293.

Vollbrecht, K. 1966. The relationship between wind records,
    energy of longshore drift and energy balance off the coast
    of a restricted body of water, as applied to the Baltic.
    Mar. geol. 4, 119-147.

Pachycerianthus multiplicatus CARLGREN — BIOTOPE OR BIOCOENOSIS?

B. O'Connor, G. Könnecker, D. McGrath and B. F. Keegan

Zoology Department, University College, Galway, Ireland

ABSTRACT

Part of the upper reaches of a shallow bay on the Irish west coast (Kilkerrin Bay; lat. 53'. 20"N, long. 9'.39"W) has been colonised by the cerianthid anemone *Pachycerianthus multiplicatus*. This represents a surprising addition to the Irish fauna for, insofar as the authors can ascertain, the cerianthid was previously known only from the Gullmar and Trondhjem fjords and from the Kattegat. *Pachycerianthus multiplicatus* - adult body length + 30 cm. with 160-170 long marginal tentacles - occupies a vertical tube of its own manufacture. This tube, which can exceed 130 cm. in length, projects some 10 cm. free of the sediment surface. In Kilkerrin Bay, the tubes have been variously settled upon by more than 40 animal species including *Myxicola infundibulum* (Renier) and *Golfingia elongata* (Keferstein). Prompted by their known ecological requirements, by the frequency of association and by the fact that they were not otherwise found in the immediate area, particular attention has been given to the occurrence of *M. infundibulum* and *G. elongata* and to the nature of their relationships with *P. multiplicatus*.

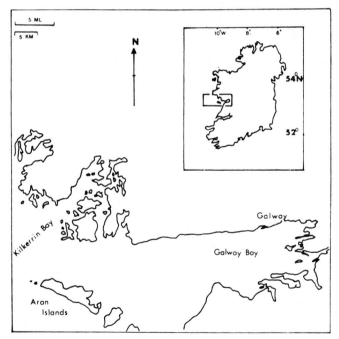

Fig. 1   Location Map of Study Area .

INTRODUCTION

The upper reaches of Kilkerrin Bay (Fig. 1) were surveyed as part of an ongoing investigation of benthic macrofauna off the Irish west coast. Remote and *in situ* sampling revealed that a semi-enclosed portion of the bay is spectacularly colonised by the cerianthid anemone, *Pachycerianthus multiplicatus*. A surprising addition to the Irish marine fauna, this anemone was previously known only from the Kattegat and the Trondheim fjord (Carlgren 1912) and from the Gullmar fjord (Molander 1928). Populating deposit substrates (85% silt/clay 15% sand), between 10 and 20 m., *Pachycerianthus multiplicatus*, *Virgularia mirabilis* (O.F. Müller), *Sagartiogeton undata* (Muller) and *Lysilla loveni* (Malmgren) are the most conspicuous elements of what is presumed to be an impoverished form of the boreo-mediterranean *Amphiura* community (Thorson 1957). A series of grab samples showed that, in numerical terms, the infauna is dominated by supposed deposit-feeders and, to a lesser degree by carnivores. A summary list of numerically prominent animals is presented in Table 1.

TABLE 1 Numerically prominent members of the infauna

Species	Max. Nos/$0.1m^2$
Thyasira fluxosa (Montagu)	211
Abra nitida (Müller)	59
Notomastus latericeus (M. Sars)	29
Scalibregma inflatum (Rathke)	12
Lysilla loveni (Malmgren)	7
Nephthys homergii (Audouin & Milne Edwards)	6
Scoloplos armiger (O.F. Müller)	5
Abra alba (W. Wood)	4
Prionospio malmgreni (Clarpareole)	4
Araea trilobata (Sars)	4

Detailed examination of a small number of *P. multiplicatus* suggested some functioning relationships with certain animals particularly the fan-worm, *Myxicola infundibulum* and with the sipunculid, *Golfingia elongata*. The authors have attempted to interpret the causality behind these suspected relationships.

Plate 1. *Pachycerianthus multiplicatus* Carlgren

RESULTS

A full description of *Pachycerianthus multiplicatus* (Plate 1) is given by Calgren (1912). With a maximum langth of 27 cm. (authors' experience), the cerianthid occupies a vertical tube of its own manufacture. This tube can be more than 130 cm. long and projects some 10 cm. above the sediment surface. It comprises two main layers:

(i) an outer layer, 3-7 mm. in thickness, which is composed of sedimentary debris, forameniferan remains and sponge spicules and is ramified by filamentous algae.

(ii) an inner layer, 2-4 mm. in thickness which is composed of mucous strings and nematocysts derived from the anemone.

Initial fears regarding the population's ability to withstand significant sub-sampling were dispelled when transectal counts across the bay yielded numbers in excess of 400 individuals with a maximum density of $3/m^2$. Forty specimens were returned to the laboratory for detailed examination. A total forty-eight macro-animals were found in or upon the anemone's tube (Fig. 2, Table 2). Smaller animals i.e. Foramenifera, Nematoda, Harpacticoida and Ostracoda were similarly located but were not identified to species level. Prompted by their known ecological requirements, by the frequency of association and by the fact that they were not otherwise found in the immediate study area, particular attention has been given to the occurrence of *Myxicola infundibulum* and *Golfingia elongata*. *Myxicola*, occuring with thirty-two of the forty specimens examined, has its gelatinous tube pocketed between parallel layers of the anemone's tube (Plate 2). It is perhaps significant that the eight *Pachycerianthus* without *Myxicola* were all small

TABLE 2 ANIMALS FOUND IN/ON THE TUBE OF P. multiplicatus.

Foramanifera spp.

Sagartiogeton sp.

Nematoda spp.

Polychaeta:

    Harmothoë imbricata L.
    Sthenelais boa Fabricius
    Pholoe minuta (Johnston)
    Eulalia viridis (O.F. Müller)
    Kefersteinia cirrata (Keferstein)
    Syllis gracilis Grube
    Syllis sp.
    Nereis fucata (Savigny)
    Platynereis dumerillii (Audouin & Milne-Edwards)
    Nephthys hombergii Lamarck
    Glycera alba (Rathke)
    Lysidice ninetta Audouin & Milne-Edwards
    Lumbriconereis latreilli Audouin & Milne-Edwards
    Lumbriconereis gracilis Ehlers
    Staurocephalus kerersteinii McIntosh
    Scoloplos armiger (O.F. Müller)
    Paraonis fulgens (Levinson)
    Tharyx marioni (Saint-Joseph)
    Cirratulid sp.
    Stylaroides plumosa (O.F. Müller)
    Scalibregma inflatum Rathke
    Notomastus latericeus M. Sars
    Melinna palmata Grube
    Polymnia nebulosa Montagu
    Thelepus cinncinnatus (Fabricius
    Sabella pavonina Savigny
    Branchiomma vesiculosum (Montagu)
    Dasychone bombyx Dalyell
    Potamilla torelli Malmgren
    Myxicola infundibulum (Renier)

Priapulida:

    Priapulus caudatus Lamark

Sipunculida:

    Golfingia elongata (Keferstein)
    Golfingia vulgare (Blainville)
    Golfingia minuta (Keferstein)

Gastropoda:

    Turitella communis Risso
    Risso sp.
    Chrysallida obtusa (Brown)

Bivalvia:

    Abra alba (W. Wood)
    Abra nitida (Müller)
    Corbula gibba (Olivi)
    Mysella bidentata (Montagu)
    Thyasira flexuosa (Montagu)

Ostracoda spp.

Copepoda spp.

Amphipoda:

    Microdeutopus anomalus Rathke
    Leucothoë liljeborgi Boeck
    Leucothoë incisa D. Robertson
    Ampelisca tenuicornis Liljeborg

---

animals in the 8-15 cm. range for overall length. The maximum number of individuals recorded with a single anemone was five. Golfingia elongata, with 100% co-occurrence and a maximum number of 17 individuals/anemone, ranges within that portion of the tube which projects above the sediment. Examination of the sipunculid's contents showed the presence of nematocysts together with non-identifiable debris. Other animals which are known to

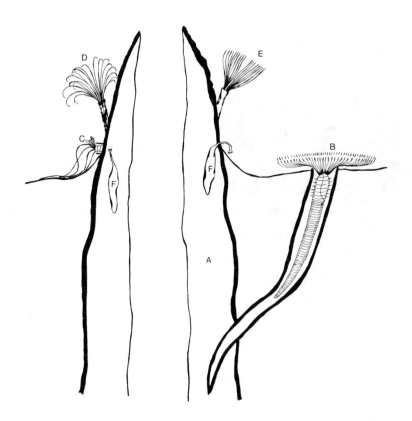

Fig. 2. Location of selected animals on/in Pachycerianthus tube.

A. Tube of Pachycerianthus multiplicatus.
B. Myxicola infundibulum
C. Thelepus cinncinnatus
D. Dasychone bombyx
E. Potamilla torelli
F. Golfingia elongata

form interspecific associations and which were taken from Pachycerianthus tubes during this survey included Mysella bidentata (Montagu), Harmothoë lunulata (Delle Chiaje) and Chrysallida obtusa (Brown). Mysella, although it can be free-living, is well-known as a commensal. It has many hosts, including Golfingia spp. (Boss, 1965), and was found in close proximity to the sipunculids in the fabric of the Pachycerianthus tubes. Harmothoë lunulata, a commensal polychaete (viz. Fauvel 1923), was found in the tube of the terebellid Thelepus cinncinnatus Fabricius which was itself embedded in the anemone's tube. The pyramidellid Chrysallida obtusa is a known parasite (Fretter and Graham, 1962) but its actual host in the Pachycerianthus 'association' has not yet been established.

Plate 2. *Myxicola* pocketed between layers of a *Pachycerianthus* tube.

DISCUSSION

There are other reported examples of interspecific partnerships which involve cerianthid anemones, e.g., *Cerianthus* sp. with *Montacuta ceriantha* Ponder and *Phoronis australis* Haswell (Ponder, 1971), and *Cerianthus maua* Calgren with *Phoronis australis* (Emig, 1971). Respectively, these have been defined as examples of commensalism and inquilinism. Whilst it seems unlikely that *Pachycerianthus* benefits from its association with *Myxicola*, possible advantages to the polychaete could relate to:

(i) the suitability of the tube as a settlement surface,

(ii) The increased food supply precipitated by the battle-effect of the anemone's tentacles,

(iii) the protection afforded by the anemone's tentacle crown.

The protection dimension is discounted as being of little importance in this association. *Myxicola* is widely and independantly distributed over the general Kilkerrin Bay area where predation must be at least as limiting as in the immediate study area. Enrichment of food-supply by baffle precipitation is similarly discounted as the dominating pressure linking the fanworm to the anemone. The epifaunal assemblages which populate sublittoral rocky outcrops and non-deposit substrates in the vicinity testify to an abundance of food material. This was further borne out when oysters *Ostrea edulis* L. and *Crassostrea gigas* Thunberg, which were held at different levely in the water body, showed large growth increments (> 2 cm.) over the period Aug. '75 to March '76. Also, *Pachycerianthus*, *Virgularia* and *Sagartiogeton*, here regarded as 'impingement' feeders (viz. Bullivant 1968), obviously depend on a plentiful supply of suspended food material. The suitability of the tube as a settlement surface is seen as the most likely basis for the association. This probably also applies to the following filter-feeding polychaetes which were here restricted in their distribution to the anemone's tube: *Branchiomma vesiculosum* (Montagu), *Dasychone bombyx* (Dalyell), *Potamilla torelli* (Malmgren). Indeed it seems that, excepting the tube, the environment is, somehow, hostile to infaunal filter-feeders. Rhoads and Young (1970) suggested that the physical instability of the reworked environment of deposit-feeders, in the presence of a weak current, may discourage settling by the larvae of many filter-feeders. Such "trophic group amensalism" might well be the cause of the large-scale exclusion of filter-feeders from the infauna. The association with *Golfingia* would seem to have the same causality - that of suitability of settlement surface. However, the presence of nematocysts in the sipunculid gut suggests that substrate suitability might not be the only basis for its relationship with *Pachycerianthus*. Clearly there are many questions to be answered concerning the precise nature of these relationships. The authors are particularly interested in the mode of settlement, growth and age of the organisms and intend to research these subjects further.

ACKNOWLEDGEMENTS

This contribution describes part of an on going programme of benthic studies which is directed by Brendan F. Keegan and financed by the National Science Council of Ireland. The authors gratefully acknowledge the assistance of Sheila Smith, Peter Gibbs nad Roger Lincoln in checking species identification. We also thank Evelyn Drinan, for her art-work, and Albert Lawless, for his co-operation as technician and diving colleague.

REFERENCES

Boss, K.J. (1965). Symbiotic Erycinacean Bivalves. Malacologia 3(2): 183-195.

Bullivant, J.S. (1968). A revised classification of suspension feeders. Tuatara 16: 151-160.

Calgren, O. (1912). Ceriantharia. Danish Ingolf Expedition 5: 1-78.

Emig, C.C., C. Herberts & B.A. Thomassin (1972). Sur l'Association de *Phoronis australis* (Phoronida) avec *Cerianthus maua* (Ceriantharia) dans les zones recifales de Madagascar. Marine Biology 15: 305-315.

Fauvel, P. (1923). Polychetes Errantes. Faune de France 5: 1-488.

Fretter, V. & A. Graham (1962). British Prosobranch Molluscs. Ray Society, London 755pp.

Molander, A.R. (1928). Animal Communities on Soft Bottom areas in the Gullmar Fjord. Kristinebergs Zoologiska Station 2: 1-90.

Ponder, W.F. (1971). *Montacuta ceriantha* n. sp., a commensal Leptonid bivalve living with *Cerianthus*. Journal de Conchyliologie Paris 109: 15-25.

Rhoads, D.C. & D.K. Young (1970). The influence of deposit feeding organisms on sediment stability and community trophic structure. Journal of Marine Research 28: 150-170.

Thorson, G. (1957). Bottom Communities Mem. geol. Soc. Amer. 67 (1): 461-534.

ON THE ECOLOGY OF A SUSPENSION FEEDING BENTHIC COMMUNITY: FILTER EFFICIENCY AND -BEHAVIOUR*

Elvira M. Ölscher and Kurt Fedra

Zoologisches Institut, Universität Wien, Lehrkanzel für Meeresbiologie, A-1090 WIEN, Währingerstrabe 17, Austria

ABSTRACT

Suspension feeding was found to be the dominant mode of feeding in a North Adriatic benthic community. 92% of the macro-epifaunal biomass is represented by suspension feeders. Active filter feeders such as sponges, ascidians, and bivalves account for 52% of the total community biomass, and passive filter feeders account for 40% (average values for the centre of the community from observations extending over a period of two years).

Filtering activity and behaviour of the dominant species of the latter compartment, namely *Ophiothrix quinquemaculata* (D. Ch.) (contributing 84% to the biomass of the compartment) and *Cucumaria planci* (BRANDT) (contributing 6%) were observed *in situ* by means of a time-lapse camera for more than 200 hours. No specific activity patterns could be observed, and the behaviour was mainly determined by the pattern of currents. *Ophiothrix quinquemaculata* was found to feed more or less continuously and to try to keep within a certain range of current velocity utilizing edaphic or biotic structures.

Concurrent laboratory experiments yielded detailed information on the feeding mechanisms. Filtration rates were determined and are mainly dependent on the type and density of suspended food (zooplankton or seston) and current velocity. The relationship of the investigated variables are given.

Based on these results, the filtering efficiency of the two species is related to their metabolic rates and to the determining environmental conditions. The role of the passive filter feeders in the trophic structure and energy budget of the community is discussed.

INTRODUCTION

Suspension feeding can be considered the most important mode of feeding among benthic invertebrates. 102 of the 205 taxa (macro-epifauna) found in the *Ophiothrix-Reniera-Microcosmus* community (Fedra *et al.*, in press) are suspension feeders. The community occupies considerable areas in the Gulf of Triest - approximately 500 km^2 have already been surveyed - and, most probably, it extends over large parts of the North Adriatic Sea. Mean standing crop of the community was evaluated at 370($\pm$ 73)g/m^2, corresponding to approximately

---

* This paper is part of a joint program between the Lehrkanzel für Meeresbiologie, University of Vienna, and the Marine Biological Station, Portoroz, University of Ljubljana.

100 kcal/m^2. An average of 92% of this biomass is contributed by suspension feeders. The predominating species is *Ophiothrix quinquemaculata* (D.Ch.) which, on average, accounts for 28% of the community biomass. For a detailed description of the structure of the community see Fedra (1977).

Suspension feeding in echinoderms, particularly brittle stars and dendrochirotid holothurians, is quite well documented (Pearse 1908, Magnus 1964, Fontaine 1965 and Fricke 1970). *Ophiothrix fragilis* (viz. Uexküll 1905, Vevers 1956, Roushdy and Hansen 1960 and Warner and Woodley 1975) and *Ophiothrix quinquemaculata* (viz. Czihak 1959, Guille 1964 and Fedra 1974) are described as suspension feeders, but this conflicts with the findings of other workers, *e.g.* Eichelbaum (1910). By comparison, the interpretation of cucumarian feeding behaviour is not in dispute. The conclusions of workers such as Dohrn 1875, Noll 1881 and Schmidt 1878 are borne out by the most recent investigations (Könnecker and Keegan 1973, Schaller 1973).

From the viewpoint of energy-flow ecology, suspension feeding can be regarded as energy transfer from the pelagos to the benthos. Energy transfer at the bottom/water interface was discussed by Riedl (1971) for porous sea beds. High-biomass benthic consumer communities may well be of comparable importance.

The importance of the macrofauna in the energetics of the sea bed, as estimated by oxygen consumption, ranges from a few percent (Pamatmat 1968) up to 40% (Banse *et al.* 1971). By comparison, the brittle-star community, dealt with below, accounts for 50% or more of the total oxygen demand of the benthos. This relationship again indicates that considerable amounts of energy are transferred from the pelagos to the benthos by this mainly suspension feeding community.

One of the most characteristic features of the North Adriatic Sea is the high nutrient level (Tusnik 1976) and, consequently, high pelagic production (Stirn 1969). Special oceanographic conditions, such as shallow depth and large river input, cause considerable fluctuations in the amount of pelagic organic matter. The pathways of this remarkable energy resource into the benthic system seem worthy of study.

MATERIALS AND METHODS

Methods and materials for community delimitation, classification and structure analysis are described in Fedra *et al.* (in press) and Fedra (1977). Long term *in-situ* observations on filtering behaviour and activity involved SCUBA diving and the use of an underwater time-lapse camera (Fedra 1974). A 16mm movie camera, capable of 4000 exposures, was used. The camera, electronic flashes, and the energy supply were sealed in plastic underwater housings. The camera is triggered electronically. Time intervals of 2 to 64 seconds can be selected, to fill a total observation time of 2.2 to 71 hours per film. For brief technical description see Fedra (1975).

An aquarium with circulating water flow was used for behavioural observations. Filtering efficiency and rates were studied in the laboratory of the Lehrkanzel für Meeresbiologie, Vienna, as well as at the Marine Biological Station, Portoroz. *Artemia salina* nauplii and deep frozen/living plankton (112 um net) were used for these purposes. Experiments in aquaria of various sizes (from 2 to 64 liters) lasted from 1 to 24 hours. *Artemia* nauplii were counted from adequate subsamples; plankton was measured as settlement volume and dry weight, to an accuracy of 0.0001 g. Energy equivalents for average plankton composition were calculated at 2735 cal/g dW (Scherübel et al., in prep,).

RESULTS

Filter techniques

The feeding posture of *Ophiothrix quinquemaculata* in directional currents (Fedra 1974) is characterized by the vertically extended arms. Two of them are extended up above the disc, which is supported above the sediment by the proximal parts of the remaining three arms. The distal parts of the latter are also extended into the current. The ambulacral sides of the arms face the currents, with the tube feet extended between the spines (cf. Warner 1971, Warner and Woodley 1975).

Food is collected by the tube feet and formed into a bolus, which is transported towards the mouth by wave-like actions of the tube feet. A detailed description of food capture and collection in *Ophiothrix fragilis*, which obviously shows the same technique, is given in Warner and Woodley (1975).

The dendrochirotid holothurian *Cucumaria planci* (BRANDT) feeds with its eight large and two smaller tentacles extended, the tentacular crown facing the currents (Fig. 1). Food particles adhere to the mucous covered branches of the tentacles. Measurements in the laboratory revealed that tentacle retraction consistently takes about 30 sec. The order of retraction is regular; seven other tentacles are retracted in more than 50% of all cases, before the same one is used again. The interval may involve 6 or 8 other tentacles in about 20% of all cases, whereas 5 or 9 intermediate retractions are very rare, and 4 or 10 never occur under normal circumstances. This rhythm can be interrupted when the density of suspended food is several orders of magnitude greater than normal.

The distal end of the food-bearing tentacle is inserted into the pharynx, while the buccal membrane contracts around it (cf. Künnecker and Keegan 1973, Schaller 1973). Immediately afterwards one of the two smaller tentacles is retracted, and also put into or across the mouth. It stays there, while the larger tentacle is extended again. The removal of attached food particles and the application of mucus from the lining epithelium of the oesophagus might be the reason for this behaviour (cf. Fish 1967).

Analysis of the stomach contents of both species yielded, in addition to expected fragments of planktonic skeletal elements of sponges, brittle stars nad holothurians as well as tiny mussels. As such particles can easily be suspended in the immediate vicinity of the sea-bottom, or might be picked up by the tube feet or tentacles from the sediment surface, their occurrence need not suggest a deposit-feeding habit.

Reactions to currents

Ophiothrix quinquemaculata and *Cucumaria planci* show acrophilic behaviour in the field. As the current velocity increases with increasing distance from the sediment surface, this can also be understood as rheophilic behaviour. This conclusion is supported by observations on the brittle stars, which relocate themselves in relation to micro-topography - edaphic or biotic - and directional changes in the current. They always favour an exposed site Fig. 2.

Fig. 1. Close up of *Cucumaria planci*. One large tentacle has been placed within the pharynx a small tentacle lies across the mouth.

Fig. 2. *Ophiothrix quinquemaculata* in typical suspension feeding position. Note food particles on the tube feet.

As with the orientation of the ambulacral side of the arms of the brittle stars towards the currents, *Cucumaria* always faces the currents with its oral side. These observations were confirmed by experiments in the laboratory. Both species migrate against directional currents in the cm/sec range.

Activity patterns

According to our long term observations, covering more than 200 hours, *Ophiothrix quinquemaculata* shows no specific pattern of activity. The feeding posture is maintained more or less continuously and the brittle stars show no spontaneous migrations. Migrations can be caused by the above mentioned changes in current direction or by motile species. Hermit crabs, snails, holothurians and the sea star *Astropecten* have been observed to disturb the spatial patterns of *Ophiothrix quinquemaculata* and cause migrations of a few decimetres. Besides the above mentioned occurrences, current velocities above a certain level (≈ 20cm/sec) result in an interruption of feeding activity, as the brittle stars are then dislodged and carried to the next convenient holdfast.

A similar pattern could be observed in the behaviour of *Cucumaria planci*. 8 individuals, observed *in-situ* for periods totalling 103 hours, were, on average, active for 97.4% of the total observation time. Individual percentages ranged from 100% to 92.9%. Inactive periods, with the tentacles retracted, were in all observed cases, caused by the proximity of motile animals such as fish (Gobiids and Blenniids), hermit crabs and *Astropecten* spp. From laboratory observation, we also concluded that activity is dependent on food supply.

Filtering efficiency and filtration rates

Filtering efficiency can be defined as the ratio of extracted suspended material to the total amount available for a passive filter feeder. The available amount depends on the density of suspended material, the rate of water exchange and on the active filtering area exposed to the currents.

The filter area of *Ophiothrix quinquemaculata*, estimated from arm length and the length of the spines, and assuming that the average length of the active tube feet is twice the length of the spines, is given as

$$FA = 0.07 \times \phi^{2.7} \qquad (r = 0.96 \quad n = 45)$$

with the filter area (FA) in cm^2 and disc diameter (ϕ) in mm. The relation to mean wet weight (wW) in g is

$$FA = 26.2 \times wW^{1.2} \qquad (r = 0.95 \quad n = 45)$$

Under ideal conditions 2/3 of this area can be used for collecting food. In dense groups and especially on very uneven substrates this ratio might decrease to 2/5.

The filtering area of *Cucumaria planci* was estimated from the diameter of the disc, formed by the tentacles. About 75% of this total area is covered by the tentacles and their branches, as deduced from *in-situ* macro photographs (Fig. 1). The relation between wet weight (wW) in g and filter area (FA) in cm^2 is estimated as

$$FA = 2.2 \times wW^{0.74} \qquad (r = 0.71 \quad n = 56)$$

Experiments in the aquarium with various food densities (ranging from 0.5 mg dW/l in intervals of 2^n up to 750 mg dW/l) also revealed a relation between efficiency and food density for brittle stars. Starting with a food density of about 400 mg dW/l, the bolus transported along the arms grows to such a size, that it cannot be held by the tube feet any longer and is swept away by the currents. In addition pseudofaeces are produced. As this phenomenon could only be observed under artificial conditions with food suspension densities far beyond the values reported from the area in question, it cannot be considered to be of ecological importance. In addition, the negatively bouyant bolus and pseudofaeces are very likely to be precipitated and are thus available to deposit feeders.

Similar experiments with *Cucumaria planci* showed a constant efficiency even for food concentrations more than three orders of magnitude above average natural values. A comparison of living with dead *Artemia salina* nauplii showed a remarkable difference in the observed efficiencies. The ratio of dead to living nauplii (taken as food) was 8:10.

From these experiments we concluded that the efficiency of both species can be approximated at 100% under natural conditions. Filtration rates are therefore assumed to be mainly a function of filtering areas, food density, and water exchange. Changes in the composition of the food might also be reflected in different rates.

The assumption of a linear relationship between food density and filtration rates, as deduced from theoretical considerations on the above results, was confirmed experimentally in the laboratory (Fig. 3). Regressions calculated for various size groups show a slope of approximately one. The relations for the size group of 7 - 12 mm disc diameter are given as

$$FR = 6.8 \times SD^{1.02} \qquad (r = 0.91 \quad n = 23)$$

with filtration rate (FR) in cal/day, and suspension density (SD) in cal/l. Substituting the values for natural plankton densities in the regression equation for the size group 7 to 12 mm disc diameter (accounting for 90% of the brittle star population's filtering capacity, measured as filter area) the filtration rates of an average square meter of the *Ophiothrix* population were calculated. Using zooplankton density only (values from November 1974), a food concentration of 0.05 cal/l will result in a filter rate of 26 cal/m^2 per day. Substituting total seston density, at 3.6 cal/l, the corresponding filter rate per square meter per day would be 2000 cal. The range for minimum (2.7 cal/l) and maximum (5.4 cal/l) seston density would result in a corresponding daily filter rate between 1500 cal/m^2 and 3040 cal/m^2 respectively.

Energy loss by respiration amounts to 426 cal/m^2 per day for the brittle star population, or 917 cal/m^2 per day for the total macro-epifaunal community. Including the sediment, the total community respiration will amount to 2100 cal/m^2 per day. The proportion of the brittle star's contribution to the total community rate is estimated to be 2/3, corresponding to the contribution to the community respiration (Fedra 1977).

Total daily community ingestion by suspension feeders will therefore range from about 2250 cal/m^2 to 4500 cal/m^2. Under the assumption of steady state conditions (Fedra et al., in press) and substituting an average assimilation efficiency of 30% for the macro-benthos, assimilation and respiration estimates are well balanced. Assuming that some 50% of the macrobenthos' egesta is

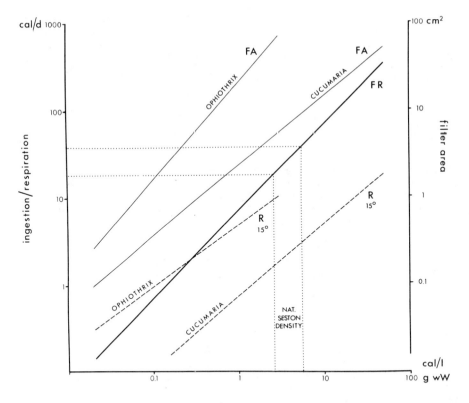

Fig. 3. Relationships between respiration (R), filter area (FA) and size of individuals; between ingestion and food availability. Filtration rates (FR) from *O. quinquemaculata* (size group 7 - 12 mm disc diameter). Dotted lines indicate observed seston density.

imported into the sediment for the meio- and microfauna components, the total community assimilation efficiency of 60% would again result in well balanced estimates for assimilation and community respiration.

DISCUSSION

Soft bottom communities, dominated by passive suspension feeders such as *Ophiothrix quinquemaculata* and *Cucumaria planci*, are known from the Adriatic Sea, where they have been recorded by Czihak (1959), Riedl (1961), and Gamulin-Brida (1974). A similar community has been described by Guille (1964) and Peres and Picard (1964) from the Mediterranean. Dense beds of *Ophiothrix fragilis* are reported from the Atlantic and around the British Isles (Vevers 1952, McIntyre 1956, Cabioch 1967, Brun 196 , Warner 1971). Könnecker and Keegan (1973) described dense aggregations of *Ophiothrix fragilis*, *Antedon bifida* and *Pseudocucumis mixta* as well as several other species of echinoderms.

These communities generally have one main environmental condition in common: more or less directional tidal currents with a high density of suspended material. For the passive suspension feeders, actual food availability depends on the density of the particulate material and on water movements. Hydrodynamic energy is therefore an important energy source for this feeding group. The importance of water movement is also reflected in rheophilic behaviour. Increasing exposure to currents means increasing food availability. The biogenic structures, preferred by the passive suspension feeders as a substrate, are one of the most conspicuous features of the community under investigation. Fedra (1977) gives a more detailed description of these multi-species aggregations.

The observed differences in the filtering efficiencies of both species for living zooplankton and dead particulate organic matter results from the fact that the living zooplankters are moving relative to the currents, whereas the dead material follows their flow. This results in deffering probabilities of contact with the food collecting apparatus.

The estimates for ingestion and respiration were found to be well balanced under the assumption of an average assimilation efficiency of about 1/3. These relations not only hold true for the total community, but also for the two species investigated in detail. Fig. 3, reveals similar proportions between filtering capacity (measured as filter area) and respiration for *Ophiothrix quinqueqmaculata* and *Cucumaria planci*.

Passive filter feeders constitute about one third of the community biomass. Although they are outweighed by active filter feeders, their metabolic activity is twice as great. This is related to their average size, which is far below the mean values measured for the sponges and ascidians. The same relation holds true for the filter rates, as the ratio of filter capacity to biomass also decreases with increasing size. The suspension feeders' filtering capacity is sufficient for the total community energy demands. Only negligible amounts of organic material seem to be imported in addition to the suspension feeders' ingestion. These relations justify the term "secondary producers" for this great pelagos-linked category of benthic organisms.

ACKNOWLEDGEMENTS

This study was supported by projects No. 1940 and 2758 of the "Fonds zur Förderung der wissenschaftlichen Forschung, Austria" with additional support from the "Boris Kidric Foundation". The authors' thanks are due to Yugoslavian authorities for research permits and their hospitality, and to the director and staff of the Marine Biological Station, Portoroz, for providing essential facilities. The authors feel deeply indepted to their colleague R. Machan for his supervision of the electronic equipment and to Dr. J. Ott for constructive criticism and the careful reading of the manuscript.

REFERENCES

Banse, K., F.H. Nichols and D.R. May 1971. Oxygen consumption by the seabed. III. On the role of the macrofauna at three stations. Vie Milieu (Suppl.) 22, 31-52.

Brun, E. 1969. Aggregation of *Ophiothrix fragilis* (Abildgaard) (Echinodermata:

Ophiuroidea). Nytt Mag. Zool. 17, 153-160.

Cabioch, L. 1967. Resultats obtenus par l'emploi de la photographie sous-marine sur les fonds du large Roscoff. Helgoländer wiss. Meeresunters. 15, 361-370.

Czihak, G. 1959. Vorkommen und Lebensweise der *Ophiothrix quinquemaculata* in der nördlichen Adria bei Rovinj. Thalassia jugosl. 1, 19-27.

Dohrn, A. 1875. Mitteilungen aus der zool. Station zu Neapel. Zool Anz. 25.

Eichelbaum, E. 1910. Über Nahrung und Ernährungsorgane von Echinodermen. Wissenschaftl. Meeresunters. K. Kommission Abteilung Kiel 11, 187-275.

Fedra, K. 1974. Filtrierverhalten und Aktivitätsrhythmik bei *Ophiothrix quinquemaculata* (Echinodermata). Wien: B.H.F.wiss. Kinematogr. Research Film p 1567.

Fedra, K. 1975. Unterwasser-Zeitrafferaufnahmen in der Meeresbiologie. Mitteilungen d. B.H.F.wiss. Kinematogr., Wissenschaftlicher Film 16, 37-43.

Fedra, K. 1977. Structural features of a North Adriatic benthic community. In Proceedings of the 11th European Symposium on Marine Biology - ed. Keegan, B.F. P.O Ceidigh and P.J.S. Boaden. Pergamon Press, Oxford.

Fedra, K., E.M. Ölscher, C. Scherübel, M. Stachowitsch and R.S. Wurzian On the ecology of a North Adriatic benthic community: Distribution, standing crop and composition of the macrobenthos. Mar. Biol. (in press).

Fish, J.D. 1967. The biology of *Cucumaria elongata* (Echinodermata: Holothuroidea). J. mar. biol. Ass. U.K. 47, 129-143.

Fontaine, A.R. 1965. The feeding mechanism of the ophiuroid *Ophiocomina nigra*. J. mar. biol. Ass. U.K. 44, 145-162.

Fricke, H.W. 1970. Beobachtungen über Verhalten und Lebensweise des im Sand lebenden Schlangensternes *Amphioplus* sp. Helgoländer wiss. Meeresunters. 21, 124-133.

Gamulin-Brida, H. 1974. Biocenoses benthiques de la Mer Adriatique. Acta Adriat 15 (9), 1-102.

Guille, A. 1964. Contribution a l'etude de la systematique et de l'ecologie d'*Ophiothrix quinquemaculata* D.Ch. Vie Milieu 15, 243-308.

Könnecker, G. and B.F. Keegan 1973. *In-situ* behavioural studies on echinoderm aggregations. Part I. *Pseudocucumis mixta*. Helgoländer wiss. Meeresunters. 24, 147-162.

Magnus, D.B.E. 1964. Gezeitenströmung und Nahrungserwerb bei Ophiuroiden und Crinoiden. Helgoländer wiss. Meeresunters. 10, 104-117.

McIntyre, A.D. 1956. The use of trawl, grab and camera in estimating the marine benthos. J. mar. biol. Ass. U.K. 35, 419-429.

Noll, F.C. 1881. Mein Seewasseraquarium. *Cucumaria planci* Brdt. Zool. Garten, Jg. 22.

Pamatmat, M.M. 1968. Ecology and metabolism of a benthic community on an inter-tidal sandflat. Int. Revue ges. Hydrobiol. Hydrogr. 53, 211-238.

Pearse, A.S. 1908. Observations on the behaviour of the holothurian *Thyone briareus* (Lesneur). Biol. Bull. mar. biol. Lab., Woods Hole 15, 259-288.

Peres, J.M. et J. Picard 1964. Nouveau manuel de bionomie benthique de la Mer Mediterranee. Recl. Trav. Stn. mar. Endoume 31(47), 1-137.

Riedl, R. 1961. Etudes des fonds vaseaux de l'Adriatique. Methodes et resultats. Recl. Trav. Stn. mar. Endoume 23, 161-169.

Riedl, R. 1971. Energy exchange at the bottom/water interface. Thalassia jugosl. 7(1). 329-339.

Roushdy, H.M. and V.K. Hansen 1960. Ophiuroids feeding on Phytoplankton. Nature, London, 188, 517-518.

Schaller, F. 1973. Uber die Tentakelbewegungen der dendrochiroten Holothurien *Cucumaria lefevrei* and *Cucumaria saxicola*. Zool. Anz., Leipzig 191, 162-170.

Schmidt, M. 1878. Nachrichten aus d. Zool. Garten i. Frankfurt, Bd. 19.

Stirn, J. 1969. The distribution of the pelagic organic matter in North Adriatic. Rapp. Comm. int. Mer. Medit., 19(4), 755-758.

Tusnik, P. 1976. Osnovne znacilnosti fizicnega okolja v obalem morju Slovenskega primorja. Magistrsko delo, MBP Portoroz, University of Ljublana.

Uexküll, J. von 1905. Studien über den Tonus. II. Die Bewegung der Schlangensterne. Z. Biol. 46, 1-37.

Vevers, H.G. 1952. A photographic survey of certain areas of the sea floor near Plymouth. J. mar. biol. Ass. U.K. 31, 215-222.

Vevers, H.G. 1956. Observations on feeding mechanisms in some ophiuroids. Proc. zool. Soc. Lond., 126, 484-485.

Warner, G.F. 1971. On the ecology of a dense bed of the brittle-star *Ophiothrix fragilis*. J. mar. biol. Ass. U.K. 51, 267-282.

Warner, G.F. and J.D. Woodley 1975. Suspension-feeding in the brittle star *Ophiothrix fragilis*. J. mar. biol. Ass. U.K. 55, 199-210.

STRATEGIES OF ENERGY TRANSFER FROM MARINE MACROPHYTES TO CONSUMER LEVELS: THE *POSIDONIA OCEANICA* EXAMPLE

Jörg Ott and Ludwig Maurer

Zoologisches Institute, Universität Wien Lehrkanzel für Meeresbiologie, A-1090 Wien, Währingerstrasse 17/VI, Austria

ABSTRACT

A stand of *Posidonia oceanica* DELILE at the island of Ischia (Gulf of Naples) was investigated as to its production and the fate of the organic matter produced. A site in 4 m depth with a mean leaf biomass of 1.15 kg dry weight had a net production of 1.52 gC per square meter and day during the period of May to July 1976. About 50 % of this production is consumed in the system, approximately 30 % exported continuously. The only potential herbivore, the sea urchin *Paracentrotus lividus* LAM. consumes about 9 - 10 % of the net production, but does not attack living *Posidonia* tissue. The hypothesis is presented that there is a strategy to limit direct herbivory on marine macrophytes, which constitutes an energy drain on the primary producer compartment without a feedback reward loop through the products of remineralization as in similar terrestrial ecosystems.

INTRODUCTION

Stands of marine macrophytes are known as sites of large biomass concentrations and high productivity (for recent reviews see MANN 1972a, b, 1973). Locally their productivity may exceed that of the plankton in the water column appreciably. Two characteristic features have been noted in all studies, which were concerned with the flow of organic energy from this primary production to consumer levels: the small amount that enters the grazing food chain in contrast to the detritus food chain, which means that herbivory as consumption of living plant biomass is quantitatively rather unimportant (TEAL 1962, MILLER and MANN 1973, MANN 1973, FENCHEL 1972, HARRISON and MANN 1975), and the high percentage of the primary production which is exported to other ecosystems, either as dissolved organic matter (KHAILOV and BURLAKOVA 1969, SIEBURTH and JENSEN 1969) or as particulate detritus (MANN 1972b, FENCHEL 1972 for recent reviews).

During a study of the productivity and the fate of this production of the large mediterranean sea grass *Posidonia oceanica* DELILE these features were again apparent. *Posidonia oceanica* DELILE is known to be responsible for enormous wrack beds, which have been described by numerous authors, and the detritus originating from its leaves may be found on all littoral bottoms (MOLINIER and PICARD 1952, MASSE 1962, PERES and PICARD 1963, POIZAT 1970). The leaves of this marine angiosperm grow from a meristeme at their base in a conveyor belt like fashion and undergo a conspicuous change from the youngest parts at the base to the oldest portion at the tip. The basal parts are green, free of epiphytic organisms, with a smooth uninjured surface and edges. Towards the tip the cover of epiphytic algae and sedentary animals increases, the leaf turns yellow to brown. The distal part of the leaf is already dead. In this region there are many signs of biological and mechanical erosion.

We want to present a hypothesis why grazing in this system is restricted to already dead parts of the plants and will attempt to generalize on the adaptive significance for aquatic macrophytes of controlling grazing on their metabolically active tissues.

To demonstrate the validity of the above two statements about marine macrophytes, namely low grazing rates and large exports, in the case of also for *Posidonia* we attempted to arrive at a crude energy budget including these compartments. Our programme, which is part of a large cooperative programme on the *Posidonia* ecosystem initiated in the Laboratorio della Ecologia Marine of the Stazione Zoologica di Napoli in Ischia, has just started. Our data cover only a short period of the year (May to July), which is furthermore the end of the growing season for *Posidonia* (MOLINIER and ZEVACO 1962). Therefore we have not attempted elaborate statistical analysis of our data and the prediction of yearly budgets.

MATERIALS AND METHODS

The Island of Ischia (Bay of Naples) is almost completely surrounded by *Posidonia* meadows, extending from the very shallow sublittoral down to 30 - 40 m. All data reported here come from a site in 4 m depth near the Castello Aragonese at Ischia Ponte. All collections and *in situ* measurements were done by SCUBA diving.

Quantitative samples were taken using a frame covering 0.25 m^2. The underground parts were sampled to the first constriction of the rhizome which was arbitrarily taken as the boundary of the metabolically active part of the root system.

In situ measurements of community metabolism employed clear acrylglass bell jars of 15 cm diameter and 1 m length, corresponding to an area of 1/63 m^2 and 15 l volume. Magnetic stirring bars in glass housings provided water mixing within the columns and sufficient agitation for the YSI polarographic oxygen electrodes. Amplifier, recorder and batteries were placed in watertight housings on the bottom.

For an estimate of the primary production of *Posidonia* the growth of individual plants was measured by punching holes (MANN 1972a) through all leaves where they appear from the bundle of fibers enclosing their base. In addition a square meter was cut repeatedly at a constant height above the ground, using two metal bars fixed to the bottom as a guide for a third, movable cutting bar

As a measure of leaf degradation and possible loss of organic matter through leaching we measured the distribution of sugars and starch as the immediate products of photosynthesis in different parts of the leaf. Leaves were dried to constant weight at $80°$ C, milled and extracted with distilled water for free glucose and fructose. Glucose and fructose from polysaccharids other than starch were estimated after extraction with dimethylsulfoxid, 8N HCl and 5N NaOH, starch as glucose after treatment with amyloglucosidase. Glucose and fructose equivalents were determined with the hexokinase glucose-6-phosphate dehydrogenase method according to the method of BÖHRINGER-MANNHEIM on a spectrophotometer at 340 nm.

The relation between the production of gO_2 and gC was 1:0.278 respectively. Dry weight was assumed to consist of 80% organic matter, which in turn was converted to gC by a factor of 0.47 (all after WESTLAKE 1963).

RESULTS

The shallow meadows of *Posidonia* around Ischia vary considerably in density, leaf length and consequently biomass. Numbers of plants ranged from 400 - 1300 per m^2, leaf biomass from 0.6 - 1.68 kg per m^2 (mean 1.154 + 0.097) dry weight, rhizome biomass from 1.1 - 5.3 kg (mean 2.788 + 0.351) dry weight. This is much higher than the values given by LARKUM (1976) for *P. australis* but compares well with the biomass in *Thalassia testudinum* beds reported by BURKHOLDER et al.(1959).

On the average a *Posidonia* plant at this time of the year has 5 leaves. The innermost is the youngest, of medium length, green throughout and with an uninjured tip. The two neighbouring leaves are the longest and show the typical zonation, their distal third is covered by epiphytic organisms and damaged. The outermost leaves are much shorter, again with dead tips. From the movement of the holes in the marked plants it could be seen, that these leaves have ceased to grow and are dying back. The longest leaves showed the fastest growth.

Table 1 gives net production estimates computed by this method, and by repeated harvesting together with the increase in mean leaf length. Regrowth after cutting compares well with the data derived from the marked plants, although most of the photosynthetic tissue was removed. A second cutting however resulted in a reduction of regrowth by more than 40 %. Only a fraction of this appears as leaf length increase (and consequently plant biomass increase). Production estimates on peak biomass or monthly leaf length increments (MOLINIER and ZEVACO 1969) will therefore underestimate production.

Whereas leaf length increase and the growth estimate on marked plants give integrated measurements over longer periods, the com-

TABLE 1 Net production estimates for *Posidonia*

Method	Nr. days	$g \cdot m^{-2} \cdot d^{-1}$	$gC \cdot m^{-2} \cdot d^{-1}$
marked plants	10	3.63	1.37
marked plants	19	4.35	1.74
marked plants	58	4.38	1.65
first harvest	61	4.39	1.65
first harvest	19	3.47	1.31
second harvest	42	1.99	0.75
leaf length increase	43	0.87	0.33
leaf length increase	40	0.82	0.31

munity net production estimates based on O_2 production over 24 hour periods may vary considerably with temperature and especially illumination. However, they may be treated as a random sample of the conditions during the period of investigation, since the measurements were taken throughout the period regardless of weather conditions (Tab. 2). We have to explain the usage of the term "community net production" here. It is taken as a measure of organic matter produced and not metabolized in the community. It consists of the matter laid down in primary producer standing crop increase and exported matter under the assumption that the consumer part of the community is in a steady state. Since the greatest part of the standing crop increase is exported at the end of the growing season through the annual leaf fall, we feel that the term "community net production" is justified.

TABLE 2 Community net production estimates

Date	Weather	Temp. °C	$gO_2 \cdot m^{-2} \cdot d^{-1}$	$gC \cdot m^{-2} \cdot d^{-1}$
760512	cloudy	18.0	1.06	0.29
760513	cloudy	17.5	1.13	0.32
760514	cl./sun.	17.0	0.95	0.26
760514	cl./sun.	17.0	3.97	1.10
760701	sunny	25.0	2.65	0.74
760701	sunny	25.0	2.84	0.79
760703	sunny	26.0	6.24	1.73
760707	sunny	26.0	3.97	1.10
mean $\pm$SE			2.85 $\pm$ 0.65	0.79 $\pm$ 0.18

The difference now, between *Posidonia* net production and the community net production corresponds to the heterotroph consumption within the system - herbivorous and detritivorous. Of the animals inhabiting the *Posidonia* meadows only the sea urchin *Paracentrotus lividus* LAM. can be considered as a herbivore occurring in sufficient numbers. Its density is 50-150 per m^2 (dry weight including skeleton 196 - 284 $g \cdot m^{-2}$). Its gut contains various epiphytic algae, pieces of bryozoan colonies and other sedentary invertebrates and numerous fragments of Posidonia leaves. However, with few exceptions no green pieces could be found, although the pigments of the algae still could be recognized at least in the first part of the gut. Inspection of the leaves showed, that the green part is almost free from any injuries which could be interpreted as grazing marks.

To test this apparent preference for the distal leaf parts, sea urchins were starved until no faeces were produced any more, which may take 2 weeks. Animals were then placed in separate dishes and presented with measured pieces of either green leaves, brown tips or a choice of both and allowed to feed for 24 hrs. From Table 3 it is apparent, that given a choice,   prefers brown tips in all cases over green leaves. If only supplied with green pieces, they consume significantly less than those presented with tips. The quantity consumed was used as a very crude estimate of the urchins' consumption rate. We are aware that sea urchin consumption rates vary greatly from day to day and between individuals (MILLER and MANN 1973). Also, only normally feeding animals should be used. However, these are the only estimates we have.

TABLE 3   Feeding experiments with Paracentrotus lividus

Nr. urchins	food	% feeding	mean consumption (mgC)
10	green	80	4.5
10	brown	100	12.5
10	choice/green	0	–
10	choice/brown	100	10.5

DISCUSSION

If we now construct a very simplified energy budget for a day during our investigation period, using the means of all parts of the equation, which could be numerically estimated, we see, that approximately a fifth of the organic matter produced is laid down in plant growth, about one half is consumed within the system and about 30% is exported. This assumes that the import of plankton and utilisation by filter feeders (e.g. bryozoa and sponges) play an insignificant role, as was substantiated by out quantitative samples for this shallow bed (Table 4).

TABLE 4   Energy budget for the production of Posidonia

	net prod.	Pos. biomass increase	export	sea urchin consumption	rest cons.
$gC \cdot m^{-2} \cdot d^{-1}$	1.52	0.31	0.48	0.14	0.59
%	100	20.4	31.6	9.2	38.8

It is again apparent, that consumption by the sea urchins even if they were truly herbivorous, is in the same order of magnitude as has been reported for other marine macrophytes. Functionally the dead tips of the *Posidonia* leaves do not differ from detritus, even though it is still attached to the plant. Therefore the sea urchins trophic position is at least partly in the detritus food chain, and herbivorous consumption will be even lower.

In comparable terrestrial ecosystems with similar structure, biomass and life expectancy of the photosynthetic parts, as e.g. grasslands, grazing may consume up to 50 % of the net production (ODUM 1971). The question is whether aquatic macrophyte vegetation could withstand appreciable grazing rates. RANDALL (1969) reports that herbivorous reef fishes keep the turtle grass *Thalassia testudinum* from growing to close to the reef's edge and that around articifical reefs placed in sea grass stands in a short time a bare strip of sand appears. As mentioned above, on the experimental square the first cutting had no significant effect on regrowth, the second cutting however reduced net production strongly (Table 1).

To find an explanation for the adaptive value to limit grazing in aquatic ecosystems one may compare the basic energy flow relationship between producers and consumers in an ecosystem. The general picture as derived from terrestrial situations shows a feed back loop from the consumers to the producers, which consists of the work done in the remineralization of organic matter taken from the primary producers (Fig. 1).

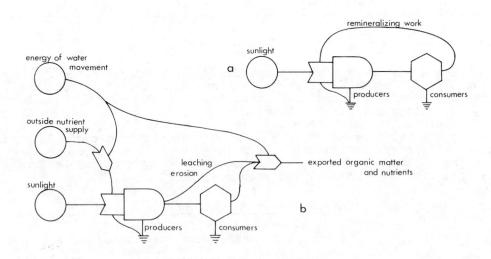

Fig. 1. Basic energy flow pathways between primary producers and consumers in (a) a terrestrial ecosystem (after H.T. ODUM 1971) and (b) in an aquatic macrophyte system.

However, this requires that the products of remineralization can be taken up at a rate, that is a function of their rate of supply. In a terrestrial situation, where the products of remineralization generally stay in the system, this is the case. In aquatic ecosystems nutrients released have only a limited residence time, so that the probability of repeated uptake by the primary producers of a quantity released will be a function of this residence time and the turnover rate of the primary producers. This may work well in the plankton, where we also find appreciable grazing rates or in enclosed water bodies. In an open system, like the benthic macrophyte vegetation, with a high rate of water exchange a regulatory function of the rate of remineralization by herbivores is highly improbable. Production will be rather dependent on external nutrient supply (Fig. 1b). Without this feedback reward herbivory acts only as an energy drain.

Aquatic macrophytes already are subject to several energy drains, which play little or no role in comparable terrestrial ecosystems: They are surrounded by a medium, which is an excellent solvent for sugars, the initial products of photosynthesis. Loss of organic matter through leaching has been demonstrated for macrophytic algae (SIEBURTH and JENSEN 1969, KHAILOV and BURLAKOVA 1969), although there has been recent doubt about the occurrence of this loss in nature (MANN 1973). It also has to be established that leaching occurs in higher plants, such as the sea grasses. In all parts of the leaf of *Posidonia* the content in free monosaccharides (glucose and fructose) is very low compared to higher sugars and starch (Table 5). The marked decrease of especially the free mono-

TABLE 5 Carbohydrate content in different regions of the leaves of *Posidonia oceanica* (mg/g dry wt)

	glucose	fructose	glucose from higher sugars	starch
leaf base	8.50	3.80	42.6	9.0
leaf blade	1.27	0.85	19.1	13.5
leaf tip	0.20	0.20	12.0	7.8

saccharids from the green part of the leaf to the tips suggests loss of these compounds during leaf degradation. However, it remains unclear, whether these substances have been actually leached or have been taken up by the microbial community developing on aging leaves. Considerable loss of inorganic nutrients has been reported for the closely related eelgrass *Zostera marina* L. by McROY et al. (1970).

Another energy drain is loss of photosynthetic surface by epiphytism. Various kinds of small algae and sedentary animals settle on the leaves and fronds of marine macrophytes. The former even profit from the leaching of nutrients through the sea grasses (McROY and GOERING 1974). Thus older parts of the plants may become completely covered and unable to photosynthesize.

A third energy drain is the constant loss of leaves or parts of the frond through the mechanical action of water movement. Especially during storms great numbers of leaves are torn off and deposited on beaches and offshore bottoms. At any time of the year these freshly torn off, still green leaves can be found on sand patches within *Posidonia* stands.

All these losses result in a net flow of organic matter and nutrients out of the system. It remains to be established, that it is energetically cheaper to counteract herbivory, e.g. by being unpalatable - which *Posidonia* may achieve through its content of tannic acid in living tissue - than to improve cell membrane properties, strengthening of mechanical tissue or to develop mechanisms against epiphytes.

In the case of our *Posidonia* bed the grazing action of the sea urchins may have been channelled to benefit the plants. The pattern of regrowth on the experimental square suggests, that *Posidonia* has a light optimum for growth, which lies between full sunlight (which seemed to produce an inhibiting effect in the middle of the square), and the natural conditions within the stand, which receives some shading from the *aufwuchs* covered tips. If these tips are not removed, their shading effect might inhibit photosynthesis. Thus the grazing action, which not only removes the fraction consumed, but also aids mechanical erosion of the tips, may serve to increase production.

ACKNOWLEDGEMENT

This study was supported by the Fonds zur Förderung der wissenschaftlichen Forschung, Austria, grant 1940. Additional support through the Stazione Zoologica di Napoli and the Gulf of Naples Ecology Programme (GONEP) is gratefully acknowledged.

REFERENCES

Burkholder, P.R. and J.A. Rivero 1959. Some chemical constituents of turtle grass, *Thalassia testudinum*. Bull Torrey bot. Cl. 86, 88-93.

Fenchel, T. 1972. Aspects of decomposer food chains in marine benthos Verh. Dtsch. Zool. Ges. 65, 14-22.

Harrison, P.G. and K.H. Mann 1975. Chemical changes during the seasonal cycle of growth and decay in eelgrass (*Zostera marina* on the Atlantic coast of Canada. J. Fish. Res. Board Can. 32, 615-621.

Khailov, K.M. and Z.P. Burlakova 1969. Release of dissolved organic matter by marine sea weeds and distribution of their total organic production to inshore communities. Limnol. Oceanogr. 14, 521-527.

Larkum, A.W. 1976. Ecology of Botany Bay. I. Growth of *Posidonia australis* (Brown) Hook f. in Botany Bay and other Bays of the Sidney Basin. Aust. J. Mar. Freshwater Res. 27, 117-127.

Mann, K.H. 1972a. Ecological energetics of the sea-weed zone in a marine bay on the Atlantic coast of Canada. II. Productivity of the sea-weed. Mar. Biol. 14, 199-209.

Mann, K.H. 1972b. Macrophyte production and detritus food chains in coastal waters. Mem. Ist. Ital. Idrobiol. Suppl. 29, 353-383.

Mann, K.H. 1973. Sea weeds: their productivity and strategy for growth. Science, 182, 975-981.

Masse, H. 1962. Cartographie bionomique de quelques fonds meubles de la partie sud-orientale du Golfe Marseille. Rec. Trav. Stn. mar. Endoume, Marseille 27, 221-259.

McRoy, C.P., R.J. Barsdate and M. Nebert 1972. Phosphorus cycling in an eelgrass (*Zostera marina* L.) ecosystem. Limn. Oceanogr. 17, 58-67.

McRoy, C.P. and J.J. Goering 1974. Nutrient transfer between the seagrass *Zostera marina* and its epiphytes. Nature 248, 173-174.

Miller, R.J. and K.H. Mann 1973. Ecological energetics of the seaweed zone in a marine bay on the Atlantic coast of Canada III. Energy transformations by sea urchins. Mar. Biol. 18, 99-114.

Molinier, R. and J. Picard 1952. Recherches sur les herbiers de Phanerogames marines du littoral mediterraneen francais. Ann. Inst. oceanogr., Paris 27, 157-234.

Molinier, R. and C. Zevaco 1962. Etudes ecologiques et biocenotiques dans la baie du Brusc (Var.). Fascicule 3: Etude statistique et physiologique de la croissance des feuilles Posidonies *Posidonia oceanica* Delile). Bull. Inst. Oceanogr. Monaco 59, 1-46.

Odum, E.P. 1971. Fundamentals of ecology. Saunders, Philadelphia.

Odum, H.T. 1971. Environment, Power, and Society. Wiley-Intersci., New York.

Peres, J.M. and J. Picard 1963. Apercu sommaire sur les peuplement marins benthique entourant l'île de Port-Cros. Terre Vie 110, 436-448.

Poizat, C. 1970. Hydrodynamisme et sedimentation dans le Golfe de Gabes (Tunesie). Téthys. 2(1), 267-296.

Randall, J.E. 1969. Grazing effect on sea grasses by herbivorous reef fishes in the West Indies. Ecology, 46, 255-260.

Sieburth, J. and A. Jensen 1969. Studies on algal substances in the sea III. The production of extracellular organic matter by littoral marine algae. J. exp. mar. Biol. Ecol. 3, 290-309.

Teal, J.M. 1962. Energy flow in the salt marsh ecosystem of Georgia. Ecology 43, 614-624.

Westlake, D.F. 1963. Comparisons of plant productivity. Biol. Rev. 38, 385-425.

THE BENTHIC ECOLOGY OF SOME SHETLAND VOES

T. H. Pearson and S.O. Stanley

Dunstaffnage Marine Research Laboratory, P.O. Box No. 3, Oban, Argyll, Scotland

ABSTRACT

The physical and chemical environment and the benthic fauna of the deeper areas of three voes (fjords) in Shetland are described and compared with those of a topographically similar basin on the west coast of Scotland. The distribution of the benthic macrofauna in the various areas can be related to a gradient of increasing organic input to the sediments. Two of the voes showed strong evidence of having potentially stagnant bottom waters with high natural organic inputs of eroded peat fragments. The faunal, chemical and physical evidence presented suggests that in these two voes the input of organic material to the sediments is approaching the natural degradative capacity of the system.

INTRODUCTION

Geographically the Shetland Isles lie 200 Kms north-east of the Scottish mainland on the same latitude as Bergen in Norway, and in the centre of the confluence of North Sea and North Atlantic waters. The coast line of the archipelago is deeply indented to form numerous fjords, locally called Voes. The development of a major oil terminal in Sullom Voe, the largest such inlet on the mainland of Shetland has prompted an assessment of the marine environment of the area in order to provide the baseline information necessary for successful control of the impact of industrialisation on that environment. As part of this assessment two surveys, in May 1974 and June 1976, have been carried out by the Scottish Marine Biological Association in which various aspects of the biology of a number of the voes were studied. This paper describes the information obtained on the sediments and benthic fauna in three of the larger voes and compares it with similar data from an intensively studied sea loch on the west coast of Scotland.

AREAS STUDIED

The three voes studied are all situated on the mainland of Shetland (Fig. 1). Ronas Voe opens to the northern coast and is approximately 9 Km long with twin sills at 2 and 5.5 Km distance from the entrance. Sullom Voe opens into Yell Sound on the east coast and shallows gradually for some 10 Km to a sill which isolates a deep inner basin about 3 x 1.5 Km in extent. Swarbacks Minn encompasses a complex of three large and two smaller inlets opening to the west, the inner most of which, Olna Firth, is described here. This is the eastern arm of the complex and consists of an outer basin about 2.4 x 1.4 Km in extent, with a narrow inner basin leading off it. These Shetland Voes are compared with an area of the inner basin of Loch Eil on the west coast of Scotland, which is subject to artificially enhanced inputs of organic material (Pearson 1970, 1975).

Fig. 1. Part of the mainland of Shetland showing the position of the sampling stations in the Voes studied.

METHODS

Grab samples, using a 0.1 m^2 Van Veen grab, and core samples, using a Craib corer (Craib 1964) having a sampling area of 0.003 m^2, were taken at various stations throughout the three voes. Grab samples were washed on a 1 mm mesh sieve and the residue preserved in 4% Formalin and returned to the laboratory for examination. Redox potential (Eh) and pH of the sediments were measured at intervals down the length of the core using a combined platinum and Ag/AgCl electrode for Eh and a combination electrode for pH. Details of the technique used are given elsewhere (Pearson and Stanley in prep.). Chemical analysis of sediments was carried out on cores which had been divided into 3 cm sections and stored frozen. Details of sediment analysis are described elsewhere. (Stanley and Pearson in prep.). Water samples from immediately above the sediment surface were obtained using a Craib corer with an extended core tube - this allowed a sample representing the 25 cms of water immediately overlying the sediment surface to be obtained. This water was used for dissolved oxygen analysis using a modification of the Winkler method and for nutrient analysis using standard techniques (Strickland and Parsons 1965). The same methods were used for nutrient analysis in the water column. Sedimentation rates were measured at 1 m above the sediment surface using a conical pot having a collecting surface area of 0.1256 m^2. This was left in situ for approximately 24 hours at each collecting site.

RESULTS

A single station from the deeper soft mud sediment of each area was chosen for comparative analysis. The position of the Shetland sampling stations is shown in Fig. 1. Table 1 a and b lists the depth and physical conditions in the bottom water and the nature of the sediment in the various areas at the time of sampling in June 1976.

Table 1(a) Composition of the sediment at various stations sampled.

AREA & STATION	SEDIMENT % COMPOSTION			
	GRAVEL >2000µ	SAND 2000-63µ	SILT 63-4µ	CLAY <4µ
SULLOM VOE D4	0	33	56	11
RONAS VOE RV1	0	13	72	14
OLNA FIRTH SM8	14	30	50	6
LOCH EIL E70	0	36	51	13

Table 1(b) Depth and physical conditions in the bottom water.

AREA & STATION	DEPTH (m)	BOTTOM WATER		
		T°C	S°/oo	O_2% Sat.
SULLOM VOE D4	44	8.8	35.2	71
RONAS VOE RV1	33	10.0	35.2	71
OLNA FIRTH SM8	41	10.0	35.2	90
LOCH EIL E70	49	8.6	29.5	104

It can be seen that the sediment in Ronas Voe was the finest, but that all the areas were essentially similar both in depth and granulometry. The salinity in Loch Eil was somewhat lower than that in the Shetland Voes, however the most striking variation lay in the values for oxygen saturation in the bottom water. Both Sullom Voe and Ronas Voe showed relatively low values of about 70% saturation whereas Olna Firth was higher but still not fully saturated. This compared with the Loch Eil basin where in the same month the bottom water was completely saturated with oxygen. Table 2(a) shows the Eh, pH and sulphide values, and Table 2(b) the Carbon/Nitrogen and Cellulose values for the four stations. Sullom Voe and Ronas Voe showed low Eh and pH values at 4 cm depth in the sediments as did Loch Eil. The Eh readings of -150 mV and lower suggest an anaerobic sulphide dominated regime in these sediments. This contention is supported by the fairly high total sulphide values found in the surface layer of the cores from these areas. No sulphide values are available from the Olna Firth station but the relatively high Eh and pH values suggest that the sediment there was much less reduced. All the stations showed high organic carbon values, but Sullom Voe was particularly high at nearly 10%. Levels of organic nitrogen and C:N ratios were similar at all the stations with the organic nitrogen varying between 0.3-0.6% and C:N between 10-19. Table 3 lists the results of

Table 2(a)   Eh, pH and sulphide values of the sediments.

AREA & STATION	Eh (mV)		pH		SULPHIDE p.p.m.
	SURFACE	4 cm	SURFACE	4 cm	0-3 cm
SULLOM VOE D4	+30	-177	7.56	7.40	64.0
RONAS VOE RV1	+173	-139	7.65	7.44	12.0
OLNA FIRTH SM8	+10	-44	7.84	7.73	-
LOCH EIL E70	+197	-158	7.25	7.18	12.8

Table 2(b)   Carbon/Nitrogen and Cellulose values of the sediments

AREA & STATION	CARBON/NITROGEN IN 0-3 cm LAYER			CELLULOSE
	% C	% N	C:N	mg/g dry wt
SULLOM VOE D4	9.6	0.6	14.9	6.2
RONAS VOE RV1	4.9	0.5	10.2	2.3
OLNA FIRTH SM8	6.4	0.3	19.5	3.5
LOCH EIL E70	6.9	0.5	14.3	5.9

Table 3(a)   Bottom Water; Nutrient Analysis

AREA & STATION	Nitrate µg-at N/l	Nitrite µg-at N/l	Ammonia µg-at N/l	DON µg-at N/l	Phosphate µg-at P/l
SULLOM VOE D4	1.70	0.04	5.10	13.44	0.66
RONAS VOE RV1	1.27	0.12	4.33	12.54	0.11
SWARBACKS MINN SM8	1.46	0.11	5.38	13.32	0.80
LOCH EIL E70	1.60	0.17	0.95	8.25	0.31

Table 3(b)   Sediments; Nutrient Analysis

AREA & STATION	Ammonia µg-at N/l	Nitrate µg-at N/l
SULLOM VOE D4	4.65	1.19
RONAS VOE RV1	2.82	0.86
SWARBACKS MINN SM8	2.72	1.00
LOCH EIL E70	4.10	1.60

nutrient analyses on the sediments and overlying bottom water. Nitrate and nitrite values in the bottom water are similar in all four areas, but at the three Shetland stations values for ammonia and dissolved organic nitrogen are much higher than in Loch Eil, and at two of the three stations phosphate values are also high. Ammonia levels in the top 0-5 cm layer of sediment are highest in Loch Eil and Sullom Voe though all the Shetland sediments are lower in extractable nitrate. The sedimentation rates in the four study areas are given in Table 4.

Table 4.   Sedimentation rates.

AREA & STATION	SEDIMENTATION	
	$g/m^2/day$	% Ash wt.
SULLOM VOE D4	5.46	65.1
RONAS VOE RV1	4.76	72.1
OLNA FIRTH SM8	1.75	67.3
LOCH EIL E70	5.20	68.1

The rates in Ronas Voe and Sullom Voe are similar to those found in Loch Eil in May 1976, but the rate in Olna Firth was much lower. The percentage ash weight of the material collected was much the same in all areas. An analysis of the macrofauna from the four areas is given in Table 5 and 6 and shows considerable variations between the stations.

Table 5(a)   Total numbers and biomass. (Sp. : species. Ind.: individuals).

AREA & STATION	BIOMASS (g/m^2 wet wt)	NUMBERS	
		Sp.	Ind.
SULLOM VOE D4	26.2	26	3939
RONAS VOE RV1	258.3	24	8235
OLNA FIRTH SM8	40.4	42	940
LOCH EIL E70	11.0	18	1200

Table 5(b)   Numbers in major faunal groups.

AREA & STATION	ANNELIDS		MOLLUSCS		ECHINODERMS		OTHERS	
	Sp.	Ind.	Sp.	Ind.	Sp.	Ind.	Sp.	Ind.
SULLOM VOE D4	16	922	5	2970	-	-	5	47
RONAS VOE RV1	12	3830	7	4315	1	15	4	75
OLNA FIRTH SM8	21	310	12	330	2	25	7	275
LOCH EIL E70	14	920	-	-	-	-	4	283

At the Sullom Voe station a relatively low total biomass composed of large numbers of individuals was found. Species numbers were low, indeed 4 species comprised 89% of the fauna and 7 over 93%. In Ronas Voe there was a very high biomass with large numbers of individuals, but the number of species was small and as in Sullom Voe 4 species made up about 90% of the fauna. In Olna Firth the biomass of 40 g/m^2 was much the same as that found on other soft mud grounds in fjordic situations in Scotland (c.f. MacIntyre 1961, Pearson 1970), the number of individuals was relatively low, but there was a considerable spread of species across all the major animal groups. In contrast at the Loch Eil station the fauna is very poor, number of species and number of individuals being very low. The dominant species types are very similar in Ronas Voe and Sullom Voe, but only have one species in common with the dominants in Olna Firth. The dominant species in Loch Eil differ completely from those in the Shetland.

Table 6. Macrofaunal dominants at each station.

AREA & STATION	DOMINANT SPECIES	PERCENTAGE OF TOTAL FAUNA
SULLOM VOE D4	Abra alba (Wood)	65
	Scalibregma inflatum Rathke	15
	Corbula gibba (Olivi)	5
	Lagis koreni Malmgren	4
RONAS VOE RV1	Abra alba	48
	Scalibregma inflatum	21
	Notomastus latericeus Sars	20
	Lagis koreni	3
OLNA FIRTH SM8	Phoronis mülleri de Selys-Longchamps	20
	Thyasira flexuosa (Montagu)	16
	Prionospio malmgreni Claparede	9
	Corbula gibba	6
LOCH EIL E70	Capitella capitata (Fabricius)	38
	Nematoda sp.	23
	Pelescolex benedeni (Udekem)	17
	Scolelepis fuliginosa (Claparede)	9

DISCUSSION

The Loch Eil basin has been shown to be an area of high organic deposition where the distribution of benthic species is primarily related to the rate of input of organic material to the sediments (Pearson 1975). The results given above suggest that the deep inner basins of both Ronas Voe and Sullom Voe are likewise dominated by high organic deposition rates.

There is a close similarity between the gross chemistry of their sediments and that of Loch Eil with high sulphide, high carbon, low Eh levels and C:N ratios of between 10-15 which are all indicators of an organically rich reduced sediment. The high levels of ammonia in all the sediments are also suggestive of biological activity under reduced conditions. The levels of extractable nitrate are low in all the Shetland Voes as would be expected under reduced conditions when nitrification would be unlikely to occur to any significant extent; the higher nitrate levels in Loch Eil may be related to the greater oxygenation of the overlying water which would favour processes leading to nitrification at the sediment water interface. This emphasises the one important difference found between Loch Eil and the Shetland Voes - in Loch Eil there is active water renewal immediately above the sediment surface which results in water high in O_2 and low in nutrients reaching the sediment. (Fig. 2).

By way of contrast in Sullom Voe there appears to be no such renewal process, at least at the time of sampling and this has resulted in a water column depleted in inorganic nutrients except at the bottom. A further consequence of the apparently poor circulation in Sullom Voe is that extensive areas of the deep basin have become oxygen deficient (Table 7), and that the most oxygen deficient waters overlie the most highly reduced sediments (Fig. 3).

Table 7. Oxygen saturation levels at stations in the inner basin of Sullom Voe.

STATION	DEPTH (m)	DISSOLVED OXYGEN % SATURATION
E7	15	98.7
B2	33	91.7
F5	37	83.1
B3	40	77.9
D4	44	70.6

Consequently any increased input of degradable organic material will result in the sediment becoming more intensely reducing and the water column becoming more oxygen depleted. High nutrient levels were also found in the bottom waters of Olna Firth, but there the sediments were less reducing and the overall sedimentation rate considerably lower, suggesting that the total organic input in that area might be lower.

The effect on the benthic fauna of these variations in the physical and chemical characteristics of the sedimentary environment can be clearly deduced. On the one hand a highly organic sediment with an active microbial flora represents a potentially rich food source for the majority of benthic species. On the other hand the twin effects of deoxygenation and sulphide accumulation will work together to limit the build up of populations in such an environment. Those opportunistic species having a greater resistance to deoxygenation and/or a higher sulphide tolerance level will be able to increase their populations rapidly under such conditions at the expense of the less resistant or 'regressive' species, (c.f. Grassle & Grassle 1974). As the organic deposition rate and deoxygenation increase in the sediments there is a progressive change in the species composition of the benthos from fairly diverse associations with low numbers of individuals through restricted species groups with high individual numbers to a very few species

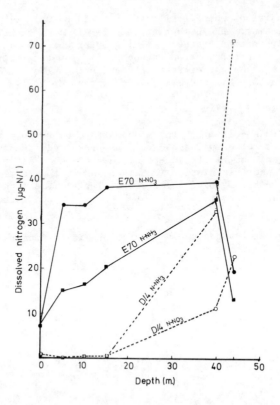

Fig. 2. Nutrient distribution through the water column at stations in Loch Eil and Sullom Voe. June 1976.

with highly fluctuating numbers, and finally to an elimination of all benthic macrofauna. Such sequential changes are fully described in Rosenberg 1973, Pearson 1975 and Pearson & Rosenberg 1975. The four areas described here each fit into different points on this gradient of faunal change. The fauna in Olna Firth is fairly typical of that end of the gradient with relatively low organic input where the populations are mixed but individual numbers are low. The Ronas Voe and Sullom Voe stations support large populations of a restricted number of species and thus relate to that part of the sequence on the border line between supporting large numbers of resistant species and the decline to afaunal conditions. The Loch Eil station where the sediment is subjected to artificially increased depositions of industrially derived organic material, has a population well into such a decline.

It is apparent that both Sullom Voe and Ronas Voe are subject to comparatively high levels of natural organic input to the sediments of their inner basins. It is probable that much of this organic material is derived from eroding peat. The surrounding land is extensively covered by blanket bog and the surface run off reaching the voes is probably rich in peat fragments. Moreover there are extensive sub-tidal peat beds in many parts of Shetland (Hoppe 1965) and exposed areas of such beds having a loose surface of

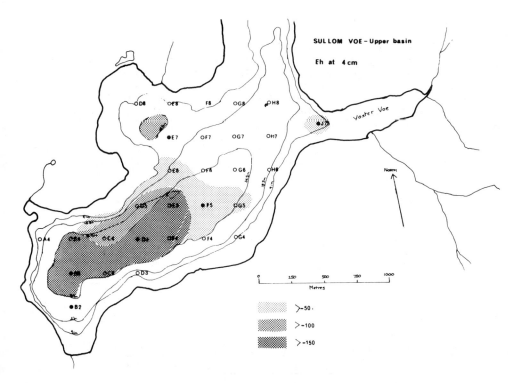

Fig. 3. Map of the distribution of Eh values at 4 cm depth in the sediments of the inner basin of Sullom Voe.

flocculent peat particles were observed on a number of occasions during diving surveys in Sullom Voe and adjacent areas during the present study. The surface of many of the core samples taken from the deep basins of Ronas Voe and Sullom Voe was covered by a flocculent layer of peat particles. Thus it seems that in these basins peat provides a rich and continuous input of organic material to the benthic environment.

CONCLUSIONS

The inner basins of the Shetland Voes studied apparently receive high input levels of organic material derived from eroding peat. In two of the three study areas these levels are apparently approaching the limit of the degradative capacity of the sedimentary ecosystem, and have significantly affected the character and distribution of the benthic macrofauna.

Should there be an increase in the actual amount of organic material reaching the sediments, or stagmation of the bottom waters under poor circulatory conditions, then rapid elimination of the benthic macrofauna in these basins might be expected.

ACKNOWLEDGEMENTS

This work was carried out as part of a survey of the marine ecology of Sullom Voe financed partly by the Natural Environment Research Council and partly by the Sullom Voe Association. The information quoted from Loch Eil was obtained as part of the Loch Eil Project, a co-operative study involving the following personnel:- from the S.M.B.A., the authors, L. Solorzano, A. Edwards, J. Blackstock, P. Tett, J. Leftley, B. Grantham, D. Edelsten, A. Beck, N. Robertson, G. Duncan, J. Nuttall, N. Pascoe, C. Wyatt, I. Vance; from the Department of Biological Sciences, University of Dundee, C. Brown, D. Blake, I. Stanley; from the Department of Civil Engineering, University of Strathclyde, G. Fleming, R. Walker.

REFERENCES

Craib, J.S. 1965. A sampler for taking short undisturbed marine cores. J. Cons. perm. int. Explor. Mer. 30, 34-39.

Grassle, J.F. and J.P. Grassle, 1974. Opportunistic life histories and genetic systems in marine benthic polychaetes. J. mar. Res. 32, 253-284.

Hoppe, G. 1965. Submarine peat in the Shetland Islands. Geogr. Annlr. 47A 195-203.

MacIntyre, A.D. 1961. Quantitative differences in the fauna of boreal mud associations. J. mar. biol. Ass. U.K. 41, 599-616.

Pearson, T.H. 1970. The benthic ecology of Loch Linnhe and Loch Eil, a sea-loch system on the west coast of Scotland. 1. The physical environment and distribution of the macrobenthic fauna. J. exp. mar. Biol. Ecol. 5, 1-34.

Pearson, T.H. 1975. The benthic ecology of Loch Linnhe and Loch Eil a sea-loch system on the west coast of Scotland. IV. Changes in the benthic fauna attributable to organic enrichment. J. exp. mar. Biol. Ecol. 20, 1-41.

Pearson, T.H. and R. Rosenberg 1976. A comparative study of the effects on the marine environment of wastes from cellulose industries in Scotland and Sweden. Ambio. 5, 77-79.

Rosenberg, R. 1973. Succession in benthic macrofauna in a Swedish fjord subsequent to the closure of a sulphite pulp mill. Oikos 24, 1-16.

Strickland, J.D.H. and T.R. Parsons 1965. A manual of sea water analysis. (2nd ed., revised). Bull. Fish. Res. Bd. Can. 125, 1-203.

PREDATION PRESSURE AND COMMUNITY STRUCTURE OF AN INTERTIDAL SOFT-BOTTOM FAUNA

Karsten Reise

Zoologisches Institut, 34 Göttingen, Germany (FR)

ABSTRACT

The macrofauna of two intertidal soft-bottom habitats, a seagrass bed and a mud flat, is compared. Although the species composition is fairly similar, species density and overall abundance is much higher in the seagrass belt. Most populations are out of phase, and differ in age structure. In the mud flat, experimental exclusion of epibenthic predators like shore crabs, shrimp and gobies caused a tenfold increase in abundance and a 4 times higher species density within cages. In the grass bed comparatively slight changes occurred, and some small sized polychaetes became even less abundant, attributable to increased numbers of infaunal predators. Spatial resistance in coarse grained and rooty sediments proved to be capable of diminishing predation pressure as compared with muddy sediments. In consequence, the mud flat community is severely overexploited by unhampered predation from summer to autumn, while the seagrass macrofauna community maintains its complex structure throughout the year.

INTRODUCTION

Organisms are adapted to sets of environmental conditions, to physical factors as well as to other organisms. Corresponding to environmental change in space and time, there is varying success in the settlements of populations, resulting in an intricate pattern of species abundance. In the present study two communities belonging to the intertidal soft-bottom fauna are compared, and the question is asked as to what determines the differences in macrofaunal abundance patterns. The complexity of the problem does not permit a strict explanation, and it is very difficult to assess the relevant initial conditions. In this paper, only one aspect, the role of predation and how it influences the structure of communities, is considered.

The study was carried out in Königshafen, a sheltered bay in the wadden sea of Sylt (German Bight). The mean tidal range is 1.7 m. Between 0 and +0.2 m elevation the small eelgrass Zostera nana Roth forms a distinct belt. In the seaward direction, an extensive sand flat takes over until an elevation of -0.5m is reached. From there on, down to low water mark, mussel beds predominate. Between them are flats composed of soft, fine silt and clay.

The macrofauna of these mud flats and that of the seagrass bed are the special topic of this investigation.

MATERIALS AND METHODS

To obtain quantitative samples of the macrofauna, a 100 cm^2 steel frame was pushed 20 cm into the sediment. The core was divided into fractions of 0-2, 2-5 and 5-20 cm depth. The animals were extracted with sieves of 0.25, 0.5 and 1.0 mm mesh size, respectively, and sorted in white dishes while still alive.

Epibenthic predators and the nekton were excluded with cages, constructed of 1 mm mesh gauze attached to iron rods. The cage dimensions were 50 by 50 by 15 cm high with the edge of the cage penetrating 10 cm into the sediment. Since periwinkles, Littorina littorea L., showed a prefence for grazing off diatoms from gauze, 20 of them were enclosed in each cage, and thus clogging of the fine mesh was prevented. Experiments with shrimp and young shore crabs as predators were carried out in aquaria with 374 cm^2 basal surface area.

RESULTS

The main aspects by which the two habitats - seagrass bed and mud flat - can be distinguished are compiled in Table 1.

TABLE 1 Comparison of the two habitats

aspects	seagrass bed	mud flat
depth at high tide (m)	0.6	1.3
hours of exposure	4 - 5	2 - 3
sediment	medium to coarse sand	soft, fine silt and clay
macrophytes	Zostera nana	none
microphytes	thin layer of diatoms	thick layer of diatoms

The macrofaunal communities differ according to habitat. The species composition is fairly similar, but the overall abundance and the relative proportions of species therein are distinct from one another. Furthermore, seasonal changes in abundance are rarely the same within one species; most of them being completely out of phase (Fig. 1). This applies not only to trends in abundances, but to age structure as well. For example, in the two populations of the polychaete Pygospio elegans Clap. the percentage of juveniles (less than 5 mm in length) decreases in the eelgrass belt from March to October 1975, while increasing in the mud flat. In 1974 a cockle population, Cerastoderma edule (L.), of 4- year old individuals inhabited

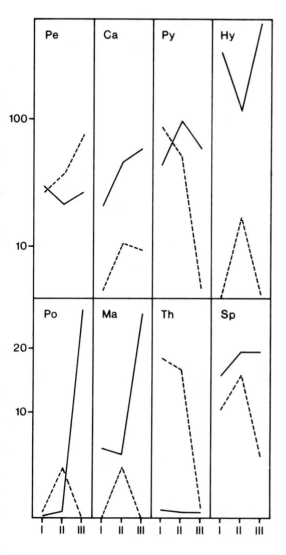

Fig. 1. Seasonal changes in abundance

No. /100 cm^2 are given in log-scale on the ordinate,
———— seagrass bed,
-------- mud flat,
I: March/April, II: June, III: Sept./Oct. 1975

Pe: <u>Peloscolex benedeni</u> (D'Ud.)
Ca: <u>Capitella capitata</u> (Clap.)
Py: <u>Pygospio elegans</u> Clap.
Hy: <u>Hydrobia ulvae</u> (Pennant)
Po: <u>Polydora</u> spec.
Ma: <u>Malacoceros fuliginosus</u> (Cl.)
Th: <u>Tharyx marioni</u> (S-J.)
sp: species density

parts of the mud flat area. Despite the general abundance of spat throughout the year, no other year classes could establish themselves. In the grass bed, however, a persistent population, composed of 1-3 year old individuals, continually receives new recruits. A prominent feature of the mud flat community is an overall decline in species density and abundance from summer to autumn with many species vanishing from the flat (Fig. 1).

The intricate differences between the two communities suggest that biological interactions are involved. One component of such interactions is predation. The most numerous and conspicuous predators on the flats from summer to autumn are young shore crabs, <u>Carcinus maenas</u> (L.), shrimp, <u>Crangon</u>

crangon (L.), and gobiid fish, Pomatoschistus microps (Kröyer). To
evaluate their impact on the infauna, enclosures were set up in summer,
and after about three months, the macrofauna within the cages was compared
with the undisturbed fauna (Table 2).

TABLE 2 Comparison between caged and uncaged infauna

No. /400 cm^2	seagrass bed$^+$		mud flat^{++}	
	uncaged	caged	uncaged	caged
total infauna	953	754	476	4932
Cerastoderma edule (L.)	4	21	3	513
Corophium volutator (Pal.)	-	-	-	196
Peloscolex benedeni (D'Ud.)	129	121	328	1222
Tharyx marioni (S-J.)	1	2	3	2129
Malacoceros fuliginosus (Cl.)	109	118	-	162
Pygospio elegans Cl.	273	257	7	140
Polydora spec.	111	30	-	213
Capitella capitata (Fabr.)	271	146	37	56
Heteromastus filiformis (Cl.)	2	1	96	89
infaunal predators	4	15	1	6
others	49	43	1	205
species density	26	24	7	28

$^+$) caged 19.6. - 16.9.1975, $^{++}$) caged 1.7. - 11.10.1975

In the eelgrass belt not much happened. Apart from some epifaunal snails,
significant increase was observed in Cerastoderma edule and infaunal
predators, the polychaete Anaitides mucosa (Oers.), and the nemertines
Lineus viridis Johns. and Amphiporus lactifloreus (Johns.). In addition, the
larger individuals of Malacoceros fuliginosus (more than 15 mm in length)
showed a significant increase. Other infaunal species tended to be less
abundant within the cage. This was significant for Polydora spec.

These results are in sharp contrast to the striking increase within the mud
flat cage. Except the deep-dwelling polychaete Heteromastus filiformis, all
species within the cage outnumbered by far the densities in the control area.
Because many species established dense settlements within the cage, but
failed to do so in the surrounding mud flat, species density increased by 4
times. The abundances of young crabs, shrimp and gobies during high tide
showed no apparent deviation between the two habitats. The only evident
difference is a shorter feeding time in the eelgrass belt (see Table 1). Thus
the profound dissimilarity in predation pressure exerted on the two
communities was not expected.

Field evidence revealed that most infaunal species are significantly more abundant under dense patches of eelgrass than where it is sparse, i.e. Polydora, Malacoceros, Capitella, and young Scoloplos armiger (Müll.) were three to four times more abundant. Experiments were designed to test the hypothesis as to whether some spatial resistance in the sediment (eelgrass-roots) serves to protect the infauna from young crabs. Undisturbed sediments from the seagrass bed (including Zostera and the infauna) and from the mud flat was placed into aquaria, and young shore crabs were added. As can be observed from Fig. 2 the crabs are much more effective in reducing the infauna in the muddy sediment, which lacks any spatial resistance like roots or shell gravel. With 13 crabs on 100 cm^2 preying for three days, 39.9% of the infauna survived in the sediment penetrated by seagrass-roots, while only 15.6% survived in the muddy sediment. Moreover, even 22 crabs in the rooty sediment were unable to reduce the infauna to such a low level as 13 did in the mud.

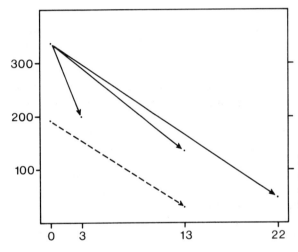

Fig. 2. Crabs reducing the infauna within three days in ——— seagrass-sediment and ----- muddy sediment.

No. /100 cm^2 of the infauna and young shore crabs (5 mm carapace width) are given on the ordinate and the abcissa, respectively.

Another experiment was set up to determine whether or not the grain size of the two sediments can alter the impact of predation. An aquarium was divided in two equal parts. One half received fine sand (0.125-0.250 mm), the other coarse sand (0.5-1 0 mm). Then 300 young cockles (less than two mm in length) were introduced into each sediment, and six shrimp were allowed to prey on the cockle spat for 39 hours. Only 22 cockles (7.3%) survived in the fine sand, while 88 (29.3%) escaped predation in the coarse sand. Apparently shrimp are less successful predators on cockle spat when the sediment is coarse grained. Thus roots and coarse grained sediments serve as a spatial resistance against predation by young crabs and shrimp. This should be regarded as an advantage for the infauna dwelling in beds of seagrass.

DISCUSSION

By the end of autumn, when shrimp, gobies and most crabs leave the intertidal zone, the mud flat community has turned into a reduced Heteromastus-Peloscolex -assemblage. Both species escape severe predation by dwelling or retracting far below the sediment-surface. While these two represent the permanent component of the community, most species are transient because predation reduced them to such an extant by the end of autumn, that recolonization during the following winter and spring has to come from adjacent habitats. Thus in the short run, predation pressure clearly acts as a destabilizing component, and causes a large part of the community to remain at the status of young colonists.

The drastic increase in species number and abundance, up to a level much higher than can be achieved in the grass flats, when the external predation is experimentally excluded, indicates that the mud flat is a benign environment to live in. According to Connell (personal communication; 1975) occasional escape from predators is typical in this type of environment. Such an escape must have occurred in 1970, when cockles managed to reach invulnerable size and thus persisted until winter 1974.

In the grass bed even in autumn a high level of abundance and species density is preserved and the exclusion of epibenthic predators revealed no striking effects. Using Connell's terminology, the seagrass belt should be regarded as an intermediate environment, where offspring survives well because predation pressure is reduced, and physical effects are not killing. Two environmental factors limiting the impact of epibenthic predators in the eelgrass community as compared with the mud flat community are recognized: (1) the seagrass belt is higher up in the intertidal zone, therefore less accessible to shrimp and gobies (barrier to the predators'dispersal) (2) coarse grained and rooty sediment provides spatial resistance to predation (refuges for the prey).
Murdoch and Oaten (1975) pointed out that such spatial heterogeneity causes stability in predator-prey relationships. As a further stabilizing component in the eelgrass community, the infaunal predators are considered. Most likely they balanced out much of the disturbance brought about by the exclusion experiment, and caused low abundances in some small sized polychaetes (see Table 2).

In conclusion, much of the differences in community structure observed in the seagrass bed and the mud flat can be attributed either directly to predation pressure or to an integrated function of spatial resistance and predation.

A similar study on biological interactions in a marine soft-sediment environment was carried out by Woodin (1974) in Mitchell Bay, San Juan Island, Washington. She found no evidence for predation determining abundance patterns. Predators analoguous to shrimp, gobies and shore crabs were largely absent. Instead, the importance of competition was demonstrated It becomes apparent that biological interactions in the soft-bottom intertidal provide quite a complex situation as is known already for the rocky intertidal.

CONCLUSIONS

1. The macrofauna of an intertidal mud flat is severely overexploited by its predators in autumn. Re-establishment of a diversified community by re-colonization from adjacent habitats has to occur during winter and spring, during which time these predators stay in the sublittoral zone.

2. In the seagrass bed such predation pressure is diminished by spatial resistance in the sediment: coarse grained sand and seagrass-roots. Species density and abundance remains high throughout the year.

3. Epibenthic predators like shore crabs, shrimp and gobies may act as a destabilizing factor, whereas infaunal predators exert a stabilizing influence on the community.

4. Much of the differences in community structure observed in the grass flat and the mud flat can be attributed either directly to predation pressure or to an integrated function of spatial resistance and predation.

ACKNOWLEDGEMENTS

I am greatly indepted to Professor Peter Ax for his helpful advice and support of this study. I also owe a great deal to Dan Whybrew for reading this paper and making helpful suggestions. The staff of the Biologische Anstalt Helgoland, Litoralstation List/Sylt and various colleagues from the 2. Zoologisches Institut Göttingen provided valuable technical assistance and friendly suggestions throughout the study.

REFERENCES

Connell, J. H. 1975 in: Ecology and Evolution of Communities, ed.: M. L. Cody and J. M. Diamond. Symp. papers (1973) Belknap, Cambr., Mass.

(has not been seen by author)

Murdoch, W. W. and A. Oaten 1975 Predation and population stability. Advances in ecol. Res. 12, 1-125.

Woodin, S. A. 1974 Polychaete abundance patterns in a marine soft-sediment environment: the importance of biological interactions. Ecol. Monogr. 44, 171-187.

ZONATION IN DEEP - SEA GASTROPODS: THE IMPORTANCE OF BIOLOGICAL INTERACTIONS TO RATES OF ZONATION

Michael A. Rex

Biology Department, University of Massachusetts-Boston, Boston, Massachusetts 02125, U.S.A.

ABSTRACT

The multivariate techniques of cluster analysis and factor analysis were used to study depth-correlated changes in the species composition of deep-sea gastropod assemblages. The most pronounced faunal change occurred at the shelf-slope transition. The fauna changed continuously from the upper slope to the abyss and the rate of change seemed proportional to the rate of change in depth, being highest on the slope, lowest on the abyss and intermediate on the abyssal rise. This pattern may reflect an increasingly uniform environment with increasing depth.

It is suggested that differing rates of zonation among components of the deep-sea benthos are determined, in part, by biological interactions. Groups dominated by croppers and predators such as the epibenthic macrofauna and gastropods are zoned more rapidly than infaunal deposit feeders. This may result because vertical ranges along the depth gradient are more compressed by interspecific competition in members of higher trophic levels. Predation may alleviate competition among infaunal groups at lower trophic levels permitting their ranges to overlap more extensively and consequently diminishing their rates of faunal change with depth. In support of this hypothesis, it is shown that rates of zonation in the epifauna, gastropods and the infaunal polychaete-bivalve fraction correspond to their relative positions in the trophic structure.

INTRODUCTION

Patterns of species replacement with depth, usually termed zonation, have been studied in a variety of deep-sea communities and taxa. Groups differ in their rates of zonation but the causes of such differences remain obscure. Most attempts to explain zonation have evoked physical factors which vary in parallel with depth gradients. In this paper, I analyze the pattern of zonation in deep-sea gastropods from the western North Atlantic. Differences between rates of zonation in gastropods and other benthos from this region suggest that biological interactions have an important role in determining rates of zonation in the deep-sea.

MATERIALS AND METHODS

The material studied was the gastropod fraction of twenty-five epibenthic sled (Hessler and Sanders 1967) samples collected from the Gay Head-Bermuda transect in the western North Atlantic (Sanders, Hessler and Hampson 1965). The samples ranged in depth from 69-4970 meters. Maps showing the station localities were given in Rex (1973, 1976). The entire collection of gastropods contained 21,895 individuals distributed among 142 species.

Since comparisons of faunal composition are sample size dependent, rarefaction methodology (Sanders 1968) was used to artificially equate sample sizes to 68 individuals. Rex (1973) discussed the species diversity of gastropods in the samples and the rationale for selecting rarefactions to 68 individuals. Sanders (1968) cautioned that this technique does not specify which of the residual species should be included in the rarefied sample. Most species in the residue can be chosen in order of their relative abundance. When it was necessary to choose which among several equally abundant species in the residue to include in the rarefied sample, I selected those which occurred most frequently in nearby samples. This criterion had the relatively minor effect of eliminating rare stenobathic or patchily distributed species from the analysis. Of the 142 species present in all 25 whole samples, 102 remained when the samples were rarefied.

Hierarchical cluster analysis and factor vector analysis were performed to determine the pattern of zonation. For the cluster analysis I used a quantified version of Jaccard's Coefficient (Sepkoski 1974) to measure the faunal similarity between stations and the unweighted pair-group method to form clusters (computer program was QUAJAC, Sepkoski 1974). The quantified Jaccard's Coefficient has the advantage of including information on relative abundance and incorporates none of the pitfalls encountered by using the mutual absence of species (Day et al. 1971). To assess the impact of relative abundance on the clustering results I also did the analysis on simple presence-absence data. Factor analysis makes it possible to observe faunal relationships between stations in a more multidimensional way than cluster analysis. I chose an oblique solution (see e.g. Gould 1967) since the meaning of any resulting groups of samples can be determined directly by inspection of the faunal make-up of the samples which serve as the oblique reference axes. This analysis was also carried out on both quantitative and presence-absence data using the computer program DUVAP (Lynts and Paris unpubl. Ms.).

RESULTS

A dendrogram based on relative abundance data is shown in Fig. 1. There are four clusters corresponding to the continental shelf, continental slope, abyssal rise and abyssal plain. Shelf samples exhibit low species richness and equitability (Rex 1973) and cluster together at relatively high levels of association mainly because they share a numerically dominant species Alvania carinata. The most pronounced faunal change along the transect occurs at the shelf-slope transition. Some species' vertical ranges include both the shelf and the upper slope, but their relative abundance patterns change markedly at the transition zone. For example, A. carinata becomes a subordinate species at upper slope depths. The slope and abyssal rise are regions of high gastropod species diversity. The pyramidal shape and low levels of association in the continental slope cluster indicate a rapidly and continuously changing fauna from about 500-2200 meters. Separation of the four abyssal rise stations from both the slope and abyssal rise clusters may partly reflect depth and horizontal distance gaps in sampling. The rate of faunal change with depth on the abyssal rise appears to be somewhat less than on the slope. Station 76 is about 900 meters shallower than the other three rise stations, yet it is more similar to them than stations separated by an equivalent depth on the slope. Another indication that the rate of faunal change decreases at greater depth is that the rise stations as a whole show greater similarity to the abyssal group than to the slope stations. The abyss, an area of comparatively low gastropod diversity, is characterized by an unusually monotonous gastropod assemblage. Station 125 has a poor association with other abyssal samples

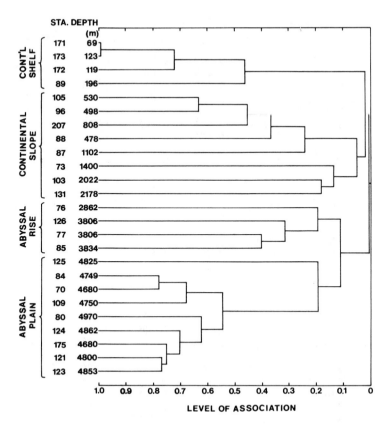

Fig. 1. Cluster analysis of rarefied gastropod samples.

because of a reversal in dominance between what are generally the two most abundant abyssal species: <u>Cithna tenella</u>, otherwise the most dominant, and <u>Adeorbis umbilicatus</u>, usually the second most abundant.

The cluster analysis using presence-absence data (not presented but available from the author) produced basically the same four groups, the main difference being lower average levels of association within clusters, especially on the shelf and abyss, and slightly higher associations between clusters. Station 125 joins the abyssal cluster and stations 103 and 131, which are separated from other slope stations by over 600m, show an especially weak association with their group. These relatively minor changes occur since all species now contribute equally to the similarity coefficient. Species which are rare in at least one cluster, but whose ranges extended between clusters have relatively more influence and abundant species responsible for tight clustering within groups become less significant. The basic pattern of faunal change with depth is, however, retained.

A reordered oblique projection matrix based on presence-absence data is shown in Table 1 (an unrotated solution explained 66% of the variance). Each station is clustered out with the reference vector on which it has the highest projection, that is, the highest faunal affinity. Relationships to other

TABLE 1 Reordered oblique projection matrix for rarefied gastropod samples.

		STA 70	STA103	STA171	STA 77	STA 96
ABYSSAL PLAIN	STA 70	1.000	0.000	0.000	0.000	0.000
	STA123	0.925	-0.005	0.002	0.093	-0.001
	STA 84	0.921	-0.039	-0.003	0.203	0.011
	STA175	0.804	-0.050	-0.006	0.208	0.017
	STA125	0.792	-0.062	-0.006	0.302	0.019
	STA124	0.775	-0.011	0.002	0.129	-0.001
	STA109	0.772	0.076	0.010	0.226	-0.024
	STA121	0.670	-0.010	0.003	0.368	-0.001
	STA 80	0.578	-0.107	-0.013	0.470	0.035
LOWER SLOPE	STA103	0.000	1.000	0.000	0.000	0.000
	STA 73	0.163	0.940	-0.069	-0.363	0.199
	STA131	0.008	0.850	-0.050	-0.056	0.143
CONT'L SHELF	STA171	0.000	0.000	1.000	0.000	0.000
	STA173	0.000	0.000	1.000	0.000	0.000
	STA 89	-0.020	-0.078	0.695	0.046	0.167
	STA172	-0.029	-0.118	0.562	0.065	0.309
ABYSSAL RISE	STA 77	0.000	0.000	0.000	1.000	0.000
	STA 76	-0.164	0.182	0.023	0.820	-0.059
	STA 85	0.466	0.011	0.001	0.759	-0.002
	STA126	0.191	0.208	-0.002	0.698	0.009
UPPER SLOPE	STA 96	0.000	0.000	0.000	0.000	1.000
	STA 88	-0.008	-0.022	-0.011	0.025	0.975
	STA207	0.027	0.175	-0.194	-0.070	0.887
	STA105	0.044	0.308	0.061	-0.108	0.854
	STA 87	0.103	0.535	-0.150	-0.230	0.660

groups are indicated by projections on other axes. In Table 1 we find essentially the same relationships revealed by the dendrogram (Fig. 1) except that the slope stations are now split into two groups representing the upper (478-1102m) and lower (1400-2178m) slope. An analysis based on percentage data produced the same five groups when five reference vectors were extracted (available from author). Table 1 shows the transitional nature of the fauna from one region to the next somewhat more clearly than the dendrogram. Abyssal stations have their highest projections on the station 70 reference vector, but also show affinities with the abyssal rise group. Abyssal rise stations show similarity to both the abyssal and lower slope regions. Associations between the upper and lower slope are evident and station 87, the deepest member of the upper slope group at 1102m, shows an especially high projection on the lower slope vector. Two shelf stations show projections on the upper slope vector showing again that some species do transgress the shelf-slope boundary.

Analyses based on relative abundance and simple presence-absence data both revealed essentially the same patterns of zonation. Differences in relative abundance can have an important impact on quantitative measures of faunal similarity (e.g. station 125). However, this was probably minimized in the present study for several reasons: (1) samples throughout most of the depth range have fairly high species evenness (Rex 1973), (2) samples with low evenness on the abyss and continental shelf tended to be dominated by the same species and share subordinate ones, (3) even though the samples were collected over a span of several years, the relative abundances of species in the deep sea are probably not as subject to spatio-temporal variation as in other environments.

DISCUSSION

The relationships indicated by the cluster and factor analyses can be summarized as follows. The greatest change in composition of the gastropod fauna occurs at the edge of the continental shelf. Similar marked shifts in composition at this depth were reported for infaunal polychaetes and bivalves (Sanders and Hessler 1969) and the epibenthic macrofauna (Haedrich et al. 1975). Sanders (1968) related this change to the differences in life history strategies that would be most adaptive to either the environmentally unstable shelf or the more stable slope. The gastropod fauna changes rapidly with depth on the continental slope and there is some indication of an especially rapid transition zone at mid-slope depths. Haedrich et al. (1975) noted a similar change in the epifauna at this depth south of New England and pointed out that it coincides with a change in the vertical gradient of the slope. The rate of change in the gastropod fauna appears to lessen on the abyssal rise and abyssal stations show the highest levels of similarity. The largest differences in similarity below the shelf, the slope to rise and rise to abyss, are partly artifacts of the sampling gaps of 684m and 846m, respectively. As with the shelf-slope transition, the vertical distributions of some species extend across these sampling gaps making the pattern of faunal change a continuous one throughout the transect. Below the shelf the rate of faunal change in gastropods seems to correspond roughly to the rate of change in depth with distance from land, being greatest down the slope and progressively less on the rise and abyss. This pattern may result from the deep-sea environment becoming more uniform at greater depths.

Attempts to explain causes of zonation in the deep-sea benthos have focused mainly on the possible effects of physical factors. These include, for example, temperature, sedimentation rates and substrate types (Haedrich et al. 1975), and the effect of deep boundary currents on larval dispersal and detrital food supplies (Rowe and Menzies 1969). It is very probable that physical factors do influence zonation in the deep-sea though their relative significance is still uncertain. Recent work in benthic ecology has brought out the importance of biological interactions in the form of competition and predation in structuring communities. In what follows, I propose a hypothesis based on biological interactions to account for different rates of zonation in groups of organisms that occupy different trophic positions in the deep-sea benthic community. Rates of zonation in gastropods, infaunal taxa and the epifauna are used to test it on a comparative basis.

The hypothesis combines aspects of two existing ecological theories, Menge and Sutherland's (1976) recent synthesis of factors affecting species diversity and Terborgh's (1971) theory of distribution on environmental gradients. Menge and Sutherland pointed out that competition and predation are both influential in structuring communities, but that their relative importance varies with trophic position. Predation is the most important factor acting to organize and diversify prey populations at lower trophic levels. At higher levels, competition for prey becomes an increasingly significant factor limiting the diversity of predator populations. Rex (1976) related this theory to patterns of diversity in deep-sea gastropods. In

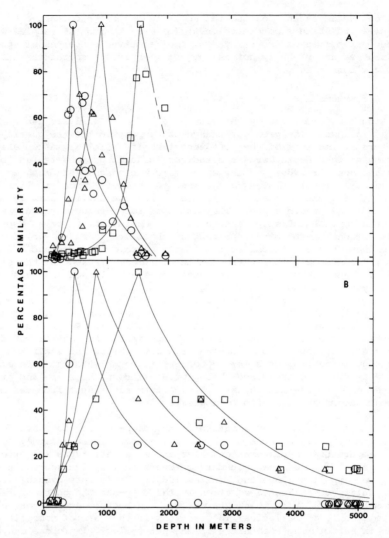

Fig. 2. Comparisons of percentage faunal similarity of six reference samples with other samples collected along a depth gradient. In A, similarity values for the epifauna from 24 stations are compared with three reference stations at depths of 475 (circles), 933 (triangles) and 1543 (squares) meters (raw data provided by R. Haedrich, used in Haedrich et al. 1975). In B, similarity values for infaunal polychaetes from 20 stations are compared with three reference stations at depths of 487 (circles), 824 (triangles) and 1500 (squares) meters (data are midpoints of intervals from Fig. 3 in Sanders and Hessler 1969). Lines are fit by eye. Faunal replacement with depth is much more rapid in the epifauna (A) than in the infaunal polychaetes (B).

Terborgh's (1971) model, the distribution of species along an environmental gradient, such as with elevation or depth, can be regulated by the availability of resources which vary in parallel with the gradient or by competition between species arrayed along the gradient. In the first case, where interspecific competition is absent, species' ranges are expected to overlap extensively along the gradient and the rate of species replacement measured away from any reference point along the gradient will be gradual. In the second case, species will tend to repulse one another competitively at the borders of their ranges and, hence, occupy mutually exclusive zones along the environmental gradient. As a consequence of this repulsion interaction, assemblages of competing species will attenuate more rapidly along the gradient than non-competing species.

Combining the two theories permits us to predict relative rates of zonation in benthos occupying different trophic levels. If members of upper trophic levels are under greater competitive pressure than members of lower levels, then they should show more rapid faunal change with depth than members of lower trophic levels. Predation from upper levels should alleviate competition among prey species allowing their ranges to overlap more extensively and their distributions to parallel other biotic resources or limiting physical factors along the depth gradient to a greater extent. This prediction can be tested by comparing rates of faunal change with depth in groups that have predominantly predatory or deposit feeding habits.

The epibenthic macrofauna, which relies principally on the infauna as a food resource, probably exerts the most significant predatory impact in deep-sea benthic communities. It seems likely that if "Foundation" or "Keystone" species (Dayton 1975) occur in the deep sea they will be components of the epifauna. The epifauna has lower density, lower species diversity and is zoned more rapidly than the infauna (Haedrich et al. 1975). The biomasses of the two groups appear to be comparable, possibly because the epifauna derives part of its food from pelagic sources (Haedrich et al. 1975). Fig. 2 compares the patterns of faunal change along a depth gradient for the epifauna and a dominant component of the infauna, the polychaetes. The curves in Fig. 2 indicate the percentage faunal similarity (Whittaker and Fairbanks 1968) of six reference samples with other samples along the depth gradient. This measure was chosen because it was common to the studies from which data were taken for comparison. Clearly, the epifauna (Fig. 2A) attenuates much more rapidly away from the reference samples than do infaunal polychaetes (Fig. 2B).

In Fig. 3 percentage faunal similarity with selected reference samples is plotted against difference in depth for both epibenthos and infaunal elements. The epifaunal data (squares) are from trawl stations (Haedrich et al. 1975) and direct observations using the deep-sea submersible Alvin (Grassle et al. 1975). These were the most accurate data available for the epifauna of this particular region. The infaunal data (circles) represent polychaetes and bivalves which together make up about 80% of the infauna in terms of abundance (Sanders and Hessler 1969). On the average, the epifauna are much more rapidly zoned than the infauna (i.e. at greater depth differences, epifaunal samples show less similarity to their reference samples than infaunal samples show to their reference samples). I have also included in Fig. 3 data on gastropods (triangles). Similarity values for gastropods tend to lie between those of the infauna and epifauna. Gastropods for the most part probably live near the surface of the sediment and some have been observed to be epifaunal (Grassle et al. 1975) They are dominated by predators (at least 54% of the species) and polychaetes and bivalves probably serve as their major prey (Rex 1973). Because of their low density and small size, the predatory impact of gastropods is probably considerably less than for the epifauna as a

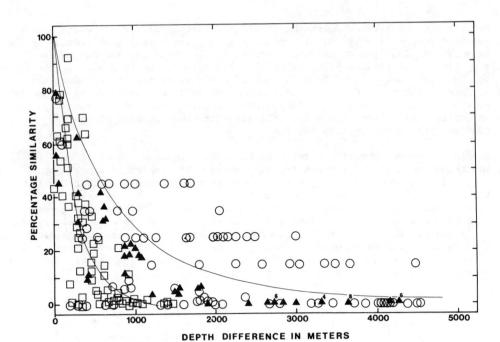

Fig. 3. Percentage faunal similarity values for samples of epifauna (squares), infaunal polychaetes and bivalves (circles) and gastropods (triangles) plotted against difference in depth to selected reference stations. Data on bivalves from Sanders and Hessler (1969, Table 2); polychaetes from Sanders and Hessler (midpoints of intervals in Fig. 3 compared to reference stations D, E, F and HH3); epifauna provided by R. Haedrich used in Haedrich et al. (1975, compared to reference samples at depths of 475, 933 and 1543m), and Grassle et al. (1975, based on density values in Table 2 compared to reference stations 281 and 279); gastropods from the 17 largest samples in the present study rarefied to 100 individuals and compared to reference stations 87, 103, 77 and 105. Lines are fit by eye and meant to indicate the general pattern in the infauna (above) and epifauna (below). Values for gastropods tend to fall between the two lines.

whole. Their rate of zonation is less than for the epifauna and greater than that of the infauna. Among these groups then, the rate of zonation does appear to be related to trophic position.

Sanders (1976) and Sanders and Grassle (1971) showed that there are sizable differences in rates of zonation between bivalves, peracarid crustaceans, polychaetes, gastropods and ophiuroids in the deep-sea. These differences were related to life history strategies. They may also reflect, in part, differences in trophic roles. Gastropods appeared to be the most rapidly zoned. Polychaetes, which also contain predatory elements (about 35% of the species in one study, Hessler and Jumars 1974) were also highly zoned. Bivalves, which are infaunal deposit feeders, had lower rates of zonation. Ophiuroids provide an exception to the pattern predicted here since they are epibenthic, but are more eurybathic than the other groups. Peracarid crustaceans are evidently also an exception since most are infaunal deposit feeders but are relatively stenobathic.

I would like to emphasize that the hypothesis relating rate of zonation to trophic role is simplified in some ways. It assumes that competition along the depth gradient is frequently manifested as broad scale spatial exclusion rather than temporal, behavioral or more subtle microhabitat distinctions. It also overlooks variation in prey selectivity and important aspects of life history stressed by Sanders (1976) and Sanders and Grassle (1971). Many life history parameters and feeding characteristics interact to regulate vertical distributions and the relative importance of these factors varies from group to group. I have compared higher taxonomic categories and whole subdivisions of the community each of which contain a variety of life styles. The limited data available on different groups provide only a very incomplete picture of community organization in the deep sea. Moreover, I have treated the depth gradient as a gradual and continuous environmental gradient, but the physical and biological factors that define this gradient are still poorly understood and quantified. The hypothesis predicts only relative rates of zonation at different trophic levels and does not attempt to explain specific qualitative differences between communities. In general, available data do suggest a relationship between trophic position and rate of zonation.

CONCLUSION

Deep-sea gastropods from the western North Atlantic show rapid and continuous faunal change with increasing depth. The rate of faunal change appears to be proportional to the rate of change in depth. This pattern may result from the deep-sea environment becoming increasingly uniform at greater depths.

It is hypothesized that rates of zonation are determined, in part, by biological interactions. Competition for prey at higher trophic levels may compress vertical ranges of predators resulting in high rates of species replacement along the depth gradient. Predation may alleviate competition at lower levels permitting vertical ranges to overlap more, thereby causing rates of zonation to be less in the deposit feeding infauna. In support of this hypothesis, it was shown that rates of zonation in the epifauna, gastropods and the infaunal polychaete-bivalve fraction do correlate with their relative positions in the trophic structure.

ACKNOWLEDGEMENTS

I thank J. F. Grassle and J. J. Hatch for reading the manuscript. R. L. Haedrich provided the similarity matrix used in Haedrich et al. (1975). The author's research is supported by N.S.F. grant BMS 75-03188. Gastropod material was collected under support of N.S.F. grants GB 6027X and GA 31105 to the Woods Hole Oceanographic Institution and was kindly made available to me by H. L. Sanders, Principal Investigator.

REFERENCES

Day, J.H., J.G. Field and M.P. Montgomery 1971 The use of numerical methods to determine the distribution of the benthic fauna across the continental shelf of North Carolina. J. Anim. Ecol. 40, 93-125.

Dayton, P.K. 1975 Experimental evaluation of ecological dominance in a rocky intertidal algal community. Ecol. Monogr. 45, 137-159.

Gould, S.J. 1967 Evolutionary patterns in pelycosaurian reptiles: A factor-analytic study. Evolution 21, 385-401.

Grassle, J.F., H.L. Sanders, R.R. Hessler, G.T. Rowe and T. McLellan 1975 Pattern and zonation: A study of the bathyal megafauna using the research submersible Alvin. Deep-Sea Res. 22, 457-481.

Haedrich, R.L., G.T. Rowe and P.T. Polloni 1975 Zonation and faunal composition of epibenthic populations on the continental slope south of New England. J. Mar. Res. 33, 191-212.

Hessler, R.R. and P.A. Jumars 1974 Abyssal community analysis from replicate box cores in the central North Pacific. Deep-Sea Res. 21, 185-209.

Hessler, R.R. and H.L. Sanders 1967 Faunal diversity in the deep sea. Deep-Sea Res. 14, 65-78.

Menge, B.A. and J.P. Sutherland 1976 Species diversity gradients: Synthesis of the roles of predation, competition, and temporal heterogeneity. Amer. Natur. 110, 351-369.

Rex, M.A. 1973 Deep-sea species diversity: Decreased gastropod diversity at abyssal depths. Science 181, 1051-1053.

Rex, M.A. 1976 Biological accommodation in the deep-sea benthos: Comparative evidence on the importance of predation and productivity. Deep-Sea Res. (in press).

Rowe, G.T. and R. J. Menzies 1969 Zonation of large benthic invertebrates in the deep-sea off the Carolinas. Deep-Sea Res. 16, 531-537.

Sanders, H.L. 1968 Marine benthic diversity: A comparative study. Amer. Natur. 102, 243-282.

Sanders, H.L. 1976 Evolutionary ecology and the deep-sea benthos. Proc. Acad. Nat. Sci. Philadelphia (in press).

Sanders, H.L. and J.F. Grassle 1971 The interactions of diversity, distribution and mode of reproduction among major groupings of deep-sea benthos. Proc. Joint Oceanogr. Assemb. (Tokyo, 1970), 260-262.

Sanders, H.L. and R.R. Hessler 1969 Ecology of the deep-sea benthos. Science 163, 1419-1424.

Sanders, H.L., R.R. Hessler and G.R. Hampson 1965 An introduction to the study of deep-sea benthic faunal assemblages along the Gay Head-Bermuda transect. Deep-Sea Res. 12, 845-867.

Sepkoski, J.J. 1974 Quantified coefficients of association and measurement of similarity. Math. Geol. 6, 135-152.

Terborgh, J. 1971 Distribution on environmental gradients: Theory and a preliminary interpretation of distributional patterns in the avifauna of the Cordillera Vilcabamba, Peru. Ecology 52, 23-40.

Whittaker, R.H. and C.W. Fairbanks 1958 A study of plankton copepod communities in the Columbia Basin, southeastern Washington. Ecology 39, 46-65.

MOLLUSCAN COLONIZATION OF DIFFERENT SEDIMENTS ON SUBMERGED PLATFORMS IN THE WESTERN BALTIC SEA[+] *

W. Richter and M. Sarnthein

*Geologisch-Paläontologisches Institut der Universität,
D 2300 Kiel, Germany (FRG)*

ABSTRACT

Gravel, sand and clay substrates were submerged in a new-style floating arrangement at depths of 11, 15 and 19 m. Over a two-year period, the substrates were colonized by 13 bivalve and 3 gastropod species. Their distribution patterns can be considered against each of the following factors: water depths, bottom types, and seasonal variations. Colonization phases occurred repeatedly and simultaneously during certain periods of the year on most experimental substrates (partly explained by spat falls), with the populations stabilizing at $20,000/m^2$. Maximum species numbers and species evenness occur at 19 m paralleling the average position of the pycnocline. *Mytilus edulis* became the dominant species with decreasing water depth, towards 15 and 11m, particularly on the gravel substrate, but also temporarily on clay. Concurrently, the proportions of the total dry weight of some other molluscs decrease and those of a third group of species remain rather constant (*Montacuta bidentata, Cardium fasciatum, Macoma baltica, Barnea candida*). The substrate factor appears to have an importance secondary to those related to water depth, except in the case of the boring clam *B. candida*.

INTRODUCTION

Benthic colonization and production are determined by many independent factors. Physical and chemical factors such as temperature, salinity, light, oxygen, turbulence, turbidity, and biological factors such as nutrition and predation, both form groups of agents, which might be dependent on water depth. Grain-size distribution, composition and structures of the sediments can be summarized as "substrate factors". Ecological interpretations often suffer from the problem of not knowing how to separate the latter from the other factors.

A field experiment, conceived to partly solve this problem, was carried out with three moored platforms, at depths of 11, 15 and 19 m, 2 to 4 m above the sea floor (Fig. 1). Each platform supported three round containers (71 cm in diameter), each filled with a different sediment: clay, sand or gravel (for details of technical layout see Sarnthein and Richter, 1974). This allowed benthonic production on one substrate type to be compared at different water depths, and from different substrates at the same depths.

Parallel studies of the environment and the macrofauna of the adjacent actual sea bottom and in bottom cages were carried out by Arntz et al. (1976), Brunswig et al. (1975) and recently, by H. Rumohr, who continues the platform

---

* The investigation is part of the interdisciplinary SFB 95 programme "Interaction Sea-Sea Bottom" at Kiel University (contribution No. 165)

experiment. Scheibel (1974) investigated its meiofauna.

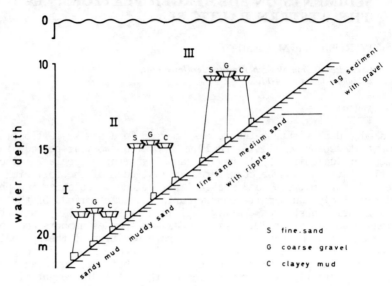

Fig. 1. Environmental positions of floating substrate platforms. Horizontal distance not to scale and greatly abbreviated (by ca. 80 m). Distribution of sediments at sea bottom from Flemming and Wefer (1973).

MATERIAL AND METHODS

The experiment ran from July 1972 until fall 1974 in a demarcated marine test area situated on the slope of one of Kiel Bay's submarine channels (Fig. 2). The channel has a maximum depth of 29 m. Its salinity varies from 15 to 27‰ and its temperature from 2.5 to 10.0°C below 10 m water depth (Arntz et al. 1976).

Samples were collected from the sediment containers at regular intervals by divers using cylindrical plastic corers 10 cm long and 5 cm in diameter. Later on, when a larger fauna had to be sampled, cylinders 15 cm long and 10 cm in diameter were employed. The samples were stored in alcohol, stained in rose Bengal and washed through a 63 micron sieve. All benthonic shells were picked out by hand after sample concentration in carbon tetrachloride. The shells were identified to species level using the keys of Joergensen (in Thorson 1946) and Ziegelmeier (1962 and 1966). Total dry weights of molluscs were determined on a Mettler microbalance.

RESULTS

The submerged floating substrates were colonized by 13 bivalve and 3 or 4 gastropod species, all (but one) having meroplanktonic larval stages. *Mytilus edulis* and *Cardium fasciatum* were the most dominant species (in terms

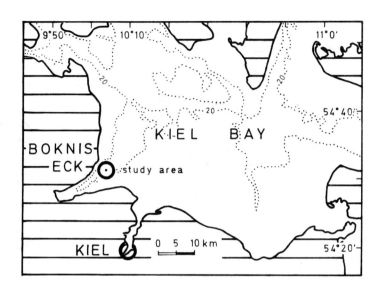

Fig. 2. Experimental site, Kiel Bay, Baltic Sea. Dotted 20 m isobaths indicate course of submarine channels.

of total dry weight) (Fig. 3). *M. edulis* dominated the gravel and sandy substrates at 11 m and the gravel substrate at 15 m (up to 100%). Sand at 15 m (and occasionally at 11 m) and gravel and sand substrates at 19 m particularly favoured the growth of *C. fasciatum* (up to 98%). *Barnea candida*, a boring clam, is another dominant form found on clay at 11, 15 and 19 m water depth (up to 80 - 98%), and occasionally on fine sand at 11 m. Finally, *Saxicava arctica* contributed an important proportion of the shell weight at 19 m, showing a slight preference for the clay substrate, although it did not occur before summer 1973. Its rapid growth (up to 30 mm long after 12 months) makes it easily monitored even in the earliest stages.

A group "miscellaneous forms" comprises mainly juvenile forms which are difficult to identify (Joergensen 1946). This includes *Abra alba*, *Montacuta bidentata*, *Aloidis gibba*, *Macoma baltica* and possibly also *Mya truncata* which was only identifiable during the second experimental study year. Small numbers of *Cyprina islandica* occurred on all sediments at 15 and 19 m, with a slight preference for the sandy substrate. The second boring clam of the Western Baltic Sea, *Zirfaea crispata*, occurred once on clay at 19 m. Generally, the sediment at 19 m, especially fine sand and clay, is characterized by the highest species diversity (species abundance as well as evenness, sensu Hurlbert 1971) of all experimental substrates, although *Mytilus edulis* and other species formed up to 70% of the dry weight for short intervals of one to four months.

The process of colonization has been clearly discontinuous as reflected, for example, by the repeated growth and decrease of specimen numbers on sand or

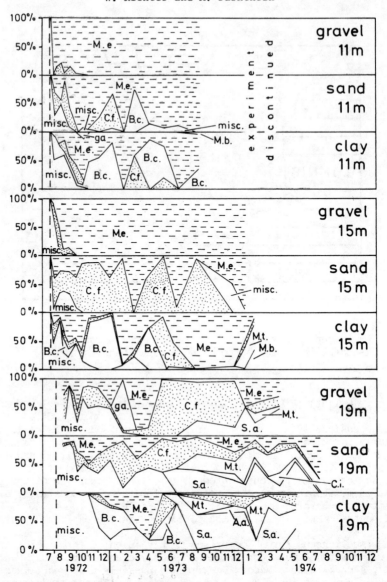

Fig. 3. Species distribution pattern. Species plotted as percentages of total dry weight. Hatched lines mark start of experiment.
A.a., *Abra alba*; B.c., *Barnea candida*; C.f., *Cardium fasciatum*; C.i., *Cyprina islandica*; ga., gastropods; M.b., *Montacuta bidentata*; M.e., *Mytilus edulis*; misc., undetermined larval stages, *Aloidis gibba*, *Macoma baltica*, *Musculus niger*, *Mya arenaria*, *Zirfaea crispata*; M.t., *Mya truncata*; S.a., *Saxicava arctica*.

clay at 15 m and 19 m (Fig. 4).

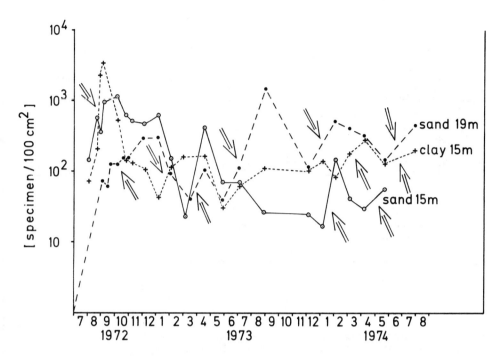

Fig. 4. Specimen numbers on different substrates and water-depth levels. Arrows mark colonization phases as indicated in Fig. 5.

The initial colonization phase started in mid July, immediately after the "set-out" of the experiment, when molluscs formed the highest numbers of meroplanktonic larvae of the year (Martens pers. comm.). It persisted on all 9 substrates until mid September, approximately 2 months (Fig. 5). It is interesting to note that further colonization phases also took place more or less simultaneously on the different sediments and water-depth levels: in early spring 1973, from late May until August 1973 as well as 1974, and in early winter 1973/74. Numbers of living individuals decreased markedly during the other parts of the years, especially during winter. Finally, they stabilized at average numbers of 10,000 to 40,000/m² after initial peak values of 500,000/m². Phases of increasing abundances are frequently linked to heavy settlement by *Mytilus* larvae (for example, in March/April 1973 on all substrate types). However, *Mytilus* has only temporary major significance on most substrates at 15 and 19 m water depth and on clay at 11 m. Several colonization pulses are related to growing numbers of *Barnea candida*, *Cardium fasciatum*, and *Saxicava artica* (various substrates at 15 and 19 m).

DISCUSSION

The experiments with submerged sediment platforms were supposed to provide new insights into the vertical and substrate related distribution patterns of

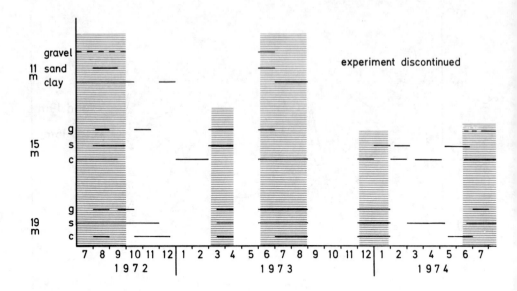

Fig. 5. Summary of colonization phases (indicated by hatching). Horizontal lines mark periods of increasing specimen numbers.

molluscan benthos. The occurrence of only 13 bivalva species might question its usefulness. However, with regard to the 23 species recorded from the S.W. Kiel Bay by Arntz et al. (1976), 4 species are restricted to the shallowest water (Cardium edule and lamarcki, Spisula subtruncata, Angulus tenuis) and 6 forms do not have meroplanktonic larvae (Musculus spp., Astarte spp., Macoma calcarea). Only the absence of Phaxas pellucidus cannot be fully understood and is possibly related to long-term variations of the annual bivalve-larvae spectrum. On the other hand, Musculus niger is present on the experimental platforms despite its lack of meroplanktonic larvae, perhaps owing to transport by drifting algae. (For details on other aspects of the experiment and the refutation of certain objections to it, see Sarnthein and Richter, 1974). Gastropods form a minor group of 3 species and in the depths under discussion are generally rare in the Western Baltic Sea.

The colonization pattern on the experimental substrates shows various trends related to water depth and independent of sediment type, e.g., the maximum of species number, species diversity and evenness at 19 m. This depth level confirms results of Arntz et al. (1976) who relate it to the influence of the thermohaline pycnocline. With an average position at 18-22 m on the channel slope, it gives rise to a number of favourable conditions.

Mytilus edulis is the only species, the relative proportion of which decreases markedly from 11 to 19 m depth on all sediment types (besides a general

decrease of biomass and shell production; Richter and Sarnthein, in prep.).
On the other hand, the relative proportion and, in some cases, the biomass of
Saxicava arctica, *Mya truncata*, *Abra alba* and *Cyprina islandica* increase
from 11 to 19 m; this also occurs in the miscellaneous group. Contrary to
previous expectations (Arntz and Rumohr, 1973), the boring clam *Barnea candida*
does not decrease with depth and shows astonishing high growth rates
(Richter and Rumohr, 1976).

The average specimen numbers are neither affected by increasing water depth
nor by different substrates.

Differences of substrate type affect benthonic colonization at 11 and 15 m
water depth. Fine sand and clay are settled by various species, with
Barnea candida dominating more frequently on clay and *Cardium fasciatum* on
sand, whereas *Mytilus edulis* covers the gravel substrate. However, it can
also settle temporarily on clay as a dominating species (Fig. 3). Only
minor differences are noted in the influence of substrate types at 19 m:
Cyprina islandica clearly prefers fine sand, *Mya truncata* is rather rare on
gravel and *Cardium fasciatum* on clay, *Barnea candida* is restricted to clay.

In summary, the conventional concept of the environmental factors "hard
bottom" and "soft bottom" can no longer be considered to be valid for
bivalves in a direct manner; the factors of the bottom type should be regarded as inferior and subject to other environmental factors related to water
depth, namely pycnocline, and nutrition, or simply as a result of bottom
morphology. For example, the rough surface of sandy gravel as used in our
experiment can provide an over-abundance of protective niches and surfaces
for adhesion of certain larvae, which will only thrive on this bottom type
if conditions are otherwise favourable. This relation holds even less truth,
when considering the data of bioproduction (Richter and Sarnthein, in prep.).

The concentration of colonization to various limited seasons is only partly
explainable (for example by temporary spat fall) owing to the fact that these
phases are generally associated with an unusual parallel increase in the
average shell weights, i.e. of benthonic bioproduction. The seasonality
had special significance for the experiment at its inception: when a larval
species arrives first it can occupy all niches and keep off all later arriving
species. Starting the experiment during July 1972 enabled the optimization
of larval species to be obtained.

CONCLUSIONS

1. A two-year field colonization experiment on a series of submerged
   floating sediment containers at 11, 15 and 19 m water depth yielded 12
   bivalve and 3 gastropod species with meroplanktic larvae, i.e. almost all
   expected in Kiel Bay.

2. Increasing water depths from 11 to 19 m are reflected by decreasing
   proportions of *Mytilus edulis* and increasing abundance of *Cyprina islandica*,
   Abra alba, *Mya truncata* and *Saxicava arctica*. The proportions of other
   species remain unchanged.

3. The maximum species number and species evenness at 19 m parallels the
   average position of the pycnocline at 18-22 m.

4. Major differences of colonization occur between the gravel, and the fine sand and clay substrates, and are limited to the shallower 15 and 11 m levels. The boring clam *Barnea candida* is almost entirely restricted to the clay substrates. The factor "bottom type" can be regarded as of less importance than the factors related to water depth and independent of "bottom softness".

5. Apart from a collective initial colonization phase, the various experimental substrates had simultaneously renewed population influxes in March-April, late May to August and December-January. These phases can be partly explained by spat falls.

ACKNOWLEDGEMENTS

The project was based on the enthusiastic help of the students' diving group in Kiel and on the valuable technical assistance of Mrs. Salomon. It was generously supported by the SFB 95 programme at Kiel University.

REFERENCES

Arntz, W.E. and Rumohr, H. 1973. Bohrmuscheln (*Barnea candida* (L.) und (*Zirfaea crispata* (L.)) in der Kieler Bucht. Kieler Meeresforsch. 29, 141-143.

Arntz, W.E., Brunswig, D., and Sarnthein, M. 1976. Zonierung von Mollusken und Schill im Rinnensystem der Kieler Bucht (Westliche Ostsee). Senckenbergiana Maritima, vol. 8, No. 4 - 6.

Brunswig, D., Arntz, W.E., and Rumohr, H. 1975. A tentative field experiment on population dynamics of macrobenthos in the Western Baltic. IV. Symp. of the Baltic Marine Biologists, Gdansk.

Flemming, B. and Wefer, G. 1973. Tauchbeobachtungen an Wellenrippeln und Abrasionserscheinungen in der Westlichen Ostsee südöstlich Bokniseck. Meyniana 23, 9-18.

Hurlbert, S.H. 1971. The nonconcept of species diversity: a critique and alternative parameters. Ecology 52, 577-586.

Joergensen, B. 1946. Lammellibranch Larvae. In Thorson, G. Reproduction and larval development of Danish marine bottom invertebrates. Medd. Komm. Havundersg. Copenhagen (Plankt.) 4, 523 pp.

Richter, W. and Rumohr, H. 1976. Untersuchungen an *Barnea candida* (L.): Ihr Beitrag zur submarinen Geschiebemergelabrasion in der Kieler Bucht. Kieler Meeresforsch. (in press).

Richter W. and Sarnthein, M. Submarine experiments on benthic colonization of sediments in the Western Baltic Sea. III. Carbonate production by mollusks. Mar. Biol. (in prep.).

Sarnthein, M. and Richter, W. 1974. Submarine experiments on benthic colonization of sediment in the Western Baltic Sea. I. Technical Layout. Mar. Biol. 28, 159-164.

Scheibel, W. 1974. Submarine experiments on benthic colonization of sediments in the Western Baltic Sea. II. Meiofauna. Mar. Biol. 28, 165-168.

Ziegelmeier, E. 1962. Die Muscheln (*Bivalvia*) der deutschen Meeresgebiete. Helgoländer wiss. Meeresunters. 6, 64 pp.

Ziegelmeier, E. 1966. Die Schnecken (*Gastropoda Prosobranchia*) der deutschen Meeresgebiete und Brackigen Küstengewässer. Helgoländer wiss. Meeresunters. 13, 1-61.

EPIFAUNAL ECOLOGY OF INTERTIDAL ALGAE

R. Seed and P.J.S. Boaden

*Department of Zoology, University of College of North Wales, Bangor
Wales and Queen's University Marine Biology Station, Portaferry, Northern
Ireland*

ABSTRACT

Algal epifaunal communities have been studied at sites with diverse physical and biological characteristics at The Dorn, Northern Ireland. Special attention is given to sessile fauna on Fucus serratus L. and the importance and extent of this habitat described. Physical site conditions influence distribution but component species are also isolated spatially by better settlement and survival on particular plant species. Even if isolation by site or plant fails, differential distribution is still possible within the plants by reaction to frond age and surface contour. Some plant parts are so favourable that isolation mechanisms fail and species are in direct competition. Evidence of this is found in bryozoans, hydroids and other organisms. Dense epifaunal growth may lead to heavy predation. The effectiveness of isolating mechanisms in pre-empting competition is discussed. It may be inferred that the "choice" of particular strategies of isolation and competition has helped increase and maintain the structural diversity of the community.

INTRODUCTION

In the study of marine community structure, algal epifaunas have been surprisingly neglected. On sheltered rocky shores, and in the sublittoral, algae are often dominant and provide an extensive habitat for epifauna. This may itself be dense and, although consisting mainly of sessile species, provide further habitat for sessile and vagile meiofauna in addition to some larger predators. Such faunal density implies that competition for resources, particularly "lebensraum" for the sessile forms, could be severe. This paper examines some of the ways in which such competition may be minimized. The effects of spatial isolation failing and of predation are also discussed.

MATERIALS AND METHODS

Our results come mostly from a study of Fucus serratus at The Dorn, Co. Down (Irish Grid J59368) a locality where several partially dammed shallow basins provide various physically and biologically contrasted sites within a quite small area. Site measurements included weed size, weed abundance, current speed, silt per unit volume of weed frond and "turbulence" (measured as weight loss of gypsum balls attached to selected plants). For fuller accounts of the area, site numbers and location, materials and methods see Boaden et al. (1975, 1976a). The area is part of a National Nature Reserve.

RESULTS

Extent of Habitat

Fucus serratus is most common in the rapids areas at The Dorn. On the main tidal sill (site 2) in May 1976 the smallest (overall length 85.9 cm) of ten F. serratus plants occurring in an area of 0.25 m^2 (i.e. average 1 plant per 250 cm^2 rock surface) had a total surface area of 564 cm^2. This plant alone could therefore be said to have increased the surface area available to epifauna by a factor of 225%. Further details of the density and diversity of its epifauna are given in Table 1.

TABLE 1 Plant parameters, metazoan diversity and density of a Fucus serratus plant, Site 2, The Dorn

Level	Segments	Plant Area (a) cm^2	Weight g	Species	Fauna Weight (f) g	Density (f/a)
Holdfast	1	0.64	0.37	6	0.10	0.16
Y_1*	1	11.65	3.26	16	8.02**	0.69
Y_2	4	39.48	9.74	26	13.17	0.33
Y_3	8	72.14	14.63	22	16.90	0.23
Y_4	14	130.99	22.52	13	6.09	0.05
Y_5	12	155.22	28.83	11	2.27	0.01
Y_6	15	112.80	20.42	1	0.18	0.002
Y_7	9	41.36	5.10	0	0	

* Y_1 = Y shaped segment between holdfast and 2nd dichotomy, Y_2 between 2nd and 4th dichotomy, Y_3 4th to 6th etc.

** Rather high value including a Grantia of 4.35 g.

Isolating Mechanisms

Species which are potentially in competition with each other have generally evolved specific mechanisms through which the severity of their interactions is mollified (Hardin 1960). Such isolating mechanisms can be of an abiotic, biotic or temporal nature. In studying such mechanisms in algal epifaunas it must be remembered that although the fauna is undoubtedly influenced directly by various environmental factors these may also act indirectly via plant response.

Choice of shore site. A comparison of some of the plant fauna from two quite dissimilar sites at The Dorn is given by Table 2. With the exception of Spirorbis, these predominantly filter feeding species were most abundant in the relatively clean fast current area. Dynamena was completely absent from the stiller silty site. When five Fucus plants bearing Dynamena were transferred to this site from site 2 for a six week period in February - March 1975 the height of erect colonies regressed by nearly 50% from 12.2 $\pm$ 0.7 mm. to 6.3 $\pm$ 0.7 mm; the control hydroids showed no such change.

TABLE 2 Fauna of Fucus serratus from two sites at The Dorn

	Fl.	Al.	El.	Gr.	Sy.	Dy.	Go.	Sp.	(no.)
Strong water movement: low silt load (Site 2)	***	**	**	***	**	***	*	*	(119)
Sluggish water movement: high silt load (Site 12)	*	*	*	*	*	O	*	***	(3913)

*** Abundant; ** Common; * Occasional; O Absent

Fl. Flustrellidra; Al. Alcyonidium; El. Electra; Gr. Grantia; Sy. Sycon; Dy. Dynamena; Go. Gonothyraea; Sp. (no.) Number of Spirorbis on one plant

It cannot be inferred from our present data whether the abundance of Spirorbis in the inner Dorn basins actually reflects better recruitment and/or survival, however its occurrence here helps isolate this species from overgrowth and other competition.

Gonothyraea, Flustrellidra, Dynamena and Didemnum share preferences for sites with high turbulence, low silt and large plants, so do Grantia and Sycon except for plant size. Membranipora, Polyclinum and Spirorbis share preferences for low turbulence and high silt; their abundance on smaller plants possibly reflects the correlation of plant size and distribution with physical site conditions (Boaden et al. 1975, 1976a).

Choice of plant species. Boaden et al. (1975) recorded 79 metazoan taxa, excluding meiofauna, from F. serratus at The Dorn, the commonest eleven were all sessile. A similar situation prevails for Ascophyllum but in the high littoral F. spiralis L. and Pelvetia only 6 of a total of 14 species were common and no attached forms were found (Hazlett & Seed 1976). Such differences do not simply reflect the relative exposure time since in this locality the lowest occurring macro-algae also have a very sparse epifauna (Table 3).

TABLE 3 Distribution of common epifauna on various intertidal algae from the Sill, The Dorn

	Fl.	Al.	El.	Ce.	Me.	Cl.	Dy.	Go.	Sp.
F. spiralis	No sessile species recorded								
F. vesiculosus	***	*	**	*	*	O	*	*	***
F. serratus	***	***	***	O	O	O	***	**	**
Ascophyllum	**	**	*	O	O	***	***	O	***
Laminaria	O	O	**	O	**	O	O	O	O

O Absent; * Present or Occasional; ** Common; *** Abundant

Fl. Flustrellidra; Al. Alcyonidium; El. Electra; Ce. Celleporella; Me. Membranipora; Cl. Clava; Dy. Dynamena; Go. Gonothyraea; Sp. Spirorbis

TABLE 4 Spirorbis populations on different substrata at The Dorn

%	Laminaria	F.serratus	F.vesiculosus	Ascophyllum	Rock
S. borealis	91	86	87	14	8
S. pagenstecheri	9	14	13	86	39
S. tridentatus	0	0	0	0	35
S. vitreus	0	0	0	0	18
Total counted	1956	1656	1942	812	1101

In most analyses we have not separated Spirorbis species. However, closely related species could not have become distinct without efficient isolating mechanisms. Choice of, and success on, different substrata is one such isolating device which would increase the success of the species group (Table 4 - from unpublished data by P. Lamont).

Choice within plants. In spite of plant and shore site choice potentially competing species are frequently encountered on the same plant. However, spatial isolation may still be possible via a number of mechanisms. For example, epifaunal species are not randomly distributed along the length of fronds but show zonal preferences (Fig. 1). The oldest and normally most

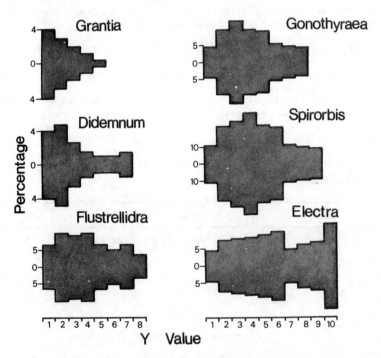

Fig. 1 Distribution of epifauna, as % of available frond-faces colonized, along the longest fronds of Fucus serratus from the Dorn.

heavily incrusted frond regions occur basally (Table 1) since Fucus grows by apical meristems. Accordingly some species such as Electra and Membranipora select younger regions at settlement thereby avoiding intense competition. On Laminaria saccharina (L.) the number of established Membranipora colonies increases distally whilst recently settled stages are predominantly basal. This does not however contradict the Fucus situation since kelps have an intercalary, not apical, meristem and the distal region is oldest.

Most F. serratus fronds are partly folded giving a recognizable concave and convex face. Preliminary analyses suggested a preference by the common species, except Electra, for the concave face. This has been statistically confirmed for Grantia and Sycon, but remarkably the preference of Polyclinum colonies changed according to plant length, (convex on short plants, concave on long plants); Grantia was not only commoner on concave surfaces but achieved greater size there (Boaden et al. 1976a, b).

Competition

Isolating mechanisms such as site choice, plant choice and within plant distribution are by no means infallible. Species therefore come into direct interspecific competition in at least part of their range. Table 5 summarises the major interactions between bryozoans on F. serratus at The Dorn. Since these observations were made in February and March when growth and interactions may be minimal, comparable summer data by Stebbing (1973) from Wembury is included. Apparently Electra is the most susceptible species. The effect of competition on restriction of the inhabitable length of frond has been documented by O'Connor et al. (1975).

TABLE 5   Bryozoan interactions on Fucus serratus at The Dorn

Type of interaction		Concurrent faces studied	Interactions	Type %	Stebbing's (1973) %
(i)	Fl. v El.	96	94	41	70
	El. v Fl.		4	2	27
	El. / Fl. meet		130	57	3
(ii)	Fl. v Al.	74	37	12	6
	Al. v Fl.		19	6	4
	Al. / Fl. meet		255	82	90
(iii)	El. v Al.	40	0	0	5
	Al. v El.		0	0	95
	Al. / El. meet		14	100	0
(iv)	Fl. v Ce.	14			
	Fl. v Me.	10	No interactions		-
	El. v Ce.	23			
	El. v Me.	8			

Fl. v El. indicates Flustrellidra overgrows Electra
Al. Alcyonidium; Ce. Celleporella; Me. Membranipora

Table 6 shows the effects of bryozoan species on hydroids and Spirorbis; of 8370 Spirorbis examined only 1.35% of the 5462 concurrent with Bryozoa were overgrown. Competitive interaction between the tunicates Polyclinum and Didemnum has been illustrated by Boaden et al. (1976). In the absence of mutual competition both species were most frequent and abundant on the basal part of F. serratus but in competition Polyclinum was displaced distally.

TABLE 6 Influence on and overgrowth by bryozoans of Dynamena, Gonothyraea and Spirorbis at The Dorn

	Dynamena	
	Height (mm)	Density (no/cm^2)
Non-encrusted areas	9.80	0.65
Encrusted areas	6.15	0.45
"t" test probability	$p < 0.05$	$p < 0.05$

	Gonothyraea		Dynamena		Spirorbis
	No. faces with both species	% of faces showing some overgrowth	No. faces with both species	% of faces showing some overgrowth	% of bryozoan overgrowth
Fl.	95	17	183	60	60
El.	71	16	100	29	24
Al.	34	0	79	37	9
Ce.	21	0	27	0	7
Me.	0	0	1	0	0

Fl. Flustrellidra; El. Electra: Al. Alcyonidium; Ce. Celleporella; Me. Membranipora

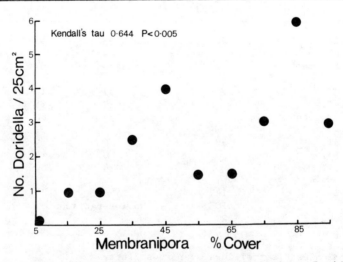

Fig. 2 Correlation between density of the predator Doridella with its prey Membranipora on Laminaria at Friday Harbor.

Predation

The dominant organisms of The Dorn F. serratus community are sessile but the vagile component includes predatory species which may assert a far larger influence than numbers suggest. The plant described in Table 1 bore 3 Doto at the Y2 level and a further 3 specimens each of this and of another hydroid-eating nudibranch Dendronotus at the Y3 level. Some idea of the importance of such predation may be gained from Seed (1976) who calculated that in summer at Friday Harbor the small nudibranch Doridella which occurred at an average density of nearly 1 per 10 sq. cm. on Laminaria fronds could have been cropping over 80% of the total daily production of Membranipora. It is to be expected that the density of such specialized predators will be strongly correlated with prey density (Fig. 2).

DISCUSSION

The major site differences at The Dorn relate to turbulence, silt and weed abundance. We have referred to "choice" and "preference" but epifaunal differences will really be largely determined by larval supply, settlement, growth and mortality. Many larvae carried across the Sill by the tidal currents will be "trapped" by settlement or predation amongst the weed fauna, but many larvae will pass through the weed beds. In stiller areas the supply of larvae will be less but settlement easier. However our Dynamena data and work by Round et al. (1961) show that lack of larval supply is not the only reason for failure of rapids fauna in other habitats.

The extremes of choice leading to settlement on a particular plant are either a entirely opportunistic with settlement on any available surface or b entirely specific to that plant. Strategy a species may have lower larval wastage and more rapid growth but will not be adapted to their host's specific milieu. They are therefore likely to fare worse in competition than b species which will have been able to evolve accurate attunement to the whole biology of the plant. Species with these extremes therefore correspond to the "r and K" selected species of ecological theory (MacArthur & Wilson 1967).

Spirorbids tend to strategy b but S. borealis "is less specific than some of its close relatives" (Knight-Jones et al. 1971) and at The Dorn avoids much competition by shore site choice and settlement on young parts of plants. Membranipora and Electra, the bryozoan which fares worst in competition, both settle on and grow toward young regions (Ryland & Stebbing 1971). Membranipora is most abundant on Laminaria whose relatively sparse frond fauna offers little competition. The widespread distribution of Electra, clearly an a strategy species, enables survival in spite of locally intense competition. It may also lessen predation both because of prey density and because there will be less environmental guides to its location.

It is not known how effectively a larva can search within plants for favourable micro-sites but it is clear that various species flourish in particular parts of plants. This must be partly due to differential survival and growth. Colonies on concave faces may avoid abrasion but be favourably placed for feeding in back eddies. Crowding may have some benefits, for example the production of collectively stronger feeding currents. However, competition often becomes sufficiently intense to restrict niche breadth or to force species into less favourable micro-sites

(O'Connor et al. 1975). This competitive displacement apparently occurs at settlement but may also be brought about by redirection of colony growth.

CONCLUSION

Clearly larval settlement is of particular importance in pre-empting adult competition amongst algal epifauna. However, differential survival and mortality caused by abiotic and biotic factors such as competition and predation still play a dynamic role in determining community structure and diversity. Through studies such as those at The Dorn we are perhaps just beginning to understand the ways in which sessile epifaunal species may attain (and be kept in) their proper place in the community.

REFERENCES

Boaden, P.J.S., O'Connor, R.J. & Seed, R. 1975 The composition and zonation of a Fucus serratus community in Strangford Lough, Co. Down. J. exp. mar. Biol. Ecol. 17, 111 - 136.

- 1976a. The fauna of a Fucus serratus L. community: ecological isolation in sponges and tunicates. J. exp. mar. Biol. Ecol. 21, 249 - 267.

- 1976b. Some observations on the size of the sponges Grantia compressa (Fabr.) and Sycon ciliatum (Fabr.) in Strangford Lough, Co. Down. Proc. R. Ir. Acad. in press.

Hardin, G. 1960 The competitive exclusion principle. Science, N.Y. 131, 1292 - 1297.

Hazlett, A. & Seed, R. 1976 A study of Fucus spiralis and its associated fauna in Strangford Lough, County Down. Proc. R. Ir. Acad. in press.

Knight-Jones, E.W., Bailey, J.H. & Issac, M.J. 1971. Choice of algae by larvae of Spirorbis, particularly of Spirorbis spirorbis. In Fourth European Marine Biology Symposium (ed. D.J. Crisp) 89 - 104, Cambridge University Press, London.

MacArthur, R., Wilson, E.O. 1967 The Theory of Island Biogeography. Princeton University Press, Princeton, N.J.

O'Connor, R.J., Boaden, P.J.S. & Seed, R. 1975 Niche breadth in Bryozoa as a test of competition theory. Nature, Lond. 256, 307 - 309.

Round, F.E., Sloane, J.F., Ebling, F.J. & Kitching, J.A. 1961. The ecology of Lough Ine. X. The hydroid Sertularia operculata (L.) and its associated flora and fauna: effects of transference to sheltered water. J. Ecol. 49, 617 - 629.

Ryland, J.S. & Stebbing, A.R.D. 1971 Settlement and oriented growth in epiphytic and epizoic bryozoans. In Fourth European Marine Biology Symposium (ed. D.J. Crisp) 105 - 123. Cambridge University Press, London.

Seed, R. 1976 Observations on the ecology of Membranipora (Bryozoa) and a major predator Doridella steinbergae (Nudibranchiata) along the fronds of Laminaria saccharina at Friday Harbor, Washington. J. exp. mar. Biol. Ecol. in press.

Stebbing, A.R.D. 1973 Competition for space between the epiphytes of Fucus serratus L. J. mar. biol. Ass. U.K. 53, 247 - 261.

THE HERMIT CRAB MICROBIOCOENOSIS - THE ROLE OF MOBILE SECONDARY HARD BOTTOM ELEMENTS IN A NORTH ADRIATIC BENTHIC COMMUNITY*

Michael Stachowitsch

I. Zoologisches Institut, Universität Wien, Lehrkanzel für Meeresbiologie, A-1090 Wien, Währingerstrasse 17/VI, Austria

ABSTRACT

A soft bottom benthic community at a depth of 23 m in the North Adriatic Sea (Gulf of Triest) was found to contain a high density of hermit crabs (1.88 individuals/m^2). This population is composed mainly of the species *Paguristes oculatus* which inhabits the shells of the gastropod species *Aporrhais pes-pelecani*, *Murex brandaris* and *M.trunculus*. Over 50 species of epifaunal as well as boring invertebrates were found to be associated with the hermit crab occupied shells. Both species of *Murex* as well as *Aporrhais pes-pelecani* were frequently found at the base of established multi-species clumps which are characteristic for this bottom and account for the large epifaunal biomass (370 g/m^2). The hermit crab microcosm is compared to these clumps in an attempt to determine whether the gastropod shell of the hermit crab forms the basis of the observed sedentary associations.
In situ experiments as well as diver taken samples show that as opposed to living gastropods, empty shells and artificial substrates, hermit crab occupancy affords a unique opportunity for the development of a fauna and flora which is capable of survival and further growth once deposited by the hermit crab.

INTRODUCTION

The Gulf of Triest has been investigated by a number of authors (Vatova 1949, Orel and Mennea 1969, Gamulin-Brida 1974) and been shown to consist of several communities, some with a well developed epifauna. One such community - the *Ophiothrix-Reniera-Microcosmus* community - with an epifaunal biomass average of 370 (±73) g/m^2 measured as wet weight is described by Fedra et al. (in press).

---

*This investigation was carried out within the frame of a joint program between the Lehrkanzel für Meeresbiologie, University of Vienna, and the Marine Biological Station Portoroz, University of Ljubljana.

The sessile as well as the motile epifaunal species of this benthic community are found in the form of multi-species clumps. This pattern results in the contagious or aggregated distribution of the biomass. It is therefore of interest to examine the origin, growth, and maintenance of the multi-species clumps which play such an important role in the distribution and high level of the biomass.

Investigations of the North Adriatic Sea have shown *Paguristes oculatus* (L.) to be constantly present and linked to detrital elements in the sediment (Picard 1965). The gastropod shells inhabited by hermit crabs in the Gulf of Triest were observed to provide a base for a large variety of benthic species. In addition, a similarity in structure and composition between the epizoics of the hermit crab and established sedentary associations was observed. Although hermit crabs have always been linked to isolated sessile invertebrates, the association of *Paguristes oculatus* to a large variety of the fauna and flora of the *Ophiothrix-Reniera-Microcosmus* community merits a more thorough investigation.

MATERIAL AND METHODS

144 $1m^2$ epifauna samples were taken by SCUBA divers off the coast of SR Slovenia in the Gulf of Triest at an average depth of 23 m. The samples were taken between 1973 and 1975 at three positions in the center of the *Ophiothrix-Reniera-Microcosmus* community (abbreviated ORM). The density and distribution of the hermit crabs were taken from this data. Numbers in parentheses indicate 95% confidence levels of the mean.

In order to examine the role of the hermit crab on the epizoics of the gastropod shell it inhabits, as well as to detect possible migrations, 190 shells of the three most commonly inhabited gastropod species - *Aporrhais pes-pelecani* (L.), *Murex brandaris* (L.), and *M. trunculus* (L.) - were tagged between August and October of 1975. A hole was drilled into the shell, and numbered polypropylene tabs were attached by means of fishing line. The average length of the fishing line was 15 cm in order to allow a diver to more easily relocate the shells. The shells were marked so that neither the knot, the line, nor the tag interfered with the activity of the crab. The marked shells were divided into three main categories:
1. Shells which were occupied by hermit crabs, some with and some without epizoics - 56 in number.
2. Clean empty shells which were taken from gastropods - 62 in number.
3. Clean shells which were taken from gastropods and the openings sealed with pebbles and plaster - 66 in number.

The clean shells were obtained by boiling gastropods in water, removing the soft parts with tweezers, and boiling the shells again in a NaOH solution to remove all organic matter.
In order to examine the fate of the epizoics upon removal of the hermit crab, a fourth group consisting of 6 shells was tagged. In this category, hermit crabs were removed from shells containing a large amount of epizoics. The openings of the shells were then

sealed with pebbles and plaster. All four groups of shells were set out at an underwater experimental site constructed approximately 2 km off the coast of SR Slovenia (position 39 in Fig. 1 of Fedra et al., in press).

77 multi-species clumps were dredged from the center of the ORM community in an attempt to determine the number of bases of these associations that were represented by gastropod shells previously inhabited by hermit crabs. These samples were examined solely for the dominant invertebrate representative and the type of particle that served as the original base. In addition, 46 $0.25m^2$ epifauna samples were gathered using SCUBA in order to obtain undamaged material as well as to gain a more quantitative picture of the size and distribution of the multi-species clumps. A multi-species clump is defined as a sedentary association containing more than one sessile invertebrate species and usually having a mollusc shell or combination of shell particles serving as a substrate.

The above experiments were accompanied by the observation of approximately 150 000 m^2 (13 transects totalling approx. 80 km) in the Gulf of Triest by means of a combined TV- and photocamera sled (Fedra et al., in press).

RESULTS

144 $1m^2$ SCUBA diver taken macro-epifauna samples showed a hermit crab density of $1.88(\pm 0.27)$ individuals/m^2, with a range of 0-8 individuals/m^2 ($s^2 > \bar{x}$). 87% of the hermit crab population was found to consist of the species *Paguristes oculatus* - $1.63(\pm 0.26)$ individuals/m^2, with the remaining 13% being largely represented by *Eupagurus cuanensis* (Thomp.). Agreement with the Poisson series is rejected at the 99% level (d=3.07) strongly indicating that the dispersion of the hermit crab population is contagious.

Observations from the transects covered with the TV camera sled and the examination of the video tapes and photos taken by the cameras on the sled, as well as diver observations over a period of two years confirmed the fact that large numbers of hermit crabs are present throughout the area under investigation, both in and outside the ORM community borders. Smaller hermit crabs (small *Paguristes oculatus* individuals as well as most *Eupagurus cuanensis*) were found to occupy the shells of *Aporrhais pespelecani*, while larger crabs were to be found almost exclusively in the shells of *Murex brandaris* and *M. trunculus*.

The gastropod shells inhabited by hermit crabs were found to provide a base for a large variety of invertebrate species found in the ORM community (Fig. 1). Of 37 species contributing at least 0.1% to the total epifaunal biomass of the above community (Table II, Fedra et al.), 22 (59%) were found to be associated with the housings of *Paguristes oculatus* and *Eupagurus cuanensis*. In addition, 42 species, or 54% of the total taxa identified (Table III, Fedra et al.) were found on hermit crab shells. This includes representatives of the biomass dominants *Ophiothrix quinquemaculata* (D.Ch.), *Reniera* ssp., and *Microcosmus* ssp., which account for 64% of the biomass of this benthic community, as well as representatives of the majority of the filter and suspension feeding organisms which make up 86% of the community biomass. Noticeably absent from the shells of hermit crabs are motile carnivorous and deposit feeding species such as gastropods, ophiuroids, and crustaceans. Not included in the list of identified taxa but nevertheless of importance in the hermit crab microbiocoenosis are the species *Cliona* ssp., *Fosliella farinosa* (Lam.), *Salmacina* sp., several species of bryozoans and polychaetes, as well as a number of algae. In most cases the species of crab seems to have little effect on the type of epifauna on the shell. No obvious differences could be observed in the epifauna of the species *Pagurus oculatus* and *Eupagurus cuanensis*. An exception seems to be *Eupagurus prideauxi* (Leach), whose shell is exclusively covered by *Adamsia palliata* (Bohadsch).

Fig. 1 Hermit crab shell with epifauna. (Left to right - *Epizoanthus aranaceus* (D.Ch.), two species of sponges, *Ophiothrix quinquemaculata*, and *Calliactis parasitica* (Couch). Eye and antennae of crab visible at bottom center.

The tagged gastropod shells were examined after a period of 10 to 11 months after the set-out date (August,September, 1975 - July, 1976). All empty shells introduced into the community were occupied by hermit crabs within two days. Of the 190 shells that were tagged, 105 (55%) were found within a 30 m radius of the experimental field in which they were set out (Table 1). This high percent of shells found indicates that there are no major migrations of the hermit crab population to different areas of the community. In addition, the species of shells found from these three major categories were in approximately the same proportion as those set out. This indicates that both large and small crabs behave similarly in that they remain in a defined area for extended periods of time.

The category in which the most shells were found was from the 66 shells with the plastered openings (52=80%). An unexpected result of this experiment, however, was the fact that 28 of the shells found from this category were inhabited by hermit crabs (52%). The plaster was evidently not strong enough to withstand the effects of the salt water and the attention of the hermit crabs. The 6 plastered shells with the large epifaunal growth were also occupied by hermit crabs within a period of two weeks. The shells with plastered openings proved to be of great interest to the hermit crabs and were continually examined and overturned.

TABLE 1 Shell Tagging Experiments

Shell category	Tagged	Aporrhais pes-pelecani	Murex brandaris	Murex trunculus	Calliostoma sp.
With epizoics	56	39	15	2	2
Shells found	19 (30%)	12	5	0	2
Clean free	63	37	11	15	0
Shells found	32 (52%)	21	4	7	0
Clean plastered	66	43	10	13	0
Shells found	52 (79%)	34	7	11	0
With crab	28	19	5	4	0
With plaster	24	15	2	7	0
Plastered shells with epizoics	6	4	1	1	0
Shells found	2 (33%)	1	0	1	0

The majority of the plastered shells that had not been taken by hermit crabs were buried in the sediment with only the plastic tabs and a portion of the fishing line showing above the sediment surface. These shells were collected and the depth to which they had sunk was measured. Several shells were grey and black in color indicating that they lay in the reduced layer of the sediment. The depth to which the shells had sunk ranged from 0.8 - 7.5 cm, with an average depth of 3.6 cm. Only two shells, that of an *Aporrhais pes-pelecani* and a half buried *Murex trunculus* showed an epifaunal growth. This epifauna consisted of the sponge *Reniera* sp. (2.5 cm in length) and a small specimen of the anthozoan *Aiptasia* sp. (1.5 cm in length in the opening of the shell) on *Aporrhais pes-pelecani*, and the sponge *Reniera* sp. with a length of 5 cm on *Murex trunculus*. Examination of the remaining shells showed isolated Foraminifera and small patches of bryozoans and calcareous algae. Despite the interest shown by the hermit crabs, the shells were covered with sediment making it impossible for epifauna to attach itself.

The majority of the clean shells, both those which were introduced as clean free shells as well as clean plastered shells which were subsequently inhabited by hermit crabs, showed a relatively small epifaunal growth after a period of 10 months. This experiment is therefore being continued, and only three shells from the remaining three categories were collected. These three tagged *Paguristes oculatus* occupied shells, 1 *Murex trunculus* and 2 *Aporrhais pes-pelecani*, were from the clean empty shell and the clean plastered shell category, respectively. The epizoics found, although small in size, totalled over 20 species and included 2 species of Foraminifera, 4 species of bryozoans, *Pomatoceras triqueter* (L.), *Salmacina* sp., *Chlamys varius* (L.), *Anomia ephippium* (L.), *Fosliella farinosa*, a colonial ascidian (family Didemnidae), as well as several species of algae. Several shells showed a larger epifauna. For example, a sponge of the genus *Reniera* with a length of 7 cm was found on one *Murex trunculus* shell inhabited by *Paguristes oculatus*. The sponge was attached at two points, forming an arc over most of the length of the shell. This *Murex* was again examined after a period of three weeks. Due to unknown causes the middle part of the sponge had broken off, resulting in two seperate colonies on both ends of the shell.

77 multi-species clumps obtained by dredging showed 23% to consist mainly of sponges, 47% mainly of ascidians, 4% of a combination of sponges and ascidians, and the remaining 26% represented largely by bivalves such as *Arca noae* (L.), and *Gastrochaena dubia* (Pennant). The 46 0.25m^2 diver taken samples, containing 73 multi-species clumps, showed 48% of the clumps to consist mainly of sponges, 21% mainly of ascidians, 5% of a combination of sponges and ascidians, with the remaining 24% largely being represented by *Arca noae* and *Gastrochaena dubia*. The type of substratum providing the base for the multi-species clumps from both the dredge and the diver taken samples is summarized in Table 2.

TABLE 2 Substratum of Multi-Species Clumps

Number of multi-species clumps		Shell particles	Former hermit crab shells	Stone	No base
Dredge	77	86%	8%	6%	X
SCUBA	73	81%	7%	5%	7%

The category "Shell particles" in Table 2 includes isolated valves and living specimens of all bivalves - for the most part *Arca noae*, *Modiolus barbatus* (L.), *Chlamys varius*, and *Pecten jacobaeus* (L.). It also includes all shells as well as shell fragments of *Aporrhais pes-pelecani*, *Murex brandaris*, and *M. trunculus* that could not with certainty be identified as being previously inhabited by a hermit crab. In many cases more than one shell particle was found at the base of the multi-species clumps.

The category "Former hermit crab shells" includes the empty *Aporrhais pes-pelecani*, *Murex brandaris*, and *M. trunculus* shells at the base of a sedentary association where previous hermit crab occupancy could be determined. In most cases this was possible. First, the majority of the hermit crab occupied shells show a characteristic aggregation of sedentary polychaetes around the opening of the shell. The presence of calcareous tubes in the openings of gastropod shells which form the base of established multi-species clumps is therefore a strong indication of previous hermit crab occupancy. Second, the epifauna of the hermit crab's shell must for the most part grow on the top side of the shell, that is, the side opposite the opening. Gastropod shells lying randomly on the sediment would be expected to show an epifauna with its point of attachment at all possible points on the shell. Multi-species clumps with their point of attachment on the bottom side of a gastropod shell were therefore not placed in this category. Third, certain species, such as the sponge *Cliona* sp., which could not be found on living gastropods or those buried in the sediment, was often found to be associated with the housings of hermit crabs. The presence of *Cliona* marks on the gastropod bases of multi-species clumps therefore indicates a former hermit crab inhabitant. The fulfillment of two of these three criteria was required before the gastropod shell was placed in this category.

The multi-species clumps obtained by the dredge that had no base were discarded. The dredge damaged a portion of the material brought up. This method is likely to rip the multi-species clump from its substrate base if the point of attachment is weak or the substrate firmly anchored in the sediment. In many cases it was not possible to distinguish between the multi-species clumps that in fact had no base and those in which the substrate base had been torn off.

DISCUSSION

Picard (1965) states that *Paguristes oculatus* is a species characteristic of more or less detrital muddy bottoms and Gamulin-Brida (1974) classifies *Paguristes oculatus* as "une espèce à large repartition écologique" and reports its regular occurence throughout the "Biocoenose Complexe des Fonds Sablo-Détritiques Plus ou Moins Envasés" (DC-E). Vatova (1949), using the Peterson grab, found no *Paguristes oculatus* in the Gulf of Triest, but estimates the density of this crab to be 1 individual/10m^2 in the Mid- and North Adriatic Sea, ranging from depths of 0 - 100 m.

The hermit crab occupied shell was found to provide a satisfactory base for a large number of sedentary invertebrates of the community. It is well known that a number of invertebrates are associated with hermit crabs. The majority of these invertebrates are isolated coelenterates, but Jenson and Bender (1973), for example, regard the snail shell inhabited by *Pagurus bernhardus* (L.) to be a mobile substratum for a specific epifauna community, with over 30 species being mentioned. *Paguristes oculatus* and *Eupagurus cuanensis* may have several advantages from such a large epifauna. It has been shown that several Octobrachia learn not to attack hermit crabs with certain coelenterates, while crabs without these coelenterates are immediately attacked and eaten (Boycott 1954, Ross 1970). The epifauna of the shells of the hermit crabs, especially *Calliactis parasitica* and *Epizoanthus arenaceus*, may offer a certain degree of protection against *Sepia officinalis* (L.) and *Loligo* sp. which are present in this area. In addition it has been shown that the hermit crab *Clibanarius vittatus* (Bosc) has a larger chance of winning an encounter with a conspecific if its shell is made to appear larger (what the rival sees) or when the shell is made heavier (what the owner feels), (Hazlett 1968). It is possible that the epifauna on the gastropod shell of *Paguristes oculatus* has a similar function. Such a mechanism could play an important role in the dense hermit crab population of this benthic community. The contagious distribution of the hermit crabs is most easily explained by the multitude of agonistic encounters that must occur in such a population, especially in aggregations around a commom food source.

Perhaps even more significant are the advantages that the epifauna and -flora obtain from the hermit crab. First, the hermit crab, through its constant movement as well as the movement of its mouth parts, provides a constant source of food particles which is independent of the local water movements. Second, the shell provides a sediment free surface for settling larvae. The plastered shell experiment as well as preliminary results from horizontally placed artificial substrates indicate that the sedimentation rate may play an extremely important role in the settlement and development of sessile organisms in this community. The top side of the artificial substrates very rapidly becomes sedimented over, making it difficult for the larvae of epifaunal species to settle. After a period of approximately one year the top will be covered by
over 1 cm of sediment and will be devoid of life. The edges and bottom of the plates, however, which are sediment free, show an epifauna with a large biomass. The shell of the hermit crab allows a direct reversal of this situation. Thirdly, the shell provides a surface that is elevated above the sediment and is predator free. The hermit crab, being a vagile carnivore as well as a scavenger, is able to protect itself and avoid contact with the majority of the predators in the community. This benthic community contains an average of 103 specimens of *Ophiothrix quinquemaculata* per m^2 and a large range of mobile consumers such as *Psammechinus microtuberculatus* (Blainville), *Ophiura lacertosa* (Pennant), and *Astropecten* ssp.. The sediment is therefore constantly being examined, making it difficult not only for planktonic larvae to settle, but also to survive the first critical period after settling. An adaptive mechanism such as the settling of larvae during the inactive phase in the feeding of breeding Ophiuroids as described by Thorson (1957) for Danish waters is as yet unknown for the North Adriatic. The shell of the hermit crab provides an ideal substrate for larvae which are ready to settle.

The epifauna on the shell of the hermit crab will develop and become larger. The final structure and composition of this epifauna may depend on the species which are first to settle, the microclimatic conditions and surface texture differences of the gastropod shell, as well as the spatial competition that results from the individual growth rates and growth patterns of the various epizoics Few epizoics can increase the volume of the shell. The crab must therefore change its shell at certain intervals as it grows larger. Undamaged shells are very seldom found on the sediment surface. The four categories of tagged shells which were introduced into the community, including those with plastered openings, were intensively examined and whenever possible occupied by hermit crabs within a very short period of time. It is therefore likely that a newly vacated undamaged shell will be reinhabited by another crab. The epifauna of such a shell may therefore grow through several changes in ownership.

Size alone does not seem to be the sole cause of final rejection of the shell by the hermit crab. The epifauna of the shell is often comprised of large ascidians and sponges that would seem to hinder the movements of the crab. Specimens of *Suberites domuncula* (Olivi) over 15 cm in diameter may occasionally be found rolled over and the opening for the hermit crab may face the surface of the water rendering the crab immobile.

Two simultaneous processes can be observed on the shell of the hermit crab - a constructive and a destructive one. The constructive one involves the growth of the above mentioned epizoics. The destructive process involves the weakening of the shell structure by the sponge *Cliona* sp., the bivalve *Gastrochaena dubia*, and possibly several polychaetes. Hermit crab shells are often severely scarred and laced with tunnels. This process can reach such a stage that a layer of calcareous algae or a colony of bryozoans may significantly strengthen the shell structure. It is this destructive process, resulting in holes or the breaking off of certain parts of the shell, which more likely causes the hermit crab to reject and deposit the shell and its epifauna as a multi-species assemblage.

A multi-species clump base consisting of more than one shell particle is the result of the ability of several sponges and ascidians (notably *Reniera* ssp. and *Microcosmus* ssp.) to attach themselves to particles they come in contact with during their growth. The most frequently found bivalve bases - *Arca noae*, *Modiolus barbatus*, and *Chlamys varius* - are not infauna species. In the living state they are attached to secondary hard bottom elements, hermit crab shells, and multi-species clumps. It is therefore likely that the epizoics attached to these bivalves settled while the bivalves were alive and elevated above the sediment - not as isolated valves lying on the sediment. The shells of living specimens of *Murex brandaris* and *M. trunculus* are associated with a very limited number of epizoics such as small patches of bryozoans, calcareous algae, and more rarely *Calliactis parasitica*. Living *Aporrhais pes-pelecani* are for the most part found in the sediment and show an even more limited epizoic growth. It is therefore unlikely that living gastropods form the base of the observed sedentary associations.

The percentages of the substrate type of the dredge and SCUBA taken multi-species clumps are in good agreement with each other. The percent of multi-species aggregations which contained a previously hermit crab inhabited *Murex* or *Aporrhais pes-pelecani* as its base was shown to lie between 7% and 8%. While this value points to the fact that hermit crabs play a role in the establishment of multi-species clumps, it may underestimate this role due to the properties of the dredge previously mentioned. In addition, the dredge is bound to be selective in the kind of multi-species clumps that it brings up, missing smaller hard bottom elements and breaking up the more delicate aggregations such as those with large sponges. This accounts for the low percentage of multi-species clumps represented by sponges using this method. The SCUBA diver picked samples, while lacking the above disadvantages, are quite time consuming and did not provide a sufficient amount of material. From the 46 $0.25m^2$ samples, 6 were null counts and many others contained only very small shell fragments with a minimum of epizoics. Not included in either set of samples were the numerous hermit crabs with large epifaunal growth - that are essentially mobile multi-species clumps.

A part of the significance of the hermit crab microbiocoenosis lies in its function as a more or less discrete and independent ecological unit. Without a doubt a large portion of the epifauna can complete its life cycle on the gastropod shell carried by the hermit crab. The favorable growth conditions on the shell, as well as the deposition of the shell at a certain stage, points to an important mechanism in the origin of established multi-species clumps and therefore in the structure of this benthic community.

ACKNOWLEDGEMENTS

This study was supported by the projects No. 1940, 2084, and 2758 of the "Fonds zur Förderung der wissenschaftlichen Forschung, Österreich" with additional support from the "Boris Kidrič Foundation". The author's thanks are due to Yugoslavian authorities for research permits and their hospitality, and to director and staff of the Marine Biological Station, Portoroz, for providing essential facilities. The author feels indepted to his colleagues K. Fedra, E.M. Ölscher, C.Scherübel, and R.S. Wurzian for their essential help during the field work and sampling program.

REFERENCES

Boycott, B.B. 1954 Learning in *Octopus vulgaris* and other cephalopods Pubbl. Staz. Zool. Napoli 25, 67.

Fedra, K., E.M. Ölscher, C. Scherübel, M. Stachowitsch, and R.S. Wurzian On the ecology of a North Adriatic benthic community: Distribution, standing crop, and composition of the macrobenthos Mar. Biol. (in press)

Gamulin-Brida, H. 1974 Biocoenoses benthiques de la Mer Adriatique Acta adriat. 15(9), 1-102.

Jenson, K. and K. Bender 1973 Invertebrates associated with the snail shells inhabited by *Pagurus bernhardus* (L.) (Decapoda) Ophelia 10, 185-192.

Hazlett, B.A. 1970 The effect of shell size and weight on the agonistic behaviour of a hermit crab Z.f. Tierpsychol. 27, 369-374.

Orel, G. e B. Mennea 1969 I popolamenti bentonici di alcuni tipi di fondo mobile del Golfo di Trieste Pubbl. Staz. Zool. Napoli 37, (2 suppl.), 261-276.

Picard, J. 1965 Recherches qualitatives sur les biocoenoses marines des substrates meubles dragables de la region marseillaise Rec. Trav. Stat. Mar. Endoume 52(36), 1-160.

Ross, D.M. 1971 Protection of hermit crabs (*Dardanus* spp.) from Octopus by commensal sea anemones (*Calliactis* spp.) Nature 230, 401-402.

Thorson, G. 1957 Bottom communities (Sublittoral or shallow shelf) Geol. Soc. Am. Mem. 67(1), 461-534.

Vatova, A. 1949 La fauna bentonica dell'Alto e Medio Adriatico Nova Thalassia 1(3), 1-110.

SUB-LITTORAL COMMUNITY STRUCTURE OF OXWICH BAY, SOUTH WALES IN RELATION TO SEDIMENTOLOGICAL, PHYSICAL OCEANOGRAPHIC AND BIOLOGICAL PARAMETERS

Paul Tyler

Department of Oceanography, University College, Singleton Park, Swansea SA2 8PP, U.K.

ABSTRACT

The sublittoral benthic community structure in Oxwich Bay (South Wales, Gower coast) is more closely related to the hydrodynamic regime of the bay than to the nature of the sediment-types which form the seabed. The sediments are both glacial and post-glacial in origin, and are undergoing redistribution and sorting according to the magnitude and direction of near-bed current-velocities and wave-orbital velocities to which they are exposed. Many seabed sediments are therefore not in equilibrium with the present hydrodynamic regime. However, the distribution of meroplanktonic larvae and newly metamorphosed adults is directly influenced by the hydrodynamic environment: there is an anticlockwise tidal gyre, during both flood and ebb, and the area of its centre is protected from the prevailing southwesterly, long-wavelength swell by refraction around Oxwich Point. Centripetal motion transports larvae to the centre of the gyre, where relatively low current velocities and low eddy viscosity values permit settlement. After settlement, the adult distribution pattern depends on the degree of migratory activity. Burrowing species (e.g. Syndosmya (Abra) alba, Cultellus pellucidus) have a distribution closely related to the hydrodynamic environment, but this relationship becomes obscured in errant species (e.g. Ophiura albida) by migration after settlement has occurred. Errant species within a community aid its development by feeding on the larvae of possible predators of the dominant species of that community.

INTRODUCTION

Marine benthic communities were first described by Petersen (1913) and have since been classified according to biocenoses (Petersen 1913; Sparck 1937; Thorson 1957) or biotypes (Jones 1950). The effect of sediment distribution on the benthic community structure have been discussed by numerous authors (Buchanan 1963; Wade 1972; Parker 1975; Lie 1968) but the effect of hydrodynamic factors including wave and tidal current action on sub-littoral marine faunas has rarely been studied in detail (Jones 1951; Holme 1953; Rasmussen 1973; Gage 1972). This paper is the result of a preliminary survey of the hydrodynamic effects on sub-littoral benthic community structure in Oxwich Bay, South Wales, and is continuing as a spatial and seasonal evaluation of the communities in this bay.

METHODS.

Benthic faunal samples were collected sublittorally at 39 stations in Oxwich Bay by Shipek grab, bucket dredge and Agassiz trawl, whilst at a further 15 stations single Shipek grabs for sediment only were taken (Fig. 1). The retrieved faunal samples were sieved through a 2mm. brass sieve and the macrofauna was fixed in 5% seawater formalin. Sediment samples were oven dried at 110°C before being sieved for particle size analysis, and for quantitative determination of carbonate content and oxidizable organic carbon content (Gaudette et al. 1974). For the sand-sized fraction, the parameters of $\bar{\emptyset}$ mean, sorting coefficient and skewness were calculated (according to Folk, 1961); for this study the silt and clay fractions were combined (Buchanan 1964, Fish 1967). For the calculation of wave refraction patterns the method given by Arthur et al. (1952) was used. Position fixing was carried out using the Decca Mk.21 Navigator (error ± 25m).

RESULTS

HYDROGRAPHY OF OXWICH BAY.

The circulation and wave refraction patterns for Oxwich Bay (Fig. 2) have already been described (Tyler and Banner 1976). It was suggested that there is an anticlockwise tidal gyre within the Bay, and that ebb flow, round Oxwich Point, had a greater duration than the corresponding flood. However, the flood gyre is more strongly developed than the ebb gyre, because of the

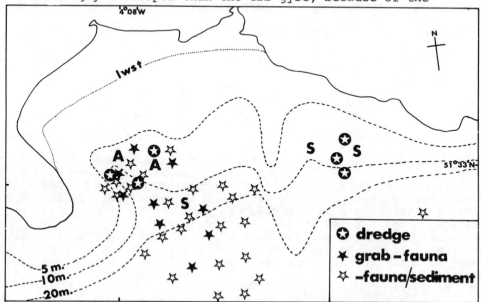

Fig.1. Oxwich Bay, South Wales. Bathymetry and Benthic sampling stations. A = Syndosmya community. S = Spisula community.

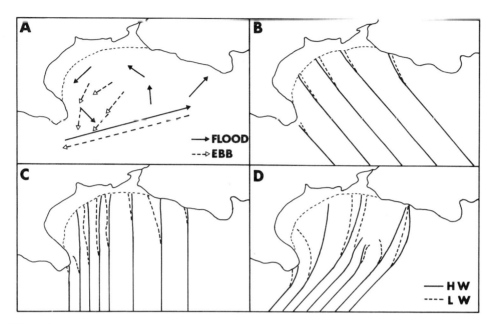

Fig. 2. Hydrography of Oxwich Bay. A. Circulation pattern. B,C,D. Wave orthogonals for B, Southeasterly swell (5 sec. period) C, Southerly swell (5 sec. period) D, Southwesterly swell (12 sec. period).

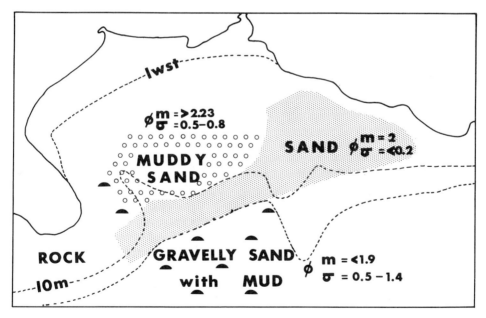

Fig. 3. Sediment distribution in Oxwich Bay to show mean grain size and sorting coefficient.

presence of a northerly directed offshoot from the WSW-ENE rectilinear reversing current immediately to the south of the gyre (Fig 2a). Maximum current velocities occur round Oxwich Point; those within the Bay are lower, due to the presence of the gyre and to loss of energy by frictional dissipation. Tidal current energies are lowest within the centre of the gyre compared to both the perimeter currents and the rectilinear reversing currents.

Wave action within the bay varies according to the strength and duration of winds from the SE, S and SW (Fig.2b, c, d). Northern quadrant winds generate little or no effective wave action within the bay. Southerly and south-easterly wind-generated waves have relatively short fetches (the maximum fetch of 50km. would give rise to fully developed waves of a maximum 5 sec. period according to Neumann (1954)). Waves of this period from the south and south-east undergo very little wave refraction until the 15m isobath is reached (Fig.2b,c). Even in more shoal water the refraction is only slight, resulting in the energy of wavetrains from the south east being evenly expended all over the bay.

However, the southwesterly winds not only account for 50% of the winds in this area but also the southwesterly fetch is up to 7000km; the generation of southwesterly wave trains can result in wave periods of 8 to 12 secs. These waves, on entering Oxwich Bay, are refracted northward around Oxwich Point giving rise to an area of low energy concentration in the west central part of the bay (Fig.2d). The eastern half of the Bay is much less affected by such wave refraction and is exposed to relatively undispersed energy of the southwesterly swell.

SEDIMENT DISTRIBUTION

Particle size analyses of sediment samples from Oxwich Bay suggest the presence of three distinct sediment populations (Fig. 3)(Tyler and Banner 1976). Across the mouth of the bay, between the 10 and 20m isobaths, is a zone of sand-sized sediment (between $\emptyset$ 0 and 3.5 (Folk 1954)). This is very well sorted sand ($\emptyset_\sigma$= 0.157 to 0.209, $\emptyset_m$ = 2.02) with a low organic matter content. The distribution of this sand zone is determined by both tidal currents and wave action, and it expands near Sir Christophers Knoll to cover most of the eastern half of the Bay. To the south of the sand zone, in water of depth greater than 20m, there is a population of coarser sediment. However, the concentration of fine material (>$\emptyset$ 3.5) and organic carbon is erratically distributed in this area. North of the sand zone, in the western central part of the Bay, is an area of muddy sand with a mean $\emptyset$ distinctly higher than elsewhere in the Bay. Concentrations of oxidizable organic matter are also high in this area. The origin and distribution of this sediment population is a function of low tidal velocities (as it lies towards the centre of the tidal gyre) and of the refraction pattern of the dominating south-westerly waves and swell.

FAUNAL DISTRIBUTION

Two distinct faunal populations appear to occur in the sublittoral of Oxwich Bay (Fig. 1). One is an 'impoverished' sandy fauna associated with the sand zone and the eastern half of the bay. This community is dominated by Spisula elliptica with occasional Nephtys hombergi and Lanice conchilega.

The second faunal population is found within the tidal gyre of the bay and to some extent in water deeper than 20m. This population would appear to be a typical rich Syndosmya community. The numerical dominants are Nucula turgida, Syndosmya alba, Owenia fusiformis, Sabella pavonina and Ampelisca spinipes with localized increases in Acrocnida brachiata, Lanice conchilega, Nephtys hombergi, Pectinaria koreni and Cultellus pellucidus. Other common species found include Venus striatula, Philine denticulata and Nereis diversicolor. Detailed quantitative evaluation of this community is under study at present.

DISCUSSION

In the Bristol Channel, IMER (1975) have referred the faunas of the muddy sand areas at the entrances to the larger bays to the Syndosmya community. This classification could include Oxwich Bay, where features typical of the Syndosmya community are found. Environmentally, these include restriction of the area to between 5 and 14m isobaths, 'soft bottoms' (Sparck 1937), and sheltered areas with mixed to muddy sediments often rich in organic matter (Thorson 1957). The characteristics of species of the Syndosmya community are rapid growth, early maturity and a short life due to predation and natural death (Thorson 1957). Increased amounts of sand lead to a Venus community (Thorson 1957), characteristically represented in Oxwich by Spisula. Features normally associated with a Spisula-dominated community include its occurrence between the 8 and 40m isobaths, on loose sand with a marked annual fluctuation in the Spisula population (Thorson 1957).

Within Oxwich Bay the distribution of the two communities can be related to the hydrodynamic regime rather than to bottom sediment type. The poor correlation which exists between sediment type alone and the faunal distribution has also been noted off the Northumberland coast (Buchanan 1963) and off the Isle of Man (Jones 1951). The obscurity of the relationship between faunal distribution and sediment type may be due to the fact that much of the seabed sediment of NW European shelf seas are relict Pleistocene and Flandrian deposits, having a grain size distribution unrelated to the present hydrodynamic regime. These sediments are still being reworked by wave and tidal-current energies. These relict sediments provide a substrate for faunal populations which would not reflect the hydrodynamic energies to which these organisms are exposed. Thus, if the hydrodynamic environment determines the settlement of metamorphosed larval forms of a population, then no direct correlation between species distribution and sediment type of the substrate can be made. However, if the sediment is modern and mobile then the grain size would reflect the present day hydrodynamic environment, and it is this which will control the transport and settlement of both sedimentary particles and newly metamorphosed meroplankton.

In Oxwich Bay, some hydrodynamic control over sediment distribution has been determined (Tyler and Banner 1976). In the area of hydrodynamic high energy, with the rectilinear reversing tidal-current between the 10 and 20m isobaths, and in the area of high wave energy on the east side of the bay (indicated by only slightly divergent orthogonals in Fig. 2d), the sediment is well sorted sand. This sand contains only minor amounts of fine material and organic matter and has a typical Spisula community. The constitution of the Spisula community may reflect the higher hydrodynamic energy values of this area by selective winnowing of settling organisms and even by the physical destruction of larvae entering this environment and of the more delicate species that do manage to settle. The winnowing out of fine material and organic matter by wave and tidal-current action may remove an important food source for deposit feeders that settle in the area. Spisula also has the ability to grow rapidly and in areas of high energy produces the more solid shell typical of Spisula on a 'Branchiostoma' bottom (Thorson 1957).

However, in the western central part of the bay there is an area of lower hydrodynamic energy near the centre of the tidal gyre and in the area where the orthogonals of the southwesterly swell diverge widely (Fig. 2d). This area appears to have a highly productive Syndosmya community which can be self-perpetuating because of its hydrodynamic environment. Many of the members of this community are short lived (Thorson 1957) but, due to the effect of centripetal motion in the tidal gyre, many of the meroplanktonic larvae released by members of this community can be retained within that gyre. This has been seen similarly to occur with plaice eggs in the Southern North Sea (Hela and Laevastu 1961) The transport of larvae to the centre of the gyre moves them into an area of still lower energy, and settlement can occur. This area of low hydrodynamic energy, whilst permitting the settlement of larvae, will also permit the settlement of fine grained sediment and organic material. Fine grained material will give some protection to the larvae once they have settled and the organic matter will provide a possible food source for them. Once the metamorphosed larva of an infaunal species has settled it will burrow; its limited range of lateral migration means that its adult distribution can still be related to the prevailing, local hydrodynamic regime. However, with errant species (e.g. Ophiura albida), although the distribution of newly settled specimens will reflect the hydrodynamic regime, this relationship will become obscured by the subsequent migration of the adults.

It can be suggested that, due to the combined effects of onshore wave and swell transport and the presence of a partially closed tidal-current gyre in Oxwich Bay, the community will be largely self-sustaining, with only limited escape of larvae from the community and, possibly, only limited entry of foreign larvae.

Once settled in the community, the metamorphosed larvae become subject to the biological controls which have been summarised by Thorson (1966), especially that of predation. However, according to the compilers of the Plymouth Marine Fauna, M.B.A. (1957), all dominant members of the Syndosmya community, with the exception of

Nucula turgida, breed during late spring and early summer; if the members of that community do not feed at the same time as they breed, this will offer some protection from predation to the settling larvae. The dominant epifauna species associated with the Syndosmya community in Oxwich Bay is the ophiuroid Ophiura albida (Tyler and Banner 1976), which appears to feed at a maximum rate during the time of settlement of larvae of the Syndosmya community (Taylor 1958, and personal observation). Its principal food consists of benthic diatoms, but crustacean larvae are also taken readily although polychaete and bivalve larvae are only rarely consumed. Thus, it is possible that O. albida exerts an influence on recruitment to the community by consuming potential predators of the larvae of the dominants of that community. It has also been noted that (Thorson 1953) the dominant bivalves of the Syndosmya community (S. alba and Cultellus pellucidus) grow rapidly and are unlikely to be taken by brittle stars which form the main predators of bivalves such as Macoma Balthica. These results however, contradict those of Thorson (1953) who has suggested that the Syndosmya community is protected from predation by brittle stars, as during settlement of Syndosmya community larvae the gonads of ophiuroids are swelling, thus reducing their stomach size and preventing feeding. This passivity may allow the development of the Syndosmya community to the exclusion of other species settling outside the passive period.

CONCLUSIONS

1. The sublittoral benthic community of Oxwich Bay is closely related to the hydrodynamic regime.
2. The distribution of sediment types is only partially related to the modern hydrodynamic regime, and, therefore, there is no strong correlation between sediment type and the distribution of the benthic communities.
3. The same hydrodynamic factors control the settlement of newly metamorphosed meroplankton and that part of the sediment which is modern and mobile.
4. The meroplanktonic larvae, fine sedimentary material and particulate organic matter all have a distribution which is affected by the same hydrodynamic factors, thus aiding successful recruitment to the community.
5. Infaunal species show a distributional pattern reflected in the hydrodynamic environment whilst the relationship of epifaunal species to the hydrodynamic regime becomes obscured by adult migrations.
6. Errant species of the epifaunal community aid community development by feeding on the larvae of possible predators of community members.

ACKNOWLEDGEMENTS

I wish to thank the Master and crew of the R.V. Ocean Crest for valuable assistance in the field; to Professor F.T. Banner for his critical reading of the manuscript and to the University of Wales for a Postgraduate Scholarship tenable during part of this study.

REFERENCES

Arthur, R.S., Munk,W.H., and Isaacs, J.D. 1972. The direct construction of wave rays. Trans. Am. Geophys. Un. 33, 855-865.

Buchanan, J.B., 1963. The bottom fauna communities and their sediment relationships off the coast of Northumberland. Oikos 14, 154-175.

Buchanan, J.B., 1964. A comparative study of some features of the biology of Amphiura filiformis and A.Chiajei (Ophiuroidea) considered in relation to their distribution, J. Mar. biol. Ass. U.K., 44, 565-576.

Fish, J.D., 1967. The biology of Cucumaria elongata (Echinodermata: Holothuroidea). J. Mar. biol. Ass. U.K. 47, 129-143.

Folk, R.L., 1954. The distinction between grain size and mineral composition in sedimentary rock nomenclature. J. Geol., 62, 344-359.

Folk, R.L., 1961. Petrology of Sedimentary Rocks, University of Texas, 154p.

Gage, J., 1972. A preliminary survey of the benthic macrofauna and sediments of Lochs Etive and Creran, sea lochs along the west coast of Scotland, J. Mar. biol. Ass. U.K., 52, 237-276

Gaudette, H.E., Flight, W.R., Toner, L., and Folger,D.W., 1974. An inexpensive titration method for the determination of organic carbon in recent sediments. J. Sedim. Petrol., 44, 249-253.

Hela, I., and Laevastu, T., 1961. Fisheries Hydrography, Fishery News (Books) Ltd., 137pp.

Holme, N.A., 1953. The biomass of the bottom fauna of the English Channel off Plymouth, J. Mar. biol. Ass. U.K., 32, 1-49.

I.M.E.R., 1975. Report 1974-75, Institute for Marine Environmental Research Plymouth, 85pp.

Jones, N.S., 1950. Marine Bottom Communities, Biol. Rev., 25, 283-313.

Jones, N.S., 1951. The bottom fauna off the south of the Isle of Man, J. Anim. Ecol., 20, 132-144.

Lie, U., 1968. A quantitative study of benthic infauna in Puget Sound, Washington, U.S.A. in 1963-64. Fiskeridir. Skr.Ser.Havunders,14(5),229-556.

Marine Biological Association, 1957. Plymouth Marine Fauna, Mar. biol. Ass. U.K., 3rd. ed. 457pp.

Neumann, G., 1954. Zur Characteristick des Seeganges, Arch. Met. Geophys. Bioklim., (A) 7, 353-377.

Parker, R.H., 1975. The Study of Benthic Communities. A Model and a Review., Elsevier, New York, 279pp.

Petersen, C.G.J., 1913. The animal communities of the sea bottom. Rep. Dan. biol. Stn., XXI, 1-44.

Rasmussen, E., 1973. Systematics and ecology of the Isefjord marine fauna (Denmark). Ophelia, 11, 1-507.

Sparck, R., 1937. The benthic animal communities of the coastal waters. Zoology Iceland, 1 (6), 45pp.

Taylor, M., 1958. Studies on the ecology of Manx Ophiuroidea,M.Sc., Thesis, University of Liverpool, 59pp.

Thorson, G.,1953. The influence of larval settlement upon the composition of marine level bottom communities,Proc.8th Pac.Sci.Cong.,III-A, 1171-1176.

Thorson, G., 1957. Bottom communities (Sublittoral or shallow shelf)., Mem. geol. Soc. Am., 1 (67), 461-534.

Thorson, G., 1966. Some factors influencing the recruitment and establishment of marine benthic communities. Neth. J. Sea Res., 3, 267-293.

Tyler, P.A., & Banner, F.T., 1976. The effect of coastal hydrodynamics on the echinoderm distribution in the sublittoral of Oxwich Bay, Bristol Channel. Estuar. & Coast. Mar. Sci. (In press).

Wade, B.A., 1972. A description of a highly diverse soft bottom community in Kingston Harbour, Jamaica. Mar. Biol. 13 (1), 57-69.

ON THE SHAPES OF PASSIVE SUSPENSION FEEDERS

G. F. Warner

Zoology Department, Reading University, U.K.

ABSTRACT

In situ observations have been made on the shapes of passive suspension feeders in currents. Particular shapes occur in different animal groups and may constitute adaptations to maximise particle capture. Flat fans are found in oscillating and tidal currents and radial or bushy arrangements occur under turbulent conditions. In persistent unidirectional currents dish-shaped fans with the concave side facing the current may be found. In other current regimes small-scale, dish-shaped filters are often found within larger supporting structures. Laboratory tests with models suggest that dish-shaped filters may capture a greater proportion of available particles than flat filters of the same area.

In a variety of suspension feeders the food-catching surfaces are located on the leeward side of the supporting structure. It is suggested that the relatively low current velocity, microturbulence and eddying which occurs to leeward probably favours particle capture. When currents change direction mobile animals such as crinoids can actively re-orientate and maintain the leeward location of their food catching surfaces. The shapes of some sessile animals allow passive re-orientation which achieves the same result. These shapes are mounted on flexible stems and include cones, spirals and shallow V-sectioned pinnate fans. These structures are only stable when their apices face upcurrent, food-catching surfaces are located on their leeward sides, within the cones or V-sections.

INTRODUCTION

Passive suspension feeders trap suspended food from exogenous currents. Purely passive forms expend no energy on the production of a feeding current other than to maintain themselves in it; they include echinoderms such as crinoids, some brittle-stars and some holothurians, and a variety of coelenterates such as hydroids, octocorals and antipatharians. In addition to these purely passive suspension feeders there are a number of active forms which, when conditions are right, behave as passive feeders: barnacles and porcelain crabs may cast their nets across currents. Lastly, active suspension feeders can augment their own feeding currents by appropriate orientation to exogenous currents; examples include sponges (Vogel 1974), bivalve molluscs (Hartnoll 1967) and ascidians (Monniot 1967). Indeed it is true to say that suspension feeders in general, whether predominantly active or passive, benefit from water circulation around them: by keeping food in suspension, by removing depleted water, by increasing the number of particles per minute that might encounter the feeding surfaces, and by providing an external source of energy which, by appropriate orientation, can increase the chances of particle capture (for reviews see Riedl 1971, Wainwright et al 1976).

A number of studies have established certain trends in orientation. Organisms in oscillating or unidirectional currents tend to be fan-shaped with the plane of the fan perpendicular to the direction of water movement. When water movement is turbulent bushy or radial orientations are found. These trends have been observed in hydroids (Svoboda 1970, Velimirov 1974), gorgonians (Theodor 1963, Grigg 1972, Schuhmacher 1973), crinoids (Magnus 1964, Meyer 1973) and brittle-stars (Warner and Woodley 1975). Fan-shapes which are perpendicular to the current are hydrodynamically more stable than when orientated parallel to the current since they are less liable to twisting (Wainwright and Dillon 1969). Gorgonian fans which are experimentally turned through $90°$ to give them a parallel orientation either die or grow accessory fans at right angles to the parent fan (Kinzie 1973). In hydroids the same experiment results in the fans being cast off and new fans regrowing from the stolon; these are orientated at right angles to the original fans, re-establishing the perpendicular orientation (Svoboda 1970). The orientation of fans perpendicular to currents is also advantageous in terms of food since a greater volume of water is filtered by a perpendicular fan than by a parallel fan (Leversee 1972). Bushy or radial shapes exploit currents from all directions.

In sessile colonies a particular shape may persist through life and can be modified only by growth. Fan-shapes therefore have little ability for short-term re-orientation. Mobile animals, however, are capable of active re-orientation. Crinoids and brittle-stars can adopt appropriate orientations whatever the current direction and in turbulent conditions can adopt radial shapes (Meyer 1973, Warner and Woodley 1975).

As a result of casual diving observations made over a number of years and, in particular, a series of observations made in Trinidad, West Indies, on the ecology of antipatharians (the bulk of which will be published elsewhere) I believe that I can add to the generalisations described above. Two points will be mentioned here since they have precedents in the literature. The first concerns the precise shape of a fan. Unidirectional currents are often tidal and reverse every six hours; flat fans are found in such situations. Where unidirectional currents are persistent, however, the fans are usually dish-shaped; the concave side of the dish faces the current. This phenomenon has been observed in gorgonians (Theodor 1963, Grigg 1972) and in stalked crinoids (Macurda and Meyer 1974). Second, it has sometimes been observed that in persistent unidirectional currents the food catching organs are located on the leeward side of the supporting structure. This has been observed in hydroids (Svoboda 1970) and in crinoids (Magnus 1964, Meyer 1973, Macurda and Meyer 1974) which, because they can actively re-orientate, are effectively feeding in persistent unidirectional currents. This paper presents some new observations on these two phenomena and suggests some explanations. In addition an as yet undescribed phenomenon is introduced: that of the shapes which allow short-term passive re-orientation in sessile colonies.

MATERIALS AND METHODS

Studies on antipatharians in Trinidad were carried out in the deep channels between the islands which occur between the north-west tip of Trinidad and the mainland of Venezuela. A prevailing surface current flows north through these channels. Below about 20 m, close to the shores of the islands and much modified by local features, a counter-current flows south. The rocky sea-bed in this area is rich in octocorals, hydroids and antipatharians. Methods of investigation included close-up photography. Notes and orientations of fans were recorded in pencil on a roughened white perspex sheet.

Laboratory experiments with models were carried out in a rectangular wooden tank. A water current was propelled round the tank by a paddle-wheel on one side of a central wooden partition. The experimental channel on the other side of the partition was about 50 cm wide and 150 cm long. The water was about 25 cm deep. Models were made of wire mesh covered with muslin to simulate the resistance to the water of living organisms. Parallel lines 12.5 mm apart were drawn on a perspex sheet which rested on the bottom of the experimental channel; the sheet was arranged such that the lines were parallel to the current. Detritus was added to the water and particles could be observed from above drifting down the paths between the parallel lines. The paths were numbered 1-13 and models were placed such that their centres were opposite the end of the central path (7). The fates of particles on different paths as they approached, encountered or passed the model could be recorded. Current speeds were measured by timing particles over measured distances. A disadvantage of the apparatus was that current speed was greater at the outer edges of the experimental channel producing a gradation of currents across all models.

RESULTS

Dish-shapes

Two species of antipatharian were commonly observed to form dish-shaped fans: Antipathes gracilis Gray and A. atlantica Gray. Two other species, A. hirta Gray and Antipathes sp (undescribed) were more bushy but frequently formed fans which tended to be dish-shaped. In all dish-shaped fans the concave sides faced the prevailing current. The form of the colonies of A. gracilis and A. atlantica is of a fan up to 40 cm in diameter formed by irregular frequent branching and anastomosis. Apart from the slight curvature necessary to form a dish the branches are mostly in one plane. Small branches and small accessory fans also develop on the convex (leeward) side of the fan giving this side a slightly bushy appearance compared to the smooth concave (windward) side. The form of the colonies of the other two species, A. hirta and A. (undescribed), is different. One, or occasionally more, main stems arise from the holdfast and these branch irregularly, infrequently, and roughly in one plane; there is no anastomosis. Large fan-shaped colonies may be 40 cm or more in diameter and a dish-shape can sometimes be discerned. The branches of these colonies bear two sets of pinnules: long and shorter. Underwater observation has shown that the long pinnules project from the windward sides of the branches and the shorter pinnules project from the leeward sides.

Observations near Discovery Bay in Jamaica where the prevailing long-shore current is from East to West provided further examples of dish-shapes. Antipathes atlantica and the large black gorgonian Iciligorgia schrammi (Duchassaing) occur on vertical cliffs below about 20 m; they both form dish-shaped colonies orientated perpendicular to the shore-line. The concave sides of the dishes face East.

Laboratory experiments were designed to test whether a curved model fan captured more particles than a flat model fan of the same area. One model, 178 mm wide and 152 mm tall, was given a radius of curvature of 121 mm in one plane (a curved model, not a dish). Another model of the same size was flat. The models were immersed one at a time in the experimental channel of the current producing machine and particles drifting down different paths (see Materials and Methods) towards the model were observed. In each experiment 130 particles, 10 in each of the 13 paths, were watched as they drifted towards the model. If they passed through the model or were trapped on the muslin they were regarded as having been caught, if they passed round either side of the model they were not caught, and if they passed over the top or underneath they were discounted. Two experiments with each model were carried out at each of three current speeds and the results of two typical experiments are illustrated in Fig. 1. The complete results are tabulated in Table 1. The experiments show first, that the approaching streamlines diverge and the outermost ones bypass the model which, therefore, filters much less water than the product of its area and the current speed might suggest. Second, it appears that the curved model catches a greater proportion of the available particles than the flat model, particularly at high current speeds. This must be because it filters more water, probably by capturing divergent streamlines on its curved edges. It is likely that both these effects obtain in the sea and that in unidirectional currents of equal speed dish-shaped suspension feeders sample more water, for their surface areas, than flat suspension feeders.

TABLE 1. Particles captured by model fans

Current speed cm/sec	flat model	curved model	total particles
4.7	38	46	130
4.7	34	44	130
10.9	21	37	130
10.9	25	40	130
15.1	24	38	130
15.1	20	38	130

Leeward food-catching

In the colonies of all antipatharian species mentioned above the polyps are born predominantly on the leeward sides. In $\underline{A}$. $\underline{gracilis}$ and $\underline{A}$. $\underline{atlantica}$ the somewhat bushy nature of the convex leeward sides provides extra space for leeward polyps. In $\underline{A}$. $\underline{hirta}$ and $\underline{A}$.(undescribed) polyps are born on the leeward sides of both windward and leeward pinnules.

Leeward hydranths have been observed in unidentified fan-shaped hydroids both in Trinidad and off British coasts. In many cases the hydranths appear to have been deflected to the leeward side by the current. Thus the current impinges on the aboral sides of the hydranths and is filtered through the extended tentacles (Fig. 2A). In some hydroids the hydranths normally arise from one side, the ventral side, and this side usually faces to leeward (Svoboda 1970, Wainwright et al 1976). This situation has been observed in $\underline{Plumularia}$ $\underline{setacea}$ (Ellis and Solander) (R.G. Hughes, personal communication)

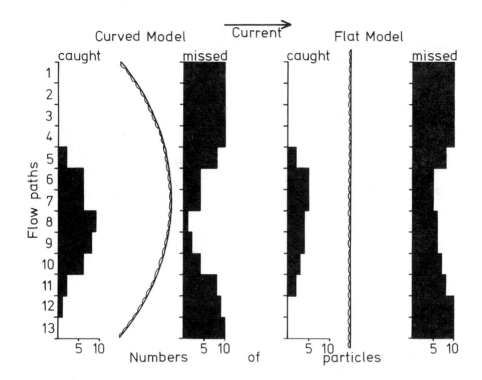

Fig. 1. The capture of particles by model fans. Mean current speed was 10.9 cm/sec but was slower in flow-path 1 than in flow-path 13 skewing the zone of peak particle capture away from the centre. The fates of 10 particles in each flow path were recorded in all cases.

which occurs as an epizoite on the hydroid Nemertesia antennina (L.) (Hughes 1975). The effect of a water current on N. antennina is to bend it over so that water flows along it from base to tip. The pinnate fans of P. setacea are always orientated perpendicular to this current with their hydranth-bearing ventral faces to leeward, facing the tip of the N. antennina colony. The hydranth-bearing branches of N. antennina are similarly arranged with the hydranths at all times facing the tip of the colony (Fig. 2B,C).

On the windward side of any object in a current there is a relatively static region bounded by divergent streamlines of faster moving water. To leeward, however, the region bounded by converging streamlines is not static but is subject to turbulence and eddying. This effect obtains whether the object is a porous sheet, such as a fan-shaped suspension feeder, or a solid strut, such as a branch of a suspension feeder. I think it likely that the main advantage of leeward feeding is that the food-catching surfaces of suspension feeders have a better chance of capturing food when they are located in leeward regions of turbulence than when located either in windward static regions or in edge regions subject to convergent streamlines and faster current flow (see Riedl and Forstner 1968). It is well known that particles both in water and air tend to accumulate on the leeward sides of objects; reduced velocity and turbulence both aid this effect. In the case of suspension feeders, particles which spin round slowly two or three times in an eddy are probably more than two or three times more likely to be caught than particles that drift rapidly past only once.

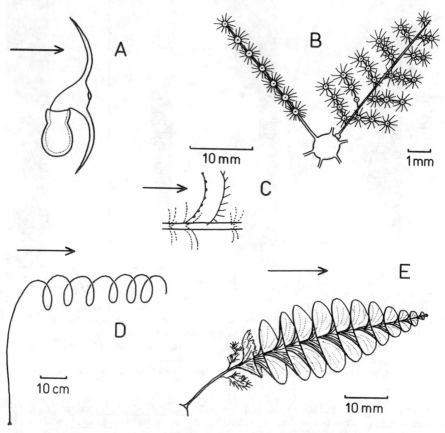

Fig. 2. Diagramatic views of suspension feeders in currents. A, a section through the branch of a hydroid showing a hydranth bent to leeward by a current. B, simplified view from the leeward of the hydroids <u>Nemertesia</u> <u>antennina</u> (main stem and left-hand branch) and <u>Plumularia</u> <u>setacea</u> (pinnate right-hand branch). C, side view of the same. D, a spiral antipatharian. E, a bryozoan consisting if a series of spiral cones. Arrows indicate current direction.

Passive re-orientation

The ability of the hydranths of fan-shaped hydroids to flop one way or the other allows leeward feeding at all times in regularly reversing tidal currents, and even, perhaps, in the oscillating movements produced by waves.

In areas subject to less predictable water flows, however, short-term re-orientation through angles other than 180° is possible in flexible colonies. Such colonies are able to swing round on their stems to adopt the same orientation whatever the direction of the current. This is only useful where the particular orientation carries some advantage - as in the case of the leeward positions of the hydranths of N. antennina when this hydroid is bent by a current. This, indeed, is an example of passive re-orientation. Another example is that of the spiral form of the antipatharian Cirrhipathes lutkeni Brook. A colony of this black coral consists of an upright stem some 50 cm high which gives rise to a horizontal spiral some 10 cm in diameter and comprising 2-12 coils (Fig. 2D). The spiral trails downcurrent, whatever its direction, rather like a wind-sock at an airport. The relatively large polyps on the spiral are invariably found to leeward; the longest tentacles of the polyps project laterally in a row along either side of the coils and the shorter tentacles project obliquely downcurrent. Although all the coils of the spiral have the same diameter it is likely that there is sufficient turbulence behind each coil to mix the water and to ensure that the most distal coils do not feed in depleted water.

Antipatharians such as A. atlantica, which are normally fan-shaped, sometimes become bushy when growing in shallow or turbulent water. The bushy growth-form consists of several flat sheets each lying in a slightly different plane. These colonies show some capacity for passive re-orientation since the flat sheets tend to curve into incomplete cones. The cones drift downcurrent and the polyps, born on the insides of the cones, are thus always situated to leeward. A similar situation was observed in Trinidad in a large (30 cm high) unidentified hydroid. The branches of this hydroid bear pinnate fans which stand out at an angle from their points of origin. These fans are not flat. The pinnules on either side arise at 90-120° to each other and the ventral, hydranth-bearing side, which usually faces distally, is on the inside of the shallow V formed by the pinnules. Whatever the direction of the current these pinnate fans are passively orientated such that their hydranths are to leeward. The V-shaped sections of the fans are responsible for this ability since, like the cones of the antipatharians, they are hydrodynamically stable only when their apices face the current.

Certain cone-shaped active suspension feeders may also benefit from passive re-orientation. When extended into a current the cone-shaped tentacular crown of a fan-worm flops over, the apex of the cone facing the current. The actively produced feeding current of the worm is drawn from the outside of the cone; propelled by lateral cilia it passes between the tentacles and out of the mouth of the cone. Food particles are deposited on the frontal (leeward) side of each tentacle. Passive orientation of the tentacular crown ensures that the exogenous current augments the current produced by the cilia. Bugula turbinata Alder is a bryozoan the colonies of which form a spiral series of cones (Fig. 2E). Here again the food-catching tentacles are located on the insides of the cones which, with passive orientation of the whole colony, are always situated to leeward. Assuming that there is an advantage associated with leeward feeding, passive re-orientation achieved by trailing spirals, cones and V-sections must also be advantageous.

DISCUSSION

If one accepts that dish-shapes with concave sides facing the current filter more water than flat shapes (and certainly more than streamlined shapes such as cones or V-sectioned pinnate fans), one might be surprised at their relatively infrequent occurrence. One reason is probably that persistent unidirectional currents are not common in the most frequently studied marine habitats. A second reason is that a dish-shape attached at one edge is only capable of limited passive re-orientation. It can accomodate to slight variations in current direction in the same manner as a flat fan, but not to changes of $90°$ or to reversals. A final reason may be that, function efficiently, a dish-shape must not bend with the current: its structural members must be strong enough to resist the force of the water. It appears from Table 1 that the relative efficiency of dish-shapes may increase with current speed, emphasising the importance of this requirement. The very strong skeletons of antipatharians (Wainwright et al 1976) no doubt play an important part in the ability of these animals to sustain dish-shaped feeding surfaces. In stalked crinoids passive re-orientation of the dish-shaped calyx is possible since the calyx is attached to the stalk at its centre. Nevertheless, the stalk must be strong to hold the calyx perpendicular to the current and the arms forming the calyx also need the strength to maintain the dish-shape (Macurda and Meyer 1974). Unstalked crinoids, despite a capacity for active re-orientation, do not exploit the advantages of dish-shaped filtering. Instead the ends of the arms are bent downcurrent such that the windward side of the fan is convex (Meyer 1973). This is probably because the arm tips simply do not have the strength to stand up to the current, let alone bend into it. Some animals capable of active re-orientation, however, do exploit dish-shapes. Examples are barnacles and porcelain crabs which have dish-shaped filtering nets with concave sides facing the current.

Small-scale dished, curved or angled orientations in which the concave sides face the current are also often found within larger filtering shapes. In the pinnate antipatharians A. hirta and A. (undescribed) the windward rows of long pinnules stand at between 90 and $120°$ to each other forming a shallow V-shaped filtering groove which faces the current. Precisely the same occurs in the brittle-star Ophiothrix fragilis (Abildgaard) in which the rows of tube feet, projecting into the current on either side of the ambulacrum, form a V-shaped filtering groove on each arm (Warner and Woodley 1975). The pinnules and tube feet of crinoids also bend into the current: each arm and each pinnule is curved in section with the concave side facing the current (Magnus 1964). The same thing can be observed in the hydranths of hydroids. Each hydranth can be likened to the calyx of a tiny stalked crinoid since it forms a dish, suspended at the centre, with the tentacles curved back into the current (Fig. 2A). On this small scale, structures such as tube feet and tentacles do not need great strength to adopt recurved orientations unless the currents are very strong.

The common location of food-catching surfaces to leeward of the filtering structure - antipatharians, crinoids, fanworms - and the occurrence of shapes which permit re-orientation such that food-catching surfaces are leeward - hydroids, some bryozoans, some antipatharians - give good evidence that there is some advantage to leeward feeding. Riedl and Forstner (1968) pointed out that in many fan-shaped gorgonians the polyps arise from the lateral sides of branches and so project into the spaces between branches through which the water flows. They suggested that this arrangement allowed the polyps to exploit a range of current speeds. At low speeds the polyps filtered the water passing between the branches; at high speeds, however, they would be bent by the current through the spaces between the branches into the turbulent, relatively low velocity regions to leeward. Thus the polyps could feed at roughly the same current speed over a range of environmental current speeds. It is possible, therefore, that leeward feeding is a device for ensuring that, although the entire animal may be advantageously located in a region of relatively great water movement (see Introduction), the food-catching organs or surfaces are situated in a relatively low velocity, protected region in which, considering their often delicate structure, they may function more effectively. An additional possibility, however, is that the turbulence and eddying found to leeward of the filtering structures may itself increse the chances of particle capture. Much work remains to be done in this field to work out exactly what happens to particles in the close vicinity of food-catching surfaces, and what properties of these surfaces are important in determining their efficiency.

CONCLUSION

There is a case for believing that dish-shaped, curved, and shallow V-shaped porous structures with concave sides facing the current filter more water than flat structures of the same composition and area. Passive suspension feeders probably gain some advantage by adopting these shapes. The common occurrence in suspension feeders of food-catching surfaces located to leeward of the filtering structures, and of shapes such as cones, shallow V-sections, and spirals on flexible bases which, through passive re-orientation, achieve leeward locations for their food-catching surfaces, suggests that there is a distinct advantage associated with leeward feeding. This advantage is probably connected with the relatively low velocity turbulent conditions which occur to leeward of objects in currents.

ACKNOWLEDGEMENTS

I should like to thank Dr D.M. Opresko for identifying my antipatharians, and Mrs R. Cleevely for carrying out the experiments on particle capture by model fans.

REFERENCES

Grigg, R.W. 1972 Orientation and growth form of sea fans. Limnol. Oceanogr. 17, 185-192.

Hartnoll, R.G. 1967 An investigation of the movement of the scallop, Pecten maximus. Helgoländer wiss. Meeresunters. 15, 523-533.

Hughes, R.G. 1975 The distribution of epizoites on the hydroid Nemertesia antennina (L.). J. mar. biol. Ass. U.K. 55, 275-294.

Kinzie, R.A. III. 1973 Coral reef project - papers in memory of Dr. Thomas F. Goreau. 5. The zonation of West Indian gorgonians. Bull. mar. Sci. 23, 93-155.

Leversee, G.J. Jr. 1972 Field and laboratory studies of the effect of water currents on morphology and feeding in the sea whip, Leptogorgia. Am. Zool. 12, 719.

Macurda, D.B. and Meyer, D.L. 1974 Feeding posture of modern stalked crinoids. Nature Lond. 247, 394-396.

Magnus, D. 1964 Gezeitenströmung und Nahrungsfiltration bei Ophiuren und Crinoiden. Helgoländer wiss. Meeresunters. 10, 104-117.

Meyer, D.L. 1973 Feeding behaviour and ecology of shallow-water unstalked crinoids (Echinodermata) in the Caribbean Sea. Mar. Biol. 22, 105-129.

Monniot, C. 1967 Problémes écologiques posés par l'observation des Ascidiens dans la zone infralittorale. Helgoländer wiss. Meeresunters. 15, 371-375.

Riedl, R. 1971 Water movement: Introduction (1085-1088) and Animals (1123-1156). In Marine Ecology, O.Kinne, ed. Vol. 1, Part 2. Wiley, New York.

Riedl, R. and Forstner, H. 1968 Wasserbewegung im Mikrobereich des Benthos. Sarsia, 34, 163-188.

Schuhmacher, H. 1973 Morphologische und ökologische Anpassungen von Acabaria-Arten (Octocorallia) im Roten Meer an verschiedene Formen der Wasserbewegung. Helgoländer wiss. Meeresunters. 25, 461-472.

Svoboda, A. 1970 Simulation of oscillating water movement in the laboratory for cultivation of shallow water sedentary organisms. Helgoländer wiss. Meeresunters. 20, 676-684.

Theodor, J. 1963 Contribution à l'étude des gorgones. 3. Vie Milieu 14, 815-818.

Velimirov, B. 1974 Orientiertes Wachstum bei Millepora dichotoma (Hydrozoa). Helgoländer wiss. Meeresunters. 26, 18-26.

Vogel, S. 1974 Current-induced flow through the sponge, Halchondria. Biol. Bull. mar. biol. Lab. Woods Hole, 147, 443-456.

Wainwright, S.A., Biggs, W.D., currey, J.D. and Gosline, J.M. 1976 Mechanical design in organisms. Arnold, London.

Wainwright, S.A. and Dillon, J.R. 1969 On the orientation of sea fans (genus Gorgonia). Biol. Bull. mar. biol. Lab., Woods Hole, 136, 130-139.

Warner, G.F. and Woodley, J.D. 1975 Suspension-feeding in the brittle-star Ophiothrix fragilis. J. mar. biol. Ass. U.K. 55, 199-210.

THE STRUCTURE AND SEASONAL FLUCTUATIONS OF PHYTAL MARINE NEMATODE ASSOCIATIONS ON THE ISLES OF SCILLY

Richard M. Warwick

NERC Institute for Marine Environmental Research, 67 Citadel Road, Plymouth, UK

ABSTRACT

The species composition of nematodes in fine intertidal seaweeds can be related directly to the growth form and texture of the weeds. Four major associations are distinguished, the species compositions of which relate to the locomotory abilities of different nematode species among algal fronds of varying coarseness, and to the availability of different food types.

Within a single weed form the abundance of each major nematode species exhibits cyclical changes of annual frequency. In the spring and early summer carnivore/omnivore and deposit feeding species with no visual mechanisms predominate, whereas species present in the latter part of the year tend to be epigrowth feeders with ocelli or visual pigment. Abundance cycles are probably caused by seasonal changes in the availability of different food types.

INTRODUCTION

The work of Wieser (1951, 1952, 1959), Ott (1967) and Moore (1971) has led to the establishment of a relationship between the morphology of seaweeds and the species structure of phytal nematode populations. These studies, however, take no account of the possibility of seasonal changes in nematode population structure, due either to seasonal variability in food supply or to reproductive activity. The opportunity to make such a temporal study arose from the author's sorting of a large series of nematode samples from fine intertidal seaweeds collected by Professor L.A. Harvey on the Isles of Scilly. This was part of a broader study being undertaken by Professor and Mrs. Harvey on the biology of the recently established weed Asparagopsis armata and its supposed asexual plant Falkenbergia rufolanosa, and on the fauna associated with these weeds.

METHODS

Sample Collection and Treatment

The collection of faunal samples has been heavily biased towards Asparagopsis and Falkenbergia, whilst smaller collections have been made from other species for comparative purposes. Samples of weed were collected between October 1969 and June 1972 at various sites on St. Mary's (see map in Harvey, 1969) and at Porth Killier on St. Agnes. Several samples of Asparagopsis and Falkenbergia were collected from various sites each month between February 1970 and June 1972 and these samples provide a good time-series for the study of seasonal and spatial variability. Samples of other weed species were collected more intermittently. All weeds were sampled in screw-top jars of sea water at LWST except for a few of the Corallina samples which were collected higher up the shore (up to MTL). Samples were generally taken from rock surfaces, except for Sphacelaria which is epiphytic on larger weeds and Jania which was attached to Zostera roots on sand. Batches of each weed species were kept as uniform as possible, but there was probably a variation of about $\pm 20\%$ about the mean. There was, however, a considerable variation in weight between one algal species and another, i.e. the Falkenbergia samples were about 1.5 g wet weight, Asparagopsis 7.5 g, etc. Formalin was subsequently added to the samples and the nematodes sorted under a binocular microscope. They were then transferred to pure glycerine using the slow evaporation technique of Seinhorst (1959) and mounted in glycerine for identification and counting under the high-power microscope.

Mathematical Treatment of Data

In view of the variability in sample size between weed species, the numerical data have been transformed to dominance values. For each nematode species, dominance is the total number of specimens of that species expressed as a percentage of the total number of nematodes in the sample. The numerical classificatory technique described by Field (1971) was used to compare faunal similarity between weed species and to compare within-species samples of Asparagopsis and Falkenbergia. Briefly, this technique involves the erection of a matrix of

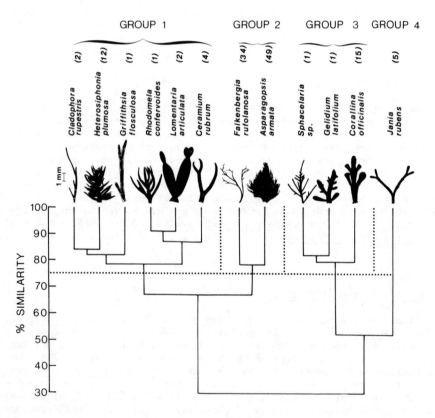

Fig. 1. Dendrogram of faunal affinities between weed types, showing four weed groups defined at the 75% similarity level. Silhouettes of frond tips drawn to scale. Number of samples used in analysis bracketed.

similarities between samples based on species attributes using the Czekanowski coefficient (Bray & Curtis, 1957) as a measure of similarity: the samples are then ordered into a hierarchy or dendrogram using group average sorting (Lance & Williams, 1967). Samples of less than ten specimens have been omitted from the numerical analysis. Whilst data on all the nematode species have been used in the analysis, only the distribution of the commoner species is discussed below.

RESULTS

Comparison of the Nematode Fauna of Different Weed Types

A dendrogram of faunal affinities between weed types, based on species dominance averaged over all samples, is given in Fig. 1. Weed species represented by single or small numbers of samples have been included for completeness, but their links in faunal similarity with other species should not be regarded as reliable in view of seasonal influences which, as will be shown below, can markedly affect species composition. A major dichotomy at the 30% level separates a left-hand group of eight weed species from a right-hand group of four. The right-hand group comprises relatively coarse weeds, all notably stiff in texture: Corallina and Jania are both coralline, Gelidium has a cartilaginous texture and the tufts of Sphacelaria are harsh to the touch. The left-hand group comprises softer textured weeds which, with the exception of Lomentaria which is represented by only two samples, are generally rather fine. At an arbitrarily selected similarity level of 75% the two very fine weeds Falkenbergia and Asparagopsis are separated from the remainder in the left-hand group. Jania is isolated at this level from the remaining three species in the right-hand group. This weed is probably richer in fine deposits since it was collected from sandy habitats and, as shown below, contains a higher percentage of deposit feeding nematode species.

TABLE 1 Percentage dominance of the major nematode species in the four weed groups.

	Mean dominance	Weed Group 1	2	3	4
Oncholaimus dujardinii	54.2	82.1	56.5	21.1	6.2
Enoplus communis	10.1	3.2	2.6	48.9	31.2
Theristus acer	8.5	0.5	12.0	0.1	13.7
Symplocostoma tenuicolle	8.2	3.8	9.9	1.8	20.0
Euchromadora striata	5.2	2.6	6.5	0.8	11.2
Cyatholaimus gracilis	4.7	1.7	5.7	2.2	10.0
Anticoma acuminata	1.4	0.3	0.6	4.8	7.5
Chromadora nudicapitata	1.2	1.6	1.2	0.9	-
Chromadorella filiformis	1.1	-	1.6	0.1	-
Monhystera refringens	0.9	1.1	0.8	1.4	-
Thoracostoma coronatum	0.8	0.1	<0.1	5.9	-
Graphonema scampae	0.7	0.1	0.3	3.6	-
Dolicholaimus marioni	0.2	-	-	1.7	-
Anticoma eberthi	0.2	-	-	1.3	-

Table 1 gives the dominance of the major nematode species in the four weed groups isolated at the 75% similarity level, and Fig. 2 shows how certain physiognomic characters of the nematodes, which have been considered by previous authors to be selectively advantageous in differing habitats, are distributed

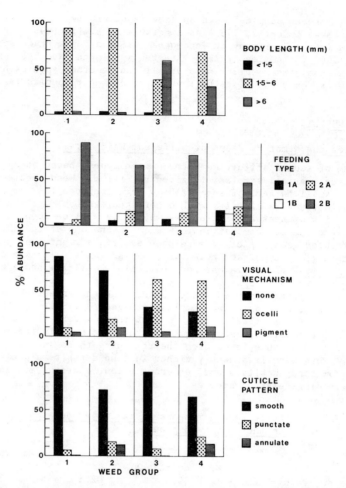

Fig. 2. Distribution of physiognomic characters as defined by Wieser (1959), based on percentage of specimens in each weed group. Feeding types deduced from structure of buccal cavity: 1A selective deposit feeders, 1B non-selective deposit feeders, 2A epigrowth feeders, 2B predators and omnivores.

among the four weed groups. The maximum length of body setae is not included, since in all cases this was less than 20 μm. The coarse rigid weeds in groups 3 and 4 have a much higher percentage of large species of nematodes (>6 mm body length) and of species with visual mechanisms, particularly true ocelli. Species in feeding group 2B (predators and omnivores) predominate in all weed types and deposit feeding forms are rare, except in group 4 (Jania) which, as stated above, was collected from sandy habitats. Nematodes with smooth cuticles are uniformly dominant, but those with punctate and annulate cuticles are slightly more abundant in weed groups 2 and 4.

Variations within a single weed type

In this part of the study the same mathematical classificatory technique has been applied to the individual samples of Asparagopsis and Falkenbergia in an attempt to relate groups of similar samples with site or season. The remaining weed species have not been investigated since the number of samples was too low for worthwhile conclusions to be drawn. Dendrograms of faunal affinities between samples are given in Fig. 3 (Asparagopsis) and Fig. 4 (Falkenbergia).

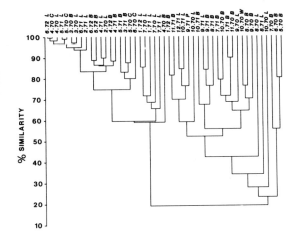 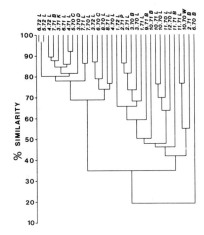

Fig. 3. Dendrogram of faunal affinities between Asparagopsis samples, designated by month, year and locality. L - Porth Loo, C - Porth Cressa, B - Bar Point, P - Priglis Cove, W - Watermill Bay.

Fig. 4. Dendrogram of faunal affinities between Falkenbergia samples. K - Porth Killier, D - Doctors Keys, P - Pelistry. Other notations as Fig. 3.

In both dendrograms a major dichotomy separates a very homogeneous left-hand group of samples from a more heterogeneous right-hand group. These groupings are independent of the locality from which the samples were collected and are entirely dependent on season. In Table 2 the number of samples in the left- and right-hand groups of the dendrogram are given on a monthly basis.

TABLE 2 Numbers of samples collected each month in the left- and right-hand groups of the dendrogram sequence.

	Months											
	J	F	M	A	M	J	J	A	S	O	N	D
Asparagopsis												
left-hand group	2	2	4	6	4	4	3					
right-hand group	2				1	1	2	4	5	6	2	1
Falkenbergia												
left-hand group			3	3	2	3	3	2				
right-hand group	2	3	2						1	4	3	1

For Asparagopsis this shows that all the samples in the left-hand group were collected between mid-winter and early summer (January-July) and those in the right-hand group mainly in the latter half of the year. For Falkenbergia the left-hand group of samples was collected a little later in the year (March-August) and the right-hand group from September to March.

Seasonal fluctuations in species abundance

The occurrence of different species associations at different times of the year led to an investigation of the seasonal distribution of the major nematode species present. For this purpose the samples of Asparagopsis and Falkenbergia have been combined since they show essentially the same pattern and the earlier analysis had shown their nematode species compositions to be very similar.

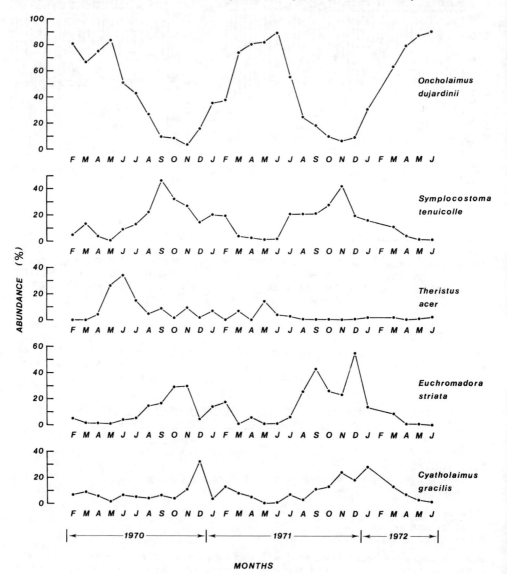

Fig. 5. Variations in abundance of the five major nematode species in Asparagopsis and Falkenbergia.

Figure 5 gives the percentage abundance of the five major nematode species, which comprise 90.6% of the total nematode fauna in the Asparagopsis/Falkenbergia weed group, at monthly intervals throughout the sampling period. Marked seasonal cycles are apparent. Oncholaimus dujardinii has peak dominance in spring and early summer and low dominance in late summer and winter. This pattern is repeated in successive years in a very regular way, so that it is very unlikely that such a pattern could have been produced by chance fluctuations in numbers. Theristus acer shows a similar pattern, although the amplitude of the peaks decreases throughout the sampling period so that in the early summer of 1972 no increase in dominance is evident. The three remaining species, Symplocostoma tenuicolle, Euchromadora striata and Cyatholaimus gracilis show the opposite trend, with peaks in the late summer and winter and low dominance in spring and early summer. It could be argued that, since dominance percentages are used, the two opposing patterns could be produced by seasonal changes in numbers of Oncholaimus dujardinii with the other species remaining at relatively constant levels of abundance. This would give rise to a marked increase in total numbers of worms in the period when Oncholaimus was dominant. However, examination of the mean number of nematodes per sample throughout the period on a monthly basis shows that seasonal changes in total numbers do not follow the same clear pattern as the dominance of Oncholaimus. There is some evidence for peaks of abundance in spring, but during the period July 1970 to April 1972 fluctuations in numbers appear to be random. It therefore seems clear that these seasonal cycles of dominance are due to real changes in species abundance.

Since the five species treated above comprise more than 90% of the total fauna they have a dominating effect on the dendrograms given in Figs. 3 and 4. The left-hand groups on the dendrograms consist of Oncholaimus and Theristus dominated samples and the right-hand groups of Symplocostoma, Euchromadora and Cyatholaimus dominated samples.

The physiognomic characters of the species in the two groups are very different, particularly with respect to feeding types and visual mechanisms. Oncholaimus is a carnivore or omnivore and Theristus is a non-selective deposit feeder; neither of these species has visual mechanisms. In the second group Euchromadora and Cyatholaimus are both epigrowth feeders. Symplocostoma shows marked sexual dimorphism in head structure; in the adult male a buccal cavity is completely lacking, suggesting a selective deposit feeding habit, whilst in the female the head is oncholaimid in appearance suggesting a carnivore or omnivore, but in this case the head is so small that only minute prey could be taken and it is possible that, in fact, these females are epigrowth feeders. All three species in this group have visual mechanisms, Symplocostoma and Cyatholaimus having true ocelli and Euchromadora concentrated ocellar pigment spots.

DISCUSSION

Differences between Weed Species

The results agree with those of previous authors in that faunal composition appears to be directly related to the coarseness, texture and silt content of the weed. Explanations of this relationship can be only tentative, but are probably related largely to the different types of epiflora and fauna growing on the weed filaments, upon which different nematode species feed preferentially. Clearly, the ability to move easily amongst the fronds is an additional factor, since the largest species (Enoplus communis, Toracostoma coronatum and Anticoma eberthi) are much more abundant in the coarser weeds whilst the smallest species (Chromadora nudicapitata, Chromadorella filiformis and Monhystera refringens) are more catholic in habitat. On the other hand, distributions of other species are more easily explained on the basis of feeding preferences. For example, Oncholaimus dujardinii (carnivore/omnivore) and a group of species comprising Theristus acer, Symplocostoma tenuicolle, Euchromadora striata and Cyatholaimus gracilis (epigrowth or deposit feeders) appear to be mutually exclusive. The former is commonest in the moderately fine group 1 weeds, whereas the latter is more abundant in the very fine group 2 weeds and in Jania but rare in group 2.

Seasonal Fluctuations within a Single Weed Form

The available evidence suggests that the seasonal cycles of species abundance are caused by seasonal changes in the availability of various food types. Epigrowth feeders dominate in late summer, autumn and winter: they either scrape off the films of bacteria and unicellular algae from the algal fronds or pierce larger objects and suck out the contents. Visual mechanisms may be necessary to maintain the worm's position near the frond tips where the light intensity is greater and the epiflora consequently richer. In spring and early summer, carnivore/omnivore or deposit feeders dominate, indicating a change in the available food. These types will have no preference for the frond tips and thus require no visual mechanisms.

Ott (1967) has shown that the juveniles of Enoplus meridionalis are found at the tips of Cystoseira growths, whereas the adults are found only at the bases, and he suggests that this species changes from an epigrowth feeder as a juvenile to a predator as an adult. A microdistribution of this kind could also be the result of body size, since the weed is much coarser at its base. If this species has an annual life cycle similar to that of Enoplus communis (Wieser & Kanwisher, 1960) this phenomenon could lead to an alternation in abundance from fine to coarse weeds with season. The state of maturity of the worms was not noted in the present study and it is not certain, therefore, whether this effect operates. This is unlikely, however, since the available evidence reviewed by Gerlach (1971) suggests that smaller nematode species have relatively short generation times with more than one generation per year. The alternation of generations between fine and coarse weeds would thus not produce cycles of annual frequency.

Further investigation is needed to test the tentative explanations of these annual cycles: an examination of the actual food preferences of the worms, the seasonal distribution of different food types and the stratification of these various types of food particle within the weeds. Whatever the cause, it is clear that a considerable degree of caution should be exercised when comparing algal nematode faunas from collections made at different times of the year, because the nematode "community" may vary considerably in different seasons both in the proportions of actual species present and in the distribution of certain physiognomic characters which have been considered by previous authors to be influenced only by the structure of the habitat.

ACKNOWLEDGEMENTS

I am indebted to Leslie Harvey for providing the nematode samples, to Clare Harvey for the collection and identification of the weed samples used to illustrate Fig. 1, and to John Field for his computer program CLASS. The practical work for this paper was done at the Dove Marine Laboratory, University of Newcastle upon Tyne. Computer analyses were run at the Institute for Marine Environmental Research, Plymouth, with the kind assistance of Jackie Forsyth.

REFERENCES

Bray, J.R. & Curtis, J.T. 1957 An ordination of the upland forest communities of Southern Wisconsin. Ecol. Monogr. 27, 325-349.

Field, J.G. 1971 A numerical analysis of changes in the soft-bottom fauna along a transect across False Bay, South Africa. J. exp. mar. Biol. Ecol. 7, 215-253.

Gerlach, S.A. 1971 On the importance of marine meiofauna for benthos communities. Oecologia (Berl.) 6, 176-190.

Harvey, L.A. 1969 The marine flora and fauna of the Isles of Scilly. The Islands and their ecology. J. nat. Hist. 3, 3-18.

Lance, G.N. & Williams, W.T. 1967 A general theory of classificatory sorting strategies. I. Hierarchical systems. Computer J. 9, 373-380.

Moore, P.G. 1971 The nematode fauna associated with holdfasts of kelp (Laminaria hyperborea) in North-east Britain. J. mar. biol. Ass. U.K. 51, 589-604.

Ott, J. 1967 Vertikalverteilung von Nematoden in Beständen nordadriatischer Sargassaceen. Helgolander wiss. Meeresunters. 15, 412-428.

Sienhorst, J.W. 1959 A rapid method for the transfer of nematodes from fixative to anhydrous glycerine. Nematologica 4, 67-69.

Wieser, W. 1951 Untersuchungen über die algenbewohnende Mikrofauna mariner Hartböden I. Zue Oekologie und Systematik der Nematodenfauna von Plymouth. Ost. Zool. Z. 3, 425-480.

Wieser, W. 1952 Investigations on the microfauna inhabiting seaweeds on rocky coasts. IV. Studies on the vertical distribution of the fauna inhabiting seaweeds below the Plymouth Laboratory. J. mar. biol. Ass. U.K. 31, 145-74.

Wieser, W. 1959 Freeliving marine nematodes. IV. General part. Reports of Lund University Chile Expedition, 1948-9. Acta Univ. Lund. N.F. Avd. 2, 55, 1-111.

Wieser, W. & Kanwisher, J. 1960 Growth and metabolism in a marine nematode Enoplus communis Bastian. Z. vergl. Physiol. 43, 29-36.

DISSOLVED ORGANICS IN THE NUTRITION OF BENTHIC INVERTEBRATES

Brian West, Maureen de Burgh and Frank Jeal

Department of Zoology, Trinity College, Dublin 2, Ireland

ABSTRACT

Kinetic analyses of transport in isolated tissue preparations support the hypothesis that soft-bodied benthic invertebrates can absorb amino acids actively from natural concentrations. The significance of this source of nutriment is discussed and the fate of the absorbed substances described. It is suggested that invertebrates play a significant role in the marine DFAA cycle.

INTRODUCTION

The average concentration of dissolved free amino acids (DFAA) in sea-water (10 µg C/l, Williams 1975) is equal to about half that of phytoplankton. Fairly typical values for an individual DFAA are those for alanine (nmoles/l): oceanic 100, coastal 200, coastal bottom 300; Stephens (1963) reported 16,000 for interstitial water from an intertidal mudflat.

In the following section we will review recent evidence for the existence in benthic invertebrates of mechanisms for absorption of DFAA from the extremely low absolute concentrations occurring in sea-water. We then consider the significance of DFAA in nutrition, their fate upon absorption, and the role of invertebrates in the flux of DFAA in the sea.

Throughout this paper amino acids exclusively are considered. Although there are fewer data relating to other compounds such as sugars and fatty acids the principles of specialised active epithelial uptake and subsequent utilisation, described here for the amino acids, are probably applicable to them also.

ABSORPTION OF DFAA FROM NATURAL CONCENTRATIONS IN SEA-WATER

The majority of experiments done on absorption of DFAA have involved the use of whole animals. Use of isolated tissues or organs, however, has two advantages: they are physiologically simpler systems with fewer tissue compartments, and therefore are more suitable for kinetic analysis of transport; and they overcome the possibility of absorption via the gut with subsequent translocation.

Isolated preparations used for investigation of alanine uptake by benthic invertebrates are listed in Table 1. The uptake of all amino acids so far studied has occurred against a concentration gradient, depended on metabolic energy, followed Michaelis-Menten kinetics, and been competitively inhibited by structural analogues: it is, therefore, an active process. Kinetic analyses yield the Michaelis-Menten constant K_t (substrate concentration at

TABLE 1 K_t Values for L-Alanine in Isolated Preparations

Species	Tissue	K_t (μmoles/l)	Reference
Cardium edule	gill	24	Bamford & McCrea 1975
Mytilus edulis	gill	33	Elliott 1974
Mya arenaria	gill	95	Stewart & Bamford 1975
Ascidia mentula	pharynx	50	West (unpublished)
Paracentrotus lividus	spine	8.4	de Burgh 1975

which uptake rate is half maximum) the importance of which is that it indicates the substrate concentration for which the transport system is specialised.

K_t values for amino acid transport by epidermal tissues typically fall in the range 10-100 μmoles/l (Table 1). They are an order of magnitude lower than those for invertebrate gut, and that for Paracentrotus is comparable to bacterial values (1-10 μmoles/l). Although the values are somewhat higher than the concentration of alanine in sea-water, they are extremely low compared with other animal absorptive tissues (Southward and Southward 1972). Thus, transport mechanisms of the epithelia which have been investigated are specialised for absorption of amino acids from μmolar concentrations, and can actively transport them at the levels in sea-water.

Uptake of various amino acids has been studied in isolated tissues of only a few invertebrates, but including representatives of several major benthic groups. However, considering the results obtained by other methods and for other dissolved substances it appears that uptake from sea-water is a widespread phenomenon among soft-bodied invertebrates. It has been reported for over 80 species including 10 cnidarians, 14 polychaetes and 27 echinoderms. It is generally associated with epithelia possessing microvilli, producing mucus and often bearing cilia -- features found in the majority of marine invertebrate groups with the important exception of the Arthropoda, in which DFAA uptake has not yet been convincingly demonstrated.

Thus it appears that benthic invertebrate epidermal epithelia are specialised for absorption of DFAA from sea-water, and that the process is widespread in the soft-bodied phyla.

THE SIGNIFICANCE OF ABSORBED DFAA IN NUTRITION

In the absence of precise information about nutritional requirements it is convenient to consider the contribution that absorbed DFAA could make to the reduced-C requirement for normal respiration. This varies greatly (Table 2), but it is significant to the gross requirements of at least some species.

There are, however, more subtle ways in which DFAA make important, perhaps vital, contributions to nutrition. For example, in many organisms it has been observed that some absorbed DFAA remain and are utilised at their sites of uptake in the epithelia, and are not translocated (e.g. Ferguson 1971, Schlichter 1974). It may be, therefore, that the epithelium itself depends on this source of nutriment whereas other tissues rely mainly on food absorbed through the gut. In echinoderms many peripheral appendages, e.g. crinoid cirri, are in poor contact with the circulatory systems; and some, e.g. echinoid spines and pedicellariae, have no circulatory supply whatsoever. Since

TABLE 2   DFAA Uptake as % of Respiratory Carbon Requirement

Organism	Substrate	umoles/l	%C*	Reference
Clymenella torquata	phe	0.6	15	Stephens 1963
	gly	0.8	27	Stephens 1963
	mixed AAs	20	108	Stephens 1963
Nereis virens	mixed AAs	20	16	Taylor 1969
N. diversicolor	1^y amines	54	70	Stephens 1975
Capitella	1^y amines	54	53	Stephens 1975
Siboglinum ekmani	phe	2.7	50	Southward & Southward 1972
S. fjordicum	phe	0.2	0	Southward & Southward 1972
S. atlanticum	mixed AAs	10	8	Southward & Southward 1972
Mytilus edulis (gill)	ala	1	8	Elliott 1974
Rangia cuneata	gly	46	4	Anderson & Bedford 1973
Paracentrotus (spine)	ala	1	15	de Burgh 1975
Strongylocentrotus	gly	0.4	9	Clark 1969
	ser	0.4	12	Clark 1969
	ala	0.2	3	Clark 1969
	asp	0.2	2	Clark 1969
Leptometra phalangium	mixed AAs	0.2	5	West (unpublished)

coelomocytes are now believed to be of little importance in nutrient translocation (e.g. Farmanfarmian 1969) it seems inescapable that direct absorption, well known to occur in echinoderm epidermis (de Burgh 1975, Ferguson 1971), is of paramount importance to these 'remote' organs.

Invertebrate eggs and larvae would seem excellent candidates for uptake of DFAA: they have large surface areas relative to volume, the gut may be absent or poorly developed, and they are often covered with 'absorptive' epithelia. Interestingly, Epel (1972) has shown that fertilisation of sea urchin eggs activates a latent amino acid transport mechanism. DFAA rapidly enter the fertilised egg and various developmental stages (Monroy and Maggio 1966), and become incorporated. The embryos or larvae of Actinia (Chia 1972), Neanthes (Reish and Stephens 1969) and Amphiura (Fontaine and Chia 1968) are known to absorb DFAA; and planktonic larvae of Nereis virens absorb at 200 times the rate achieved by the benthic adults (Bass et al. 1969).

It can be said, therefore, that DFAA are significant to the gross nutrition of some organisms, and that they play a regional role in the nutrition of others. They are probably of especial importance in the nutrition of larval forms.

THE FATE OF ABSORBED DFAA

Amino acids upon absorption enter the free pools in epithelial cells, from which they may be translocated elsewhere. Generally less than 20% of absorbed amino acids are converted into other soluble substances (Bamford and McCrea 1975, de Burgh 1975, Elliott 1974, Schick 1975, Stewart and Bamford 1975).

---

*Minimum values, in cases where ranges have been published.

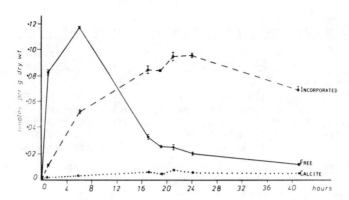

Fig. 1. Incorporation of 1 μM L-alanine by Paracentrotus spines

As time progresses a considerable proportion of the free (soluble) amino acids become incorporated into a form insoluble in 80% ethanol but soluble in KOH (Fig. 1). That such incorporation is largely into protein has been demonstrated by Fontaine and Chia (1968) for Cucumaria in which 80% of glycine incorporation was prevented by chloramphenicol, an inhibitor of protein sythesis. In Paracentrotus spines de Burgh (1975) found that all KOH-soluble label precipitated with TCA. Tritiated proline has been found to incorporate into the collagen both of Psammechinus (Pequignat and Pujol 1968) and of regenerating wound tissue in Calliactis (Young 1974). In Rangia Anderson and Bedford (1973) demonstrated that ^{14}C-glycine is incorporated mainly into protein, but also into nucleic acids and to a lesser extent into glycogen and lipids.

The incorporated radioactivity is what is seen in conventional autoradiography, and in carefully prepared autoradiographs mucus secreting cells, and later mucus itself on the epithelial surfaces, appear labelled (de Burgh 1975, Fontaine and Chia 1968, Southward and Southward 1968). Elliott (1974) ingeniously collected mucus produced by isolated gills of Mytilus incubated for one hour in 20 μM ^{14}C-alanine, and found it to contain five times the concentration of label in the sea-water. In view of the continuous (often copious) production and secretion of mucus by the epidermal epithelia of most soft-bodied marine invertebrates, it is not surprising to find that a large proportion of absorbed DFAA is directed to the synthesis of mucoid substances.

In echinoderms small amounts of amino acid C may be used in the synthesis of skeletal calcite (Fig. 1). In Aurelia polyps (Schick 1975), Clymenella (Stephens 1963), Ophiactis (Stephens and Virkar 1966), Echinocardium and Psammechinus (Pequignat 1970), and Paracentrotus (de Burgh 1975), the ^{14}C of labelled amino acids has been identified in respired CO_2. In the case of Paracentrotus spines this constituted less than 10% of alanine uptake, and about 1% of CO_2 production; in Aurelia, less than 1% of glycine uptake. It seems probable, therefore, that DFAA are relatively unimportant as respiratory substrates.

Ferguson (1975) found that in Echinaster amino acids are conserved in the free pool as an amino-N store, and are later utilised in the synthesis of gonadal products. In waters of fluctuating salinity amino acids are used for osmoregulation, presumably including DFAA. At constant salinities, too, there

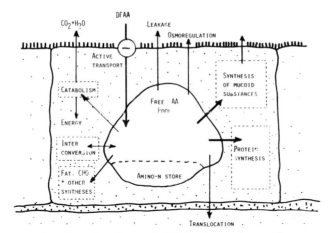

Fig. 2. The fate of DFAA after absorption

is some loss of absorbed amino acids to the surrounding sea-water, but in experiments using trace-labelled compounds (e.g. de Burgh 1975, Elliott 1974) and using gas-liquid chromatography for measuring absolute intracellular and DFAA concentrations (Ferguson 1971), net uptake has been indicated in all cases of free-living, soft-bodied invertebrates.

The main fate of absorbed DFAA is synthesis, primarily of proteins and mucoid substances. These, and other minor fates, are illustrated in Fig. 2.

THE ROLE OF INVERTEBRATES IN THE MARINE DFAA CYCLE

Since marine invertebrates of all the major groups except arthropods absorb DFAA the question arises whether uptake by them is significant in relation to the flux of DFAA in the seas (Fig. 3).

Relative to a DFAA standing crop of about 25 mg/m^3, the calculated daily input from phytoplankton exudates alone of 7 mg/m^3.day (40% of estimated released C) suggests an extremely high turnover rate in the euphotic zone.

If we imagine zooplankton at a concentration of 2 mg C/m^3, all of which remove alanine at the rate achieved by Paracentrotus spines from a concentration of 100 nmoles/l, their total rate of uptake of alanine would be about 0.5 mg/m^3. day. Alternatively, if the entire zooplankton absorbed leucine at the rate described by Bass et al. (1969) for the planktonic larvae of Nereis virens, uptake could account for some 80 mg/m^3. day of this amino acid. In relation to standing crop and input from phytoplankton exudates these figures, though crude and based on unrealistic assumptions, are noteworthy. They suggest that the invertebrates have a significant part to play in the marine DFAA cycle, and therefore in any attempt to quantify it their role should be considered.

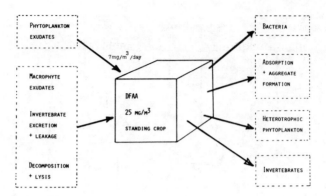

Fig. 3. Postulated flux of DFAA in the euphotic zone

REFERENCES

Anderson, J.W. and Bedford, W.B. 1973 The physiological response of the estuarine clam Rangia cuneata (Gray) to salinity. II. Uptake of glycine. Biol.Bull.mar.biol.Lab.Woods Hole 144, 229-247.
Bamford, D.R. and McCrea, R. 1975 Active absorption of neutral and basic amino acids by the gill of the common cockle, Cerastoderma edule. Comp.Biochem.Physiol. 50A, 811-817.
Bass, N., Chapman, G.J. and Chapman, J.M. 1969 Uptake of leucine by larvae and adults of Nereis. Nature,Lond. 221, 476-477.
Chia, F.-S. 1972 Notes on the assimilation of glucose and glycine from seawater by the embryos of a sea anemone, Actinia equina. Can.J.Zool. 50, 1333-1334.
Clark, M.E. 1969 Dissolved free amino acids in seawater and their contribution to the nutrition of sea urchins. Kelp Habitat Improvement Project Annual Report 1968-9, 70-93. California Institute of Technology, Pasadena.
de Burgh, M.E. 1975 Aspects of the absorption of dissolved nutrients by spines of Paracentrotus lividus (Lamarck). Ph.D. Thesis, Univ. of Dublin.
Elliott, A.J. 1974 Investigations on the uptake of dissolved amino acids by the branchial epithelium of Mytilus edulis L. M.Sc. Thesis, Univ. of Dublin.
Epel, D. 1972 Activation of a Na$^+$-dependent amino acid transport system upon fertilisation of sea urchin eggs. Expl Cell Res. 72, 74-89.
Farmanfarmian, A. 1969 Intestinal absorption and transport in Thyone. I. Biological aspects. Biol.Bull.mar.biol.Lab.Woods Hole 137, 118-131.
Ferguson, J.C. 1971 Uptake and release of free amino acids by starfishes. Biol.Bull.mar.biol.Lab.Woods Hole 141, 122-129.
Ferguson, J.C. 1975 The role of free amino acids in nitrogen storage during the annual cycle of a starfish. Comp.Biochem.Physiol. 51A, 341-350.
Fontaine, A.R. and Chia, F.-S. 1968 Echinoderms: an autoradiographic study of assimilation of dissolved organic molecules. Science, N.Y. 161, 1153-1155.
Monroy, A. and Maggio, R. 1966 Amino acid metabolism in the developing embryo. In Boolootian, R.A. Physiology of Echinodermata. Wiley, New York.
Pequignat, E. 1970 On the biology of Echinocardium cordatum (Pennant) of the Seine Estuary. New researches on skin digestion and epidermal absorption in Echinoidea and Asteroidea. Forma et Functio 2, 121-168.
Pequignat, E. and Pujol, J.P. 1968 Absorption cutanée de ^{3}H-proline à trés faible concentration et son incorporation dans le collagène chez Psammechinus miliaris. Bull.Soc.Linn.Normandie Ser.10, 9, 209-219.

Reish, D.J. and Stephens, G.C. 1969 Uptake of organic material by aquatic invertebrates. V. The influence of age on the uptake of glycine-C^{14} by the polychaete Neanthes arenaceodentata. Mar.Biol. 3, 352-355.

Schick, J.M. 1975 Uptake and utilization of dissolved glycine by Aurelia aurita scyphistomae: temperature effects on the uptake process; nutritional role of dissolved amino acids. Biol.Bull.mar.biol.Lab.Woods Hole 148, 117-140.

Schlichter, D. 1974 Aufname in Meerwasser gelöster Aminosäuren durch Anemonia sulcata Pennant. Z.Morph.Tiere 79, 65-74.

Southward, A.J. and Southward, E.C. 1968 Uptake and incorporation of labelled glycine by pogonophores. Nature, Lond. 218, 875-876.

Southward, A.J. and Southward, E.C. 1972 Observations on the role of dissolved organic compounds in the nutrition of benthic invertebrates. III. Uptake in relation to organic content of the habitat. Sarsia 50, 29-46.

Stephens, G.C. 1963 Uptake of organic material by marine invertebrates. II. Accumulation of amino acids by the bamboo worm, Clymenella torquata. Comp.Biochem.Physiol. 10, 191-202.

Stephens, G.C. 1975 Uptake of naturally occurring primary amines by marine annelids. Biol.Bull.mar.biol.Lab.Woods Hole 149, 397-407.

Stephens, G.C. and Virkar, R.A. 1966 Uptake of organic material by aquatic invertebrates. IV. The influence of salinity on the uptake of amino acids by the brittle star Ophiactis arenosa. Biol.Bull.mar.biol.Lab.Woods Hole 131, 172-185.

Stewart, M.G. and Bamford, D.R. 1975 Kinetics of alanine absorption by the gills of the soft-shelled clam, Mya arenaria. Comp.Biochem.Physiol. 52A, 67-74.

Taylor, A.G. 1969 The direct uptake of amino acids and other small molecules from seawater by Nereis virens Sars. Comp.Biochem.Phsyiol. 29, 243-250.

Williams, P.J.le B. 1975 Biological and chemical aspects of dissolved organic material in sea water. In Riley, J.P. and Skirrow, G. Chemical Oceanography 2nd Ed. vol.2, Chap.12. Academic Press, London.

Young, J.A.C. 1974 The nature of tissue regeneration after wounding in the sea anemone Calliactis parasitica (Couch). J.mar.biol.Ass.U.K. 54, 599-617.

STUDIES ON THE SHALLOW, SUBLITTORAL EPIBENTHOS OF LANGSTONE HARBOUR, HAMPSHIRE, USING SETTLEMENT PANELS

R. G. Withers and C. H. Thorp

Portsmouth Polytechnic Marine Laboratory, Ferry Road, Hayling Island, Hampshire, PO11 0DG, England

ABSTRACT

'Tufnol' panels were immersed in Langstone Harbour to investigate seasonal settlement, growth and competition amongst the epibenthos and the structure and development of the epibenthic community. The panels were suspended parallel to, and 0.5 m above, the flat, muddy sea-bed in a minimum depth of 4.0 m and current speeds of 1.7 to 4.6 km/hr. Algal settlement was largely prevented by turbidity but 151 animal species have so far been identified. The majority of sessile species settled over a broad period (spring-autumn), some exhibiting more than one settlement peak, but panels initially colonised by winter settlers subsequently attracted little summer settlement and vice versa.

Upper surface colonisation was dominated by species which tolerated silt deposition by displacing or outgrowing it (barnacles, erect bryozoans, hydroids), or forming temporary tubes within it (amphipods, polychaetes). The lower surfaces, however, although similarly colonised by barnacles and encrusting bryozoans initially, were soon characterised by compound ascidians which overgrew the early settlers and were in turn replaced by large solitary ascidians. After 4 - 8 months the lower surfaces were generally occupied by densely packed, pendulous ascidians supporting a prolific epifauna. These were major contributors to the total biomass of most well fouled panels and highest biomass values coincided with maximum development of the ascidian fauna. Encrusting sponges became significant only on panels immersed > 2 years, when peak abundance of the ascidians was passed.

INTRODUCTION

Langstone Harbour is centrally placed in an inter-connected system of 3 fully marine, shallow, natural harbours bordering the northern shore of the eastern Solent (south coast of England). It covers an area of approximately 20 km^2 at MHWS but largely dries out at low water, leaving 2 well defined channels. The dominant sediment type is mud and, with the tidal range varying between 1.5 m (neaps) and 4.5 m (springs), tidal streams are very rapid. Almost 100 Ml/day, dry weather flow, of sewage effluent are discharged in the vicinity of the harbour; 40% is secondarily treated, the remainder is macerated, raw sewage.

Since 1945, scientists at the Ministry of Defence (Navy) Exposure Trials Station (ETS), Portsmouth, have collected data on fouling organisms in Chichester and Langstone Harbours, mostly using raft-borne, vertically aligned, settlement panels, suspended within 1.5 m of the surface. A paper by Pearce and Chess (1971) outlining the use of panels fixed to a frame, anchored close to the seabed, to study epibenthic communities off North America, prompted the initiation of a similar study in Langstone Harbour. Our objectives were to extend existing information on the structure and development of the epibenthic communities of the harbour's hard surfaces, with a view to providing much needed base-line data against which to measure the effects of increasing urban pollution, and to observe settlement, growth and competition between both major and minor fouling organisms.

The present data were collected from panels exposed in the Langstone Channel, close to the ETS rafts, in a minimum depth of 4.0 m. There, maximum current speeds (Houghton, 1959) ranged from 2.59 km/h (neaps) to 4.64 km/h (springs) and the water was turbid with visibility rarely exceeding 3.0 m. Sea temperatures in Langstone Harbour have been monitored since 1974. Mean monthly values ranged from 4^0C, (February, 1976) to 20^0C, (August, 1975; July and August, 1976).

MATERIALS AND METHODS

The panel holder was constructed from galvanised steel and comprised a rectangular frame (2.6 x 2.2 m) of tubing (o.d. 50 mm), supported on short legs (0.5 m) with distal plate-like metal shoes (0.3 x 0.3 m) to prevent undue sinking into soft sediments. Lengths of thinner (o.d. 25 mm), but similar, tubing were welded across the top of the main frame to support horizontal settlement panels. The panels (260 x 260 x 6 mm), cut from 'Tufnol' sheets (a paper laminate impregnated with phenol-formaldehyde resin), were attached to the frame by means of plastic pipe-clips bolted to the underside at each corner. This greatly facilitated rapid attachment and removal of the panels by SCUBA diving. Panels were immersed for periods of time ranging from 7 days to > 2 years. On removal from the frame, they were immediately placed in fine-mesh bags, before transfer to the surface, in order to minimise the loss of free-swimming species. Panels could be maintained in their mesh bags under running sea water in the laboratory until their examination was completed. Routine assessment of the panel fauna was made by detailed examination, identification and counting under a dissection microscope and a photographic record was made. For biomass assessments, groups of panels were immersed at different seasons and, over periods exceeding 2 years, 2 - 3 panels at a time were removed. Biomass values (ash-free dry weight) were determined by carefully scraping the panels, drying the fauna thus removed to constant weight at 85^0C and ashing in a muffle furnace at 425^0C.

RESULTS

151 animal and 10 plant taxa have been identified from the panels (Appendix 1). Little attention has been given to the few algae occurring on the benthic panels but, with the exception of *Laminaria* sporelings, recorded in 1976, they were all small members of the Rhodophyceae. Further information on the fouling algae of the harbour is given by Fletcher (1974). Animal species occupied three broad groupings, those which attached permanently, those which formed temporary attachments, often in the form of soft tubes, and those which were wholly free-living. While information has been collected on all three, this paper concentrates mainly on those in the first group. The data are presented under the following headings: 1) settlement and growth; 2) community development and structure.

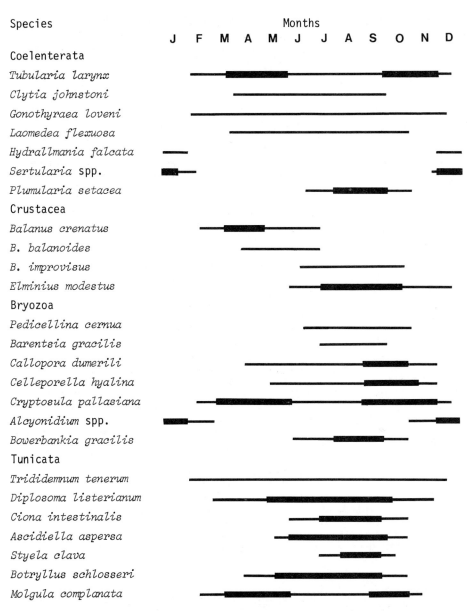

Fig. 1. Settlement Periods of selected sessile species. Thick lines indicate periods of maximum settlement.

Settlement and Growth

This was largely investigated between November 1972 and December 1974, using panels immersed for up to eight weeks both on the benthic frame and within 1.5 m of the surface on one of the ETS rafts. Species settlement periods were subsequently monitored in the course of other experiments using benthic panels only. The settlement periods of some of the more common species are indicated in Fig. 1. Although most species settled between spring and autumn, sometimes with more than one period of peak settlement, a few settled only in winter e.g. the large hydroids *Sertularia* spp. and *Hydrallmania falcata* (L.), and the bryozoans *Alcyonidium* spp. The settlement of these hydroids and to a lesser extent that of *Alcyonidium* spp., was confined to the benthic panels but colonisation was slight on those immersed for less than 4 weeks. Heaviest settlements occurred on those immersed for more than 8 weeks, e.g., a panel immersed from 19.10.73 to 29.01.74 collected 169 stalks of *Sertularia* spp., 55 stalks of *H. falcata* and 676 colonies of *Alcyonidium* spp.

The hydroid *Tubularia larynx* Ellis and Solander, the bryozoan *Cryptosula pallasiana* (Moll) and the solitary ascidian *Molgula complanata* Alder and Hancock frequently exhibited settlement peaks in both spring and autumn. Autumn settlements, at times very heavy (maxima of 100 colonies, 250 colonies and 80 individuals/8 week panel, respectively), extended into December, while spring settlements usually commenced in March. Very occasionally, a very few colonies both of the above species and the compound ascidians, *Trididemnum tenerum* Verrill and *Diplosoma listerianum* (Milne Edwards), settled throughout the winter. Most other species preferred to settle between June and September and some were restricted to this period. The following species settled abundantly on raft-borne, near-surface panels but were relatively scarce on benthic panels: the sponges *Grantia compressa* (Fabricius) and *Sycon ciliatum* (Fabricius), the crustaceans *Balanus balanoides* (L.) and *Jassa falcata* (Montagu), and the bryozoan *Bugula stolonifera* Ryland. *J. falcata* was always abundant and colonised new panels throughout the year, but especially in spring, while *B. stolonifera* settled, sometimes abundantly, between July and September. On benthic panels *Corophium* spp. replaced *J. falcata* as the dominant tubicolous amphipod while *Bicellariella ciliata* (L.), absent on near-surface panels, replaced the similar and closely related *B. stolonifera*.

Maximum growth of the common sessile species occurred in summer, declining almost to zero in winter. Autumn and winter-settling species exhibited little growth until the following spring when they accelerated very rapidly. Common endemic ascidians and most other species normally approached their maximum size within a single growth season and certainly within a year of settlement. The ascidian, *Styela clava* Herdman, introduced from the Far East in 1954, however, grew relatively slowly, reaching full development only after 18 months to 2 years.

Community Development and Structure

Between spring and autumn, the sessile macrofauna initially colonising benthic panels comprised hydroids, barnacles and encrusting forms of bryozoans and compound ascidians. The more common species were *Clytia hemisphaerica* (L.), *Obelia* spp., *Gonothyraea loveni* (Allman), *Laomedea flexuosa* Hincks, *Plumularia setacea* (Ellis and Solander), *Elminius modestus* Darwin, *Balanus crenatus* Bruguière, *C. pallasiana, Callopora dumerili* (Audouin), *T. tenerum, D. listerianum* and *Botryllus schlosseri* (Pallas). Although some smaller hydroids quickly declined in importance, the majority of initial colonisers became major components of the developing community. Most other major sessile species settled after 6 - 8 weeks, in summer, e.g. *Pedicellina cernua* (Pallas),

Bowerbankia gracilis Leidy, *Ascidiella aspersa* (O. F. Müller), *Ciona intestinalis* (L.) and *Molgula* spp. along with many minor species which were largely determined by the date of immersion. Panels immersed in late autumn and early winter attracted heavy settlements of winter-settling species, e.g. *Sertularia* spp., *H. falcata* and *Alcyonidium* spp., while those already well fouled by summer-settling species were subjected to much lighter settlements. Similarly, heavy winter settlement discouraged subsequent colonisation by most other sessile species. Benthic copepods and other small active species arrived on panels within a few days of immersion. Small numbers of other semi-sessile and free-living organisms appeared in the first 2 months but their greatest incidence normally occurred only after considerable sessile epibenthos had developed. From an early stage, differences became manifest between the fauna of the upper and lower surfaces because silt, suspended in the water column, settled out on the former surface but not on the latter.

Upper surface. Silting, to a depth of 10 mm on the more heavily colonised panels, was greatest at slack water following rough weather. Tidal currents, however, tended to cleanse the panels although a thin layer, adhering to the primary film, remained even on those recently immersed.

Algae, largely confined to the upper, illuminated surface, were erect forms; silt deposition presumably excluding crustose species. The fauna was normally dominated by barnacles, especially *E. modestus* (50 - 80% cover), while medium and large hydroids (*P. setacea, Sertularia* spp.) were common, and occasionally dominant on panels immersed in winter. Numerous digitate lobes of *Alcyonidium gelatinosum* (L.) also grew on the latter panels and, during the summer, they were joined by *B. ciliata*, colonising both the panel surface and old hydroid stalks. Routine laboratory observations suggested that while barnacles probably dispersed silt by their feeding movements, the other species simply outgrew the deposit. Small Bryozoa e.g. *Bowerbankia gracilis*, *P. cernua* and *Barentsia gracilis* (Sars), frequently abundant on the upper surface, avoided the silt by growing over or settling on, larger erect species. The only common solitary ascidian was the small species, *M. complanata*, and although compound ascidians settled frequently, their subsequent development was generally poor; larger colonies adopted a lobe like, upward growth form around hydroid stalks. The small, sessile species mentioned above, together with others such as the sponge *Leucosolenia botryoides* (Ellis and Solander), formed the upper layer of the many organisms living on and amongst the barnacles. Many tubicolous forms colonised crevices between barnacles while nematodes and other free-living species dwelt between and beneath the tubes.

Lower surface. The faunal community developed rapidly through a number of stages before reaching a relatively stable state, persisting for at least 12 months, in which compound and/or solitary ascidians were dominant. Initial colonisation was similar to that on the upper surface and, although *E. modestus* was less common, greater numbers of the other barnacles resulted in a 30% cover. During the early development of the community, encrusting bryozoan colonies expanded rapidly, smothering some of the barnacles. Even greater success was achieved by compound ascidians, especially *D. listerianum* and *B. schlosseri*, which outgrew both barnacles and bryozoans to reach, within a few weeks in summer, a cover in excess of 60%. Most species recorded from the upper surface also occurred on the lower surface but were generally outcompeted by the encrusting bryozoans and ascidians. Hydroids, such as *P. setacea* and, on panels which had received winter settlements, *Sertularia* spp., were an exception, often being abundant, especially towards the panel edges where they provided substrata for settlement of small erect bryozoans. After 4 - 12 months, depending on season, solitary ascidians, especially *A. aspersa* and *C. intestinalis*, became dominant or co-dominant. Their annual abundance varied considerably but, when great, reduced the water flow close to the undersides of the panel. While this resulted in a considerable reduction of previously common, encrusting species on the panel surface, the tunics of the solitary ascidians not only provided a greatly increased surface area for recolonisation, but also elevated the major encrusting species clear of the panel surface, and into increased water flow. *A. aspersa* and *S. clava*, with relatively rigid tunics, were colonised by many phyla but the more contractile *C. intestinalis* and *M. manhattensis*, supported mainly hydroids and *D. listerianum*.

After 1.5 - 2 years immersion, the numbers of large *A. aspersa* and *C. intestinalis* on heavily fouled panels decreased. This resulted from both individual mortality and groups of live animals being detached, either through drag (a consequence of an ever-increasing epizoite load) or predators. Juvenile recruitment of the above species together with development of the slower growing *S. clava* and massive encrustaceans of compound forms, e.g. *T. tenerum*, normally ensured continued ascidian dominance although encrusting sponges, especially *Halichondria* spp., occupied 5 - 40% of the panel surface. Small sponge colonies were observed, after 4 - 6 months immersion, growing successfully beneath densely packed, solitary ascidians, where faster growing species survived with difficulty. The significance of the subsequent slow expansion of the sponge colonies, often over the panel edges, was enhanced by the fact that very few other sessile organisms were able to colonise their surfaces although many free-living and tubicolous species occurred in, on and under them.

The major contribution to the biomass of heavily fouled panels was made by large erect or pendulous species such as *A. gelatinosum*, *S. clava*, *S. aspersa* and *C. intestinalis*, because they not only occupied a relatively small attachment site, but also usually supported a considerable epifauna. The highest recorded biomass values of 49.5, 40.6 and 37.03 g ash-free dry wt/0.1 m² resulted from 3 panels immersed from May, 1974, to June, 1975, on which *A. aspersa* and *C. intestinalis* were abundant (mean total of both species = 85/panel). *In situ* observation, in June, 1975, of the remaining panels in the group indicated the presence of similarly abundant ascidian populations but, on the panels' removal, some 3 - 12 months later, the number of ascidians had declined drastically and the biomass values were in the range of 12.5 - 27.8 g ash-free dry wt/0.1 m².

DISCUSSION

There is little doubt that this continuing study will greatly extend our knowledge of the epibenthos of the Solent area: some 30 species, previously unrecorded in published data, have already been identified (Appendix 1). Indeed, *Obelia bidentata* Clarke, although recorded from Norfolk (U.K.), Belgium and Holland (Cornelius, pers. comm.), is identified from the south coast of England for the first time.

Previous settlement data for *T. larynx*, barnacles and ascidians in Langstone and Chichester Harbours (Stubbings and Houghton 1964) agree broadly with those presented here. On the lower surface of our benthic panels *D. listerianum* and *B. schlosseri* achieved an almost 100% cover within 38 days of immersion during July and August. This confirms the data of Stubbings and Houghton who showed the major settlement period of *D. listerianum* to be, usually, June-August, beginning when sea temperatures exceeded 14 - 16°C. Such a complete ascidian cover was surely partly responsible for the low incidence of some other common encrusting species at this time; their larvae soon being excluded from the panel surface and their young colonies often being overgrown and killed. In autumn, as the settlement and growth of compound ascidians declined, large settlements of encrusting bryozoans (*C. pallasiana, C. dumerili* and *Celleporella hyalina* (L.)) frequently occurred.

Settlement data for the winter-settling species *Sertularia* spp., *H. falcata* and *Alcyonidium* spp., are few but their presence on benthic panels from Langstone Harbour is not surprising in view of their local abundance. SCUBA diving has shown them to be the dominant epiziotes on the multitudes of *Crepidula fornicata* (L.) shells which carpet large areas of the harbour. While silt deposition on the upper panel surface undoubtedly resulted in a restricted fauna, certain animals, e.g. *P. ciliata*, erect bryozoans and nematodes, flourished: this observation confirms the findings of Nair (1962) and Persoone (1971) that *P. ciliata* and nematodes became abundant on panels in very turbid conditions. Similarly Maturo (1959), appreciating that silt deposition discouraged settlement of many bryozoans on horizontal panels, noted that *Bowerbankia gracilis* and *P. cernua* were unaffected, the latter probably displacing the silt by its active bowing and swaying movements.

Gulliksen and Skjaeveland (1973) reported the predation of *C. intestinalis* by *Asterias rubens* L. and, more recently, Gulliksen (1975) noted a similar predation of *Ascidia conchilega* O.F. Müller by *Trivia arctica* (Montagu). Although both predators have been recorded locally, neither has been observed on benthic panels in Langstone Harbour. It would appear that the crabs, *C. pagurus* and *P. puber*, most frequently consume or accidently dislodge the larger ascidians from the panels, although other predators include the wrasses *Labrus bergylta* Ascanius and *Crenilabrus melops* (L.), and the pouting *Gadus luscus* L.

CONCLUSIONS

Settlement panels immersed in Langstone Harbour, parallel and close to the seabed, were colonised by an insignificant number of algae and a very large number and variety of animals. A maximum biomass of 49.5 g ash-free dry wt/0.1 m^2 has been recorded. Although some of the macrofauna settled exclusively, but abundantly, in winter, most settled from spring to autumn and matured within one year.

The upper surfaces of the panels received a deposition of silt from the turbid water column and the fauna was dominated by species which either displaced (barnacles) or outgrew (hydroids, erect bryozoans) the deposit or dwelt within it (nematodes, polychaetes, amphipods). The lower surfaces were dominated by compound and/or solitary ascidians. The latter were heavily colonised by epizoites and made the major contribution to the biomass on many panels. After > 2 years immersion, panels often collected a significant covering of sponges.

ACKNOWLEDGEMENTS

We wish to acknowledge the assistance of D. R. Houghton and the staff of the Exposure Trials Station, Portsmouth, Hampshire.

REFERENCES

Fletcher, R.L. 1974 Results of an international co-operative research programme on the fouling of non-toxic panels by marine algae. Bulletin de liaison du comite international permanent pour la recherche sur la preservation des materiaux en milieu marin, 14, 7 - 31.

Gulliksen, B. 1975 The prosobranch *Trivia arctica* as predator on the solitary ascidian *Ascidia conchilega*. Proc. malac. Lond. 41, 377-378.

Gulliksen, B. and Skjaeveland, S.H. 1973 The sea-star, *Asterias rubens* L., as predator on the ascidian, *Ciona intestinalis* (L.), in Borgenfjorden, North-Tröndelag, Norway, Sarsia 52, 15 - 20.

Houghton, D.R. 1959 Tidal measurements in Langstone Harbour, Hampshire. The Dock and Harbour Authority 468, vol. 40, 172-179.

Marine Biological Association 1957 The Plymouth Marine Fauna, 3rd Ed. Plymouth.

Maturo, F.J.S. 1959 Seasonal distribution and settling rates of estuarine Bryozoa. Ecology 40, 116 - 127.

Nair, N.B. 1962 Ecology of marine fouling and wood-boring organisms of Western Norway. Sarsia 8, 1 - 88.

Parke, M. and Dixon, P.S. 1976 Check-list of British marine algae - third revision. J. mar. biol. Ass. U.K. 56, 527 - 594.

Pearce, J.B. and Chess, J.R. 1971 Comparative investigations of the development of epibenthic communities from Gloucester Massachusetts, to St. Thomas, Virgin Islands. Pp. 55 - 61 in 4th European Marine Biology Symposium (D. J. Crisp. ed.). Cambridge University Press.

Personne, G. 1971 Ecology of fouling on submerged surfaces in a polluted harbour. Vie Milieu suppl. 22 (3rd European Marine Biological Symposium) 613 - 636.

Stubbings, H.G. and Houghton, D.R. 1964 The ecology of Chichester Harbour, S. England, with special reference to some fouling species. Int. Revue gen. Hydrobiol.49, 233 - 279.

APPENDIX 1   Species List for the Benthic Settlement Panels
             in Langstone Harbour

Except where an authority is given, nomenclature follows Parke & Dixon (1976) and the Plymouth Marine Fauna (Marine Biological Association 1957).
  * Species new to published data from the Solent area.
 ** Species new to published data from the south coast of England.

ALGAE

Laminaria saccharina
Antithamnion plumula
Callithamnion tetragonum
Ceramium rubrum
Griffithsia floculosa

Spermothamnion sp.
Cryptopleura ramosa
Hypoglossum woodwardii
Polysiphonia nigrescens
P. urceolata

PORIFERA

Leucosolenia botryoides
Sycon ciliatum
Grantia compressa
Hymeniacidon perleve

Halisarca sp.
Halichondria panicea
H. bowerbanki

COELENTERATA

Tubularia indivisa
T. larynx
Coryne sp.
Bougainvillia sp.
Eudendrium ramosum
Clytia hemisphaerica (L.)
Obelia dichotoma
O. geniculata
O. flabellata Hincks
O. bidentata Clarke**
Gonothyraea loveni
Laomedea flexuosa
*Phialella quadrata**

Opercularella lacerta
Dynamena pumila
*Sertularella polyzonias**
Hydrallmania falcata
Sertularia argentea
S. cupressina
Kirchenpaueria pinnata
Plumularia setacea
*P. halecioides**
Tealia felina
*Diadumene cincta**
Metridium senile

PLATYHELMINTHES

*Fecampia erythrocephala**
Leptoplana tremellaris
Rhabdocoela: unidentified

*Prostheceraeus vittatus**
*Cycloporus papillosus**

NEMATHELMINTHES

Nematoda: unidentified

NEMERTINI

Amphiporus lactifloreus

ANNELIDA

Lepidonotus squamatus
Harmothoë lunulata
Lagisca extenuata
Phyllodoce maculata
Kefersteinia cirrata
Syllis gracilis
S. prolifera
Autolytus sp.
*Leptonereis glauca**
Nereis pelagica

Platynereis dumerili
*Ophryotrocha puerilis**
Polydora ciliata
Sabellaria spinulosa
Amphitrite johnstoni
Polymnia nebulosa
Nicolea venustula
Sabella pavonina
Hydroides norvegica
Pomatoceros triqueter

CRUSTACEA

*Harpacticus gracilis**
Atleutha interrupta
*A. depressa**
Tisbe furcata (Baird)*
T. tenera (Sars)*
*Thalestris longimana**
Parathalestris harpacticoides
Ameiropsis brevicornis (Sars)*
*Verruca stroemia**
Balanus crenatus
B. balanoides
B. improvisus
Elminius modestus
Limnoria lignorum
Idotea granulosa
*Nannonyx goësi**
*Lysianassa ceratina**
Orchomene humilis
*Leucothoë spinicarpa**
*Stenothoë spinimana**
*S. valida**
Nototropis swammerdami

Melita palmata
Gammarus locusta
Aora typica
Microdeutopus anomalus
Jassa falcata
*Corophium insidiosum**
C. acherusicum
C. sextoni
*C. acutum**
Phthisica marina
Caprella aequilibra
Athanas nitescens
Palaemon serratus
P. elegans
Galathea squamifera
Porcellana longicornis
Cancer pagurus
*Portunus puber**
Carcinus maenas
Pilumnus hirtellus
Macropodia rostrata

ACARINA

Halacaridae: unidentified

PYCNOGONIDA

Nymphon gracile
N. rubrum
*Anoplodactylus pygmaeus**

*Endeis spinosa**
E. laevis Grube
Achelia echinata Hodge

MOLLUSCA

Crepidula fornicata
*Onchidoris muricata**
Doto coronata
*Trinchesia aurantia**

*Tergipes despectus**
Mytilus edulis
Anomia ephippium

BRYOZOA

Pedicellina cernua
Barentsia gracilis
Scruparia chelata
Electra pilosa
Conopeum reticulum
Callopora dumerili
Bicellariella ciliata

Bugula stolonifera Ryland
Celleporella hyalina
Cryptosula pallasiana
Alcyonidium gelatinosum
A. polyoum
Bowerbankia gracilis Leidy

TUNICATA

Clavelina lepadiformis
Distaplia rosea
Polyclinum aurantium
Morchellium argus
Trididemnum tenerum
Diplosoma listerianum
Ciona intestinalis
Perophora listeri
Ascidiella aspersa

Ascidia conchilega
Styela clava Herdman
Polycarpa rustica
P. fibrosa
Dendrodoa grossularia
Botryllus schlosseri
Botrylloides leachi
Molgula manhattensis
M. complanata

VERTEBRATA

Blennius pholis
Pholis gunnellus

Cottus bubalis
Cyclopterus lumpus

MODIFICATION OF ASSOCIATION AND SWARMING IN NORTH ADRIATIC *MYSIDACEA* IN RELATION TO HABITAT AND INTERACTING SPECIES

Karl J. Wittmann

I. Zoologisches Institut der Universitat Wien, Lehrkanzel fur Meeresbiologie, A-1090 Wien, Wahringerstrasse 17/VI, Austria

ABSTRACT

The distribution of *Mysidacea* has been studied in Strunjan Bay and islands off Rovinj (Istria, Yugoslavia). Underwater observations were made by diving, TV, and time-lapse-camera.

Different species show a preference for certain bathymetric zones and, within a zone, occupy different substrates and distinct microhabitats or differ in period of activity, so that they are usually aggregated in monospecific swarms. Occurrence of polyspecific swarms is referable to overlapping requirements and to interaction of species.

Swarm formation is controlled by two patterns: habitational and inter-individual responses. The relative influence of both patterns differs in species - this is referable to different substrate-relations.

INTRODUCTION

Swarming of mysids has won increasing interest in recent years. The chief attention has been given to the distributional, behavioural, and physiological aspects of swarming (Clutter 1967, 1969, Macquart-Moulin 1971, 1973, 1973a, Mauchline 1971, Zelickman 1974). The general task of this report is to describe and compare modifications of swarming in relation to the habitat, where mysids occur, and to estimate the patterns which might control and influence swarm formation. This report is mainly concerned with swarming in bright light and with the effects of distribution on swarming. Possible causes and functions of distribution are not discussed - more detailed data will be provided by Wittmann (in prep.) at a future time. The author did not follow the previous definitions of grouping in mysids (Clutter 1969, Mauchline 1971, Zelickman 1974). The following definitions are primarily geometric and secondarily related to social interaction:
swarm: Close groups showing regularity in spatial arrangement (inter-individual distance). Discontinuities are referable to structuring in subswarms.
school: Swarms consisting of individuals swimming in the same direction.
subswarm: Internal structures of a swarm differing in abundance of distinguishable individual types (species, age groups, colour).
aggregation: No or reduced regularity in spatial arrangement. Internal density changes are not referable to abundance of distinguishable individual types. Outside, the density often becomes continuously thinner until dispersion.
dispersion: The inter-individual distance is so great that social interaction can be excluded.
guests: The non-dominant species in an association.

MATERIALS AND METHODS

Field collections and observations were made on the eastern coast of the North Adriatic Sea, off the Istrian peninsula, during the period 1972-75. The stations were Strunjan Bay and islands off Rovinj: I. Banjole, I. Figarola, I. Katarina, I. Rossa. Recent studies at Ischia (Gulf of Naples, Italy) are partly included. Collections were made mainly along transect lines perpendicular to shore. This paper mainly deals with the most abundant genus *Leptomysis*: *Leptomysis mediterranea* G.O. Sars, *Leptomysis lingvura* G.O. Sars, *Leptomysis sardica* (G.O. Sars), *Leptomysis gracilis* G.O. Sars, *Leptomysis apiops* (G.O. Sars), and the undescribed species *Leptomysis* sp.A, and *Leptomysis* sp.B. The paper deals with the following species of other genera: *Siriella armata* (Milne-Edwards), *Siriella jaltensis* (Czerniavsky), *Mesopodopsis slabberi* (van Beneden), *Anchialina agilis* (G.O. Sars), *Neomysis sp.*, and *Acanthomysis* sp. Some species exhibited intraspecific differences in distribution and behaviour. These differences are linked to variations in numbers of setae. In regard to the foregoing, the author is solely concerned with the most abundant variety of each species - detailed information shall be available in Wittmann (in preparation).

451 samples were obtained by diver operated nets at and near the bottom, and by plankton nets near the surface and at the surface itself (hyponeuston). These methods allowed a semi-quantitative reconstruction of macrodistribution and its diurnal changes. Monthly collections were made at Strunjan Bay during the period July 1973 - July 1974.

Underwater observations were made by diving, TV, and time-lapse-camera (intervals of 16 sec.). Microdistribution has been observed by these methods combined with field photographs.

Adult mysids could tolerate being out of water for up to 6 seconds and were marked, on the carapace, with spots of Ripolin paint. Following this, healthy specimens were liberated in the field where they could subsequently be found for up to 12 days (females of *Leptomysis lingvura* moult every 12-14 days).

DIURNAL CHANGES AND MAINTAINANCE OF DISTRIBUTION

Swimming activity in mysids is photokinetically controlled by light intensity (Macquart-Moulin 1971, 1973a). All observed mysids were active at night, in caves, inactive mysids sit on or cling to the substrate. Usually they do not swarm, but in caves however they may swarm or be guests in swarms (*Siriella jaltensis, Anxhialina agilis*). Semi-active and active mysids occur in dispersion, aggregations and/or swarms.

At Strunjan Bay swarms of *Leptomysis lingvura* and *Leptomysis* sp.A have been observed by TV (2 daytime sequences) and time-lapse-camera (2 night sequences). The observations are combined with results obtained by photographs, plankton samples, and near bottom samples taken at Strunjan Bay, I. Figarola and Ischia.

Both species show nearly the same microdistributional behaviour, but they are separated by bathymetric zonation. At Strunjan Bay *Leptomysis lingvura* occurs from 0.5 to 3 m, *Leptomysis* sp.A from 5 to 8 m. At daytime they swim pendulously to and fro, the main axes of movement oriented to discontinuities of bottom structure (parallel to rocks, recesses, cave entrances). The swarms remain motionless, while all mysids move to and fro. Each moving object (greater than about 3 cm and moving relative to the water movement) causes escape behaviour. After a disturbance, mysids return exactly to the place where they were before, so that no new adaption of the camera was necessary, even when the greatest zoom was used.

At sunset, pendula become longer and the inter-individual distances increase. The swarm (adults) expands slowly out of the recesses to a layer of 10 to 30 cm above the bottom. At a surface light of 9.6 mW/cm^2 the swarms start to loose monospecificity. Mysids, which have been inactive in daytime, rise up from their substrate and the cavernicolous mysids come out. They migrate into the near-bottom layers, surface plankton and hyponeuston (compare Macquart-Moulin 1971). In bright moon light a part of the swarm remains in a slightly expanded state in shallow water. When surface light falls below 0.8 mW/cm^2 mysids start to sink down to a layer of 0 to 7 cm above the *lingvura* population rises up into the surface plankton (compare Macquart-Moulin 1973). The remaining mysids are now members of a polyspecific night-aggregation near the bottom. The expansion continues. Where dense populations occur, the bottom is covered with mysids nearly everywhere about one hour after sunset, if the water is quiet. At the place where the original swarm stood, mysids occur in greater densities until sunrise. At places where rocks border on sand bottom, polyspecific swarms of *Leptomysis lingvura* and *Leptomysis mediterranea* occur in daytime. They also maintain substrate-preferences during night expansion, the former species mainly covering rocks and algae, the latter mainly sandy bottoms.

At sunrise the mysids rise up and concentrate in expanded swarms. Being now monospecific, most of the swarms contract to the same places where they stood the day before. This happens so precisely that the swarm is sharply focused by the camera again.

This return is not caused by a homing behaviour *sensu stricto*, but is probably caused by a precise habitat-preference. At I. Figarola in a rocky area of 140 m^2 places, where swarms of *Leptomysis lingvura* occurred, were marked and mapped. Within 16 days the number of swarms varied from 12 up to 26. They only occupied 31 different places during this places. 9 places were occupied every day (average swarm diameter: 20 cm). Three years later 11 swarms occurred in this district, 5 of them at places where swarms had been before. At the marked places sometimes a single specimen was found alone. These were mainly adult females, which swam pendulously to and fro in a similiar manner as they do in swarms. The single specimen differed from swarmers in showing reduced return after artificial disturbance.

Marked specimens showed a daily return-rate ranging fron 0 to 93%, mainly related to the amount of swarms occurring in the surroundings. This is due to the fact that the next day these specimens often occurred in neighbouring swarms. Therefore swarms return to the same place more exactly than their members do.

DISTRIBUTIONAL SEPARATION

Interspecific Separation

Interspecific separation seems to be a world-wide phenomenon in swarming mysids Swarms are well known to be usually monospecific in bright light and the various species to be separated by bathymetric zonation (Clutter 1967, 1969, Mauchline 1971). Zonation and substrate-preferences of *Leptomysis* species are independent. On one hand the species only occur where their substrates occur, on the other hand the zones do not expand where the substrate overlaps neighbouring zones (Table I). The zones were nearly identical in different regions and show no seasonal or diurnal changes with the exception of *Leptomysis apiops*, which migrated into shallow waters (4-28 m) in the period December - March. Similar results were obtained by Clutter (1967). It seems to be a rule, that swarms which occur in overlapping bathymetric zones often prefer different substrates while swarmers preferring the same substrate often occur in separate zones.

Intraspecific Separation

Intraspecific separation is quite common in mysids. Swarms show a different composition of age groups and within the swarms, age groups separate into subswarms. The subswarms are uniform in regard to the length of their members and inter-individual distances (Clutter 1969, Macquart-Moulin 1973, Mauchline 1971, Zelickman 1974). Subswarms of *Leptomysis lingvura* show nearly the same return-behaviour as do swarms. After each disturbance, swarms restructure themselves into subswarms again, which usually stay at the same places where they stood before. After artificial or natural changes in the population of subswarms, they can join to intimately neighbouring swarms. Neighbouring swarms can connect to form a common swarm. Subswarms are generally more cohesive than swarms. After capture of a subswarm the remaining mysids organize themselves in subswarms again. The place where the captured subswarm was located is now completely or partly filled by another subswarm. These place-changes indicate that arrangement of subswarms is partly referable to avoidance of other age groups and is also

TABLE I Bathymetric Zonation and Substrate-Preferences of the Genus *Leptomysis*

species	substrates	SW	SE	F	B	R	K
lingvura	*Anemonia sulc.* rocks, algae	0.5-3	0.5-2	0.5-4	-	0.5-6	0.8-2.5
sp.A	as in *lingvura*	4-8	-	-	6-10	6-12	-
sp.B	rocks, algae	-	-	-	14-30	-	-
gracilis	mud, biogenic structures	21-28	-	-	-	-	-
apiops	as in *gracilis*	23-28	18-28	-	30-34	-	-
medit.	sand	-	1.5-5	1-5	-	2-8	1-8
sardica	*Zostera* beds	-	1-12	-	-	-	-
greatest depth observed		28	28	15	34	15	10

Depths in meters below low water level.
SW = Strunjan Bay west SE = Strunjan Bay east F = I.Figarola
B = I.Banjole   R = I.Rossa   K = I.Katarina

partly referable to habitat-preferences. At I. Figarola juveniles of *Leptomysis lingvura* swam at distances of I to 7 cm above the substrate and adults at distances of I to 28 cm. Where *Amemonia sulcata* populations occur, juveniles are dominant in most of the associated swarms, while adults are dominant in other swarms.

ASSOCIATIONS

The occurrence of polyspecific swarms in bright light is considered in relation to the two assumed causes:

The Habitat-Association

All species occur at their usual habitat, but the habitats are near enough and the swarms are large enough, so that both swarms overlap and connect to subswarms of one common swarm. After artificial reduction of the population, the author often succeeded in separating the subswarms and in installing two monospecific swarms, which then stood at a distance of I0 to 20 cm from each other. These habitat-associations are characterized by a zonal overlap and by closely bordering of overlapping substrates. The subswarms stay at the same substrate the species prefer, as monospecific swarms. All possible combinations of *Leptomysis* species which showed zonal overlapping (Table I) were found to be associated. In entrance of microcaves, cave-insiders and cave-outsiders associate in swarms consisting of up to 5 species belonging to different genera of the family *Mysidae*.

The Gregarious Association

The guest-species occur at the usual habitat of the dominant species, and not where they usually occur as a monospecific swarm. Therefore it can be assumed, that those associations are a result of species interaction. The animals cannot identify others exactly like themselves and prefer grouping with other species rather than being alone.

These associations are only present for short time periods. The number of guests range from I to a maximum of 8 specimens. If guests form subswarms, they often do not follow the returning of the dominant species after disturbance (*Leptomysis lingvura*, *Leptomysis mediterranea*, *Leptomysis* sp.A). In *Leptomysis lingvura* mainly adult males and immatures of both sexes are guests in gregarious associations. Species interaction also is visible in habitat-associations. Intimately neighbouring swarms often have a thin connection consisting of mysids. Associated species tolerate one another, and *in situ* there is no visibly increased aggression.

SWARMING AND SUBSTRATE

In situ observations in bright light on adult mysids at Strunjan Bay showed that swarming of mysids is influenced by substrate-relations. There is a broad spectrum ranging from substrate-specialists, which do not swarm, to swarm-specialists, which show but little substrate-relations. A distinct place in this spectrum is characterized by combinations of colour, substrate-preference, substrate-distance, activity, antipredatory behaviour and evidence of swarm formation (Table 2). These findings are exclusively related to the occurrence of mysids in bright light - as mentioned above, inactive species were found to be active in caves and to be active at night.

This broad spectrum is not only characterized by differing species for recent studies showed that those patterns also may differ in relation to races, generations, and age groups.

TABLE 2   The Spectrum Ranging from Substrate-Specialists to Swarm-Specialists

A. *Anchialina agilis, Siriella jaltensis, S. armata*:
   inactive, colourless; the body is extremely elongated; they cling to the leaves of *Zostera* and show no reaction to predators, and to being touched by a diver.

B. *Leptomysis sardica* (Summer):
   semi-active, colourless; they partly cling to *Zostera* leaves; partly swim dispersed, occasionally swim in loose aggregations of 2 to 8 specimens; when disturbed, they sit on leaves. Substrate-distances up to 3 cm.

C. *Leptomysis apiops, Acanthomysis* sp., *Neomysis* sp.:
   active, red ('deep' species), they are dispersed over mud bottoms and aggregate at biogenic structures; when disturbed they escape by swimming down over the substrate at a distance of 1 to 2 cm. Substrate-distances up to 8 cm.

D. *Leptomysis* sp.A, *Leptomysis* sp.B, *L. lingvura*:
   active, grey or black; they prefer *Cystoseira,* rocks, and recesses of rocks; they swarm and swim pendulously to and fro in quiet water. When a possible predator approaches, they swim near their substrate (10 cm); when the predator attacks, the attacked part of the swarm escapes by swimming down over the substrate at a distance of 1 to 4 cm. Then they return to the same spot. Substrate-distances up to 30 cm.

E. *Leptomysis sardica* (February), *L. mediterranea*:
   active, yellowish-grey spots; they prefer sandy bottoms and also group in schools in quiet water. They escape in a similar manner to *Leptomysis lingvura,* but the return is often not precise. Swarms do not move without visible disturbance by possible predators. Substrate-distances up to 50 cm.

F. *Mesopodopsis slabberi*:
   active, colourless; they show no visible substrate preferences. The schools sometimes move with changing leadership without visible external stimuli. They do not return, apart from staying at major discontinuities such as cave-entrances. Escaping swarms remain in the same layer. as long as the predator does not come from above. Substrate-distances up to 50 cm.

DISCUSSION

Like swarms of fishes and other crustaceans, swarms of mysids show evidence of monospecificity, internal structure and cohesiveness (see also Clutter 1969, Macquart-Moulin 1973, Mauchline 1971, Steven 1961, Zelickman 1974).

According to Zelickman (1974) it is assumed, that these patterns are
controlled by the combined affect of habitational (external stimuli) and
collective (inter-individual) responses.

Monospecificity and separation of age groups are controlled by distributional
separation such as bathymetric zonation, substrate-preference and activity
periods.   They are also controlled by active avoidance of individuals of other
lengths and/or preference for the same length.   Clutter (1969) assumes that
mysids respond to the visual image of their neighbours.   The effect of
distributional separation is not great enough to prohibit associations.
Species are not capable of identifying others exactly like themselves.
There may be a capability of a general mysid-identification.   Associations with
young fishes are rare but occur (young gobiids observed at Ischia).

Cohesiveness of swarms is controlled by preference for the same habitat and
probably by a non-specific gregariousness.   Gregariousness seems to be the
mechanism of species interaction.

The precise return-behaviour of *Leptomysis lingvura* seems to be a finely
modified active orientational behaviour.   This assumption is based on the
fact that microhabitats did not change over long periods, while water
movement, illumination, and temperature changed with season and weather.
Mysids try to resist increased water movement until the swarms break down and
their members are swept away.   The precise habitat-preference may be a
strategy combining cohesiveness of swarms and protection by the substrate.
Gregariousness is also evident in this species but is confined to males.
Females seem to have more precise hahitat-prefenerences than males.   This
distributional dimorphism seems to exist in other species of the genus
Leptomysis.

The relative importance of inter-individual responses was found to increase
with the decrease of substrate-relations (Table 2).   These patterns occur
combined with colour and antipredatory behaviour.   It seems conclusive that
these combinations are antipredatory adaptations of mysids occurring in bright
light.   Mysids show two patterns of reducing predation in bright light.   One
is to camouflage on the substrate, the second pattern to swarm.   Both patterns
are spatially separated.   Swarming demands swimming in the water layer,
camouflage demands a proximity to the substrate.   Therefore both patterns are
conflicting.   As shown above, in some species a combination of the two
patterns is possible, swarming in the water layer and escaping close to the
substrate when a predator attacks.   These species show a middle position in
the spectrum ranging from substrate-specialists to swarm-specialists.

The adaptational spectrum seems to be selectively important only in bright
light;   its patterns show changes with changing light intensity.   It is not
implied that swarm formation is exclusively an antipredatory adaption - there
are also other purposes such as aggregation for breeding (Clutter 1969,
Mauchline 1971).

It seems that mysids have evolved two different methods of bringing the
stability and cohesiveness of swarms to an optimum.   One shown by
Mesopodopsis slabberi, is to loose substrate relations and to intensify social
interaction.   The second method, shown by *Leptomysis lingvura*, is to
maintain a relatively low level of collective responses and to intensify
habitat-relations in such a way that the population only occurs at a few

distinct places within its area.   This second method allows the species to profit by the protection of the substrate.

REFERENCES

Clutter, R.I. 1967 Zonation of nearshore mysids.   Ecology 48(2), 200-208.

Clutter, R.I. 1969 The microdistribution and social behaviour of some pelagic mysid shrimps.   J. exp. mar. Biol. Ecol. 3, 125-155.

Macquart-Moulin, Cl. 1971 Modifications des reactions photocinetiques des peracarides de l'hyponeuston nocturne en fonction de l'importance de l'eclairement.   Tethys 3(4), 897-920.

Macquart-Moulin, Cl. 1973 Le comportement d'essaim chez les mysidaces. Influence de l'intensite lumineuse sur la formation, le maintien et la dissociation des essaims de Leptomysis lingvura.   Rapp. Comm. int. Mer Medit. 21(8), 499-501.

Macquart-Moulin, Cl. 1973a L'activite natatoire rythmique chez les peracarides bentho-planctoniques.   Determinisme endogene des rythmes nycthemeraux.   Tethys 5(1), 209-231.

Mauchline, J. 1971 Seasonal occurence of mysids (Crustacea) and evidence of social behaviour.   J. mar. biol. Ass. U.K.   51, 809-825.

Steven, D.M. 1961 Shoaling behaviour in a mysid.   Nature 191(4799), 280-281.

Wittmann, K.J.   Biotop- und Standortbindung einiger Mysidacea.   (in preparation).

Zelickman, E.A. 1974 Group orientation in Neomysis mirabilis (Mysidacea: Crustacea).   Marine Biology 24, 251-258.

PREDATOR-PREY INTERACTION BETWEEN THE CRAB *Pilumnus hirtellus* (LEACH) AND THE BRITTLE STAR *Ophiothrix quinquemaculata* (D. CHIAJE) ON A MUTUAL SPONGE SUBSTRATE

Rolf Stephan Wurzian

*I. Zoologisches Institut, Universität Wien, Lehrkanzel für Meeresbiologie,
A-1090 Wien, Währingerstrasse 17/VI, Austria*

ABSTRACT

A benthic soft bottom community in the Gulf of Triest at a depth of 20 - 25 m is dominated by suspension feeding organisms such as brittle stars, sponges and ascidians. The spatial distribution of the sedentary organisms is obviously clumped. These aggregations serve as a substrate for the most abundant species, *Ophiothrix quinquemaculata*, which accounted, on average, for 102 g/m^2 wet weight, corresponding to 28% of the total macro-epifaunal biomass. *Ophiothrix quinquemaculata* occurs in low densities on the muddy sediment, but is strongly aggregated on the multi-species clumps. This pattern is influenced by preferences for special kinds of substrates and by numerous interactions. *O. quinquemaculata* is only found on certain sponges and seems to avoid the more mucilaginous species and those inhabited by *Pilumnus hirtellus*, which feeds to some extent, on the brittle star. Low densities of *O. quinquemaculata*, on and around clumps which were occupied by *P. hirtellus*, were observed by diving and photographically documented.

Diver taken samples show a negative correlation between the two species, *O. quinquemaculata* even avoids a distinct area of sediment around clumps containing *P. hirtellus*. On larger aggregations *O. quinquemaculata* is restricted to the peripheral areas. Associations of sessile invertebrates which have been abandoned by the crab, remain unattractive to the brittle star for several hours, although recolonisation of clumps, which were cleaned of *O. quinquemaculata*, normally proceeds within a short period of time.

INTRODUCTION

The soft bottom community in the south of the Gulf of Triest, has been described several times (Vatova 1949, Gamulin-Brida 1967, Orel and Mennea 1969, Fedra et al., in press). Fedra et al. showed that this community is characterized by the group *Ophiothrix-Reniera-Microcosmus* which contribute 238(± 43 = 95% confidence limits of the mean) g/m^2, i.e. 64% of the total biomass which amounted to 370 (± 73) g/m^2, measured as wet weight. The high macro-epifaunal biomass is obviously aggregated. Clumps of sponges, ascidians, the alga *Rhodimenia corallicola*, bryozoans and hydrozoans colonise secondary hard-bottom elements formed by bivalves, sedentary polychaetes with tubes and calcareous algae. These multi-species aggregations serve as a substrate for semi-sessile animals such as *Cucumaria planci* and especially

---

This paper is part of a joint program between the Lehrkanzel für Meeresbiologie, University of Vienna, and the Marine Biological Station, Portoroz, University of Ljubljana.

O. quinquemaculata. The fact that the brittle star is found on sponges, ascidians and other sedentary species, and that it can be densely aggregated on these substrates has previously been described (Peres and Picard 1958, Monniot 1961, Guille 1964). Guille (1964) mentioned the importance of sponges and ascidians for the development and growth of *O. quinquemaculata*. Juvenile brittle stars have been found only on such substrates while adults also live on the sediment. Detailed observations on the aggregations of sessile invertebrates and on brittle star distribution showed that curiously, not all clumps are covered by *O. quinquemaculata*.

Another species inhabiting these clumps is *Pilumnus hirtellus* - a crab that feeds, at least to some extent, on the brittle star (Stevcic 1971). The interaction between *P. hirtellus* and *O. quinquemaculata*, vis-a-vis these aggregations, is described in this paper.

MATERIAL AND METHODS

A 10x10 m area in the centre of the community was marked out to facilitate a rough survey of the spatial and size frequency distribution of the aggregations. This test area was divided into single metre squares. Each m^2 was examined using SCUBA. The location of each clump was noted and it was categorised gravimetrically as belonging to 1 of 4 weight classes. This information and a list of dominant species from each clump were recorded against the reference grid. The aggregations were examined for phirtellus or *O. quinquemaculata* and these animals' size frequency distributions were plotted. The frequency distribution of clumps/m^2 was established, for all clumps with a weight exceeding 40 g ww, and this was tested against 6 mathematical distribution models. The percentage of aggregations inhabited by *P. hirtellus* was calculated for all 4 weight classes together as well as for each weight class individually. Confidence intervals were taken from Clopper and Pearson (Weber 1967).

27 quantitative 0.25 m^2 random samples were taken by means of a 50 x 50 cm frame using SCUBA. If a clump was in the sample frame, it was placed so that the clump was at its centre before the epifauna was collected. Using the 27 samples, a correlation between clump size and *O. quinquemaculata* numbers was shown by linear regression. The influence of *P. hirtellus* on the density of the brittle star (on aggregations) and the relationship between clump size and the number of *P. hirtellus* individuals was determined. In addition, 7 random samples of 0.25 m^2 were gathered from the *O. quinquemaculata* inhabiting the sediment. Several experiments were carried out *in situ* to determine the influence of the crab on the brittle star. 10 clumps were numbered and their *O. quinquemaculata* or *P. hirtellus* inhabitants removed. These aggregations were checked at intervals of approximately 24 hours. At each check, the *O. quinquemaculata* or *P. hirtellus* individuals on the clump were counted and again removed.

RESULTS

In this paper the expressions "clump" or "aggregation" are used to refer to concentrations made up of different sedentary species. 79.9% of these clumps consist mainly of sponges and the remaining 20.1% are represented mainly by *Microcosmus* ssp. As *Pilumnus hirtellus* was found only once in a *Microcosmus* dominated clump, it seems justified to speak of a mutual sponge substrate.

Sessile species, which build up these aggregations, represent 62% (230 g wet weight/m^2) of the total biomass (Fedra et al., in press).

229 aggregations were found in the 10x10 m test area. The weights of the clumps, estimated in situ, were divided into 4 weight classes. The 100 m^2 had 171 aggregations in weight class I (0-80 g wet weight), 43 clumps in weight class II (80-240 g ww), 12 aggregations in the weight class III (240-560 g ww) and 3 clumps in weight class IV (more than 560 g ww) (see Fig. 2). The clumps were randomly distributed in the 100 m^2 test area and fitted well in a Poisson-series with p 0.025.

These aggregations serve as a substrate for *O. quinquemaculata* and are very important for the recruitment of this brittle star. Further to Guille's (1964) observations on the importance of sponges and ascidians in the development and growth of the brittle star, Fedra (in press) found that *O. quinquemaculata* settles almost exclusively on several species of sponges. The comparison of a mean size frequency distribution for *O. quinquemaculata* on the sediment (established from 7 samples of 0.25 m^2) with the mean size frequency distribution for the brittle star on aggregations (established from 10 samples of 0.25 m^2), clearly shows that specimens of *O. quinquemaculata* with a disc diameter of less than 5 mm can only be found on clumps. However, adults with a disc diameter of more than 5 mm can be found on the sediment as well as on these aggregations. The high density of the brittle star on the clumps as compared to that on the sediment is shown in Fig. 1.

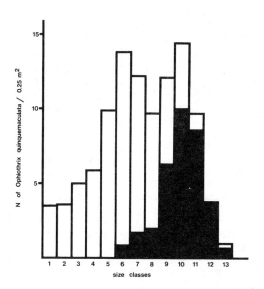

Fig. 1. A mean size frequency distribution of *O. quinquemaculata* on an aggregation (non-shaded area) in comparison to a mean size frequency distribution of the brittle star on the mere sediment (shaded area). Size class interval = 1 mm.

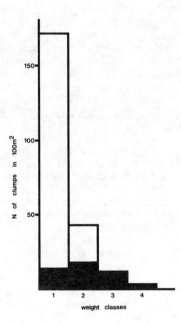

Fig. 2. Frequency distribution of aggregations in 100 m^2 divided into 4 weight classes (non-shaded area). Frequency distribution of aggregations in 100 m^2 inhabited by P. *hirtellus* (shaded area).

The 0.25 m^2 samples containing clumps show an increase in the numbers and weight of *O. quinquemaculata* with increasing clump size. Clump size was established using wet weights. A linear regression for the increase in *O. quinquemaculata* numbers with increase in clump weight was calculated and found to be highly significant (see Fig. 3).

The crab *Pilumnus hirtellus* also inhabits aggregations and depend on them for shelter, at least during the day time. Nothing is known of this crabs nocturnal activity. During the day, P. *hirtellus* is found almost exclusively in these aggregations. In over 200 hours of diving on this biotope, P. *hirtellus* was observed on the sediment only on two occasions.

The preference of P. *hirtellus* for large aggregations can be seen in the percentage of clumps occupied by the crab in each weight class. 14 clumps of weight class I were inhabited by P. *hirtellus*. This is equal to 8.2% (with a confidence interval of 5.7 - 15.9%) of the clumps in this weight class found in the 100 m^2 test area. 18 aggregations of weight class II equalling 41.9% (confidence interval 27.9 - 57.1%) of the 43 clumps of this weight class contained the crab. In weight classes III and IV, all clumps - 12 and 3 respectively - were occupied by the crab (see Fig. 2).

The examination of the aggregations showed that not all serve as a substrate for *O. quinquemaculata*. The brittle star, and P. *hirtellus*, never associated with certain sponges, such as the more mucilaginous species e.g. *Dysidea* ssp. and *Tedania anhelans* (Lieberk) which are very rare in this community.

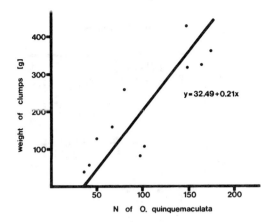

Fig. 3. A linear regression showing the increase of *O. quinquemaculata* numbers with an increase of clump weight.

Other clumps, however, consisting of species that are often settled upon, were found without brittle stars. Such aggregations were always inhabited by *P. hirtellus*. All clumps in the 100 m² test area which were not populated by brittle stars, and did not contain sponges such as were mentioned above, were occupied by the crab. On the other hand, *P. hirtellus* was not found on clumps that were populated by brittle stars. An exception to this observation were two large clumps which were estimated to weigh approximately 1000 g ww. These were inhabited by one *P. hirtellus* each, and *O. quinquemaculata* was restricted to their edges. This agrees with the observation that the brittle star avoids an area, with a radius of approximately 0.2 m, around a smaller clump inhabited by the crab. If the aggregation is larger that the area around a *P. hirtellus* which is normally devoid of *O. quinquemaculata*, the brittle stars can be found on the edge of the clump. A 0.25 m² sample was used in order to get a better picture of the effect of the crab on the brittle star as well as to take the *O. quinquemaculata* free area, around clumps with *P. hirtellus*, into account. The area free of brittle stars was in most cases smaller than the 0.25 m² frame and explains why samples with *P. hirtellus* also contained *O. quinquemaculata*. As previously mentioned, the 27 0.25 m² samples showed an increase in the number of brittle stars with increasing clump size, as well as a drastic decrease in the number of *O. quinquemaculata* individuals when *P. hirtellus* occurred. In addition, *P. hirtellus* became more numerous with increasing clump size. The majority of clumps weighing over 240 g ww (weight class III and IV) are inhabited by *P. hirtellus* (see Fig. 4).

Additional observations show that *P. hirtellus* changes its shelter. Clumps abandoned by the crab were settled upon by few if any brittle stars. Experiments confirm the fact that clumps without *O. quinquemaculata*, and from which *P. hirtellus* is removed, show a lower brittle star density after 24 hours when compared to clumps without the crab and from which brittle stars were removed. After a period of 48 hours no difference could be observed between the repopulation of clumps that were and those that were not inhabited by *P. hirtellus*. The influence of *P. hirtellus* on repopulation by *O. quinquem-*

aculata seems to persist, although in weakened form, for a period of over one day.

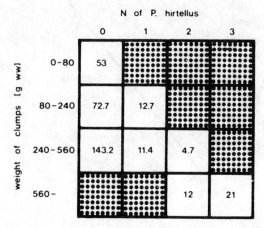

Fig. 4. Relationship between clump size and the presence of *P. hirtellus* and *O. quinquemaculata*. The mean brittle star number is shown in squares.

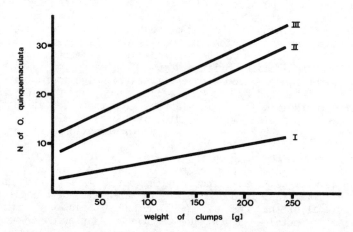

Fig. 5. Recolonisation Experiment - Linear regression I showing the number *O. quinquemaculata* in relation to clump weight 24 hours after *P. hirtellus* was removed from the clump. Regression II shows the same relationship 48 hours after the crab was removed. Regression III shows the same relationship 24 hours after *O. quinquemaculata* was removed from clumps which had not been inhabited by *P. hirtellus*.

An interesting observation was made during the repopulation experiments. Aggregations from which *P. hirtellus* was removed were frequently reinhabited by another individual. Whether the preference for such clumps depends on the structure of the aggregation, i.e. the type of shelter offered, or whether it depends on previous occupation by *P. hirtellus* could not be determined. The daily check showed that even if *O. quinquemaculata* resettled a clump which had been abandoned by the crab, the brittle stars could not be found the next day if the clump was reinhabited by a crab.

DISCUSSION

The fact that a clump will be repopulated by *O. quinquemaculata* following the removal of *P. hirtellus*, and that several hours after a *P. hirtellus* newly occupies a clump no brittle star will be found there, leads to the conclusion that the interaction between *P. hirtellus* and *O. quinquemaculata* cannot be explained by predation alone. It is not possible for the crab to devour all the brittle stars on a clump in the 24 hours interval between the observation periods. It is more likely that *O. quinquemaculata* flees from *P. hirtellus*. What triggers this flight reaction could not be determined with certainty. The observation that clumps from which *P. hirtellus* has been removed are more slowly reinhabited by *O. quinquemaculata* than clumps without the crab from which brittle stars have been removed, suggests that the flight reaction is, to some degree, a response to a chemical stimulus. The *Ophiothrix*-free sediment area around clumps containing *Pilumnus* seems to confirm this assumption. These regions are assumed to lie within the area of influence of the crab. Whether the presence of *P. hirtellus* in a clump hinders the settling of *O. quinquemaculata* larvae by chemical exclusion or whether the larvae and juvenile brittle stars are eaten by *P. hirtellus* remains to be investigated. The settlement of brittle star larvae on adult individuals (Guille 1964) or the presence of larvae on young individuals on the sediment was not observed in this biotope. The multi-species clumps therefore represent the only substrate for the settlement and development of *O. quinquemaculata*. More than 20% of these aggregations, especially the larger ones, are inhabited by the crab. 1/5 of the possible substrates are unsuitable for the recruitment of *O. quinquemaculata*.

CONCLUSIONS

1. *Ophiothrix quinquemaculata* and *Pilumnus hirtellus* use sponges as a mutual substrate.
2. The presence of *P. hirtellus* in a clump prevents settlement by *O. quinquemaculata*.
3. The number of brittle star individuals increases with increase in clump size, when the clump is not inhabited by the crab.
4. The number of *P. hirtellus* and the probability of finding the crab in an aggregation increase with clump size.
5. A strong effect of *P. hirtellus* on the population and the distribution of *O. quinquemaculata* is suggested.

ACKNOWLEDGEMENTS

This study was supported by the projects Nos. 1940, 2084 and 2758 of the 'Fonds zur Förderung der wissenschaftlichen Forschung', Austria, with additional support from the Boris Kidric Foundation. The author's thanks are due to Yugoslavian authorities for research permits and their hospitality, and to the director and staff of the Marine Biological Station, Portoroz, for providing essentail facilities. The author feels indebted to his colleagues K. Fedra, E.M. Olscher, C. Scherübel, and M. Stachowitsch for their essential help during the field work and sampling program. I am also grateful to Dr. J. Ott for carefully reading the manuscript.

REFERENCES

Fedra, K. 1977. Structural Features of a North Adriatic Benthic Community. In Proc. 11th Europ. Sym. on Mar. Biol. eds. Keegan, B.F., P. O Ceidigh and P.J.S. Boaden.

Fedra, K., E.M. Ölscher, C. Scherübel, M. Stachowitsch and R.S. Wurzian on the ecology of a North Adriatic benthic community: Distribution, standing crop and composition of the macrobenthos. Mar. Biol. (in press).

Gamulin-Brida, H. 1974. Biocenoses benthiques de la Mer Adriatique. Acta adriat. 15(9), 1-102.

Guille, A. 1964. Contribution a l'etude de la systematique et de l'ecologie d'*Ophiothrix quinquemaculata* d. Ch. Vie et Milieu 1, 243-308.

Monniot, C. 1961. Un connexe ecologique: les *Microcosmus* de Banyuls. These de 3^e cycle d'oceanographie biologique. Fac. Sciences, Paris.

Orel, G. and B. Mennea 1969. I popolamenti bentonici di alcuni tipi di fondo mobile del Golfo di Trieste. Pubbl. Staz. Zool. Napoli Suppl. 37, 261-276.

Peres, J.M. and J. Picard 1958. Manuel de bionomie benthique de la Mer Mediterranee. Rev. Trav. Stat. Mar. Endoume 23 (14): 78-80.

Stevcic, Z. 1971. Contribution aux recherches dur la nourriture du Crabe *Pilumnus hirtellus* L. Rapp. Comm. int. Mer. Medit. 20 (3), 255-257.

Vatova, A. 1949. La fauna bentonica del'Alto e Medio Adriatico. Nova Thalassia 1(3).

Weber, E. 1967. Grundrib der biologischen Statistik. G. Fischer, Stuttgart.

ANALYSIS OF ECOLOGICAL EQUIVALENTS AMONG LITTORAL FISH

C. Dieter Zander and Armin Heymer

Zoologisches Institut und Museum, D-2000 Hamburg 13
Laboratoire d'Ecologie Générale, F-91800 Brunoy

ABSTRACT

The vertical distribution of blenniid, cliniid and gobiid fishes was investigated in the Mediterranean. Observable adaptions in these species to distinct littoral habitats prompted a study of the types and dimensions of their ecological niches. The speleophils Blennius nigriceps and Tripterygion melanurus, show similar colour patterns and, together with other species, occupy the same habitats in caves. They can be considered as syntopic ecological equivalents. Their ecological niches are not totally coincident, for there are slight differences in their abiotic demands and food requirements. The same applies to Blennius rouxi and Gobius vittatus. from a coralligeneous biotope, which show a strickingly identical pattern. Comparing the occurrences of Tripterygion xanthosoma and Pomatoschistus pictus in different geographical locations, it was learned that secondary limitations of ecological niches are possible. On the other hand distinct factors, e.g. deficiency of food, lead to an enlargement of habitats and ecological niches, as was shown in the case of speleophil fishes. On the basis of these results, the limitation of species to narrow habitats and conditions for "ecological speciation" are discussed.

INTRODUCTION

Remane (1943) first used the term ecological equivalent for organisms which posses identical structures because of similar ways of life in identical biotopes. Structure and way of life affect very decisively the role an organism plays in the eco-system or which ecological niche it occupies. Generally, ecological equivalents with identical ecological demands are not sympatric but are separated geographically. Hutchinson (1957), however, mentions 3 ways by which this rule can be modified: (1) in colonizing species which live in unstable environments that never reach equilibrium, (2) in species that do not compete for resources, and (3) in fluctuating environments that reverse the direction of competition before extinction is possible. On the other hand, a narrow sympatry - syntopy - of almost identical ecological equivalents will compress the fundamental niche to the realized niche in the sense of Hutchinson (1957). Therefore we examined the types and dimensions of niches of ecological equivalents which live together and the evolution of competing species, considering especially closely related ones. In the following, some demersal fishes which live in the upper littoral or in caves are investigated. These 2 biotopes can be compared because in both of them a reduction of light and water movement occurs; in the open littoral in a vertical, in caves in a horizontal direction (Riedl 1966).

MATERIAL AND METHODS

The investigations were carried out in the Mediterranean Sea by SCUBA dives, especially in the open littoral of Banyuls-sur-Mer (France) and in the grotto of Banjole at Rovinj (Yugoslavia). The abundances of fishes were determined;

water depths and light intensities were measured. Most of the fishes were caught and fixed in formaline. For food investigations, the intestinal tracts were removed and the contents counted and assigned to taxonomic groups.

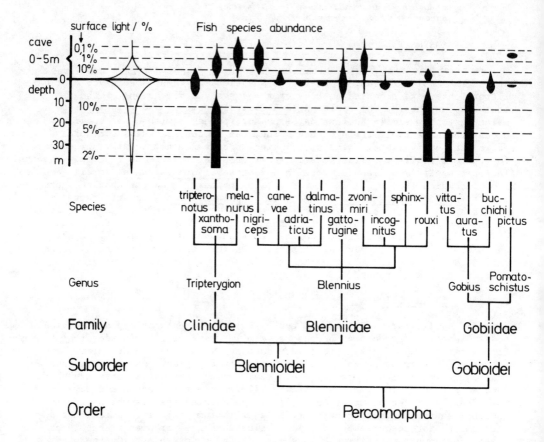

Fig. 1. The distribution and abundance of some demersal fishes related to light intensities in caves and in the open littoral of the Mediterranean Sea.

RESULTS

Types of Ecological Niches
Fig. 1 shows the vertical decrease of light in the open littoral of the Mediterranean Sea and the horizontal decrease in caves (Riedl 1966). These light values are plotted against the abundance of fish belonging to the families Blenniidae, Clinidae, and Gobiidae, whose relationships are also presented in Fig. 1. It can be seen that closely related species inhabit very different biotopes, whereas distantly related species may live together in the same

habitats. The degree to which the latter constitute true ecological equivalents is considered in relation to 2 biotopes.

Grotto of Banjole. The Grotto of Banjole at Rovinj shows a continual light decrease up to the rear of the cave (section 5, Riedl 1966), where a light shaft interrupts the dimly illuminated section. Eleven demersal fish species belonging to 4 families were found in the grotto (Zander and Jelinek 1976): Blennius rouxi Cocco, B. gattorugine Brunnich, B. zvonimiri Kolombatovic, B. nigriceps Vinciguerra; Tripterygion tripteronotus (Risso), T. xanthosoma Zander and Heymer, T. melanurus Guichenot; Gobius bucchichi Steindachner, Pomatoschistus pictus adriaticus Miller, Speleogobius trigloides Zander and Jelinek; Scorpaena porcus L. These species can be assigned to different cave sections with changing abundances (Fig. 1). Whereas most of the species are photophobic or speleophilic, T. tripteronotus, B. rouxi, G. bucchichi, and S. porcus are euryphotic species which are also distributed in the open littoral.

The colouration of the cave-dwelling species is not uniform. Some species show a brown body colour with bright spots on the back, e.g. B. zvonimiri. A similar pattern, in which dark and light bars alternate, is also present in B. gattorugine, P. pictus, and T. xanthosoma. At spawning time ♂♂ of T. xanthosoma, however, are coloured a brilliant yellow and show a black head mask (Zander and Heymer 1970). The patterns of B. nigriceps and T. melanurus, which dominate in the hind cave sections, are more uniform. They have a marbled head and a bright red body (Fig. 2), which is sometimes interrupted on the back by white spots. they resemble other speleophilic fishes such as Apogon imberbis L. and Anthias anthias L., which, however, follow a supra-demersal way of life. At spawning time the ♂♂ of B. nigriceps have black crowns and yellow cheeks (Abel 1964, Zander 1975), whereas T. melanurus has a totally black head mask (Zander and Heymer 1976).

The relationships of the speleophilic species to dimly lit biotopes are measured by their preference for caves as spawning, hiding, resting, and feeding places (Zander and Jelinek 1976). With this method only habitats with a light reception below 1% were considered. It turned out that the resting and feeding places of B. gattorugine, B. zvonimiri, and P. pictus, but also the spawning and hiding places of T. xanthosoma, are not as strongly bound to habitats of lower light intensities as those of B. nigriceps or T. melanurus (Fig. 3). In these 2 species the relationships to caves are very close; their resting or feeding places are rarely located outside of the diffused light section.

Aside from the fact that single T. melanurus are found in the vacant section (section 6, Riedl 1966), the habitats of this species and those of B. nigriceps can be considered as identical. Thus these 2 species represent syntopic ecological equivalents, since they agree in shape and colouration. Abel (1959) describes the differences in their spatial niches: whereas B. nigriceps, like related Blennius species, seek narrow holes as hiding and spawning places, T. malanurus behaves as do its relatives; spawning on the open substrate and taking refuge in deeper water layers.

In spite of these differences a common exploitation of the food resources in the caves is to be expected. Besides the sedentary stocks of algae, Porifera, Hydrozoa, Balanidae, and Bryozoa, a plentiful vagile fauna occurs between the algae (Table 1). Comparing the pre-cave with the cave-entrance section,

TABLE 1 Numbers of the vagile fauna in 25 cm^2 phytal-samples of 3 shadowy biotopes at Rovinj/YU

	Banjole-grotto		Red Island
	Pre-cave section	Cave-entrance section	"Cave 8"
Nematoda	56	37	53
Gastropoda	8	-	24
Polychaeta	55	43	230
Pantopoda	-	-	2
Halacaridae	8	4	3
Ostracoda	69	27	37
Harpacticoida	476	282	565
Cumacea	2	2	-
Anisopoda	3	5	20
Gammaridea	10	9	141
Caprellidae	1	-	7
Isopoda	1	-	9
Brachyura	-	-	1
Insecta-larvae	-	-	5
Ophiuroidae	-	-	2

Fig. 2. Blennius nigriceps (left) and Tripterygion melanurus with identical patterns: Marbled head and red body

Fig. 3. Relationships of cave-dwelling species to shadowy biotopes, expressed by black filling of the boxes. B.g., B.n., B.z.: Blennius gattorugine, nigriceps, zvonimiri; P.p.: Pomatoschistus pictus; T.m., T.x.: Tripterygion melanurus, xanthosoma.

whilst one finds a reduction in the colonization of vagile fauna, a large number of Harpacticidae is found in both habitats. An important part is played by Nematoda, Ostracoda, and Polychaeta, but gammarid amphipods, the main food of the Blennius species of the littoral, seldom occur (Table 1).

The food consumed by B. nigriceps and T. melanurus corresponds to these findings, but the 2 species differ in the selection of some components (Table 2). In B. nigriceps the phytal (algae, Balanidae, Hydrozoa) is the main food (60%), whereas in T. melanurus the vagile meiofauna, especially Harpacticidae, dominates (74%). Vagile macrofauna (Amphipoda, Isopoda, Anisopoda, Polychaeta) makes up only a small portion (11-13%) of the food of both species (Table 2). In comparison with other Blennius species living in the open littoral, the percentage of meiofauna (27%) consumed by B. nigriceps is relatively high. It should be mentioned that 4% of the food found in T. melanurus derives from suprabenthal (Calanoidea, Mysidacea), a food resource which is exploited by 21% of the specimens (Fig. 4).

Summing up, we may assert that whilst the syntopic ecological equivalents B. nigriceps and T. melanurus inhabit identical habitats, they are adapted to different ecological niches with different reproductive and feeding behaviour. This behaviour corresponds with that of their relatives respectively (Abel 1964, Zander and Haymer 1970, Gordina et al. 1972, Zander and Bartsch 1972).

Coralligene of Banyuls-sur-Mer. The coralligeneous biotope is a secondary hard-bottom formation which is built up chiefly by Lithothamniacea and Squamariacea. At Banyuls-sur-Mar it ranges from depths of 20 to 38 m. Through the colonization of other crust-forming organisms (Anthozoa, Polychaeta, Bryozoa), a cleft cavity system is created and this is used by vagile animals for refuge. Under consideration of vertical distribution in Fig. 1, there are 4 demersal fish species which can inhabit this biotope: Tripterygion xanthosoma, Blennius rouxi, Gobius auratus Risso, and G. vittatus Vinciguerra, beside the euryecious predator Scorpaena porcus. With the exception of T. xanthosoma the above-mentioned species are regularly found there (Heymer and Zander 1975). The population density was between 10 (G. auratus), 4 (B. rouxi), and 1 (G. vittatus) in 10 m^2.
Since the investigated depths between 20 and 38 m are located in the area of the reflected light section (Fig. 1), we ask if the species are restricted to this biotope. Only G. vittatus is found exclusively in the coralligene, whereas B. rouxi and G. auratus also live in the upper light section (below 2 and 5 m respectively).

The colouration of G. auratus is similar to other Gobiidae, but the patterns of B. rouxi and G. vittatus, which are identical, are strickingly different from those of their relatives: on a light body, a dark horizontal stripe runs from the head to the caudal peduncle (Fig. 5). Though the meaning of this pattern is yet unknown, we can state that the conditions for syntopic ecological equivalents are also given for these fishes.

B. rouxi, like related Blennius species, inhabits narrow holes which are at the same time hiding and spawning places - at the centres of the males' territories. G. vittatus, however, takes refuge within the cavity system of the coralligene. Nothing is yet known about its spawning habits. The main food of B. rouxi in the coralligene is algae; of secondary importance are Harpacticidae, Polychaeta, and Porifera. Only a small proportion of Amphipoda are consumed (Heymer and Zander 1975). In G. vittatus the Harpacticidae dominate, whereas algae are not important. A comparative

TABLE 2 Percentage of food in different populations of Blennius nigriceps and Tripterygion melanurus in comparison with related species

Species	B. nigriceps		B. zvo-nimiri		T. melanurus			T. xanthosoma
Population	Banjole	Red Island	Banjole	Banjole	Banyuls	Finike		Banjole
Number	28	6	5	33	17	10		9
Porifera		1,7	13,0					
Algae	28,9	21,6	38,8		0,7			
Balanidae	14,5	13,4	9,2	1,6	0,7			1,6
Hydrozoa	12,7	1,7	11,1					
Bryozoa					0,7			
Detritus	4,1		7,4	3,8	2,9	6,2		
Fish-eggs			3,8	5,4	4,3	5,5		4,8
Phytal	60,2	38,4	83,3	10,8	9,3	11,7		6,4
Gastropoda	0,3			0,3				3,2
Bivalvia		3,3	1,9					
Polychaeta	4,1	3,3	5,6	3,8	3,6	0,8		3,2
Amphipoda	4,6	3,3		1,0	7,9	2,3		4,8
Anisopoda	2,3	3,3		2,5	4,3	18,7		6,5
Isopoda		6,7		1,3	1,4	0,8		3,2
Decapoda	1,7			1,9				9,7
Insecta-larv.		1,7			1,4	18,7		
Macrofauna	13,0	21,6	7,5	10,8	18,6	41,3		30,6
Harpacticoida	22,0	36,6	7,4	51,7	65,7	27,4		43,5
Ostracoda	2,3	3,3	1,9	10,2	5,0	10,2		14,6
Halacaridae	2,3			12,1	1,4	7,8		1,6
Nematoda	0,3							
Meiofauna	26,9	39,9	9,3	74,3	72,1	45,4		59,7
Calanoida				3,5				
Mysidacea				0,6				3,2
Suprabenthal				4,1				3,2
Insecta						1,6		

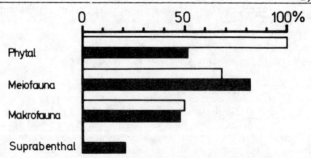

Fig. 4. Comparative frequency analysis of food resources used by Blennius nigriceps (white) and Tripterygion melanurus (black).

frequency analysis of both species brought further differences on the phytal and less on the vagile fauna; G. vittatus, on the other hand, behaves reversely. The decisive difference is the use of the suprabenthal as a food resource by G. vittatus only, 30% of the individuals making use of it. A more detailed description has been prepared by Heymer and Zander.

Fig. 5. Blennius rouxi (left) and Gobius vittatus with identical patterns: Light body and dark horizontal stripe.

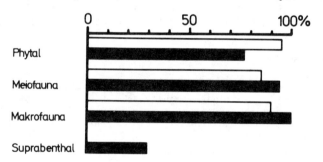

Fig. 6. Comparative frequency analysis of food resources used by Blennius rouxi (white) and Gobius vittatus (black).

Summing up, we may state that the syntopic ecological equivalents, B. rouxi and G. vittatus have developed adaptations to partially different ecological niches niches which are not affected by their habitats but by phylogenetically inherited behaviour patterns, analogous to the situation found in the speleophilic B. nigroceps and T. melanurus.

Dimension of Ecological Niches

It could be shown that the ecological equivalents of caves, B. nigriceps and T. melanurus, and those of the coralligene, B. rouxi and G. vittatus, have few differences in their abiotic demands, but greater ones in their food resources. Therefore, we must ask whether one of them may play the part of the other if one is absent from the biotope.

This situation could be studied near the Banjole grotto at Red Island, where T. melanurus is absent almost totally but B. nigriceps is found very frequently, as well as in Banyuls-sur-Mer, where until now only 1 B. nigriceps was found but T. melanurus exists in greater abundance. The food of the populations without interspecific competition differ indeed from those of the Banjole grotto (Table 2). B. nigriceps of the Red Island population feed more frequently on macro- and meiofauna than in the Banjole grotto, but less on phytal. On the other hand, T. melanurus from Banyuls-sur-Mer consume more from the macrofauna than in the Banjole grotto, whereas the proportions of meiofauna remain identical.

Generally the main food of B. nigriceps may be the phytal and the meiofauna, that of T. melanurus primarily the meiofauna. Therefore the meiofauna is the object of competition between both the species. It must however be taken into consideration that the supply of this food resource is very rich; this is true also at Red Island, where T. melanurus is quite rare (Table 1). Moreover, both species are able to substitute other components for their main food (Table 2). On the other hand there is more competition with related species. Thus the main food of Blennius zvonimiri is the phytal, that of Tripterygion xanthosoma the meiofauna (Table 2).

An enlargement of the ecological niche for feeding is frequently observed. B. zvonimiri normally leaves its resting places in the extremely shaded areas and swims to the open littoral for feeding (Fig. 3). The same was observed in T. melanurus of Finike (Turkey). As known from cave samples, the food of this population has a different composition, meiofauna and macrofauna making up almost identical proportions (Table 2).

DISCUSSION

If 2 or more species compete for space and food a restriction of the fundamental niche of one or all partners results (Wallace and Srb 1961). An analysis of the syntopic ecological equivalents shows that these compete not so much for habitats as for food resources. On the other hand it was determined that the competition for food is still greater in related species: between Blennius nigriceps and B. zvonimiri in caves, between Gobius vittatus and G. auratus in the coralligene. Females of Tripterygion xanthosoma swimming around in all sections of the Banjole grotto also compete with T. melanurus (Zander and Jelinek 1976). But, in most cases, the light demands and therefore the habitats of the 3 Tripterygion species of the Mediterranean Sea are clearly different: T. tripteronotus inhabits the upper light section, T. xanthosoma the reflected, and T. melanurus the diffused-light section (Zander and Heymer 1970, 1976)(Fig. 1). This strict distribution is modified by the T. melanurus population of Finike (se above) and other sites from the southern coast of Turkey between Silifke and Mersin, the last ones being found exclusively in the upper light section (Zander and Heymer 1976).

T. xanthosoma is the only Tripterygion species which is also found in the Atlantic Ocean. Observations at 2 sites on Teneriffa Island show that T. xanthosoma at least lives in the upper light section (spawning place), and partly in the diffused light section (resting place), which are typical for T. tripteronotus and T. melanurus in the Mediterranean Sea (Zander and Heymer 1976). As the phylogeny of all Mediterranean Tripterygion species probably originated from T. xanthosoma populations of the Atlantic Ocean, the Mediterranean population of this species must have developed a niche restriction owing to the competition pressure of the 2 relatives. After these conditions had been fixed genetically, the realized niche became the new fundamental niche in the Mediterranean Sea. T. Tripteronotus and T. melanurus would have adapted to their special niches in the same way.

The enlargement of niches of the T. melanurus populations mentioned above, is a secondary one, for it is restricted to single geographical areas where a food deficiency exists in shaded biotopes and competitors are missing. The same may be true for the T. melanurus population in the Banjole grotto which even exploits the suprabenthal, since B. nigriceps competes for the meiofauna as food resource.

Whereas allopatric ecological equivalents developed along different taxonomical lines under identical environmental conditions (Remane 1943), the evolution of syntopical ecological equivalents was an adaptation of the habitus of one partner to that of the other: mimicry. During this development the shapes and colourations were especially imitated, while behaviour patterns and special structures of related species were maintained. It is still the speleophilic T. melanurus and the coralligene inhabiting B. rouxi which have greater ecological (and maybe geographical) distributions than their partners; these species may be the models for B. nigriceps and G. vittatus. The adaptations of the speleophilic species are so perfect that even geographically restricted colour variations appear. In the southern Mediterranean Sea a T. melanurus sub-species exists which is marked by a black spot on the caudal peduncle. In exactly the same areas the B. nigriceps populations also posses this pattern (Zander and Heymer 1976).

CONCLUSIONS

In the fishes studied we have observed that a sympatric occurrence of close relatives excludes a syntopic existence. This indicates that the separation of their ecological niches is effected by a separation of habitats, the different preferences for distinct abiotic factors being thereby decisive (Fig. 7). Hence their speciation was primarily an adaptation to different habitats and spatial niches, respectively. On the other hand, the evolution of syntopic ecological equivalents was an adaptation to the habitats of a model. In this case biotic factors caused the separation of niches due to different behaviour patterns and special structures, which are fixed phylogenetically in each species (Fig. 7).

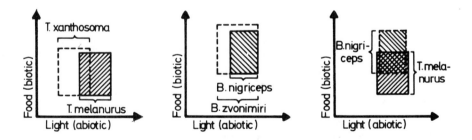

Fig. 7. Separation of the ecological niches of related species and syntopic ecological equivalents owing to abiotic and biotic factors.

ACKNOWLEDGEMENTS

This work was supported by grants from the Deutsche Forschungsgemeinschaft. I am very obliged to Prof. Drach and Mrs. A. Fiala (Banyuls-sur-Mer), Dr. Z. Stevcic and Dr. D. Zavodnik (Rovinj) for their assistance. I thank Miss M. Hänel for the drawings.

REFERENCES

Abel, E.F. 1959. Zur Kenntnis der Beziehungen der Fische zu Höhlen im Mittelmeer. Pubbl. Staz. Zool. Napoli 30 Suppl., 519-528.

Abel, E.F. 1964. Freiwasserstudien zur Fortpflanzungsethologie zweier Mittelmeerfische, Blennius canevae Vinc. und Blennius inaequalis C.V. Z. Tierpsychol 21, 205-222.

Gordina, A.D., L.A. Duka and L.S. Oven 1972. Sexual dimorphism, feeding and spawning in the "black-headed blenny" (Tripterygion tripteronotus Risso) of the Black Sea. J. Ichthyol. 12, 401-407.

Heymer, A. and C.D. Zander 1975. Morphologische und ökologische Untersuchungen an Blennius rouxi, Cocco 1833 (Pisces, Perciformes, Blenniidae). Vie Milieu 25 A, 311-333.

Hutchinson, G.E. 1957. Concluding remarks. Cold Spring Harbor Symp. Quant. Biol. 22, 415-426.

Remane, A. 1943. Die Bedeutung der Lebensformtypen für die Ökologie. Biol. Gen. 17, 164-182.

Riedl, R. 1966. Biologie der Meereshöhlen, Parey, Hamburg and Berlin.

Wallace, B. and A.M. Srb 1961. Adaptation Prentice-Hall, Englewood Cliffs, New Jersey.

Zander, C.D. 1975. Secondary sex characteristics of Blennoid fishes (Perciformes). Pubbl. Staz. Zool. Napoli 39 Suppl., 717-727.

Zander, C.D. and I. Bartsch 1972. In situ-Beziehungen zwischen Nahrungsangebot und aufgenommener Nahrung bei 5 Blennius-Arten (Pisces) des Mittelmeeres. Marine Biol. 17, 77-81.

Zander, C.D. and A. Heymer 1970. Tripterygion tripteronotus (Risso, 1810) und Tripterygion xanthosoma n. sp. - eine ökologische Speziation (Pisces, Teleostei). Vie Milieu 21 A, 363-394.

Zander, C.D. and A. Heymer 1976. Morphologische und ökologische Untersuchungen an den speleophilen Schleimfischartigen Tripterygion melanurus Guichenot, 1850 und T. minor Kolombatovic, 1892 (Perciformes, Blennioidei, Tripterygiidae). Z. Zool. Syst. Evolut.-forsch. 14, 41-59.

Zander, C.D. and H. Jelinek 1976. Zur demersen Fischfauna im Bereich der Grotte von Banjole (Rovinj/YU) mit Beschreibung von Speleogobius trigloides n. gen. n. sp. (Gobiidae, Perciformes). Mitt. Hamburg. Zool. Mus. Inst. 73, in press.